ANALYTICAL METHODS IN VIBRATIONS

M A C M I L L A N   S E R I E S   I N

A P P L I E D   M E C H A N I C S

F R E D   L A N D I S ,   E D I T O R

The Analysis of Stress and Deformation
  by George W. Housner and Thad Vreeland, Jr.
Analytical Methods in Vibrations by Leonard Meirovitch
Continuum Mechanics by Walter Jaunzemis
Statics of Deformable Bodies by Nils O. Myklestad

ANALYTICAL METHODS IN VIBRATIONS

LEONARD MEIROVITCH

PREFACE

This book has been written for analytically oriented engineers. It should prove useful to engineering students enrolled in formal graduate-level programs, although sufficient detail has been furnished to allow for independent study.

Many oscillatory systems exhibit a striking degree of similarity which, in the author's opinion, has not been sufficiently emphasized in engineering texts on vibrations. The similarity between discrete and continuous vibratory systems, on the one hand, and between various types of discrete as well as continuous systems, on the other hand, demands a unified treatment. A unified approach to problems described by similar mathematical models enables one to gain deeper insight into the associated physical systems. Mathematical tools allowing for a broad treatment of such similar systems are introduced. This book represents an attempt to close the gap between the practicing engineer and the applied mathematician. To this end, the text uses extensively matrix methods, calculus of variations, operator notation, integral transform methods, and statistical methods, all of which are becoming standard mathematical tools in a modern engineering curriculum.

The book had its beginnings in an open-ended course offered at the International Business Machines Corporation during the years 1961–1962. The material forms the basis of a two-semester graduate course in vibrations.

The author wishes to acknowledge the many valuable suggestions from his former students and colleagues. In particular, I thank Mr. Frank B. Wallace, Jr., and Dr. Daniel F. Jankowski for their thorough proofreading of the entire manuscript as well as Dr. John D. Spragins for his suggestions concerning Chapter 11. Special thanks are also due Mr. I. Edward Garrick and Dr. David A. Evensen of the National Aeronautics and Space Administration, Langley Research Center, for their valuable comments. Last but not least, I am eager to express my appreciation to Mrs. Dee Meade for her expert typing of the manuscript.

L. M.

CONTENTS

Contents

INTRODUCTION

The last several decades have witnessed tremendous changes in the study and practice of engineering sciences. As technology advances, the problems the engineer of today must face are getting progressively more sophisticated, and it is safe to predict that this trend will continue. As a result, in many engineering colleges throughout this country greater emphasis than ever before is being placed on the study of the fundamentals of the mathematical and physical sciences. It is in recognition of this trend that this book has been written.

Because of the complexity of current engineering problems, assumptions taken for granted for a long time must be reviewed and possibly revised. This process requires deeper insight into the problems, which in turn demands more precise formulations and exact solutions. All this points to a basic and mathematically rigorous approach to the problems. Fortunately mechanics, of which the study of vibrations is an integral part, is one of the fields in which the physical evidence agrees well with the mathematical formulations. This text is concerned exclusively with analytical methods.

One of the purposes of this book is to present the study of vibrations in a mathematically rigorous manner. At the same time this text is written with the idea in mind that the mathematical language is a means of obtaining information about a physical system and not an end in itself. Finding a proper balance between the mathematics and the physics of a problem is a very delicate task; one must always ask to what extent the rigor of the mathematical formulation must be preserved without clouding the physical picture. To this end the problems of uniqueness and existence of solutions have been avoided. The question of uniqueness is brought up only when it is absolutely necessary.

Another objective of this text is to present a unified approach to the problems of vibrating systems. Although the discussion is centered around problems in mechanics, many of the methods shown can be easily applied to electrical systems as well. An attempt has been made to present the material in a systematic fashion by gradually increasing the level from the very simple to the more complex. The intimate relation between discrete and continuous systems is emphasized throughout the text. The common features of discrete and continuous systems, as well as the similarities between various types of discrete and continuous systems, are stressed. An attempt is made to list the fundamental assumptions, which brings into sharper focus the limitations of the theories. A large variety of approximate and numerical methods are presented. An attempt is made to examine in depth the reasons the various methods work, which in turn points out the generalities and limitations of the methods.

Perhaps one of the most dramatic technical developments in our century is the high-speed electronic computer. This factor alone is having a profound effect on the present and future course of research. The incredible speed of the computations makes it possible to obtain numerical solutions to problems not dreamed of years ago. At the same time it must be realized that, in general, a computing machine cannot formulate problems and choose the proper method of solution; in fact, the computer must be given specific instructions on the procedure to be followed, after which it carries out the specified operations very rapidly and without error. Furthermore, there is nothing sacred about the numerical results produced by a high-speed computer; they can be just as wrong as some of the assumptions made in the formulation of the problem. Hence the entrance of the computer into the scientific arena only reinforces the need for more rigorous and complete mathematical treatment of problems. The present text recognizes these facts and therefore many problems are formulated in matrix and series form, both of which are ideally suited for numerical evaluation by high-speed computer. However, the language through which a human being communicates with the computer is in a constant state of flux, so no attempt is made to specify the exact form of the instructions to be accepted by a computer. The formulations are in a form that is bound to be suitable for computer evaluation for many years.

The text is intended for a two-semester course on the graduate level. As such, students are expected to be familiar with the material covered in the undergraduate engineering curriculum in the areas of vibrations and solid mechanics. Furthermore, the student should be equipped with a working knowledge of such modern mathematical tools as matrix algebra and integral transform methods (the elements of which are presented in Appendixes A, B, and C), in addition to the more traditional material covered in undergraduate mathematics. It is hoped that the mathematical rigor will provide a base on which the interested student can build research capabilities in the field of vibrations.

The practicing engineer should also find this book a useful tool. The various methods have not been "rated," and it is left to the reader to establish their merits according to his particular needs. Furthermore, the rigorous treatment should increase the confidence of the user in the validity of the various methods presented and allow for an easy inclusion of additional factors. One must continuously re-examine the traditional assumptions and methods, with regard to new needs, and adopt more refined ones if these prove more suitable. The re-evaluation process must never cease, as it is an integral part of modern engineering; one cannot tell what the demands of tomorrow will be.

The problems at the end of each chapter are an integral part of the text and quite often are designed to illustrate a specific idea. Their solution is mandatory to the understanding of the material.

The material has been selected to reflect current interests in the area of vibrations. The choice of topics can be best discussed by reviewing the material chapter by chapter.

Chapter 1 is essentially a review of some of the fundamentals of vibration analysis covered in undergraduate courses. This chapter provides an opportunity to introduce a modern treatment of the subject by discussing the harmonic analysis and integral transform methods in conjunction with the transient response of linear systems.

Chapter 2 discusses selected topics from "analytical mechanics." Basic concepts such as work and energy, degrees of freedom, generalized coordinates, virtual displacements, and virtual work are discussed. The variational principles of mechanics are presented in a rigorous manner. Because the text is concerned primarily with multi-degree-of-freedom discrete systems and continuous systems, the methods of analytical mechanics are particularly suitable for the formulation of problems associated with such systems.

Chapter 3 introduces some special concepts such as influence coefficients and functions, coordinate transformations, and coupling. The groundwork for a modern treatment of multi-degree-of-freedom systems is prepared by introducing the methods of matrix algebra. The matrix formulations provide clear and concise treatments of very complex problems and are ideally suited for high-speed computation.

Chapter 4 deals with the free vibration of discrete systems. The general characteristic-value problem for a multi-degree-of-freedom system is formulated in matrix form and several matrix methods of obtaining the natural modes are presented. Several concepts that will prove useful in obtaining approximate solutions of the free vibration of continuous-system problems are introduced.

Chapter 5 discusses the free vibration of continuous systems. Particular attention is given to the subject of boundary conditions. The operator notation is used extensively to provide a unified treatment of continuous systems, and the concepts of admissible functions, comparison functions, and eigenfunctions are introduced.

Chapter 6 provides a rigorous treatment of the approximate methods for obtaining the natural modes of continuous systems, for which exact solutions are difficult if not impossible to obtain. The reasons why and conditions under which the individual methods work are discussed. The question of the geometric and natural boundary conditions in conjunction with the series solution of continuous-system problems is brought into sharp focus. Various lumped-parameter methods, particularly suitable for nonuniform systems, are also presented. Matrix methods are used extensively.

Chapter 7 discusses responses of discrete and continuous systems to various types of excitations. The presentation of modal analysis reveals the importance of the normal modes in the solution of vibration problems. The problem of systems with time-dependent boundary conditions is discussed. Matrix and series solutions of forced vibration problems are given.

Chapter 8 presents solutions of continuous-system problems by means of integral transform methods. In particular, the versatility of transform methods in dealing with time-dependent boundary conditions and in producing solutions in terms of traveling waves is illustrated.

Chapter 9 treats the response of damped discrete and continuous systems. The conditions under which modal analysis can be used to uncouple the differential equations of motion are discussed. The solution of damped systems for which modal analysis cannot be used is presented in great detail. Advantage is taken of the inherent expediency of matrix formulations.

Chapter 10 discusses the application of the methods of this text to selected complex problems of current interest. The effect of axial forces upon the flexural modes of bars receives particular attention. This effect is important in stability-of-motion problems. Approximate solutions are presented in matrix form.

Chapter 11 presents a statistical approach to random vibration problems.

The various concepts of statistical and harmonic analysis are presented without assuming previous knowledge of these subjects on the part of the reader. Responses of linear systems to random excitations are obtained.

Elements of matrix algebra and transform methods are discussed in Appendixes A, B, and C. It is assumed that a student who graduated recently from a four-year engineering college is familiar with the material discussed in the appendixes. For those readers the appendixes can serve as a reference. Readers not familiar with these subjects should view the appendixes as stepping stones toward acquiring proficiency in the use of these techniques.

CHAPTER 1

BEHAVIOR OF SYSTEMS

1-1 INTRODUCTION

A *system* is defined as an assemblage of components acting as a whole. To describe the behavior of a physical system one must conceive a mathematical model. Various components of a system are identified and ascribed characteristics according to their physical properties. Using laws of mechanics, mathematical expressions describing the behavior of the mathematical model are derived. If the behavior of the mathematical model agrees within a desired range of accuracy with the observed behavior of the physical system, the model is considered reliable over that range. Otherwise refinements of the assumptions must be made to bring about closer agreement between the predicted and observed behavior of the system.

There are two types of mathematical models, classified according to the manner in which physical properties of a system are described: *discrete systems* and *continuous systems*.

(a) Discrete Systems

The physical properties of system components are discrete quantities. The system behavior is described by ordinary differential equations.

As an example consider the spring-mass-damper system (Figure 1.1). The components of the system are identified as a spring with a spring constant k,

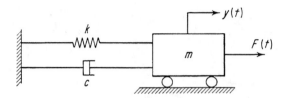

FIGURE 1.1

a damper with a coefficient of viscous damping c, and a mass m. There is just one independent variable, the time t. The system is acted upon by a force $F(t)$. The system behavior is defined by the displacement $y(t)$ of mass m; using Newton's second law it can be shown that this displacement must satisfy the ordinary differential equation

$$m \frac{d^2y(t)}{dt^2} + c \frac{dy(t)}{dt} + ky(t) = F(t), \tag{1.1}$$

where the coefficients m, c, and k are independent of time. One must recognize the high degree of idealization of this model, because in reality springs and dampers possess some mass and a mass possesses some deformability and damping capabilities. Quite often, however, it is possible to justify such a model on physical grounds.

(b) Continuous Systems

The physical properties of system components are functions of spatial coordinates. The system behavior is described by partial differential equations and less frequently by integral equations.

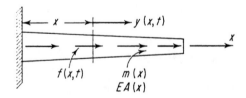

FIGURE 1.2

A bar subjected to axial forces is such an example (Figure 1.2). The mass $m(x)$ is a function of the spatial coordinate x and so is the stiffness $EA(x)$. The force per unit length $f(x, t)$ and the displacement $y(x, t)$ are functions of the spatial coordinate x and time t. The partial differential equation describing the system is (and see Chapter 5)

$$\frac{\partial}{\partial x}\left[EA(x)\frac{\partial y(x,\,t)}{\partial x}\right] + f(x,\,t) = m(x)\frac{\partial^2 y(x,\,t)}{\partial t^2}. \tag{1.2}$$

In conclusion it can be said that a system subjected to some excitation displays a certain behavior. This behavior is called its *response*. The relation between the excitation and response in the case described by (1.1) is given by

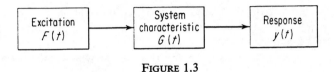

FIGURE 1.3

the block diagram of Figure 1.3, where the system characteristic $G(t)$ is in the form of the differential operator

$$G(t) = m\frac{d^2}{dt^2} + c\frac{d}{dt} + k. \tag{1.3}$$

Perhaps the most important thing to know about a system is whether it is linear or nonlinear, because this has a profound effect upon the choice of methods of solution. In the case of a linear system the dependent variables describing the system must be either of first or zero power (absent). Obviously there cannot be any cross products of the dependent variables present, and it follows that the system is described by linear differential equations. As a consequence, the superposition principle holds true and a tremendous number of mathematical tools are available. On the other hand, in the case of nonlinear systems, the dependent variables may be raised to any power, and cross products of the dependent variables may be present, yielding nonlinear differential equations. In this case the superposition principle no longer holds, which prohibits the use of many mathematical methods.

There is a simple way of testing whether a system is linear or nonlinear. Let $y_1(t)$ be the response of the system in question to the excitation $F_1(t)$ and $y_2(t)$ be the response to the excitation $F_2(t)$. In terms of the operator notation, the above can be written

$$F_1(t) = G[y_1(t)], \qquad F_2(t) = G[y_2(t)], \tag{1.4}$$

where G is a differential operator that reflects the properties of the system only. Let the system be excited now by $F_3(t) = c_1 F_1(t) + c_2 F_2(t)$, where c_1 and c_2 are arbitrary constants, and observe the response. If

$$F_3(t) = G[y_3(t)] = c_1 G[y_1(t)] + c_2 G[y_2(t)], \tag{1.5}$$

the system is linear. If, on the other hand,

$$F_3(t) = G[y_3(t)] \neq c_1 G[y_1(t)] + c_2 G[y_2(t)], \tag{1.6}$$

the system is nonlinear.

In terms of the operator notation the expression

$$G[c_1 y_1 + c_2 y_2] = c_1 G[y_1] + c_2 G[y_2] \tag{1.7}$$

represents the statement that the operator G is linear, implying that the superposition principle holds true for the system whose characteristics are described by G.

The linear theory is well developed and on a good mathematical foundation. Many analytical solutions are available. The nonlinear theory is relatively new, although much of the present state of the art was developed by Poincaré in the latter part of the nineteenth century. Analytical solutions are very few. Solutions are either qualitative in nature and obtained by means of topological methods or are quantitative, in which case they are obtained by some sort of series expansion in terms of a small parameter which in essence assumes that the nonlinearity is a second-order effect. In some cases, in which only the stability of motion is of interest, it may be possible to predict stability (or instability) without actually obtaining the solution of the differential equations. The method for investigating stability was developed by Liapounov and is known as *Liapounov's method*.

The fact that there has been so much more work done in linear theory than nonlinear theory is not an indication that most systems are linear. Actually the opposite is true: All systems are nonlinear. Over a certain range, however, the system behavior may be linear and if the system is restricted to this range, one can safely use linear theory. This restriction is often called the *small-motion assumption*.

Only linear systems will be treated here, in particular those systems for which the physical properties (system parameters) do not depend on time.

The methods of solution depend to a large extent upon the nature of the excitation. The excitation can be divided into two major types: *deterministic* and *nondeterministic* (or random). In the case of deterministic excitation one knows the value of the excitation at any time. Deterministic excitation can be further classified as *periodic* (harmonic excitation is a special case of periodic excitation) and *nonperiodic* (or transient). Periodic excitation can be represented by the Fourier series, which is an infinite series of harmonic components. A nonperiodic function can be looked upon as a periodic function of infinite period. As we increase the period more and more harmonics will participate, and in the limit the Fourier series becomes a Fourier integral. Hence a nonperiodic excitation can be represented by a Fourier integral. The response to a deterministic excitation is deterministic. The random excitation is characterized by indeterminacy in expected behavior, so it is impossible to predict the value of the excitation at a given time. A large collection of records of the excitation, however, may exhibit statistical regularity, enabling one to estimate averages, such as the mean and mean square values. Examples of random excitations are wind velocities or road roughness. In this case the response cannot be given in a deterministic form and will also be in terms of statistical quantities.

Both deterministic and random excitation will be treated here.

This chapter serves a dual purpose. Although it introduces some new concepts, it basically serves as a review of many of the concepts of vibrations with which the reader is assumed to be familiar. In addition, to make the review more meaningful, the material is presented in a relatively modern form by means of mathematical techniques such as the harmonic analysis and integral transforms.

1-2 HARMONIC OSCILLATOR

Consider the simple pendulum free to oscillate in the xy plane (Figure 1.4).

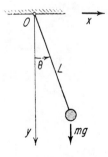

FIGURE 1.4

The equation governing the motion of the pendulum can be obtained by writing the moment equation about point O,

$$M_O = -mgL \sin \theta = mL^2 \ddot{\theta}, \tag{1.8}$$

where dots indicate differentiations with respect to time. Rearrangement yields

$$\ddot{\theta} + \frac{g}{L} \sin \theta = 0. \tag{1.9}$$

Equation (1.9) is a nonlinear ordinary differential equation, because $\sin \theta = \theta - (1/3!)\theta^3 + (1/5!)\theta^5 - \cdots$ is a nonlinear function of θ. If the angle θ remains relatively small, one can linearize (1.9) by letting $\sin \theta \cong \theta$. In using the results, of course, one must always be sure that this assumption is not violated. Hence for small values of θ the pendulum equation becomes

$$\ddot{\theta} + \frac{g}{L} \theta = 0. \tag{1.10}$$

Equation (1.10) represents the differential equation of a harmonic oscillator. Let ω^2 be defined by

$$\omega^2 = \frac{g}{L}, \tag{1.11}$$

where ω is the *circular natural frequency* of the system and is measured in radians per second. The solution of (1.10) is

$$\theta(t) = A \cos \omega t + B \sin \omega t, \tag{1.12}$$

where A and B are constants depending on the initial conditions. Let the initial angular displacement and initial angular velocity be θ_0 and $\dot{\theta}_0$, respectively, so that A and B become

$$A = \theta_0, \qquad B = \frac{\dot{\theta}_0}{\omega}. \tag{1.13}$$

Introducing the relations

$$C^2 = A^2 + B^2 = \theta_0^2 + \left(\frac{\dot{\theta}_0}{\omega}\right)^2,$$

$$\varphi = \tan^{-1} \frac{B}{A} = \tan^{-1} \frac{\dot{\theta}_0}{\omega \theta_0}, \tag{1.14}$$

(1.12) takes the form

$$\theta(t) = C \cos(\omega t - \varphi), \tag{1.15}$$

which represents simple harmonic motion. The constant C is known as the *amplitude* and the constant φ is called the *phase angle*. The motion repeats itself after a *period T* given by

$$\omega T = 2\pi. \tag{1.16}$$

The *natural frequency f* is the reciprocal of the period

$$f = \frac{1}{T} = \frac{\omega}{2\pi} \tag{1.17}$$

and is measured in cycles per second.

Let us now relax the small-motion assumption and adopt a different approach to the problem. Multiplying (1.8) by $\dot{\theta}$ and integrating with respect to time, we obtain

$$\int mL^2 \ddot{\theta}\dot{\theta}\, dt = -\int mgL \sin\theta\dot{\theta}\, dt + E, \tag{1.18}$$

where E is a constant of integration. But

$$\int mL^2 \ddot{\theta}\dot{\theta}\, dt = \int mL^2 \frac{d}{dt}\left(\frac{1}{2}\dot{\theta}^2\right) dt = \frac{1}{2} mL^2 \dot{\theta}^2 = T, \tag{1.19}$$

$$\int mgL \sin\theta\dot{\theta}\, dt = -\int mgL \frac{d}{dt}(\cos\theta)\, dt = -mgL \cos\theta = V, \tag{1.20}$$

where T and V are the kinetic and potential energy, respectively. Hence (1.18) to (1.20) yield

$$T + V = E, \tag{1.21}$$

where the constant E is recognized as the total energy of the system. Equation (1.21) implies that the total energy of the system is constant and is the statement of the conservation of energy. It is, obviously, not restricted to small angles θ.

A more detailed discussion of potential and kinetic energy will be presented in Chapter 2.

1–3 SPRING-MASS-DAMPER SYSTEM. FREE VIBRATION

Lumped mechanical systems are composed of springs, dampers, and masses. The excitation may be of displacement† type or external forces. The spring, damper, and mass are denoted by k, c, and m and have units FL^{-1},

---

† This could consist either of initial conditions or moving supports.

FTL^{-1}, and FT^2L^{-1}, respectively.† Their values are constant in time and the relation between the forces across the elements and the displacement difference across the terminals,‡ the velocity difference across the terminals, and the acceleration, respectively, are linear, as shown in Figure 1.5. As indicated

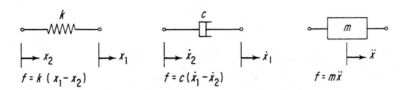

FIGURE 1.5

previously, k is called the *spring constant* or *stiffness* and c is the *coefficient of viscous damping*.

If two springs are connected in parallel or in series, the force-displacement differential relations are as shown in Figure 1.6, where k_{eq} denotes an

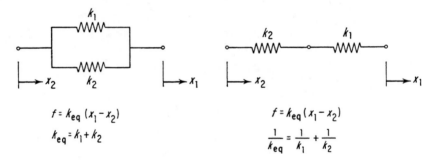

FIGURE 1.6

equivalent spring constant. The above result can be extended to any number of springs. Analogous relations are obtained for equivalent damping coefficients.

Figure 1.7(*a*) represents a spring-mass-damper system and Figure 1.7(*b*) represents the corresponding free-body diagram, where $x(t)$ denotes the displacement from the position of static equilibrium. In this position the spring is compressed by an amount δ_{st}, so there is a force in the spring balancing the weight: $k\delta_{st} = mg$. Measuring $x(t)$ from this equilibrium position

---

† Note that F, T, and L indicate units of force, time, and length, respectively.
‡ End points.

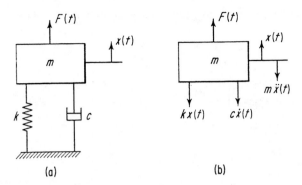

(a)   (b)

FIGURE 1.7

allows us to simplify the equations of motion by canceling the force in the spring $k\delta_{st}$ and the weight mg. Consequently, Newton's second law gives

$$F(t) - c\dot{x}(t) - kx(t) = m\ddot{x}(t), \tag{1.22}$$

which can be rearranged in the form

$$m\ddot{x}(t) + c\dot{x}(t) + kx(t) = F(t). \tag{1.23}$$

Next introduce the notation†

$$\frac{c}{m} = 2\zeta\omega_n, \qquad \frac{k}{m} = \omega_n^2, \qquad \frac{F(t)}{k} = f(t), \tag{1.24}$$

where ζ is called the *viscous damping factor* and ω_n is the circular natural frequency. Equation (1.23), divided through by m, in view of (1.24), yields

$$\ddot{x}(t) + 2\zeta\omega_n\dot{x}(t) + \omega_n^2 x(t) = \frac{1}{m}F(t) = \omega_n^2 f(t). \tag{1.25}$$

The case in which $F(t) = 0$ is called *free vibration* and is characterized by the homogeneous differential equation

$$\ddot{x} + 2\zeta\omega_n\dot{x} + \omega_n^2 x = 0, \tag{1.26}$$

where it is understood that x is a function of t. The solution of (1.26) is of the form

$$x = Ae^{\alpha t}, \tag{1.27}$$

† Note that $f(t)$ has units of displacement.

which upon substitution in (1.26) yields the quadratic equation in α,

$$\alpha^2 + 2\zeta\omega_n\alpha + \omega_n^2 = 0 \tag{1.28}$$

with two roots,

$$\alpha_1 = -\zeta\omega_n + \omega_n\sqrt{\zeta^2 - 1}, \qquad \alpha_2 = -\zeta\omega_n - \omega_n\sqrt{\zeta^2 - 1}. \tag{1.29}$$

Hence the solution of (1.26) can be written

$$x = A_1 e^{\alpha_1 t} + A_2 e^{\alpha_2 t}, \tag{1.30}$$

where A_1 and A_2 are constants depending on the initial displacement and velocity.

When $\zeta > 1$, both roots α_1 and α_2 are negative, so the motion x is decreasing monotonically with increasing time t. This is the *overdamped case* characterized by aperiodic motion given by

$$x = e^{-\zeta\omega_n t}(A_1 e^{\sqrt{\zeta^2 - 1}\omega_n t} + A_2 e^{-\sqrt{\zeta^2 - 1}\omega_n t})$$

$$= e^{-\zeta\omega_n t}(B_1 \cosh \sqrt{\zeta^2 - 1}\omega_n t + B_2 \sinh \sqrt{\zeta^2 - 1}\omega_n t). \tag{1.31}$$

When $\zeta < 1$, we can introduce the notation

$$\omega_d = \omega_n\sqrt{1 - \zeta^2}, \tag{1.32}$$

where ω_d is known as the *circular frequency of the damped free vibration*. In this case, called the *underdamped case*, the solution is

$$x = e^{-\zeta\omega_n t}(A_1 e^{i\omega_d t} + A_2 e^{-i\omega_d t})$$

$$= e^{-\zeta\omega_n t}(C_1 \cos \omega_d t + C_2 \sin \omega_d t)$$

$$= C_0 e^{-\zeta\omega_n t} \cos (\omega_d t - \varphi), \tag{1.33}$$

where the constant C_0 and the phase angle φ depend on the initial conditions. Equation (1.33) indicates damped oscillatory motion. The response oscillates within an envelope defined by $x = \pm C_0 e^{-\zeta\omega_n t}$, and as t increases the response dies out.

The case for which $\zeta = 1$ is known as the *critically damped case* and is the limit between the region for which $\zeta > 1$, in which the system is overdamped and the motion aperiodic, and the region for which $\zeta < 1$, in which the system is underdamped and the motion consists of a damped oscillation. The value of the critical damping is simply

$$c_{cr} = 2m\omega_n = 2\sqrt{km}. \tag{1.34}$$

Equation (1.33) is not valid for $\zeta = 1$ (see Problem 1.2).

1-4 SYSTEM RESPONSE. TRANSFER FUNCTION

It was pointed out in Section 1-1 that a system excited by a function $F(t)$ has a response $x(t)$ which depends on the system characteristic $G(t)$. In the system of Figure 1.1, $x(t)$ denoted the displacement and $G(t)$ was a linear differential operator with constant coefficients. If the response sought is the velocity $v(t) = \dot{x}(t)$ instead of the displacement, the system characteristic will be in the form of a linear differential-integral operator which is just the integral of $G(t)$.

If the differential equation describing the system is transformed by means of the Laplace transformation (see Appendix B), the relation between the excitation and response (also called *input* and *output*, respectively) will be a simple algebraic expression. Transforming (1.23) and setting the initial conditions equal to zero, we obtain

$$ms^2\bar{x}(s) + cs\bar{x}(s) + k\bar{x}(s) = \bar{F}(s), \tag{1.35}$$

where

$$\bar{x}(s) = \mathscr{L}x(t) = \int_0^\infty e^{-st}x(t)\,dt, \tag{1.36}$$

$$\bar{F}(s) = \mathscr{L}F(t) = \int_0^\infty e^{-st}F(t)\,dt \tag{1.37}$$

are the Laplace transformations of $x(t)$ and $F(t)$, respectively, and s is called a subsidiary variable and, in general, is a complex quantity.

The ratio of the transformed input to the transformed output is

$$\frac{\bar{F}(s)}{\bar{x}(s)} = ms^2 + cs + k \tag{1.38}$$

and is an algebraic expression. This ratio is called the *generalized impedance of the system* and is denoted

$$Z(s) = \frac{\bar{F}(s)}{\bar{x}(s)}. \tag{1.39}$$

Its reciprocal is called the *admittance of the system*,

$$Y(s) = \frac{1}{Z(s)}. \tag{1.40}$$

In the study of systems one encounters a more general expression relating the input† and output at various points. This is the *system function*, better known as the *transfer function*, defined by

$$G_{ij}(s) = \frac{\mathcal{L}(\text{output at point } j \text{ due to input at point } i)}{\mathcal{L}(\text{input at point } i)}. \tag{1.41}$$

In the case of the spring-mass-damper system the transfer function reduces to the simple relation

$$G(s) = \frac{\bar{x}(s)}{\bar{F}(s)} = \frac{1}{ms^2 + cs + k} = \frac{1}{m(s^2 + 2\zeta\omega_n s + \omega_n^2)}. \tag{1.42}$$

The Laplace transform counterpart of the block diagram of Figure 1.3 is as shown in Figure 1.8. Note, however, that $G(s)$ is an algebraic expression

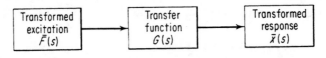

FIGURE 1.8

rather than a differential operator.‡ The transformed response is

$$\bar{x}(s) = G(s)\bar{F}(s), \tag{1.43}$$

and the response $x(t)$ is the inverse Laplace transformation of $\bar{x}(s)$,

$$x(t) = \mathcal{L}^{-1}\bar{x}(s) = \mathcal{L}^{-1}G(s)\bar{F}(s). \tag{1.44}$$

We note that (1.43) contains three quantities, $\bar{x}(s)$, $G(s)$, and $\bar{F}(s)$. A large number of problems consist of determining one of the quantities having advance knowledge of the other two. Accordingly, one can distinguish three types of problems, depending on the quantity sought.

1. The most straightforward type of problem is the one in which $\bar{F}(s)$ and $G(s)$ are known and $\bar{x}(s)$ is unknown. The problem of obtaining the response of a system when the excitation and the system characteristics are given is known as *analysis*.

2. A different type of problem consists of seeking $\bar{F}(s)$ when $\bar{x}(s)$ and $G(s)$ are known. This is the equivalent of knowing the response and the system characteristics and attempting to determine the excitation function. *Instrumentation, measurement,* or *calibration* fall in this category. In such cases the system is used to record the response, from which the excitation can be deduced.

---

† The initial conditions can be included in the transformed input.
‡ The reciprocal of $G(s)$ is related to $G(t)$ in Figure 1.3.

3. A more flexible problem consists of determining $G(s)$ provided $\bar{x}(s)$ and $\bar{F}(s)$ are given. The problem of determining the system characteristics when the excitation and the response are given is called *synthesis*. This amounts to designing a system which, for a given excitation, exhibits a desired response. Of course, there may be many systems able to produce the desired response. The problem of obtaining the most satisfactory system, when all pertinent factors are taken into account, is known as *optimum design*.

In this text we shall be concerned exclusively with analysis.

1-5' INDICIAL RESPONSE. UNIT STEP FUNCTION

The *unit step function* (see Figure 1.9) is defined as

$$u(t - a) = \begin{cases} 0, & t < a, \\ 1, & t > a. \end{cases} \tag{1.45}$$

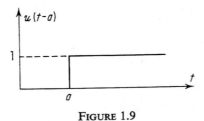

FIGURE 1.9

The unit step function applied at $t = 0$ is denoted $u(t)$. Note that in our usage the unit step function is dimensionless.

The *indicial response*, denoted $g(t)$, is the response of a system with zero initial conditions to a unit step function applied at $t = 0$. The Laplace transformation of the excitation is, therefore,

$$\mathscr{L}u(t) = \frac{1}{s}, \tag{1.46}$$

and using the definition (1.41) for the case when points i and j are identical, we obtain

$$\mathscr{L}g(t) = \bar{g}(s) = G(s)\mathscr{L}u(t) = \frac{G(s)}{s}. \tag{1.47}$$

The indicial response is simply

$$g(t) = \mathscr{L}^{-1}\bar{g}(s) = \mathscr{L}^{-1}\frac{G(s)}{s}. \tag{1.48}$$

1-6 IMPULSIVE RESPONSE. UNIT IMPULSE

The *unit impulse* or *Dirac's delta function* (see Figure 1.10) is defined as

$$\delta(t - a) = 0, \qquad t \neq a,$$

$$\int_0^\infty \delta(t - a)\, dt = 1. \tag{1.49}$$

The unit impulse applied at $t = 0$ is denoted $\delta(t)$. The unit impulse has units of reciprocal seconds.

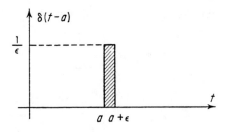

FIGURE 1.10

The *impulsive response*, denoted $h(t)$, is the response of a system with zero initial conditions to a unit impulse applied at $t = 0$. The Laplace transformation of the excitation is

$$\mathscr{L}\delta(t) = 1, \tag{1.50}$$

and it follows that

$$\mathscr{L}h(t) = \bar{h}(s) = G(s)\mathscr{L}\delta(t) = G(s). \tag{1.51}$$

Hence the transformed impulsive response is equal to the transfer function

$$\bar{h}(s) = G(s). \tag{1.52}$$

The impulsive response is, therefore,

$$h(t) = \mathscr{L}^{-1}\bar{h}(s) = \mathscr{L}^{-1}G(s). \tag{1.53}$$

The unit impulse is just the time derivative of the unit step function, so one may expect that there is a relationship between the indicial and impulsive responses. Indeed, comparing (1.47) and (1.51), we conclude that

$$\bar{h}(s) = s\bar{g}(s). \tag{1.54}$$

The Laplace transformation of the derivative of $g(t)$ is

$$\mathscr{L}\,\frac{dg(t)}{dt} = s\bar{g}(s) - g(0) = \bar{h}(s) - g(0), \tag{1.55}$$

from which it follows that

$$h(t) = \mathscr{L}^{-1}\left(\mathscr{L}\,\frac{dg(t)}{dt} + g(0)\right) = \frac{dg(t)}{dt} + g(0)\delta(t), \tag{1.56}$$

because $g(0)$ is a constant.

1-7 DUHAMEL'S INTEGRAL. CONVOLUTION INTEGRAL

Consider a system with an indicial response $g(t)$. At a given time τ the exciting function has the value $f(\tau)$, and corresponding to an increment of time $\Delta\tau$ there is an increase in the amplitude $\Delta f(\tau)$ as shown in Figure 1.11.

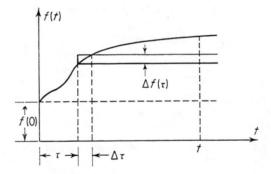

FIGURE 1.11

The contribution to the response of a step function of amplitude $\Delta f(\tau)$ applied approximately at $t = \tau$ is

$$\Delta x(t, \tau) \cong \Delta f(\tau)g(t - \tau) = \frac{\Delta f(\tau)}{\Delta\tau}\,g(t - \tau)\,\Delta\tau. \tag{1.57}$$

If $f(0)$ is the amplitude of a step function applied at $t = 0$, the response at any time t is given by

$$x(t) \cong f(0)g(t) + \sum \frac{\Delta f(\tau)}{\Delta\tau}\,g(t - \tau)\,\Delta\tau. \tag{1.58}$$

In the limit as $\Delta\tau \to 0$ we obtain

$$x(t) = f(0)g(t) + \int_0^t \frac{df(\tau)}{d\tau} g(t - \tau) \, d\tau. \tag{1.59}$$

The integral in (1.59) is known as *Duhamel's integral*. Equation (1.59) can be integrated by parts to obtain

$$x(t) = f(0)g(t) + g(t - \tau)f(\tau)|_0^t - \int_0^t f(\tau) \frac{dg(t - \tau)}{d\tau} \, d\tau$$

$$= g(0)f(t) + \int_0^t g'(t - \tau)f(\tau) \, d\tau, \tag{1.60}$$

where the prime indicates differentiation with respect to $t - \tau$.

Following the Leibnitz rule for differentiation under the integral sign,[†] (1.60) is reduced to

$$x(t) = \frac{d}{dt} \int_0^t g(t - \tau)f(\tau) \, d\tau. \tag{1.61}$$

Recalling (1.56) and replacing t by $t - \tau$ we can write

$$g'(t - \tau) = h(t - \tau) - g(0)\delta(t - \tau), \tag{1.62}$$

which, upon substitution in (1.60), gives

$$x(t) = g(0)f(t) + \int_0^t [h(t - \tau) - g(0)\delta(t - \tau)]f(\tau) \, d\tau$$

$$= \int_0^t f(\tau)h(t - \tau) \, d\tau. \tag{1.63}$$

Letting $t - \tau = \lambda$ and adjusting the limits accordingly, the above takes the form[‡]

$$x(t) = \int_0^t f(t - \tau)h(\tau) \, d\tau, \tag{1.64}$$

indicating that (1.63) and (1.64) are symmetric in the excitation $f(t)$ and the impulsive response $h(t)$. These are known as *convolution* (or *Faltung*) *integrals* and denoted

$$x(t) = f(t) * h(t) = \int_0^t f(\tau)h(t - \tau) \, d\tau = \int_0^t f(t - \tau)h(\tau) \, d\tau. \tag{1.65}$$

[†] See I. S. Sokolnikoff and R. M. Redheffer, *Mathematics of Physics and Modern Engineering*, McGraw-Hill, New York, 1958, p. 262.

[‡] Note that after the above substitution, the dummy variable of integration, λ, was replaced by τ.

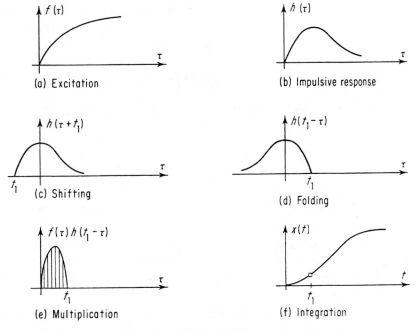

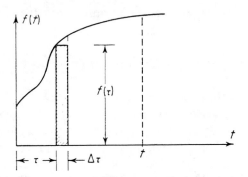

FIGURE 1.12

The convolution consists of shifting, folding, multiplication, and integration, as shown in Figure 1.12.

Figure 1.12(f) shows the response as a function of time. The circle indicates the value of the convolution integral for any particular time t_1 and is equal to the shaded area in Figure 1.12(e). The above is actually the procedure that should be followed if a graphical evaluation of the integral is desired.

Equation (1.63) can be easily proved by regarding the excitation as a series of impulses of magnitudes $f(\tau)\,\Delta\tau$, as shown in Figure 1.13. The

FIGURE 1.13

response to a unit impulse $\delta(t - \tau)$, applied at $t = \tau$, is the impulsive response delayed by the time interval $t = \tau$; that is, $h(t - \tau)$. Hence an impulse of magnitude $f(\tau)\,\Delta\tau$, applied at $t = \tau$, contributes to the response an amount

$$\Delta x(t, \tau) = f(\tau)\,\Delta\tau\, h(t - \tau). \tag{1.66}$$

Letting $\Delta\tau \to 0$ and replacing the summation by integration, we obtain

$$x(t) = \int_0^t f(\tau)h(t - \tau)\, d\tau; \tag{1.67}$$

which is identical to (1.63).

The convolution integral, (1.65) or (1.67), is of considerable importance in the study of transient response of linear systems problems. These formulas are quite general and their only limitations are that the system must be linear and the excitation functions $f(t)$ defined only for $t > 0$. The convolution integral will be used repeatedly to obtain general expressions for the response of vibrating systems.

In Chapter 11 we will show a slightly different version of the convolution integral, obtained by modifying the limits of integration to accommodate excitation functions $f(t)$ defined for any value of t.

1–8 RESPONSE TO HARMONIC EXCITATION

Referring to the spring-mass-damper of Section 1–3, let the excitation be given by the *real part* of

$$f(t) = Ae^{i\omega t}, \tag{1.68}$$

where A is the complex amplitude, with units of displacement and contains information about the phase angle.† Consequently the response will be the real part of $x(t)$ as obtained from

$$\ddot{x}(t) + 2\zeta\omega_n\dot{x}(t) + \omega_n^2 x(t) = \omega_n^2 Ae^{i\omega t}. \tag{1.69}$$

The solution of the homogeneous differential equation (obtained by letting A equal zero) was treated in Section 1–3. This solution is known as the *starting transient solution*, because it eventually dies out as t increases indefinitely. The particular solution of (1.69) does not vanish for large t and is called the

---

† The representation of the force as a complex vector rotating in the complex plane is discussed in more detail in Section 9–4.

steady-state response of the system to harmonic excitation. It is given by

$$x(t) = \frac{\omega_n^2 A e^{i\omega t}}{\omega_n^2 - \omega^2 + i2\zeta\omega_n\omega} = \frac{A e^{i\omega t}}{1 - (\omega/\omega_n)^2 + i2\zeta(\omega/\omega_n)}. \tag{1.70}$$

We introduce now the concept of *complex frequency response* (or *magnification factor*), given by

$$H(\omega) = \frac{1}{1 - (\omega/\omega_n)^2 + i2\zeta(\omega/\omega_n)}, \tag{1.71}$$

and it follows from (1.24), (1.68), (1.70), and (1.71) that

$$H(\omega) = \frac{x(t)}{A e^{i\omega t}} = \frac{x(t)}{f(t)} = \frac{kx(t)}{F(t)} = \frac{F_{sp}(t)}{F(t)}. \tag{1.72}$$

In words, the complex frequency response is the ratio between the force in the spring (ignoring the force balancing the gravity force) and the excitation force, where both forces are harmonic.

Denote the complex conjugate of $H(\omega)$ by

$$H^*(\omega) = \frac{1}{1 - (\omega/\omega_n)^2 - i2\zeta(\omega/\omega_n)}, \tag{1.73}$$

so the magnitude of $H(\omega)$ can be obtained from

$$|H(\omega)|^2 = H(\omega)H^*(\omega) = \frac{1}{[1 - (\omega/\omega_n)^2]^2 + [2\zeta(\omega/\omega_n)]^2}. \tag{1.74}$$

The maximum of $|H(\omega)|$ is obtained by differentiating (1.74) with respect to ω and letting the result be zero. If this is done, we conclude that the maximum occurs at

$$\omega = \omega_n(1 - 2\zeta^2)^{1/2}, \tag{1.75}$$

hence at a value lower than the natural frequency ω_n (Figure 1.14).

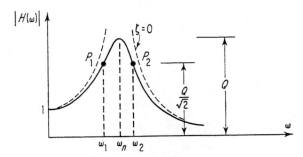

FIGURE 1.14

It is easy to see that for light damping ($\zeta < 0.05$) the maximum occurs approximately at the natural frequency ω_n. Denote the maximum value $|H(\omega)|_{\max} = Q$ and note that for light damping,

$$Q \cong \frac{1}{2\zeta},$$ (1.76)

and the curve $|H(\omega)|$ vs. ω is nearly symmetrical with respect to the vertical $\omega = \omega_n$. The points P_1 and P_2, where the amplitude of $|H(\omega)|$ falls to $(1/\sqrt{2})Q$, are called *half power points*, because the power absorbed by a damper in simple harmonic motion at a given frequency is proportional to the amplitude squared. The frequency difference between the half power points P_1 and P_2 is generally called the *bandwidth* of the system, and for small damping its value can be shown to be

$$\Delta\omega = \omega_2 - \omega_1 = 2\zeta\omega_n.$$ (1.77)

In many problems in vibrations only the amplitude of vibration and not the phase is of interest. The *mean square value* provides a measure of the amplitude. The mean square value of the response over an interval T_1 is defined as

$$\overline{x^2} = \frac{1}{T_1} \int_0^{T_1} x^2(t)\, dt.$$ (1.78)

It was mentioned previously that the response is the real part of $x(t)$; again considering only the steady-state response we write

$$\begin{aligned}
x(t) &= \mathrm{Re}\, \frac{Ae^{i\omega t}}{1 - (\omega/\omega_n)^2 + i2\zeta(\omega/\omega_n)} \\[2mm]
&= \frac{[1 - (\omega/\omega_n)^2]\, \mathrm{Re}\, Ae^{i\omega t} + 2\zeta(\omega/\omega_n)\, \mathrm{Im}\, Ae^{i\omega t}}{[1 - (\omega/\omega_n)^2]^2 + [2\zeta(\omega/\omega_n)]^2} \\[2mm]
&= |H(\omega)|^2 \left\{ \left[1 - \left(\frac{\omega}{\omega_n}\right)^2\right] \mathrm{Re}\, Ae^{i\omega t} + 2\zeta\, \frac{\omega}{\omega_n}\, \mathrm{Im}\, Ae^{i\omega t} \right\},
\end{aligned}$$ (1.79)

where

$$\mathrm{Re}\, Ae^{i\omega t} = \frac{1}{2}(Ae^{i\omega t} + A^* e^{-i\omega t}),$$

$$\mathrm{Im}\, Ae^{i\omega t} = \frac{1}{2i}(Ae^{i\omega t} - A^* e^{-i\omega t}).$$ (1.80)

In the above A^* is the complex conjugate of A.

Squaring (1.79) in conjunction with (1.80), we obtain

$$x^2(t) = \frac{|H(\omega)|^4}{4} \left[A^2 e^{2i\omega t} \left\{ \left[1 - \left(\frac{\omega}{\omega_n}\right)^2\right]^2 - i4\zeta \frac{\omega}{\omega_n} \left[1 - \left(\frac{\omega}{\omega_n}\right)^2\right] - \left(2\zeta \frac{\omega}{\omega_n}\right)^2 \right\} \right.$$

$$+ 2AA^* \left\{ \left[1 - \left(\frac{\omega}{\omega_n}\right)^2\right]^2 + \left(2\zeta \frac{\omega}{\omega_n}\right)^2 \right\}$$

$$\left. + A^{*2} e^{-2i\omega t} \left\{ \left[1 - \left(\frac{\omega}{\omega_n}\right)^2\right]^2 + i4\zeta \frac{\omega}{\omega_n} \left[1 - \left(\frac{\omega}{\omega_n}\right)^2\right] - \left(2\zeta \frac{\omega}{\omega_n}\right)^2 \right\} \right].$$

$$(1.81)$$

If the interval T_1 coincides with the period $2\pi/\omega$, (1.78) in conjunction with (1.81) gives

$$\overline{x^2} = \frac{\omega}{2\pi} \int_0^{2\pi/\omega} x^2(t)\, dt$$

$$= \frac{|H(\omega)|^4}{2} AA^* \left\{ \left[1 - \left(\frac{\omega}{\omega_n}\right)^2\right]^2 + \left(2\zeta \frac{\omega}{\omega_n}\right)^2 \right\} = \tfrac{1}{2}|A|^2 |H(\omega)|^2. \quad (1.82)$$

On the other hand,

$$f^2(t) = (\mathrm{Re}\, A e^{i\omega t})^2 = \tfrac{1}{4}(A^2 e^{2i\omega t} + 2AA^* + A^{*2} e^{-2i\omega t}), \quad (1.83)$$

and the mean square value over the interval $2\pi/\omega$ is

$$\overline{f^2} = \frac{1}{T} \int_0^T f^2(t)\, dt = \frac{\omega}{2\pi} \int_0^{2\pi/\omega} f^2(t)\, dt = \tfrac{1}{4}AA^* = \tfrac{1}{2}|A|^2. \quad (1.84)$$

If the time interval over which the mean square values are calculated is T_1, which is different than the period $T = 2\pi/\omega$, the results obtained for $\overline{x^2}$ and $\overline{f^2}$ will be different than the ones given by (1.82) and (1.84). As T_1 becomes very large, however, $\overline{x^2}$ and $\overline{f^2}$ will converge to the values given by (1.82) and (1.84). Hence (1.82) and (1.84) yield

$$\lim_{T_1 \to \infty} \frac{\overline{x^2}}{\overline{f^2}} = |H(\omega)|^2. \quad (1.85)$$

In words, the response ratio of the mean square output to the mean square input (adjusted to give units of displacement) converges to the square of the absolute value of the frequency response $H(\omega)$ as the interval of integration T_1 increases indefinitely.

The importance of mean square values will become more obvious in Chapter 11, where random vibration is discussed.

1-9 RESPONSE TO PERIODIC EXCITATION. FOURIER SERIES

Consider the case in which $f(t)$ is periodic and of period T. If the fundamental frequency (the lowest frequency) is $\omega_0 = 2\pi/T$, it follows that all other frequencies are its multiples: $\omega_p = p\omega_0$ $(p = 1, 2, \ldots)$. Such a periodic function may be represented by a Fourier series of the complex form

$$f(t) = \sum_{p=-\infty}^{\infty} C_p e^{ip\omega_0 t}, \tag{1.86}$$

where the complex coefficients C_p contain the information regarding the phase angles of the various harmonics. The coefficients can be evaluated by means of the formula†

$$C_p = \frac{1}{T} \int_{-T/2}^{T/2} f(t) e^{-ip\omega_0 t} \, dt. \tag{1.87}$$

The same periodic function $f(t)$, if its average value is zero, can be represented by the *real part* of the series

$$f(t) = \sum_{p=1}^{\infty} A_p e^{ip\omega_0 t}, \qquad \omega_0 = \frac{2\pi}{T}, \tag{1.88}$$

where again the phase angle is contained in the complex coefficient A_p. In practice, only a finite number of terms is required for a good approximation to $f(t)$.

The steady-state harmonic response of the spring-mass-damper of Section 1-3 to $f(t)$, as given by (1.88), becomes

$$x(t) = \mathrm{Re} \sum_{p=1}^{\infty} \frac{A_p e^{ip\omega_0 t}}{[1 - (p\omega_0/\omega_n)^2 + i2\zeta(p\omega_0/\omega_n)]} = \mathrm{Re} \sum_{p=1}^{\infty} H_p A_p e^{ip\omega_0 t}, \tag{1.89}$$

where

$$H_p = \frac{1}{1 - (p\omega_0/\omega_n)^2 + i2\zeta(p\omega_0/\omega_n)}. \tag{1.90}$$

† See R. Courant and D. Hilbert, *Methods of Mathematical Physics*, Interscience, New York, Vol. 1, 1961, p. 70.

If we are interested in the amplitudes but not the phase angles, we calculate the mean square excitation and mean square responses. The mean square excitation is†

$$\overline{f^2} = \frac{1}{T} \int_0^T \left(\mathrm{Re} \sum_{p=1}^{\infty} A_p e^{ip\omega_0 t} \right)^2 dt$$

$$= \frac{1}{4T} \int_0^T \left(\sum_{p=1}^{\infty} A_p e^{ip\omega_0 t} + \sum_{p=1}^{\infty} A_p^* e^{-ip\omega_0 t} \right) \left(\sum_{r=1}^{\infty} A_r e^{ir\omega_0 t} + \sum_{r=1}^{\infty} A_r^* e^{-ir\omega_0 t} \right) dt$$

$$= \frac{1}{4T} \int_0^T \sum_{p=1}^{\infty} \sum_{r=1}^{\infty} [A_p A_r e^{i(p+r)\omega_0 t} + A_p A_r^* e^{i(p-r)\omega_0 t} + A_p^* A_r e^{-i(p-r)\omega_0 t}$$

$$+ A_p^* A_r^* e^{-i(p+r)\omega_0 t}] \, dt \cong \frac{1}{4} \sum_{p=1}^{\infty} \sum_{r=1}^{\infty} \delta_{pr}(A_p A_r^* + A_p^* A_r)$$

$$= \frac{1}{2} \sum_{p=1}^{\infty} A_p A_p^* = \frac{1}{2} \sum_{p=1}^{\infty} |A_p|^2. \tag{1.91}$$

In (1.91) T must be long enough so that the averaging process will converge to the value obtained by averaging over the period. Also, δ_{pr} is the Kronecker delta defined by

$$\delta_{pr} = \begin{cases} 0 & \text{when } p \neq r, \\ 1 & \text{when } p = r. \end{cases} \tag{1.92}$$

In a similar way we calculate the mean square of the steady-state response in the form

$$\overline{x^2} = \frac{1}{T} \int_0^T x^2(t) \, dt = \frac{1}{T} \int_0^T \left(\mathrm{Re} \sum_{p=1}^{\infty} H_p A_p e^{ip\omega_0 t} \right)^2 dt$$

$$\cong \frac{1}{2} \sum_{p=1}^{\infty} H_p H_p^* A_p A_p^* = \frac{1}{2} \sum_{p=1}^{\infty} |H_p|^2 |A_p|^2. \tag{1.93}$$

The contribution of each individual frequency component to the mean square excitation and mean square response can be displayed by means of the *excitation mean square spectral density* and the *response mean square spectral density*, defined by

$$G_f(\omega) = \frac{1}{2} \sum_{p=1}^{\infty} |A_p|^2 \delta(\omega - p\omega_0), \tag{1.94}$$

$$G_x(\omega) = \frac{1}{2} \sum_{p=1}^{\infty} |H_p|^2 |A_p|^2 \delta(\omega - p\omega_0), \tag{1.95}$$

† This equality is called *Parseval's relation*, or *formula, for periodic functions*

where $G_f(\omega)$ is the excitation spectral density and $G_x(\omega)$ is the response spectral density. The symbol $\delta(\omega - p\omega_0)$ represents a Dirac's delta function in the frequency domain with units rad^{-1} sec, whereas $G_f(\omega)$ and $G_x(\omega)$ have units $\text{in.}^2 \, \text{rad}^{-1}$ sec, because the amplitudes A_p have units of displacement and H_p is nondimensional. The functions $G_f(\omega)$ and $G_x(\omega)$ are plotted in Figure 1.15.

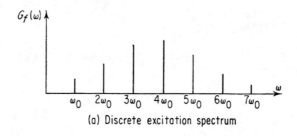

(a) Discrete excitation spectrum

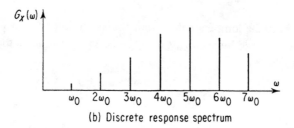

(b) Discrete response spectrum

FIGURE 1.15

It is also customary to use units of acceleration for the excitation amplitude A_p instead of displacement and to plot the spectral densities as functions of the cyclic frequency f. In this case the units of $G_x(f)$ and $G_f(f)$ are g^2 cycle^{-1} sec, where g has units of acceleration.

Although Dirac's delta functions have infinite amplitudes, their integrals are equal to 1,

$$\int_{p\omega_0 - \epsilon}^{p\omega_0 + \epsilon} \delta(\omega - p\omega_0) \, d\omega = 1, \tag{1.96}$$

so the "spikes" in Figure 1.15 can be regarded as representing finite quantities. The functions $G_f(\omega)$ and $G_x(\omega)$ are also called *power spectral density functions*, for reasons that will be discussed in Section 11–7.

The mean square spectral density provides a convenient way to describe random variables. The harmonic analysis will be used extensively in Chapter 11, where we shall deal primarily with continuous rather than discrete spectral densities.

1-10 RESPONSE TO NONPERIODIC EXCITATION. FOURIER INTEGRAL

A periodic function can be represented by a Fourier series as indicated in Section 1–9. If the period T of a periodic function increases indefinitely the function becomes nonperiodic, and the Fourier series representation becomes a Fourier integral representation.

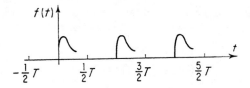

FIGURE 1.16

Consider a periodic function of period T as shown in Figure 1.16, and let it be represented by a Fourier series

$$f(t) = \sum_{p=-\infty}^{\infty} A_p e^{ip\omega_0 t}, \qquad \omega_0 = \frac{2\pi}{T}, \qquad (1.97)$$

where the coefficients are simply

$$A_p = \frac{1}{T} \int_{-T/2}^{T/2} f(t) e^{-ip\omega_0 t}\, dt. \qquad (1.98)$$

The response to this excitation is

$$x(t) = \sum_{p=-\infty}^{\infty} A_p H_p e^{ip\omega_0 t}. \qquad (1.99)$$

Let

$$p\omega_0 = \omega_p, \qquad (p+1)\omega_0 - p\omega_0 = \omega_0 = \frac{2\pi}{T} = \Delta\omega_p,$$

so (1.97) and (1.98) lead to

$$f(t) = \sum_{p=-\infty}^{\infty} \frac{1}{T}(TA_p)e^{i\omega_p t} = \frac{1}{2\pi} \sum_{p=-\infty}^{\infty} (TA_p)e^{i\omega_p t}\, \Delta\omega_p, \qquad (1.100)$$

$$TA_p = \int_{-T/2}^{T/2} f(t) e^{-i\omega_p t}\, dt. \qquad (1.101)$$

Letting $T \to \infty$, dropping the subscript p (so $\omega_p \to \omega$), and replacing the summation by integration, we obtain

$$f(t) = \lim_{\substack{T \to \infty \\ \Delta\omega_p \to 0}} \frac{1}{2\pi} \sum_{p=-\infty}^{\infty} (TA_p)e^{i\omega_p t}\,\Delta\omega_p = \frac{1}{2\pi} \int_{-\infty}^{\infty} F(\omega)e^{i\omega t}\,d\omega, \quad (1.102)$$

$$F(\omega) = \lim_{\substack{T \to \infty \\ \Delta\omega_p \to 0}} (TA_p) = \int_{-\infty}^{\infty} f(t)e^{-i\omega t}\,dt. \quad\quad (1.103)$$

Equation (1.102) is the Fourier integral representation of $f(t)$. The function $F(\omega)$ is the Fourier transformation of $f(t)$, so, in effect,†

$$F(\omega) = \int_{-\infty}^{\infty} f(t)e^{-i\omega t}\,dt, \quad\quad (1.104)$$

$$f(t) = \frac{1}{2\pi} \int_{-\infty}^{\infty} F(\omega)e^{i\omega t}\,d\omega \quad\quad (1.105)$$

represent a Fourier transform pair; $F(\omega)$ can be regarded as the continuous excitation spectral density and $F(\omega)\,d\omega$ as the contribution of the harmonics in the frequency interval ω to $\omega + d\omega$ to the excitation $f(t)$. This derivation is not mathematically rigorous‡ but is sufficient for our purposes; the derived expressions, however, are correct.

Similar reasoning leads to the Fourier integral representation

$$x(t) = \frac{1}{2\pi} \int_{-\infty}^{\infty} F(\omega)H(\omega)e^{i\omega t}\,d\omega, \quad\quad (1.106)$$

which is the counterpart of (1.99) for periodic functions.

The Fourier transform pair corresponding to the response is

$$X(\omega) = \int_{-\infty}^{\infty} x(t)e^{-i\omega t}\,dt, \quad\quad (1.107)$$

$$x(t) = \frac{1}{2\pi} \int_{-\infty}^{\infty} X(\omega)e^{i\omega t}\,d\omega, \quad\quad (1.108)$$

leading us to conclude that

$$X(\omega) = F(\omega)H(\omega). \quad\quad (1.109)$$

† Note the change in the sign of $i\omega t$ as compared with (C.5).
‡ For a more rigorous discussion see I. A. Sneddon, *Fourier Transforms*, McGraw-Hill, New York, 1951, p. 7.

In words, the Fourier transform of the response is the product of the Fourier transform of the excitation and the complex frequency response.

We showed in Section 1–7 that the response can be written in the form of the convolution integrals,

$$x(t) = \int_0^t h(\tau)f(t - \tau)\, d\tau = \int_0^t f(\tau)h(t - \tau)\, d\tau. \tag{1.110}$$

Whereas (1.106) represents the response as a superposition of responses to a continuous spectrum of harmonic excitations, (1.110) represents the response as a superposition of responses to a continuous series of impulsive excitations.

Assume for the moment that the excitation is just the unit impulse

$$f(t) = 1 \cdot \delta(t). \tag{1.111}$$

It follows from (1.104) that

$$F(\omega) = 1 \tag{1.112}$$

and from (1.109) that

$$X(\omega) = H(\omega). \tag{1.113}$$

But the response to the unit impulse is just the impulsive response, so combining (1.108) and (1.113) we can write

$$h(t) = \frac{1}{2\pi} \int_{-\infty}^{\infty} X(\omega)e^{i\omega t}\, d\omega = \frac{1}{2\pi} \int_{-\infty}^{\infty} H(\omega)e^{i\omega t}\, d\omega. \tag{1.114}$$

It follows that the complex frequency response is just the Fourier transform of the impulsive response

$$H(\omega) = \int_{-\infty}^{\infty} h(t)e^{-i\omega t}\, dt. \tag{1.115}$$

Hence a system can be described by either the impulsive response or the complex frequency response (Figure 1.17).

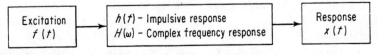

FIGURE 1.17

Problems

1.1 Calculate the response of the system shown in Figure 1.7 if the initial displacement is zero and the initial velocity is v_0. Plot the response for $\zeta = 0.25$ and $\zeta = 2.0$ over the interval $0 \le t \le 6\pi/\omega_n$.

1.2 Calculate the indicial response of the system shown in Figure 1.7 using (1.48). Plot the indicial response for $\zeta = 0.25$ and $\zeta = 2.0$ over the interval $0 \le t \le 6\pi/\omega_n$.

1.3 Calculate the impulsive response of the system shown in Figure 1.7 using (1.53). Check the validity of (1.56) using the result of Problem 1.2. Plot the impulsive response for $\zeta = 0.25$ and $\zeta = 2.0$ over the interval $0 \le t \le 6\pi/\omega_n$.

1.4 Using the impulsive response curves of Problem 1.3 for $\zeta = 0.25$ and $\zeta = 2.0$, follow the sequence indicated in Figure 1.12 and obtain graphically the corresponding indicial responses. Plot the indicial responses over the interval $0 \le t \le 6\pi/\omega_n$ and compare the curves with the curves obtained in Problem 1.2.

1.5 Plot $|H(\omega)|$ vs. ω for the spring-mass-damper system of Figure 1.7. Use $\zeta = 0$, $\zeta = 0.1$, and $\zeta = 2.0$, and plot $|H(\omega)|$ over the interval $0 \le \omega \le 2.5\omega_n$. For $\zeta = 0.1$ calculate $|H(\omega)|_{\max}$ and find the value of ω at which this maximum occurs. Calculate the half power points and the bandwidth for $\zeta = 0.1$.

1.6 Prove that any periodic function $f(t)$ with zero average value can be represented by a Fourier series as given by (1.88). Expand the periodic function $f(t)$, as shown in Figure 1.18,

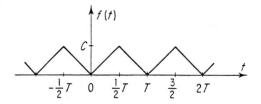

FIGURE 1.18

(a) In a Fourier series of the form

$$f(t) = \tfrac{1}{2}a_0 + \sum_{p=1}^{\infty} \left(a_p \cos \frac{2\pi p t}{T} + b_p \sin \frac{2\pi p t}{T} \right).$$

(b) In a complex Fourier series as given by (1.86).
(c) State the correction to be introduced in (1.88) for it to be the equivalent of (1.86).

1.7 Let $f(t)$ as shown in Problem 1.6 be the excitation function and calculate the mean square excitation. For the spring-mass-damper calculate the mean square response. Plot the excitation and response spectral densities for $\omega_n = 2.9\omega_0$ and $\zeta = 0.01$. Use $0 \le p \le 8$. Draw conclusions as to the relation between the excitation and response spectra.

1.8 Calculate the Fourier transform $F(\omega)$ of the pulse

$$f(t) = \begin{cases} C, & |t| < a, \\ 0, & |t| > a, \end{cases}$$

and plot it as a function of ω for $-2.5\pi/a \le \omega \le 2.5\pi/a$. Let $2Ca = k = $ const and calculate the value of $F(\omega)$ as $a \to 0$. Draw conclusions as to the relation between $f(t)$ and $F(\omega)$ under these conditions.

1.9 Let the excitation of the spring-mass-damper system be

$$F(t) = kf(t) = k\,\delta(t)$$

and calculate the impulsive response $h(t)$. Introduce the result in (1.115) and calculate the complex frequency response $H(\omega)$. Compare $H(\omega)$ with the transfer function $G(s)$ of the system and draw conclusions.

Selected Readings

Aseltine, J. A., *Transform Method in Linear System Analysis*, McGraw-Hill, New York, 1958.

Bendat, J. S., L. D. Enochson, G. H. Klein, and A. G. Piersol, The Application of Statistics to the Flight Vehicle Vibration Problems, *ASD TR 61–123*, Aeronautical Systems Division, Air Force Systems Command, Wright-Patterson AFB, Ohio, 1961.

Crandall, S. H., ed., *Random Vibration*, M.I.T. Press, Cambridge, Mass., 1958.

Myklestad, N. O., *Fundamentals of Vibration Analysis*, McGraw-Hill, New York, 1956.

Thomson, W. T., *Laplace Transformation*, Prentice-Hall, Englewood Cliffs, N.J., 1960.

CHAPTER 2

ADVANCED
PRINCIPLES
OF DYNAMICS

2-1 GENERAL CONSIDERATIONS

Newton's laws were formulated for a single particle. The second law leads to a differential equation which, upon integration, yields the motion of the particle. In the case of a system of several particles, a differential equation must be written for each particle. The resulting system of equations contains interacting forces and the solutions may be difficult to obtain. The principal concepts in this approach are force and momentum, both vector quantities.

A different approach, called *analytical mechanics* and attributed to Leibnitz and Lagrange, considers the system as a whole and requires knowledge of two scalar functions, the kinetic energy and the work function. Quite often there are present constraint forces, reflecting the action of the surroundings upon a particle, which cause certain kinematical conditions. Whereas the Newtonian approach requires knowledge of these forces, the analytical mechanics approach requires only the kinematical relations resulting from the forces. The analytical approach leads to unifying formulations such as Hamilton's principle and Lagrange's equation, which do not depend on the coordinate system used.

When dealing with multi-degree-of-freedom systems, it is often more expedient to derive the equations of motion by using the analytical rather than the Newtonian approach. The analytical approach can also be used for continuous systems, in which case one obtains not only the differential equations of motion but also the associated boundary conditions. In future discussions the advantages of the analytical approach will be amply illustrated.

The present chapter presents the general ideas behind analytical mechanics.

2–2 WORK AND ENERGY. SINGLE PARTICLE

Consider a particle m moving along curve C under the action of force $\mathbf{F}$ (Figure 2.1). The position of the particle at any time is given by the position

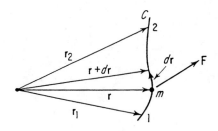

FIGURE 2.1

vector $\mathbf{r}$. If the particle is moved over an element of distance $d\mathbf{r}$, the element of work is, by definition, given by the dot (or scalar) product

$$dW = \mathbf{F} \cdot d\mathbf{r}, \tag{2.1}$$

where W denotes the work.

If the particle travels from point 1 to point 2, along curve C, the work done is

$$W_{12} = \int_{\mathbf{r}_1}^{\mathbf{r}_2} \mathbf{F} \cdot d\mathbf{r}. \tag{2.2}$$

For a particle of constant mass, Newton's second law gives $\mathbf{F} = m(d\dot{\mathbf{r}}/dt)$, and, as $d\mathbf{r} = \dot{\mathbf{r}}\,dt$, we can write

$$W_{12} = \int_{\mathbf{r}_1}^{\mathbf{r}_2} \mathbf{F} \cdot d\mathbf{r} = \int_{t_1}^{t_2} m \frac{d\dot{\mathbf{r}}}{dt} \cdot \dot{\mathbf{r}}\, dt = \frac{1}{2} \int_{t_1}^{t_2} m \frac{d}{dt}(\dot{\mathbf{r}} \cdot \dot{\mathbf{r}})\, dt$$

$$= \tfrac{1}{2}m\dot{\mathbf{r}}_2 \cdot \dot{\mathbf{r}}_2 - \tfrac{1}{2}m\dot{\mathbf{r}}_1 \cdot \dot{\mathbf{r}}_1 = \tfrac{1}{2}m\dot{r}_2^2 - \tfrac{1}{2}m\dot{r}_1^2 = T_2 - T_1, \tag{2.3}$$

where $\dot{\mathbf{r}}$ is the velocity at any point and T is the kinetic energy,

$$T = \tfrac{1}{2}m\dot{\mathbf{r}} \cdot \dot{\mathbf{r}}. \tag{2.4}$$

Define a conservative force field as a field in which the work done in moving a particle from point 1 to point 2 depends only on the positions $\mathbf{r}_1$ and $\mathbf{r}_2$ and is independent of the path of integration,

$$W_{12c} = \int_{\mathbf{r}_1}^{\mathbf{r}_2} \mathbf{F} \cdot d\mathbf{r} = \int_{\mathbf{r}_1}^{\mathbf{r}_2} \mathbf{F} \cdot d\mathbf{r}, \tag{2.5}$$

$$\text{path I} \qquad\qquad \text{path II}$$

where the subscript c denotes the work done by conservative forces only (see Figure 2.2).

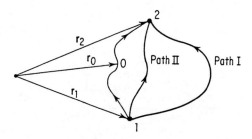

FIGURE 2.2

The potential energy $V(\mathbf{r}_1)$ associated with position $\mathbf{r}_1$ is defined as the work done in a conservative force field in moving a particle from position $\mathbf{r}_1$ to a *reference position* $\mathbf{r}_0$,

$$V(\mathbf{r}_1) = \int_{\mathbf{r}_1}^{\mathbf{r}_0} \mathbf{F} \cdot d\mathbf{r}. \tag{2.6}$$

Next let us calculate the work done in a conservative force field in moving a particle from position $\mathbf{r}_1$ to position $\mathbf{r}_2$. The path is immaterial, so choose a path passing through the reference position $\mathbf{r}_0$. Hence

$$W_{12c} = \int_{\mathbf{r}_1}^{\mathbf{r}_2} \mathbf{F} \cdot d\mathbf{r} = \int_{\mathbf{r}_1}^{\mathbf{r}_0} \mathbf{F} \cdot d\mathbf{r} + \int_{\mathbf{r}_0}^{\mathbf{r}_2} \mathbf{F} \cdot d\mathbf{r}$$

$$= \int_{\mathbf{r}_1}^{\mathbf{r}_0} \mathbf{F} \cdot d\mathbf{r} - \int_{\mathbf{r}_2}^{\mathbf{r}_0} \mathbf{F} \cdot d\mathbf{r} = -[V(\mathbf{r}_2) - V(\mathbf{r}_1)]. \tag{2.7}$$

Equation (2.7) states that the work done in a conservative field is the negative of the change in the potential energy, which is expressed in differential form by

$$dW_c = -dV. \tag{2.8}$$

In general there are both conservative and nonconservative forces acting upon a particle. The nonconservative forces are energy-dissipating forces, such as friction forces, or forces imparting energy to the system, such as external forces. Nonconservative forces are recognized as forces that do not depend on position alone and cannot be derived from a potential function. In contrast, the conservative forces do depend on position and can be derived from the potential energy expression. Equation (2.8) gives

$$dW_c = \mathbf{F}_c \cdot d\mathbf{r} = -dV = -\nabla V \cdot d\mathbf{r}, \tag{2.9}$$

where ∇ is an operator called *del* or *nabla* and in Cartesian coordinates has the form

$$\nabla = \frac{\partial}{\partial x}\mathbf{i} + \frac{\partial}{\partial y}\mathbf{j} + \frac{\partial}{\partial z}\mathbf{k}, \tag{2.10}$$

where $\mathbf{i}$, $\mathbf{j}$, and $\mathbf{k}$ are the corresponding unit vectors. The symbol ∇V means the gradient of the function V and is a vector. Equation (2.9) implies that

$$\mathbf{F}_c = -\nabla V, \tag{2.11}$$

or the conservative force vector is the negative of the gradient of the potential energy function.

Now form

$$\nabla \times \mathbf{F}_c = -\nabla \times \nabla V, \tag{2.12}$$

and, because for any function ϕ such that $\partial^2\phi/\partial x\,\partial y = \partial^2\phi/\partial y\,\partial x$, etc., we obtain

we conclude that
$$\nabla \times \nabla \phi = 0, \tag{2.13}$$

$$\nabla \times \mathbf{F}_c = \text{curl } \mathbf{F}_c = 0. \tag{2.14}$$

Thus for a conservative force field the *curl* of $\mathbf{F}_c$ is zero.

Equation (2.5) implies that

$$\oint \mathbf{F}_c \cdot d\mathbf{r} = 0, \tag{2.15}$$

or the work done by conservative forces around a closed path is zero. Either (2.14) or (2.15) can be used to determine whether a force is conservative or not. Quite often (2.14) is to be preferred, because it involves differentiation instead of integration. More often than not, however, we shall determine whether the force is conservative on the basis of physical considerations— whether the force brings about a change in the energy of the system or not.

The work expression can be divided into conservative and nonconservative work,

$$W_{12} = W_{12c} + W_{12nc}. \tag{2.16}$$

Equations (2.3), (2.7), and (2.16) combined give the nonconservative work expression

$$W_{12nc} = W_{12} - W_{12c} = (T_2 - T_1) + (V_2 - V_1)$$

$$= (T_2 + V_2) - (T_1 + V_1) = E_2 - E_1, \tag{2.17}$$

where E denotes the total energy, which is the sum of the kinetic and potential energies,

$$E = T + V. \tag{2.18}$$

Equation (2.17) indicates that the work done by nonconservative forces is responsible for the change in the total energy of the particle.

In the case of a conservative force field,

$$W_{12nc} = 0, \tag{2.19}$$

from which it follows that

$$E_2 = E_1 = E = \text{const}, \tag{2.20}$$

which means that the total energy is a constant. Equation (2.20) is the statement of the conservation of energy principle.

2–3 STRAIN ENERGY. ELASTICITY

One type of potential energy of considerable interest is the *elastic potential energy*, or *strain energy*.

Consider a spring stretched from an initial length L by an amount δ (Figure 2.3). Assume the spring is in the linear range, so that the force in the spring is proportional to the displacement, and for an intermediate position ζ, such that $0 < \zeta < \delta$, the restoring force (the force that tends to return the spring to the unstretched configuration) is

$$F_\zeta = -k\zeta. \tag{2.21}$$

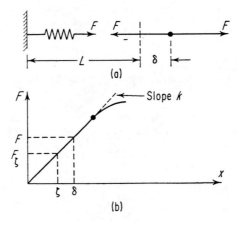

FIGURE 2.3

The force is negative, because it acts in the direction opposite to the displacement.

The potential energy $V(\delta)$ in the position δ is

$$V(\delta) = \int_{\delta}^{0} F_{\zeta}\, d\zeta = -k \int_{\delta}^{0} \zeta\, d\zeta = \tfrac{1}{2}k\delta^2 = \tfrac{1}{2}F\delta. \tag{2.22}$$

This is the energy stored in the spring and is equal in magnitude to the work done by the external force.

Analogous behavior is exhibited by any piece of elastic material. The counterpart of the force-deflection relation is the stress-strain curve. The latter relation must be used in the case of continuous systems (Figure 2.4).

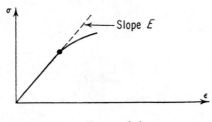

FIGURE 2.4

As an illustration consider a rod of variable cross section $A(x)$ under tension (or compression) and examine an infinitesimal element of length dx (Figure 2.5). Using the standard procedure of calculus we consider the cross

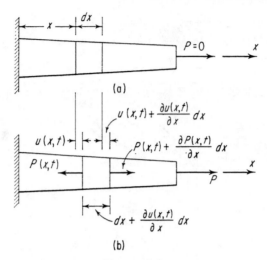

FIGURE 2.5

section as uniform over the interval dx, so the elongation of the element dx under the force $P(x, t)$ is

$$\delta(dx) = \frac{\partial u(x, t)}{\partial x} dx = \epsilon(x, t) dx, \tag{2.23}$$

where $u(x, t)$ is the displacement and $\epsilon(x, t)$ is the strain at point x.

The assumption made here is that displacements, strains, and stresses are uniform at a given cross section, which is equivalent to saying that plane cross sections remain plane during deformation. This assumption is justified when the cross-sectional dimensions of the bar are small in relation to the length of the bar.

The strain energy associated with the element of volume $A(x) dx$ is

$$dV(x, t) = \tfrac{1}{2}P(x, t)\delta(dx) = \tfrac{1}{2}P(x, t)\epsilon(x, t) dx. \tag{2.24}$$

The axial stress $\sigma(x, t)$ is defined by

$$\sigma(x, t) = \frac{P(x, t)}{A(x)}, \tag{2.25}$$

and as long as the material remains in the linear portion of the σ vs. ϵ curve the stress is also given by

$$\sigma(x, t) = E\epsilon(x, t). \tag{2.26}$$

A combination of (2.24) to (2.26) yields

$$dV(x, t) = \frac{1}{2}\frac{P^2(x, t)}{EA(x)} dx = \frac{1}{2} EA(x)\epsilon^2(x, t) dx. \tag{2.27}$$

Integrating over the entire length of the rod, we obtain the expression of strain energy for axial tension (or compression),

$$V(t) = \tfrac{1}{2} \int_0^L \frac{P^2(x, t)}{EA(x)}\, dx = \tfrac{1}{2} \int_0^L EA(x)\epsilon^2(x, t)\, dx. \qquad (2.28)$$

2–4 SYSTEMS WITH CONSTRAINTS. DEGREES OF FREEDOM

Consider a simple pendulum consisting of a mass m suspended by an inextensible string of length L and free to oscillate in the xy plane (Figure 2.6).

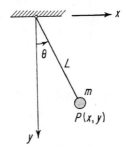

FIGURE 2.6

The position of mass m can be given by the coordinates x and y. But these coordinates are not independent, because

$$x^2 + y^2 = L^2 . \qquad (2.29)$$

Equation (2.29) is a constraint equation and indeed in this case we need only one coordinate to express the position of m. It can be either x or y, but it is even more expedient to use the angle θ, as shown in Figure 2.6.

The minimum number of independent coordinates needed to describe the motion of a system is called the *degree of freedom* of the system. Hence our pendulum is a *single-degree-of-freedom system*.

The position of a particle free to move in three dimensions is described by three coordinates. If a system of N particles must satisfy c constraint equations, the number of independent coordinates needed to describe the system is

$$n = 3N - c. \qquad (2.30)$$

because each constraint equation reduces the degree of freedom of a system by 1. Such a system is called an *n-degree-of-freedom system.*

The constraint equation, (2.29), can be written

$$f(x, y, t) = f(x, y) = c = \text{const.} \tag{2.31}$$

A system in which the constraints are functions of the coordinates or coordinates and time is called *holonomic.*

In many cases the constraint equations relate the velocities or the velocities and coordinates. If these equations can be integrated to yield relations involving coordinates and time only, the system is still holonomic. As an example, the equation

$$x\dot{x} + y\dot{y} = 0 \tag{2.32}$$

can be integrated to yield the condition (2.29).

In other cases one may have relations such as

$$A_1\dot{x} + A_2\dot{y} + A_3\dot{z} = 0, \tag{2.33}$$

where A_1, A_2, and A_3 are functions of x, y, and z, and there is no way of integrating this expression to yield a relation containing the coordinates and time. Such a system is called *nonholonomic.*

In this text we shall be concerned only with holonomic systems.

2–5 GENERALIZED COORDINATES

The minimum number of coordinates necessary to describe a system constitutes a set of *generalized coordinates.* The generalized coordinates are all independent and their number corresponds to the number of degrees of freedom of a system. They are commonly denoted $q_1, q_2, \cdots, q_n$ and do not necessarily represent Cartesian coordinates. Quantities such as the amplitudes of a Fourier series expansion can play the role of generalized coordinates.

The double pendulum is a two-degree-of-freedom system (Figure 2.7). As a set of generalized coordinates we can use the angular displacements

$$q_1 = \theta_1, \qquad q_2 = \theta_2. \tag{2.34}$$

Although these are not the only possible generalized coordinates for this problem, they are probably the most convenient ones.

The selection of generalized coordinates is quite obvious in some cases, but in other cases, in which the coordinates are related by constraint equations,

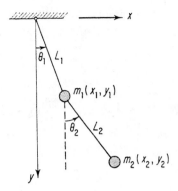

FIGURE 2.7

one must use a coordinate transformation to arrive at a set of generalized coordinates. An example of such a coordinate transformation is given in Section 4–9.

2–6 PRINCIPLE OF VIRTUAL WORK. STATIC CASE

The principle of virtual work is essentially a statement of the equilibrium definition of mechanical systems and is due to Johann Bernoulli.

We define virtual displacements δx_1, δy_1, δz_1, δx_2, ..., δz_n as infinitesimal, arbitrary changes in the coordinates of a system. These are small variations from the true position of the system and must be compatible with the constraints of the system. As an example, consider a particle that lies in equilibrium on a surface and imagine the particle in a displaced position in the neighborhood of the true position. For the virtual displacements to be compatible with the system constraints, the particle in the displaced position must still lie on the given surface. The virtual displacements are not true displacements, so there is no time change associated with them. The symbol δ was introduced by Lagrange to emphasize the virtual character of the variations as opposed to the symbol d, which designates differentials. Hence if the actual coordinates are related by the constraint equation

$$f(x_1, x_2, \ldots, y_1, y_2, \ldots, z_1, z_2, \ldots, z_n, t) = c, \qquad (2.35)$$

the virtual displacements must be such that

$$f(x_1 + \delta x_1, \ldots, y_1 + \delta y_1, \ldots, z_1 + \delta z_1, \ldots, z_n + \delta z_n, t) = c. \quad (2.36)$$

Note that the time was held constant.

The operations concerning δ follow the rules of elementary calculus. Expanding (2.36) in a Taylor series and neglecting higher order terms in δx_i, δy_i, and δz_i, we obtain

$$f(x_1, x_2, \ldots, y_1, y_2, \ldots, z_1, z_2, \ldots, z_n, t)$$

$$+ \sum_{i=1}^{n} \left(\frac{\partial f}{\partial x_i} \delta x_i + \frac{\partial f}{\partial y_i} \delta y_i + \frac{\partial f}{\partial z_i} \delta z_i \right) = c. \qquad (2.37)$$

In view of (2.35), however, we conclude that

$$\sum_{i=1}^{n} \left(\frac{\partial f}{\partial x_i} \delta x_i + \frac{\partial f}{\partial y_i} \delta y_i + \frac{\partial f}{\partial z_i} \delta z_i \right) = 0, \qquad (2.38)$$

which is the relation the virtual displacements δx_i, δy_i, and δz_i must satisfy to be compatible with the constraints of the system.

Now consider a particle i acted upon by some forces with resultant vector $\mathbf{R}_i$. If the system is in equilibrium, the resultant force is zero and the work done over the virtual displacement $\delta \mathbf{r}_i$ must be zero; that is,

$$\delta W_i = \mathbf{R}_i \cdot \delta \mathbf{r}_i = 0. \qquad (2.39)$$

If there are constraints in the system, then

$$\mathbf{R}_i = \mathbf{F}_i + \mathbf{f}_i = 0, \qquad (2.40)$$

where $\mathbf{F}_i$ is the resultant vector of externally applied forces and $\mathbf{f}_i$ is the resultant of the constraint forces. Hence (2.39) becomes

$$\mathbf{R}_i \cdot \delta \mathbf{r}_i = \mathbf{F}_i \cdot \delta \mathbf{r}_i + \mathbf{f}_i \cdot \delta \mathbf{r}_i = 0. \qquad (2.41)$$

But, in general, constraint forces do not perform work, because the displacements do not have any components in the direction of the constraint forces. As an example, consider a particle moving on a smooth surface. The constraint force is normal to the surface and the displacements are parallel to the surface. Hence

$$\mathbf{f}_i \cdot \delta \mathbf{r}_i = 0, \qquad (2.42)$$

and it follows that

$$\mathbf{F}_i \cdot \delta \mathbf{r}_i = F_{x_i} \delta x_i + F_{y_i} \delta y_i + F_{z_i} \delta z_i = 0. \qquad (2.43)$$

In general, for a system of N particles, the sum of the virtual work over all particles must be zero, or

$$\delta W = \sum_{i=1}^{N} \delta W_i = \sum_{i=1}^{N} \mathbf{R}_i \cdot \delta \mathbf{r}_i = 0. \qquad (2.44)$$

Again eliminating the constraint forces, including the internal forces in rigid bodies, one obtains

$$\sum_{i=1}^{N} \mathbf{F}_i \cdot \delta \mathbf{r}_i = 0, \tag{2.45}$$

which is the expression of the *virtual work principle*. The virtual work principle can be stated as: If a system of forces is in equilibrium, the work done by the externally applied forces through virtual displacements compatible with the constraints of the system is zero.

The variation in the potential energy associated with the virtual displacements δx, δy, and δz of a single particle is

$$\delta V = \frac{\partial V}{\partial x} \delta x + \frac{\partial V}{\partial y} \delta y + \frac{\partial V}{\partial z} \delta z. \tag{2.46}$$

In a conservative field the work is the negative of the change in potential

$$\delta W = -\delta V. \tag{2.47}$$

But for equilibrium the virtual work is zero,

$$\delta W = \mathbf{F} \cdot d\mathbf{r} = -\frac{\partial V}{\partial x} \delta x - \frac{\partial V}{\partial y} \delta y - \frac{\partial V}{\partial z} \delta z = 0. \tag{2.48}$$

If δx, δy, and δz are independent, (2.48) is satisfied only if each partial derivative is zero. But

$$\frac{\partial V}{\partial x} = 0, \qquad \frac{\partial V}{\partial y} = 0, \qquad \frac{\partial V}{\partial z} = 0 \tag{2.49}$$

are precisely the conditions for the function V to have an extremal value which could be a minimum, a maximum, or a stationary value (neither a minimum nor a maximum). A function is said to have a stationary value at a given point if the rate of change of the function in any direction vanishes at that point.

According to a theorem by Dirichlet,† a position of stable equilibrium is obtained only if V has a minimum in the position in which equations (2.49) are satisfied.

2–7 D'ALEMBERT'S PRINCIPLE

The principle of virtual work was derived for the case of static equilibrium. One can extend it to the dynamic case and speak of *dynamic equilibrium*.

† See S. W. McCuskey, *An Introduction to Advanced Dynamics*, Addison-Wesley, Reading, Mass., 1962, p. 46.

If there are some unbalanced forces acting upon a particle m_i, then according to Newton's second law the force resultant vector must be equal to the rate of change of the linear momentum vector,

$$\mathbf{F}_i + \mathbf{f}_i = \dot{\mathbf{p}}_i. \tag{2.50}$$

One can think of an *inertia force*, which is a force whose magnitude is equal to that of the rate of change of the momentum vector, is collinear with it, but acts in the opposite direction. If such a force is applied to the particle, one can express dynamic equilibrium in the form

$$\mathbf{F}_i + \mathbf{f}_i - \dot{\mathbf{p}}_i = 0. \tag{2.51}$$

Hence *D'Alembert's principle* states that the resultant force is in equilibrium with the inertia force.

In that fashion the principle of virtual work can be extended to cover the dynamic case. Following the previous line of reasoning we have

$$(\mathbf{F}_i - \dot{\mathbf{p}}_i) \cdot \delta \mathbf{r}_i = 0. \tag{2.52}$$

For a system of N particles we write

$$\sum_{i=1}^{N} (\mathbf{F}_i - \dot{\mathbf{p}}_i) \cdot \delta \mathbf{r}_i = \sum_{i=1}^{N} (F_{x_i} - m_i \ddot{x}_i)\, \delta x_i$$

$$+ \sum_{i=1}^{N} (F_{y_i} - m_i \ddot{y}_i)\, \delta y_i + \sum_{i=1}^{N} (F_{z_i} - m_i \ddot{z}_i)\, \delta z_i = 0. \tag{2.53}$$

2–8 VARIATIONAL PRINCIPLES. HAMILTON'S PRINCIPLE

D'Alembert's principle gives a complete formulation of the problems of mechanics. A different formulation, based on D'Alembert's principle, leads to Hamilton's principle, which is perhaps the most advanced variational principle of mechanics. Hamilton's principle considers the entire motion of the system between two instants t_1 and t_2 and is therefore an integral principle. The problems of dynamics are reduced to the investigation of a scalar integral. One remarkable advantage of this formulation is that it is invariant with respect to the coordinate system used.

Consider a system of N particles. The system may be subject to kinematical conditions. The virtual work expression in conjunction with D'Alembert's principle is

$$\sum_{i=1}^{N} (m_i \ddot{\mathbf{r}}_i - \mathbf{F}_i) \cdot \delta \mathbf{r}_i = 0, \tag{2.54}$$

and note that

$$\sum_{i=1}^{N} \mathbf{F}_i \cdot \delta \mathbf{r}_i = \delta W \tag{2.55}$$

is the virtual work done by the forces.

Now assume that the operations d/dt and δ are interchangeable, and consider

$$\frac{d}{dt} (\dot{\mathbf{r}}_i \cdot \delta \mathbf{r}_i) = \ddot{\mathbf{r}}_i \cdot \delta \mathbf{r}_i + \dot{\mathbf{r}}_i \cdot \delta \dot{\mathbf{r}}_i = \ddot{\mathbf{r}}_i \cdot \delta \mathbf{r}_i + \delta(\tfrac{1}{2} \dot{\mathbf{r}}_i \cdot \dot{\mathbf{r}}_i)$$

or

$$\ddot{\mathbf{r}}_i \cdot \delta \mathbf{r}_i = \frac{d}{dt} (\dot{\mathbf{r}}_i \cdot \delta \mathbf{r}_i) - \delta(\tfrac{1}{2} \dot{\mathbf{r}}_i \cdot \dot{\mathbf{r}}_i). \tag{2.56}$$

Multiply (2.56) by m_i and sum over the entire system to obtain

$$\sum_{i=1}^{N} m_i \ddot{\mathbf{r}}_i \cdot \delta \mathbf{r}_i = \sum_{i=1}^{N} m_i \frac{d}{dt} (\dot{\mathbf{r}}_i \cdot \delta \mathbf{r}_i) - \delta \sum_{i=1}^{N} \tfrac{1}{2} m_i (\dot{\mathbf{r}}_i \cdot \dot{\mathbf{r}}_i)$$

$$= \sum_{i=1}^{N} m_i \frac{d}{dt} (\dot{\mathbf{r}}_i \cdot \delta \mathbf{r}_i) - \delta T, \tag{2.57}$$

where T is the kinetic energy of the system.

Introducing (2.55) and (2.57) in (2.54), one obtains

$$\delta T + \delta W = \sum_{i=1}^{N} m_i \frac{d}{dt} (\dot{\mathbf{r}}_i \cdot \delta \mathbf{r}_i). \tag{2.58}$$

The instantaneous configuration of a system is given by the values of n generalized coordinates. These values correspond to a point in an n-dimensional space known as the *configuration space*. The configuration of the system changes with time, tracing a path known as the *true path* (also called *Newtonian* or *dynamical path*) in the configuration space. A slightly different path, known as the *varied path*, is obtained if at any given instant one allows a small variation in position $\delta \mathbf{r}_i$ with no associated change in time:

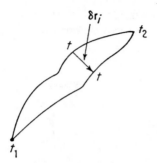

FIGURE 2.8

$\delta t = 0$ (Figure 2.8). The stipulation is made, however, that at two instants t_1 and t_2 the true and varied paths coincide:

$$\delta \mathbf{r}_i = 0 \quad \text{at} \quad t = t_1 \quad \text{and} \quad t = t_2. \tag{2.59}$$

Multiply (2.58) by dt and integrate between t_1 and t_2. The result is

$$\int_{t_1}^{t_2} (\delta T + \delta W) \, dt = \int_{t_1}^{t_2} \sum_{i=1}^{N} m_i \frac{d}{dt} (\dot{\mathbf{r}}_i \cdot \delta \mathbf{r}_i) \, dt$$

$$= \sum_{i=1}^{N} \int_{t_1}^{t_2} m_i \frac{d}{dt} (\dot{\mathbf{r}}_i \cdot \delta \mathbf{r}_i) \, dt = \sum_{i=1}^{N} m_i \dot{\mathbf{r}}_i \cdot \delta \mathbf{r}_i \Big|_{t_1}^{t_2}. \tag{2.60}$$

Recalling (2.59), one obtains

$$\int_{t_1}^{t_2} (\delta T + \delta W) \, dt = 0. \tag{2.61}$$

In the case in which the forces are conservative we can write

$$\delta W = -\delta V, \tag{2.62}$$

so that, introducing the *Lagrangian L* in the form

$$L = T - V, \tag{2.63}$$

and, assuming that the system is holonomic, (2.61) reduces to

$$\delta \int_{t_1}^{t_2} L \, dt = 0, \tag{2.64}$$

which is the mathematical statement of *Hamilton's principle*. It may be stated in words as follows: The actual path (true path) renders the value of the integral $\int_{t_1}^{t_2} L \, dt$ stationary with respect to all possible neighboring paths that the system may be imagined to take between two instants t_1 and t_2, provided the initial and final configurations of the system are prescribed. The stationary value is actually a minimum.

Equation (2.61) can be looked upon as a generalization of Hamilton's principle† so nonconservative forces are included by letting W consist of both conservative and nonconservative forces.

Hamilton's principle is an example of a variational principle which reduces the problems of dynamics to the investigation of a scalar integral that does not depend on the coordinates used. The condition rendering the value of the integral stationary leads to all the equations of motion. Hamilton's principle is a formulation and not a solution of the problems of dynamics. The principle belongs to a broader class of principles called *principles of least action*.

Example 2.1   Equation of Motion of a Thin Rod in Tension

As an application of Hamilton's principle let us derive the equation for the longitudinal motion of a thin rod.

In Section 2.3 we derived an expression for the potential or strain energy of a slender rod deforming longitudinally. The expression is

$$V(t) = \tfrac{1}{2} \int_0^L EA(x) \left[\frac{\partial u(x,\, t)}{\partial x} \right]^2 dx, \tag{2.65}$$

where $EA(x)$ denotes the stiffness and $u(x, t)$ is the longitudinal displacement. Under the same assumption made there, the kinetic energy is simply

$$T(t) = \tfrac{1}{2} \int_0^L m(x) \left[\frac{\partial u(x,\, t)}{\partial t} \right]^2 dx, \tag{2.66}$$

where $m(x)$ is the mass per unit length of rod.

Equations (2.65) and (2.66), when introduced in the Hamilton principle expression, yield

$$\delta \int_{t_1}^{t_2} (T - V) \, dt = \delta \int_{t_1}^{t_2} \left[\tfrac{1}{2} \int_0^L m \left(\frac{\partial u}{\partial t} \right)^2 dx - \tfrac{1}{2} \int_0^L EA \left(\frac{\partial u}{\partial x} \right)^2 dx \right] dt$$

$$= \int_{t_1}^{t_2} \left[\int_0^L m \frac{\partial u}{\partial t} \, \delta \left(\frac{\partial u}{\partial t} \right) dx - \int_0^L EA \frac{\partial u}{\partial x} \, \delta \left(\frac{\partial u}{\partial x} \right) dx \right] dt = 0.$$

$$\tag{2.67}$$

† We shall refer to (2.61) as the *extended Hamilton principle*.

Assuming, on the one hand, that the operators δ and $\partial/\partial t$, as well as δ and $\partial/\partial x$, are commutative, and, on the other hand, that integrations with respect to t and x are interchangeable, we can integrate by parts and obtain

$$\int_{t_1}^{t_2} m \frac{\partial u}{\partial t} \delta\left(\frac{\partial u}{\partial t}\right) dt = \int_{t_1}^{t_2} m \frac{\partial u}{\partial t} \frac{\partial}{\partial t} (\delta u) \, dt = \left(m \frac{\partial u}{\partial t}\right) \delta u \Big|_{t_1}^{t_2}$$

$$- \int_{t_1}^{t_2} \frac{\partial}{\partial t}\left(m \frac{\partial u}{\partial t}\right) \delta u \, dt = - \int_{t_1}^{t_2} m \frac{\partial^2 u}{\partial t^2} \delta u \, dt, \quad (2.68)$$

because δu vanishes at $t = t_1$ and $t = t_2$.

In addition we can write

$$\int_0^L EA \frac{\partial u}{\partial x} \delta\left(\frac{\partial u}{\partial x}\right) dx = \int_0^L EA \frac{\partial u}{\partial x} \frac{\partial}{\partial x} (\delta u) \, dx$$

$$= \left(EA \frac{\partial u}{\partial x}\right) \delta u \Big|_0^L - \int_0^L \frac{\partial}{\partial x}\left(EA \frac{\partial u}{\partial x}\right) \delta u \, dx. \quad (2.69)$$

Equations (2.68) and (2.69), when introduced in (2.67), yield

$$\int_{t_1}^{t_2} \left\{ \int_0^L \left[\frac{\partial}{\partial x}\left(EA \frac{\partial u}{\partial x}\right) - m \frac{\partial^2 u}{\partial t^2}\right] \delta u \, dx - \left(EA \frac{\partial u}{\partial x}\right) \delta u \Big|_0^L \right\} dt = 0. \quad (2.70)$$

But (2.70) must be satisfied for those δu which vanish at $x = 0$ and $x = L$ and, because otherwise δu is arbitrary throughout the domain $0 < x < L$, we must have

$$\frac{\partial}{\partial x}\left(EA \frac{\partial u}{\partial x}\right) - m \frac{\partial^2 u}{\partial t^2} = 0 \qquad (2.71)$$

throughout the domain. Furthermore, if we write

$$\left(EA \frac{\partial u}{\partial x}\right) \delta u \Big|_0^L = 0, \qquad (2.72)$$

we take into account the possibility that either δu or $EA(\partial u/\partial x)$ vanishes at either end.

Equation (2.71) can be identified as the differential equation of motion, and (2.72) represents the conditions that u must satisfy at the boundaries. The physical significance of the boundary conditions will be explored in detail later (see Sections 5–2, 5–4, 5–8, 5–9, and 6–4).

2-9 LAGRANGE'S EQUATION (Holonomic Systems)

In the case of a holonomic system one can express the dependent variables $\mathbf{r}_i$ in terms of n generalized coordinates q_k and time t in the form

$$\mathbf{r}_i = \mathbf{r}_i(q_1, q_2, \ldots, q_n, t), \tag{2.73}$$

where n is the number of degrees of freedom of the system.

The velocities are obtained by differentiating (2.73),

$$\dot{\mathbf{r}}_i = \frac{d\mathbf{r}_i}{dt} = \frac{\partial \mathbf{r}_i}{\partial q_1}\dot{q}_1 + \frac{\partial \mathbf{r}_i}{\partial q_2}\dot{q}_2 + \cdots + \frac{\partial \mathbf{r}_i}{\partial q_n}\dot{q}_n + \frac{\partial \mathbf{r}_i}{\partial t}. \tag{2.74}$$

Using (2.74) the kinetic energy of the system is written

$$T = \tfrac{1}{2} \sum_{i=1}^{N} m_i \dot{\mathbf{r}}_i \cdot \dot{\mathbf{r}}_i$$

$$= \tfrac{1}{2} \sum_{i=1}^{N} m_i \left[\sum_{r=1}^{n} \sum_{s=1}^{n} \frac{\partial \mathbf{r}_i}{\partial q_r} \cdot \frac{\partial \mathbf{r}_i}{\partial q_s} \dot{q}_r \dot{q}_s + 2 \frac{\partial \mathbf{r}_i}{\partial t} \cdot \sum_{r=1}^{n} \frac{\partial \mathbf{r}_i}{\partial q_r} \dot{q}_r + \frac{\partial \mathbf{r}_i}{\partial t} \cdot \frac{\partial \mathbf{r}_i}{\partial t} \right], \tag{2.75}$$

and note that

$$T = T(q_1, q_2, \ldots, q_n, \dot{q}_1, \dot{q}_2, \ldots, \dot{q}_n, t). \tag{2.76}$$

In the following, the generalized Hamilton principle will be used to derive Lagrange's equation. Recalling that the time is not varied, the variation in kinetic energy is

$$\delta T = \sum_{k=1}^{n} \frac{\partial T}{\partial q_k} \delta q_k + \sum_{k=1}^{n} \frac{\partial T}{\partial \dot{q}_k} \delta \dot{q}_k. \tag{2.77}$$

Integrating δT with respect to time from t_1 to t_2, one obtains

$$\int_{t_1}^{t_2} \delta T \, dt = \sum_{k=1}^{n} \int_{t_1}^{t_2} \frac{\partial T}{\partial q_k} \delta q_k \, dt + \sum_{k=1}^{n} \int_{t_1}^{t_2} \frac{\partial T}{\partial \dot{q}_k} \delta \dot{q}_k \, dt. \tag{2.78}$$

But

$$\delta \dot{q}_k = \delta \frac{dq_k}{dt} = \frac{d}{dt} \delta q_k, \tag{2.79}$$

because the order of variation and differentiation is interchangeable. Hence

$$\int_{t_1}^{t_2} \frac{\partial T}{\partial \dot{q}_k} \delta \dot{q}_k \, dt = \int_{t_1}^{t_2} \frac{\partial T}{\partial \dot{q}_k} \frac{d}{dt} (\delta q_k) \, dt. \tag{2.80}$$

Integrating by parts, (2.80) can be put in the form

$$\int_{t_1}^{t_2} \frac{\partial T}{\partial \dot{q}_k} \, \delta \dot{q}_k \, dt = \int_{t_1}^{t_2} \frac{\partial T}{\partial \dot{q}_k} \frac{d}{dt} (\delta q_k) \, dt = \frac{\partial T}{\partial \dot{q}_k} \, \delta q_k \Big|_{t_1}^{t_2} - \int_{t_1}^{t_2} \delta q_k \frac{d}{dt} \left(\frac{\partial T}{\partial \dot{q}_k} \right) dt$$

$$= - \int_{t_1}^{t_2} \frac{d}{dt} \left(\frac{\partial T}{\partial \dot{q}_k} \right) \delta q_k \, dt, \tag{2.81}$$

because δq_k is assumed to vanish at $t = t_1$ and $t = t_2$.

Introducing (2.81) in (2.78), one obtains

$$\int_{t_1}^{t_2} \delta T \, dt = \delta \int_{t_1}^{t_2} T \, dt = \sum_{k=1}^{n} \int_{t_1}^{t_2} \frac{\partial T}{\partial q_k} \, \delta q_k \, dt - \sum_{k=1}^{n} \int_{t_1}^{t_2} \frac{d}{dt} \left(\frac{\partial T}{\partial \dot{q}_k} \right) \delta q_k \, dt$$

$$= - \sum_{k=1}^{n} \int_{t_1}^{t_2} \left[\frac{d}{dt} \left(\frac{\partial T}{\partial \dot{q}_k} \right) - \frac{\partial T}{\partial q_k} \right] \delta q_k \, dt. \tag{2.82}$$

If p forces act upon the system, the virtual work is

$$\delta W = \sum_{j=1}^{p} \mathbf{F}_j \cdot \delta \mathbf{r}_j. \tag{2.83}$$

But $\delta \mathbf{r}_j$ can be obtained from (2.73) in the form

$$\delta \mathbf{r}_j = \frac{\partial \mathbf{r}_j}{\partial q_1} \, \delta q_1 + \frac{\partial \mathbf{r}_j}{\partial q_2} \, \delta q_2 + \cdots + \frac{\partial \mathbf{r}_j}{\partial q_n} \, \delta q_n, \tag{2.84}$$

so (2.83) can be rewritten

$$\delta W = \sum_{j=1}^{p} \left(\mathbf{F}_j \cdot \frac{\partial \mathbf{r}_j}{\partial q_1} \, \delta q_1 + \mathbf{F}_j \cdot \frac{\partial \mathbf{r}_j}{\partial q_2} \, \delta q_2 + \cdots + \mathbf{F}_j \cdot \frac{\partial \mathbf{r}_j}{\partial q_n} \, \delta q_n \right). \tag{2.85}$$

On the other hand, one can express the virtual work as the product of n "generalized forces" Q_k acting over n generalized virtual displacements δq_k. The directions of the generalized forces coincide with the corresponding directions of the generalized displacements, so one writes

$$\delta W = Q_1 \, \delta q_1 + Q_2 \, \delta q_2 + \cdots + Q_n \, \delta q_n = \sum_{k=1}^{n} Q_k \, \delta q_k. \tag{2.86}$$

The generalized forces Q_k take the place of the single forces acting upon each particle. These forces form the components of an n-dimensional vector in the configuration space.

Note that the generalized force Q_k does not necessarily represent a force. Its units, however, must be such that $Q_k \, \delta q_k$ has units of work. For example,

Q_k could be a moment, in which case δq_k represents an angle change. Comparing (2.85) and (2.86), one concludes that

$$Q_k = \sum_{j=1}^{p} \mathbf{F}_j \cdot \frac{\partial \mathbf{r}_j}{\partial q_k}. \tag{2.87}$$

The ingredients required for the application of Hamilton's principle consist of (2.82) and (2.86). Introducing them in (2.61), one obtains

$$\int_{t_1}^{t_2} (\delta T + \delta W)\, dt = -\sum_{k=1}^{n} \int_{t_1}^{t_2} \left[\frac{d}{dt}\left(\frac{\partial T}{\partial \dot{q}_k}\right) - \frac{\partial T}{\partial q_k} - Q_k \right] \delta q_k\, dt = 0. \tag{2.88}$$

It was pointed out, however, that virtual displacements are arbitrary, hence can be assigned values at will. In addition, the q_k, being generalized coordinates, are, by definition, independent of each other. It follows that the generalized virtual displacements δq_k are both arbitrary and independent. Therefore, one can arbitrarily choose

$$\begin{aligned} \delta q_k \neq 0 \quad & \text{if } k = r, \\ \delta q_k = 0 \quad & \text{if } k \neq r. \end{aligned} \tag{2.89}$$

If this is done, from the series in (2.88) only the coefficient of δq_r survives. But δq_r is arbitrary and can be assigned arbitrary values. It follows that (2.88) can be satisfied only if the integrand is zero, or

$$\frac{d}{dt}\left(\frac{\partial T}{\partial \dot{q}_r}\right) - \frac{\partial T}{\partial q_r} - Q_r = 0, \qquad r = 1, 2, \ldots, n, \tag{2.90}$$

which are Lagrange's equations of motion. The satisfaction of these equations is equivalent to the statement that the integral $\int_{t_1}^{t_2} (\delta T + \delta W)\ dt$ is zero.

The above equations are very general in the sense that Q_r includes both conservative and nonconservative forces. One can distinguish between the conservative force Q_{rc}, which can be derived from the potential energy expression V, and the nonconservative force Q_{rnc}, which may include internal dissipative forces or any external forces and is not derivable from a potential function, so that

$$Q_r = Q_{rc} + Q_{rnc}. \tag{2.91}$$

But for conservative forces

$$W_c(q_k) = -V(q_k), \tag{2.92}$$

$$\delta W_c(q_k) = \sum_{k=1}^{n} Q_{kc}\, \delta q_k = -\delta V(q_k) = -\sum_{k=1}^{n} \frac{\partial V}{\partial q_k}\, \delta q_k, \tag{2.93}$$

from which it follows that

$$Q_{rc} = -\frac{\partial V}{\partial q_r}.$$ (2.94)

Equations (2.91) and (2.94) introduced in (2.90) yield

$$\frac{d}{dt}\left(\frac{\partial T}{\partial \dot{q}_r}\right) - \frac{\partial T}{\partial q_r} + \frac{\partial V}{\partial q_r} = Q_{rnc}, \qquad r = 1, 2, \ldots, n,$$ (2.95)

and noting that V is not a function of the generalized velocities $\dot{q}_r$, one can use (2.63) and rewrite (2.95) in the form

$$\frac{d}{dt}\left(\frac{\partial L}{\partial \dot{q}_r}\right) - \frac{\partial L}{\partial q_r} = Q_{rnc}, \qquad r = 1, 2, \ldots, n,$$ (2.96)

where $L = T - V$ is the Lagrangian.

In the special case of a conservative system, Lagrange's equations reduce to

$$\frac{d}{dt}\left(\frac{\partial L}{\partial \dot{q}_r}\right) - \frac{\partial L}{\partial q_r} = 0, \qquad r = 1, 2, \ldots, n.$$ (2.97)

Example 2.2

A uniform link of length $2a$ and mass m has the upper end connected to a spring k and the lower end subjected to a force F (Figure 2.9). The entire

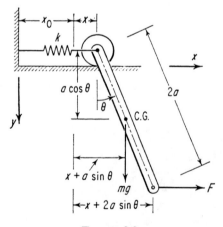

FIGURE 2.9

link is constrained to move in the xy plane with the upper end constrained to move in the horizontal direction only. Let the mass moment of inertia of the link about the mass center be $J = m\rho^2$, where ρ is the radius of gyration, and derive the equations of motion.

Choose as generalized coordinates

$$q_1 = x, \qquad q_2 = \theta,$$

where x is measured from the equilibrium position and write the kinetic and potential energy expressions

$$T = \tfrac{1}{2}mv_c^2 + \tfrac{1}{2}J\dot\theta^2 = \tfrac{1}{2}m\left\{\left[\frac{d}{dt}(x + a\sin\theta)\right]^2 + \left[\frac{d}{dt}(a\cos\theta)\right]^2 + \rho^2\dot\theta^2\right\}$$

$$= \tfrac{1}{2}m(\dot x^2 + 2a\dot x\dot\theta\cos\theta + a^2\dot\theta^2\cos^2\theta + a^2\dot\theta^2\sin^2\theta + \rho^2\dot\theta^2)$$

$$= \tfrac{1}{2}m[\dot x^2 + 2a\dot x\dot\theta\cos\theta + (a^2 + \rho^2)\dot\theta^2], \tag{a}$$

$$V = mga(1 - \cos\theta) + \tfrac{1}{2}kx^2. \tag{b}$$

The virtual work expressions are

$$\delta W = F\,\delta(x + 2a\sin\theta) = F\,\delta x + 2aF\cos\theta\,\delta\theta = X\,\delta x + \Theta\,\delta\theta. \tag{c}$$

Hence the generalized forces become

$$Q_1 = X = F, \qquad Q_2 = \Theta = 2aF\cos\theta. \tag{d}$$

The equations of motion are written using the Lagrange's equations, as given by (2.95), which in this case become

$$\frac{d}{dt}\left(\frac{\partial T}{\partial \dot x}\right) - \frac{\partial T}{\partial x} + \frac{\partial V}{\partial x} = X, \qquad \frac{d}{dt}\left(\frac{\partial T}{\partial \dot\theta}\right) - \frac{\partial T}{\partial \theta} + \frac{\partial V}{\partial \theta} = \Theta. \tag{e}$$

With T and V given by (a) and (b), we have

$$m(\ddot x + a\ddot\theta\cos\theta - a\dot\theta^2\sin\theta) + kx = F,$$

$$m[a\ddot x\cos\theta + (a^2 + \rho^2)\ddot\theta] + mga\sin\theta = 2aF\cos\theta, \tag{f}$$

which are the desired equations of motion.

Problems

2.1   For the system shown in Figure 2.10, calculate the potential and kinetic energy expressions. Assume that the springs are linear. Draw a free-body diagram

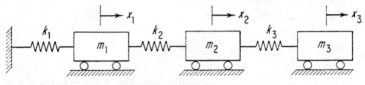

FIGURE 2.10

for each mass and calculate the force resultant acting upon each mass. Verify (2.11).†

2.2   A circular elastic shaft of length L has one end clamped and a rigid disk attached at the other end, Figure 2.11. If G denotes the shear modulus, J the area

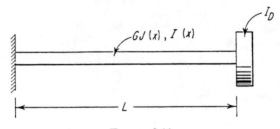

FIGURE 2.11

polar moment of inertia of the shaft, I the mass polar moment of inertia of the shaft per unit length, and I_D the mass polar moment of inertia of the disk, calculate the potential and kinetic energy expressions associated with the rotational motion.

2.3   A linear spring k is restricted so that it can move only vertically, as shown in Figure 2.12. A massless link of length L has one end hinged at the end of the spring and a point mass m attached at the other end.

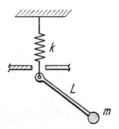

FIGURE 2.12

† Note that the displacements x_i $(i = 1, 2, 3)$ are the only components of the displacement vectors $\mathbf{r}_i(x_i, y_i, z_i)$.

(a) Identify two possible sets of generalized coordinates describing the motion.

(b) Choose one set of coordinates and write the potential and kinetic energy expressions.

2.4 The mechanism shown in Figure 2.13 consists of two uniform bars of length

FIGURE 2.13

L and mass m, and a massless roller, free to move horizontally while attached to a linear spring k. The bars are hinged at both ends. When the bars are horizontal the spring is unextended and has length B. Determine an expression for the equilibrium position in terms of the angle θ by using the virtual work principle.

2.5 For the system of Problem 2.1, use D'Alembert's principle and write the equations of motion.

2.6 For the system of Problem 2.1, use Hamilton's principle and derive the equations of motion.

2.7 For the system of Problem 2.2, use Hamilton's principle and derive the differential equation of motion and the associated boundary conditions.

2.8 Let masses m_1, m_2, and m_3 in the system of Problem 2.1 be acted upon by forces F_1, F_2, and F_3 parallel to directions x_1, x_2, and x_3, respectively, and derive the equations of motion by means of

(a) The extended Hamilton principle.

(b) Lagrange's equations.

2.9 For the system of Problem 2.4, derive Lagrange's equation of motion *about the equilibrium position.*

2.10 For the system of Problem 2.3, derive Lagrange's equations of motion *about the equilibrium position.* Assume small motions.

2.11 The system shown in Figure 2.14 consists of two linear springs k_1 and k_2 of infinite lateral stiffness and a uniform bar of mass m and length $2L$. If a force F is acting at a distance a, as shown, derive Lagrange's equations of motion. Assume small motions.

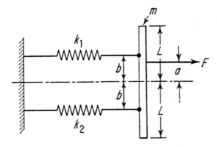

FIGURE 2.14

Selected Readings

Goldstein, H., *Classical Mechanics*, Addison-Wesley, Reading, Mass., 1959.

Lanczos, C., *The Variational Principles of Mechanics*, Univ. of Toronto Press, Toronto, Canada, 1964.

McCuskey, S. W., *An Introduction to Advanced Dynamics*, Addison-Wesley, Reading, Mass., 1962.

Thomson, W. T., *Introduction to Space Dynamics*, Wiley, New York, 1961.

CHAPTER 3

SPECIAL
CONCEPTS FOR
VIBRATION
STUDY

3–1 INTRODUCTION

In Chapter 2 we explored variational methods for the formulation of the dynamical equations of a system. These methods are very efficient, especially for multi-degree-of-freedom systems. The formulation, commonly, is in the form of differential equations, and the next step is to solve these equations.

In the case of a discrete system we obtain a finite set of coupled ordinary differential equations of motion. To solve these equations it will prove convenient to introduce matrix methods.

A continuous system possesses an infinite number of degrees of freedom and is described by partial differential equations. Some approximate methods reduce such a system to a system with a finite number of degrees of freedom, so even in this case matrix techniques are useful.

The matrix formulations provide clear and concise treatment of very complex problems. The compact notation proves its power when dealing with a large number of variables. Furthermore, the use of matrix operations provides very clear and systematic methods of solution. The matrix methods are made even more appealing by the fact that the matrix solutions are ideally suited for numerical evaluation by high-speed electronic computer. In future discussions matrix methods will be used extensively.

The present chapter prepares the groundwork for the solution of the differential equations of motion. To this end such concepts as influence coefficients, coordinate transformation, and coupling, which are particularly pertinent to the study of vibrations, will be discussed in detail.

3-2 INFLUENCE COEFFICIENTS AND FUNCTIONS

When the ends of a linear spring are subjected to a force, the elongation of the spring is proportional to the force. The constant of proportionality is the spring constant, as indicated previously. In more complex systems it is of interest to express the relation between the displacement at a point and the forces acting at various points of the system. This is done by means of influence coefficients. A simple discrete system acted upon by a number of forces F_j ($j = 1, 2, \ldots, n$) will serve to illustrate the concept.

Suppose for the moment that just one of the forces, F_j, is acting, and denote the displacement at point $x = x_i$ due to the force F_j acting at point $x = x_j$ by u_{ij} (Figure 3.1).

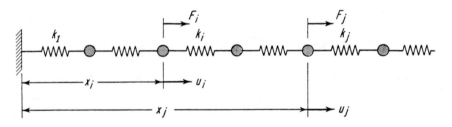

FIGURE 3.1

The *flexibility influence coefficient* a_{ij} is defined as the deflection at point $x = x_i$ due to a unit force applied at $x = x_j$. For a linear system the deflection increases proportionately with the load, so we have the relation

$$u_{ij} = a_{ij}F_j. \tag{3.1}$$

But for a linear system the superposition principle holds true, and the total deflection at $x = x_i$, denoted u_i, is obtained by summing up the contributions of all forces F_j ($j = 1, 2, \ldots, n$)

$$u_i = \sum_{j=1}^{n} u_{ij} = \sum_{j=1}^{n} a_{ij}F_j, \qquad i = 1, 2, \ldots, n. \tag{3.2}$$

The index j is called a *dummy index* and can be replaced by any other index letter without affecting the result.

Now define the *stiffness influence coefficient* k_{ij} as the force at $x = x_i$ producing a unit displacement at $x = x_j$, the points for which $x \neq x_j$ are fixed. It follows that

$$F_i = \sum_{j=1}^{n} k_{ij}u_j, \qquad i = 1, 2, \ldots, n. \tag{3.3}$$

Note that for a single spring with one end fixed and the other end free the stiffness influence coefficient is just the spring constant.

In general, for discrete systems, the displacements are evaluated at the same points where the forces are applied so that $i, j = 1, 2, 3, \ldots, n$.

Equations (3.2) and (3.3) may be represented by the matrix form (see Appendix A),

$$\{u\} = [a]\{F\}, \tag{3.4}$$

$$\{F\} = [k]\{u\}, \tag{3.5}$$

in which $\{u\}$ and $\{F\}$ are column matrices representing the displacement and force vectors, respectively. The matrices $[a]$ and $[k]$ are square matrices known as the flexibility influence coefficients matrix (or *flexibility matrix*) and stiffness influence coefficients matrix (or *stiffness matrix*), respectively. Evidently, (3.4) and (3.5) represent symbolically the algebraic operations (3.2) and (3.3).

Equation (3.4) represents a linear transformation and it appears that (3.5) represents the inverse linear transformation. We must conclude, therefore, that matrices $[a]$ and $[k]$ are related. Indeed, if we substitute (3.5) in (3.4) we obtain

$$\{u\} = [a]\{F\} = [a][k]\{u\}, \tag{3.6}$$

from which it follows that

$$[a][k] = [I], \tag{3.7}$$

where $[I]$ is a unit matrix that has all its diagonal elements equal to one and all its off-diagonal elements zero. Hence

$$[a] = [k]^{-1}, \qquad [k] = [a]^{-1}, \tag{3.8}$$

or the flexibility and the stiffness matrices are the inverse of one another.

The same concept can be used in the case in which the points x_i and x_j assume continuous rather than discrete values, as in the case of distributed load $p(x)$ (lb in.$^{-1}$). The load associated with the element of length $d\xi$ at a distance ξ from the left end (Figure 3.2) is $p(\xi)\,d\xi$.

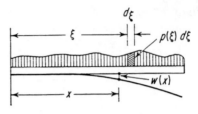

FIGURE 3.2

The *flexibility influence function* $a(x, \xi)$ is defined as the deflection at any point x due to a unit load at a point ξ. It follows that the increment of deflection due to the load $p(\xi) \, d\xi$ is

$$dw(x, \xi) = a(x, \xi)p(\xi) \, d\xi. \tag{3.9}$$

The total deflection at x is obtained by integrating over the length of the beam,

$$w(x) = \int_0^L a(x, \xi)p(\xi) \, d\xi. \tag{3.10}$$

Note that the flexibility influence function $a(x, \xi)$ is just a one-dimensional Green's function. The kernel $a(x, \xi)$ in the integral transformation (3.10) represents the same concept as the coefficient a_{ij} in the linear transformation (3.2).

Evidently one can have a two-dimensional flexibility influence function $a(x, y; \xi, \eta)$. Accordingly, for the two-dimensional case one would have the relation

$$w(x, y) = \iint_{\text{area}} a(x, y; \xi, \eta)p(\xi, \eta) \, d\xi \, d\eta. \tag{3.11}$$

Example 3.1

Three rigid circular disks of mass polar moments of inertia I_1, I_2, and I_3 are mounted on a sectionally uniform shaft clamped at both ends as shown in Figure 3.3. The torsional rigidity of the massless shaft sections are GJ_1, GJ_2, GJ_2, and GJ_4, respectively. It is required to calculate the flexibility and stiffness influence coefficients.

First let k_i be spring constants associated with each section,

$$k_i = \frac{GJ_i}{L_i}, \qquad i = 1, 2, 3, 4. \tag{a}$$

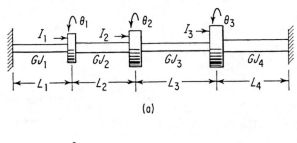

(a)

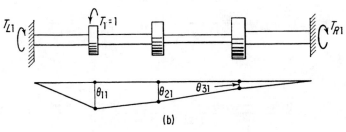

(b)

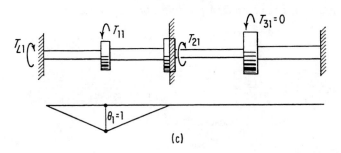

(c)

FIGURE 3.3

To evaluate the flexibility influence coefficients a_{11}, a_{21}, and a_{31}, we apply a unit torque on disk 1 as shown in Figure 3.3(b), and calculate the corresponding angular rotations θ_{11}, θ_{21}, and θ_{31}. We note that left of disk 1 the spring constant is k_1, whereas right of disk 1 we have an equivalent spring constant $1/\sum_{i=2}^{4} (1/k_i)$. Because of a torque $T_1 = 1$, we have the reactions

$$T_{L1} = \theta_{11}k_1, \qquad T_{R1} = \frac{\theta_{11}}{\sum_{i=2}^{4} (1/k_i)}, \tag{b}$$

and as

$$T_{L1} + T_{R1} = T_1 = 1, \tag{c}$$

we obtain

$$a_{11} = \theta_{11} = \frac{1}{k_1 + [\sum_{i=2}^{4} (1/k_i)]^{-1}} = \frac{\sum_{i=2}^{4} (1/k_i)}{k_1 \sum_{i=1}^{4} (1/k_i)}. \tag{d}$$

It follows that

$$a_{31} = \theta_{31} = \frac{T_{R1}}{k_4} = \frac{1}{k_1 k_4 \sum_{i=1}^4 (1/k_i)}, \tag{e}$$

$$a_{21} = \theta_{21} = \frac{T_{R1}}{k_3} + \frac{T_{R1}}{k_4} = \frac{k_3 + k_4}{k_3 k_4} T_{R1} = \frac{k_3 + k_4}{k_1 k_3 k_4 \sum_{i=1}^4 (1/k_i)}. \tag{f}$$

Similarly, by applying a torque $T_2 = 1$ we find the corresponding reactions

$$T_{L2} = \frac{\theta_{22}}{(1/k_1) + (1/k_2)}, \qquad T_{R2} = \frac{\theta_{22}}{(1/k_3) + (1/k_4)}, \tag{g}$$

where

$$T_{L2} + T_{R2} = 1, \tag{h}$$

so that

$$a_{22} = \theta_{22} = \frac{[(1/k_1) + (1/k_2)][(1/k_3) + (1/k_4)]}{\sum_{i=1}^4 (1/k_i)} = \frac{(k_1 + k_2)(k_3 + k_4)}{\prod_{i=1}^4 k_i \sum_{i=1}^4 (1/k_i)}, \tag{i}$$

where $\prod$ represents the standard product notation. Also,

$$a_{12} = \theta_{12} = \frac{T_{L2}}{k_1} = \frac{(1/k_3) + (1/k_4)}{k_1 \sum_{i=1}^4 (1/k_i)} = \frac{k_3 + k_4}{k_1 k_3 k_4 \sum_{i=1}^4 (1/k_i)}, \tag{j}$$

$$a_{32} = \theta_{32} = \frac{T_{R2}}{k_4} = \frac{(1/k_1) + (1/k_2)}{k_4 \sum_{i=1}^4 (1/k_i)} = \frac{k_1 + k_2}{k_1 k_2 k_4 \sum_{i=1}^4 (1/k_i)}. \tag{k}$$

Similarly, by interchanging subscripts we obtain

$$a_{33} = \theta_{33} = \frac{\sum_{i=1}^3 (1/k_i)}{k_4 \sum_{i=1}^4 (1/k_i)}, \tag{l}$$

$$a_{23} = \theta_{23} = \frac{k_1 + k_2}{k_1 k_2 k_4 \sum_{i=1}^4 (1/k_i)} = a_{32}, \tag{m}$$

$$a_{13} = \theta_{13} = \frac{1}{k_1 k_4 \sum_{i=1}^4 (1/k_i)} = a_{31}. \tag{n}$$

The stiffness influence coefficients are considerably simpler to obtain. We note from Figure 3.3(c) that to produce a displacement $\theta_1 = 1$ we need a torque T_{11} with reactions T_{L1} and T_{21} given by

$$T_{L1} = 1 \times k_1, \qquad T_{21} = 1 \times k_2. \tag{o}$$

It follows that

$$k_{11} = T_{L1} + T_{21} = k_1 + k_2,$$
$$k_{21} = -T_{21} = -k_2, \tag{p}$$
$$k_{31} = T_{31} = 0.$$

It is easy to check that

$$k_{12} = -k_2, \qquad k_{22} = k_2 + k_3, \qquad k_{32} = -k_3,$$
$$k_{13} = 0, \qquad k_{23} = -k_3, \qquad k_{33} = k_3 + k_4. \tag{q}$$

The flexibility influence coefficients can be arranged in the matrix form

$$[a] = \frac{1}{\prod_{i=1}^{4} k_i \sum_{i=1}^{4} (1/k_i)}$$

$$\times \begin{bmatrix} k_3 k_4 + k_2(k_3 + k_4) & k_2(k_3 + k_4) & k_2 k_3 \\ k_2(k_3 + k_4) & (k_1 + k_2)(k_3 + k_4) & k_3(k_1 + k_2) \\ k_2 k_3 & k_3(k_1 + k_2) & k_1 k_2 + k_3(k_1 + k_2) \end{bmatrix}. \tag{r}$$

Similarly,

$$[k] = \begin{bmatrix} k_1 + k_2 & -k_2 & 0 \\ -k_2 & k_2 + k_3 & -k_3 \\ 0 & -k_3 & k_3 + k_4 \end{bmatrix}. \tag{s}$$

It is obvious that in this particular example it is much easier to obtain the matrix $[k]$ than to obtain the matrix $[a]$. In fact, it may be easier to obtain the matrix $[a]$ by first obtaining the matrix $[k]$ and then inverting it. The reader is urged to evaluate $[a]$ in this manner.

3-3 STRAIN ENERGY IN TERMS OF INFLUENCE COEFFICIENTS AND FUNCTIONS. PROPERTIES OF INFLUENCE COEFFICIENTS AND FUNCTIONS

Consider a system acted upon by n forces F_i $(i = 1, 2, \ldots, n)$. (As an example see Figure 3.1). Denote by u_i the displacement at point i in the direction of the force F_i. For a linear system, if F_i is the only force present, the strain energy has the value

$$V_i = \tfrac{1}{2} F_i u_i. \tag{3.12}$$

Hence the total strain energy considering all n forces is

$$V = \sum_{i=1}^{n} V_i = \tfrac{1}{2} \sum_{i=1}^{n} F_i u_i, \tag{3.13}$$

and recalling (3.2) we can write

$$V = \tfrac{1}{2} \sum_{i=1}^{n} F_i \sum_{j=1}^{n} a_{ij} F_j = \tfrac{1}{2} \sum_{i=1}^{n} \sum_{j=1}^{n} a_{ij} F_i F_j. \tag{3.14}$$

If u_j denotes the displacement and F_j the force at point j, we may write the strain energy expression in the form

$$V = \tfrac{1}{2} \sum_{j=1}^{n} F_j u_j. \tag{3.15}$$

Equation (3.2) with indices i and j interchanged is

$$u_j = \sum_{i=1}^{n} a_{ji} F_i, \tag{3.16}$$

so that a combination of (3.15) and (3.16) yields

$$V = \tfrac{1}{2} \sum_{j=1}^{n} F_j \sum_{i=1}^{n} a_{ji} F_i = \tfrac{1}{2} \sum_{i=1}^{n} \sum_{j=1}^{n} a_{ji} F_i F_j, \tag{3.17}$$

because the order of summation is immaterial. Comparing (3.14) and (3.17) we conclude that

$$a_{ij} = a_{ji}. \tag{3.18}$$

Equation (3.18) is the statement of *Maxwell's reciprocity theorem*.

Using the same procedure, the reader can show that a similar relation exists for the stiffness influence coefficients,

$$k_{ij} = k_{ji}. \tag{3.19}$$

Equation (3.13) may be written in matrix form as follows:

$$V = \tfrac{1}{2}\{F\}^T\{u\} = \tfrac{1}{2}\{u\}^T\{F\}, \tag{3.20}$$

and as

$$\{u\} = [a]\{F\}, \qquad \{u\}^T = \{F\}^T[a]^T, \tag{3.21}$$

we can write

$$V = \tfrac{1}{2}\{F\}^T[a]\{F\} = \tfrac{1}{2}\{F\}^T[a]^T\{F\}. \tag{3.22}$$

Similarly, it is easy to show that

$$V = \tfrac{1}{2}\{u\}^T[k]\{u\} = \tfrac{1}{2}\{u\}^T[k]^T\{u\}. \tag{3.23}$$

Equations (3.22) and (3.23) are matrix representations of the quadratic form V.

Equations (3.18) and (3.19) can be written in the matrix form

$$[a] = [a]^T, \tag{3.24}$$

$$[k] = [k]^T, \tag{3.25}$$

so that (3.24) and (3.25) are just the matrix formulations of the reciprocity relations (3.18) and (3.19).

If the load is distributed, the strain energy expression for the one-dimensional case can be written

$$V = \tfrac{1}{2} \int_0^L w(x)p(x)\, dx = \tfrac{1}{2} \int_0^L p(x) \left[\int_0^L a(x,\, \xi)p(\xi)\, d\xi \right] dx, \tag{3.26}$$

where use has been made of (3.10).

Following a procedure similar to that for the discrete case, we can easily show that

$$a(x,\, \xi) = a(\xi,\, x), \tag{3.27}$$

which states that the flexibility influence function is symmetric in x and ξ.

Note that nothing was said about the stiffness influence function $k(x,\, \xi)$. Although the concept exists, it has no practical value, because the process of obtaining it is prohibitive, owing to the amount of work involved.

3–4 LAGRANGE'S EQUATIONS OF MOTION IN MATRIX FORM

Equation (2.95) represents a set of n Lagrange's equations of motion. Recalling that the derivative of a matrix is equal to the matrix of the derivatives of the elements, we can write (2.95) in the matrix form

$$\frac{d}{dt}\left\{\frac{\partial T}{\partial \dot{q}_r}\right\} - \left\{\frac{\partial T}{\partial q_r}\right\} + \left\{\frac{\partial V}{\partial q_r}\right\} = \{Q_r\}, \tag{3.28}$$

where $\{Q_r\}$ is a column matrix representing the nonconservative generalized forces.

A case frequently encountered in the field of vibrations is the one in which the kinetic energy depends on velocities alone and not on coordinates. In this case we can write the kinetic energy expression in terms of generalized velocities,

$$T = \tfrac{1}{2}\{\dot{q}\}^T[m]\{\dot{q}\}, \tag{3.29}$$

where $[m]$ is called the *inertia* or *mass matrix*. Equation (3.29) represents a quadratic form and, because the kinetic energy, by definition, cannot be negative and vanishes only when all the velocities vanish, the form is called a *positive definite quadratic form*.

A similar expression was obtained for the potential energy V, which in terms of generalized coordinates becomes

$$V = \tfrac{1}{2}\{q\}^T[k]\{q\}, \tag{3.30}$$

which also has a quadratic form.

The inertia matrix $[m]$ and the stiffness matrix $[k]$ depend on the set of generalized coordinates used.

The derivative of a continuous product of matrices is formed in the same manner used for scalar expressions, except that the order of the matrix positions must be preserved. Hence, differentiating (3.29) with respect to $\dot{q}_r$, we obtain

$$\frac{\partial T}{\partial \dot{q}_r} = \tfrac{1}{2}\{\delta_{ir}\}^T[m]\{\dot{q}\} + \tfrac{1}{2}\{\dot{q}\}^T[m]\{\delta_{ir}\}$$

$$= \{\delta_{ir}\}^T[m]\{\dot{q}\} = \lfloor m_r \rfloor \{\dot{q}\}, \qquad r = 1, 2, \ldots, n, \tag{3.31}$$

where the symmetry of $[m]$ has been accounted for. The matrix $\{\delta_{ir}\}$ is defined as the Kronecker delta column matrix whose elements in the rows for which $i \neq r$ are equal to zero and whose element in the row $i = r$ is equal to 1. The matrix $\lfloor m_r \rfloor$ is a row matrix that is identical to the rth row of matrix $[m]$. It follows that

$$\left\{ \frac{\partial T}{\partial \dot{q}_r} \right\} = \{\lfloor m_r \rfloor \{\dot{q}\}\} = [m]\{\dot{q}\}, \tag{3.32}$$

and, furthermore,

$$\frac{d}{dt} \left\{ \frac{\partial T}{\partial \dot{q}_r} \right\} = [m]\{\ddot{q}\}. \tag{3.33}$$

Since T depends on velocities alone, it follows that

$$\left\{ \frac{\partial T}{\partial q_r} \right\} = \{0\}. \tag{3.34}$$

Following the same procedure, we obtain

$$\frac{\partial V}{\partial q_r} = \tfrac{1}{2}\{\delta_{ir}\}^T[k]\{q\} + \tfrac{1}{2}\{q\}^T[k]\{\delta_{ir}\} = \{\delta_{ir}\}^T[k]\{q\}$$

$$= \lfloor k_r \rfloor \{q\}, \qquad r = 1, 2, \ldots, n, \tag{3.35}$$

which leads to

$$\left\{\frac{\partial V}{\partial q_r}\right\} = [k]\{q\}. \qquad (3.36)$$

Introducing (3.33), (3.34), and (3.36) in (3.28) we obtain the differential equations of motion in matrix form,

$$[m]\{\ddot{q}\} + [k]\{q\} = \{Q\}, \qquad (3.37)$$

where the subscript r has been dropped.

In the special case in which the generalized coordinates q_r are the actual displacements u_r, the inertia matrix $[m]$ is a diagonal matrix.

Example 3.2

Consider the system of Example 3.1 and use as generalized coordinates the angular motions

$$q_1 = \theta_1, \qquad q_2 = \theta_2, \qquad q_3 = \theta_3. \qquad (a)$$

The kinetic energy can be written

$$T = \tfrac{1}{2}I_1\dot{\theta}_1^2 + \tfrac{1}{2}I_2\dot{\theta}_2^2 + \tfrac{1}{2}I_3\dot{\theta}_3^2 = \tfrac{1}{2}\{\dot{\theta}\}^T[I_p]\{\dot{\theta}\}, \qquad (b)$$

where

$$[I_p] = \begin{bmatrix} I_1 & 0 & 0 \\ 0 & I_2 & 0 \\ 0 & 0 & I_3 \end{bmatrix} \qquad (c)$$

is the inertia matrix. The potential energy has the form

$$
\begin{aligned}
V &= \tfrac{1}{2}k_1\theta_1^2 + \tfrac{1}{2}k_2(\theta_2 - \theta_1)^2 + \tfrac{1}{2}k_3(\theta_3 - \theta_2)^2 + \tfrac{1}{2}k_4\theta_3^2 \\
&= \tfrac{1}{2}[(k_1 + k_2)\theta_1^2 - 2k_2\theta_1\theta_2 + (k_2 + k_3)\theta_2^2 - 2k_3\theta_2\theta_3 + (k_3 + k_4)\theta_3^2] \\
&= \tfrac{1}{2}\{\theta\}^T[k]\{\theta\},
\end{aligned} \qquad (d)
$$

where

$$[k] = \begin{bmatrix} k_1 + k_2 & -k_2 & 0 \\ -k_2 & k_2 + k_3 & -k_3 \\ 0 & -k_3 & k_3 + k_4 \end{bmatrix}. \qquad (e)$$

In writing (e) we made use of the fact that $k_{ij} = k_{ji}$. We can define a column matrix of applied torques,

$$\{T\} = \begin{Bmatrix} T_1 \\ T_2 \\ T_3 \end{Bmatrix}, \qquad (f)$$

so the equations of motion of the system become

$$[I_p]\{\ddot{\theta}\} + [k]\{\theta\} = \{T\}. \tag{g}$$

We must recall that the stiffness coefficients are given by

$$k_i = \frac{GJ_i}{L_i}. \tag{h}$$

3–5   LINEAR TRANSFORMATIONS. COUPLING

Consider an n-degree-of-freedom system for which the kinetic and potential energy expressions have the quadratic forms

$$T = \tfrac{1}{2} \sum_{i=1}^{n} \sum_{j=1}^{n} m_{ij}\dot{q}_i\dot{q}_j, \tag{3.38}$$

$$V = \tfrac{1}{2} \sum_{i=1}^{n} \sum_{j=1}^{n} k_{ij}q_iq_j, \tag{3.39}$$

where the inertia coefficients m_{ij} and the stiffness coefficients k_{ij} depend on the coordinates used. They are symmetric, because $m_{ij} = m_{ji}$ and $k_{ij} = k_{ji}$.

The coordinates q_i can be related to a different set of coordinates η_r by the *linear transformation*

$$q_i = \sum_{r=1}^{n} \beta_{ir}\eta_r, \tag{3.40}$$

where the coefficients β_{ir} are constant. It follows that the velocities $\dot{q}_i$ are related to the velocities $\dot{\eta}_r$ by

$$\dot{q}_i = \sum_{r=1}^{n} \beta_{ir}\dot{\eta}_r. \tag{3.41}$$

Introducing the linear transformation (3.41), in the kinetic energy expression (3.38), we obtain

$$T = \tfrac{1}{2} \sum_{i=1}^{n} \sum_{j=1}^{n} m_{ij}\dot{q}_i\dot{q}_j = \tfrac{1}{2} \sum_{i=1}^{n} \sum_{j=1}^{n} m_{ij} \sum_{r=1}^{n} \beta_{ir}\dot{\eta}_r \sum_{s=1}^{n} \beta_{js}\dot{\eta}_s$$

$$= \tfrac{1}{2} \sum_{r=1}^{n} \sum_{s=1}^{n} \dot{\eta}_r\dot{\eta}_s \sum_{i=1}^{n} \sum_{j=1}^{n} m_{ij}\beta_{ir}\beta_{js} = \tfrac{1}{2} \sum_{r=1}^{n} \sum_{s=1}^{n} M_{rs}\dot{\eta}_r\dot{\eta}_s, \tag{3.42}$$

where $$M_{rs} = \sum_{i=1}^{n} \sum_{j=1}^{n} m_{ij}\beta_{ir}\beta_{js} \tag{3.43}$$

is an inertia coefficient depending on the coefficients β and implicitly on the coordinates η.

Similarly, the potential energy becomes

$$V = \tfrac{1}{2} \sum_{i=1}^{n} \sum_{j=1}^{n} k_{ij} q_i q_j = \tfrac{1}{2} \sum_{r=1}^{n} \sum_{s=1}^{n} K_{rs} \eta_r \eta_s, \tag{3.44}$$

where

$$K_{rs} = \sum_{i=1}^{n} \sum_{j=1}^{n} k_{ij} \beta_{ir} \beta_{js} \tag{3.45}$$

is the corresponding stiffness coefficient.

Note that the inertia coefficients M_{rs} and the stiffness coefficients K_{rs} are symmetric; that is,

$$M_{rs} = M_{sr}, \tag{3.46}$$

$$K_{rs} = K_{sr}. \tag{3.47}$$

These results can be expressed most conveniently in matrix form. The linear transformation in matrix form becomes

$$\{q\} = [\beta]\{\eta\}, \tag{3.48}$$

and, similarly,

$$\{\dot{q}\} = [\beta]\{\dot{\eta}\}. \tag{3.49}$$

The transposed matrices are

$$\{q\}^T = \{\eta\}^T[\beta]^T, \tag{3.50}$$

$$\{\dot{q}\}^T = \{\dot{\eta}\}^T[\beta]^T. \tag{3.51}$$

Correspondingly, (3.42) and (3.44) have the matrix form

$$T = \tfrac{1}{2}\{\dot{q}\}^T[m]\{\dot{q}\} = \tfrac{1}{2}\{\dot{\eta}\}^T[\beta]^T[m][\beta]\{\dot{\eta}\} = \tfrac{1}{2}\{\dot{\eta}\}^T[M]\{\dot{\eta}\}, \tag{3.52}$$

$$V = \tfrac{1}{2}\{q\}^T[k]\{q\} = \tfrac{1}{2}\{\eta\}^T[\beta]^T[k][\beta]\{\eta\} = \tfrac{1}{2}\{\eta\}^T[K]\{\eta\}, \tag{3.53}$$

where

$$[M] = [\beta]^T[m][\beta], \tag{3.54}$$

$$[K] = [\beta]^T[k][\beta], \tag{3.55}$$

are the corresponding inertia and stiffness matrices, respectively. It is easy to see that $[M]$ and $[K]$ are symmetric, because $[m]$ and $[k]$ are symmetric.

The equations of motion (3.37) can be written in terms of the coordinates η. To this end we must express the forces associated with the new system of coordinates. The virtual work expression is

$$\delta W = \sum_{i=1}^{n} Q_i \, \delta q_i. \tag{3.56}$$

But the virtual displacements δq_i are related to the virtual displacements $\delta \eta_r$ by

$$\delta q_i = \sum_{r=1}^{n} \beta_{ir} \, \delta \eta_r, \tag{3.57}$$

so

$$\delta W = \sum_{i=1}^{n} Q_i \, \delta q_i = \sum_{i=1}^{n} Q_i \sum_{r=1}^{n} \beta_{ir} \, \delta \eta_r = \sum_{r=1}^{n} \delta \eta_r \sum_{i=1}^{n} \beta_{ir} Q_i = \sum_{r=1}^{n} N_r \, \delta \eta_r, \tag{3.58}$$

where

$$N_r = \sum_{i=1}^{n} \beta_{ir} Q_i \tag{3.59}$$

are the generalized forces corresponding to the coordinates η_r. In matrix form we write

$$\delta W = \{\delta q\}^T \{Q\} = \{\delta \eta\}^T [\beta]^T \{Q\} = \{\delta \eta\}^T \{N\}, \tag{3.60}$$

where
$$\{N\} = [\beta]^T \{Q\}. \tag{3.61}$$

The equations of motion in terms of the coordinates η_r are simply

$$[M]\{\ddot{\eta}\} + [K]\{\eta\} = \{N\}. \tag{3.62}$$

If the kinetic energy expression T contains cross products, the system is said to be *inertially coupled*. If the potential energy expression V contains cross products, it is said to be *elastically coupled*. In either of these cases the equations of motion form a set of n coupled differential equations.

If one can choose the coefficients β_{ij} such that

$$M_{rs} = 0 \qquad \text{for } r \neq s, \tag{3.63}$$

the kinetic energy T contains no cross products and the system becomes inertially uncoupled. In this case $[M]$ is a diagonal matrix.

Similarly, if the coefficients β_{ij} are such that

$$K_{rs} = 0 \qquad \text{for } r \neq s, \tag{3.64}$$

the potential energy V contains no cross products and the system becomes elastically uncoupled, in which case $[K]$ is a diagonal matrix. It follows that by adjusting the coefficients β_{ij} one can change the coupling of the system. It is, therefore, obvious that coupling is not an inherent property of the system but depends on the coordinate system used.

It will be shown that it is possible to find a set of coefficients β_{ij} which, when used, renders the system both inertially and elastically uncoupled. In this case $[M]$ and $[K]$ are diagonal matrices, so the equations of motion become

$$[M]\{\ddot{\eta}\} + [K]\{\eta\} = \{N\}. \tag{3.65}$$

This equation represents a set of n uncoupled differential equations of the type

$$M_{rr}\ddot{\eta}_r(t) + K_{rr}\eta_r(t) = N_r(t), \qquad r = 1, 2, \ldots, n, \tag{3.66}$$

which is similar to the equation of a single-degree-of-freedom system. Its solution can be readily obtained by the methods indicated in Chapter 1.

The coordinates η_r are called *natural coordinates*,† and the linear transformation matrix $[\beta]$ is the *modal matrix*, which consists of the *characteristic vectors* or *modal vectors* of a system.

It is obvious that it is considerably easier to solve the equations of motion in the uncoupled form, (3.65), than in the coupled form, (3.37). In fact, a widely used method of solving equations (3.37) is to bring them into the uncoupled form, equations (3.65), by using the modal matrix as a transformation matrix. For that reason the method is called *modal analysis*. The methods of obtaining the modal vectors, or natural modes, of a system will be discussed in Chapter 4.

Note that the coordinate transformation (3.48) does not change the character of the solution; it simply facilitates it. Such a transformation is called a *similarity transformation*. This subject will be discussed in Chapter 4.

Example 3.3

Consider the system of Example 3.2, for which the kinetic energy has the form

$$T = \tfrac{1}{2}\{\dot{q}\}^T[m]\{\dot{q}\}, \tag{a}$$

---

† In some references they are called *principal coordinates*. They become *normal coordinates* if a certain normalizing procedure is used. The reason for the name "principal coordinates" can be found in Section 4–7.

where the inertia matrix is

$$[m] = \begin{bmatrix} m_1 & 0 & 0 \\ 0 & m_2 & 0 \\ 0 & 0 & m_3 \end{bmatrix} \qquad \text{(b)}$$

and the potential energy is given by

$$V = \tfrac{1}{2}\{q\}^T[k]\{q\}, \qquad \text{(c)}$$

where the stiffness matrix is

$$[k] = \begin{bmatrix} k_1 + k_2 & -k_2 & 0 \\ -k_2 & k_2 + k_3 & -k_3 \\ 0 & -k_3 & k_3 + k_4 \end{bmatrix}. \qquad \text{(d)}$$

Letting

$$
\begin{aligned}
m_1 = m_2 &= m, & m_3 &= 2m, \\
k_1 = k_2 = k_3 &= k, & k_4 &= 2k,
\end{aligned}
\qquad \text{(e)}
$$

the inertia matrix reduces to

$$[m] = m\begin{bmatrix} 1 & 0 & 0 \\ 0 & 1 & 0 \\ 0 & 0 & 2 \end{bmatrix} \qquad \text{(f)}$$

and the stiffness matrix becomes

$$[k] = k\begin{bmatrix} 2 & -1 & 0 \\ -1 & 2 & -1 \\ 0 & -1 & 3 \end{bmatrix}. \qquad \text{(g)}$$

Next consider the linear transformation

$$\{q\} = [\beta]\{\eta\}, \qquad \text{(h)}$$

where the transformation matrix $[\beta]$ has the form

$$[\beta] = m^{-1/2}\begin{bmatrix} 0.4959 & 0.6074 & 0.6209 \\ 0.6646 & 0.1949 & -0.7215 \\ 0.3954 & -0.5446 & 0.2171 \end{bmatrix}. \qquad \text{(i)}$$

Introducing (h) in the kinetic energy expression and using (f) and (i), we obtain

$$T = \tfrac{1}{2}\{\dot{q}\}^T[m]\{\dot{q}\} = \tfrac{1}{2}\{\dot{\eta}\}^T[\beta]^T[m][\beta]\{\dot{\eta}\} = \tfrac{1}{2}\{\dot{\eta}\}^T\{\dot{\eta}\}$$

$$= \tfrac{1}{2}(\dot{\eta}_1^2 + \dot{\eta}_2^2 + \dot{\eta}_3^2), \tag{j}$$

which contains no cross products. Similarly, the potential energy expression reduces to

$$V = \tfrac{1}{2}\{q\}^T[k]\{q\} = \tfrac{1}{2}\{\eta\}^T[\beta]^T[k][\beta]\{\eta\} = \tfrac{1}{2}\{\eta\}^T\lceil\omega^2\rfloor\{\eta\}$$

$$= \tfrac{1}{2}(\omega_1^2\eta_1^2 + \omega_2^2\eta_2^2 + \omega_3^2\eta_3^2), \tag{k}$$

where

$$\omega_1^2 = 0.6595\,\frac{k}{m}, \qquad \omega_2^2 = 1.6791\,\frac{k}{m}, \qquad \omega_3^2 = 3.1627\,\frac{k}{m}. \tag{l}$$

It must be noted that the potential energy (k) is also free of cross products, so the coordinate transformation (h) and (i) renders the system both inertially and elastically uncoupled. It follows that the coordinates $\eta_1(t)$, $\eta_2(t)$, and $\eta_3(t)$ are the natural coordinates of the system. This should come as no surprise, because the matrix $[\beta]$ is the modal matrix of the system (see Example 4.2).

Problems

3.1 Given the system shown in Figure 3.4,

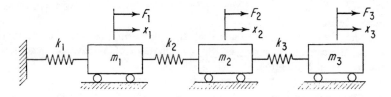

FIGURE 3.4

(a) Calculate the matrix of the flexibility influence coefficients and check its symmetry.
(b) Calculate the matrix of the stiffness influence coefficients by using the definition and check its symmetry.
(c) Calculate the stiffness matrix by inverting the flexibility matrix and compare your results.
(d) Calculate the displacement matrix if $F_1 = F_0$, $F_2 = 2F_0$, and $F_3 = 4F_0$.

3.2 Calculate the flexibility influence function for a cantilever beam, fixed at $x = 0$ and free at $x = L$, by using

(a) The area-moment method.

(b) The double-integration method.

(c) Castigliano's theorem.

3.3 Given the system of Problem 3.1,

(a) Write the kinetic and potential energy expressions in matrix form in terms of the coordinates x_i.

(b) Instead of the coordinates x_i use the coordinates z_i, where the relation between the two sets of coordinates is

$$x_1 = z_1, \qquad x_2 - x_1 = z_2, \qquad x_3 - x_2 = z_3,$$

and write the kinetic and potential energy expressions in terms of the coordinates z_1.

(c) Derive the generalized forces corresponding to the coordinates z_i by means of the virtual work method.

3.4 If the cantilever beam of Problem 3.2 is subjected to a distributed load $p(x) = p_0[1 - (x/L)]$, calculate

(a) The deflection at any point x by means of (3.10).

(b) The potential energy expression by means of (3.26).

3.5 For the system of Problem 3.1,

(a) Derive Lagrange's equations of motion by using x_1, x_2, and x_3 as generalized coordinates.

(b) Derive Lagrange's equations of motion by using as generalized coordinates

$$z_1 = x_1, \qquad z_2 = x_2 - x_1, \qquad z_3 = x_3 - x_2.$$

(c) Cast the equations of motion in matrix form for both sets of coordinates x_i and z_i.

3.6 Write the linear transformation

$$z_1 = x_1, \qquad z_2 = x_2 - x_1, \qquad z_3 = x_3 - x_2$$

in the matrix form $\{x\} = [\beta]\{z\}$. The system of Problem 3.1 is under consideration.

(a) Write the kinetic and potential energy expressions in matrix form in terms of the coordinates x_i.

(b) Use the linear transformation $\{x\} = [\beta]\{z\}$ and write the kinetic and potential energy expressions in terms of the coordinates z_i. Compare the results with the results of Problem 3.3.

(c) What conclusions can you draw about coupling?

(d) Derive Lagrange's equations of motion in matrix form for both sets of coordinates x_i and z_i.

(e) How do the equations of motion compare with the ones obtained in Problem 3.5?

Selected Readings

Bisplinghoff, R. L., H. Ashley, and R. L. Halfman, *Aeroelasticity*, Addison-Wesley, Reading, Mass., 1957.

Frazer, R. A., W. J. Duncan, and A. R. Collar, *Elementary Matrices*, Cambridge Univ. Press, New York, 1957.

CHAPTER 4

NATURAL
MODES OF
VIBRATION.
DISCRETE
SYSTEMS

4-1 GENERAL DISCUSSION

In Chapter 3 we have shown how to derive the differential equations of motion of an undamped discrete system. For an n-degree-of-freedom system the equations of motion consist of a set of n ordinary differential equations of the second order. In general these equations are coupled, which implies that one must solve the set of equations simultaneously, not an easy task. It was pointed out, however, that by using a certain linear transformation it is possible to express the set of equations in terms of a different system of coordinates, for which the equations of motion become uncoupled (see Section 3–5). An uncoupled system of equations of motion is, in essence, a set of n independent equations, every one of which resembles, in structure, the equation of motion of a single-degree-of-freedom system. The advantage of such a transformation is obvious, because a set of n independent differential equations is considerably easier to solve than a set of n coupled equations. This, of course, involves finding the desired coordinate transformation.

The linear transformation uncoupling the system of differential equations is conveniently represented in matrix form and, as pointed out in Section 3–5, the columns of the transformation matrix are the *modal vectors*, also called *natural modes*, of the system. For this reason the transformation matrix is called the *modal matrix*. The response of an n-degree-of-freedom system can be expressed in terms of n time-dependent coordinates describing the motion of the system at any time. These coordinates can be arranged in a column matrix and regarded as the components of an n-dimensional vector. The linear transformation uncouples the system of equations, which implies that the response of the system can be obtained as a linear superposition of its modal vectors.

The modal vectors of a system are obtained by solving the so-called *characteristic-value* or *eigenvalue problem*. Its solution yields a set of n natural modes and n associated natural frequencies. With the exception of special cases, all the natural frequencies are distinct. The method of obtaining the response of a system by using the modal matrix as the transformation matrix is called *modal analysis* and is based on the important property of the natural modes, called the *orthogonality property*.

In this chapter we shall first discuss the free vibration problem of an n-degree-of-freedom system, which leads to the eigenvalue problem. Several methods for the solution of the eigenvalue problem will be presented. Concepts such as the positive definite and the semidefinite systems will be introduced. A geometric interpretation of the eigenvalue problem will be discussed. The treatment of the eigenvalue problem associated with symmetric as well as nonsymmetric matrices will be presented.

The study of discrete systems gains further importance because of the fact that many of the approximate methods of solution of continuous systems reduce the problem of solving the partial differential equations associated with the continuous systems to the problem of solving a set of ordinary differential equations resembling, in structure, the equations describing a discrete system.

The problems are formulated exclusively in matrix form, to allow expedient solutions by means of high-speed electronic computer. Nevertheless, the reader is encouraged to first attempt a solution by means of a slide rule or desk calculator, to familiarize himself with the methods.

The natural modes of vibration play a predominant role in the field of vibrations. This chapter and the next two are devoted almost exclusively to this subject.

4-2 SMALL OSCILLATIONS OF CONSERVATIVE SYSTEMS. FREE VIBRATION

In Chapter 2 we defined a conservative system as a system subjected to forces depending only on position and derivable from a potential energy function V. These forces do not dissipate (or add) energy, so the total energy of the system is conserved.

To stay within the linear theory we must stipulate that the displacements are sufficiently small that the linear force-displacement relations are not violated and products of the variables vanish from the differential equations of motion.

The potential energy depends on a reference position, so in general it is defined within an arbitrary additive constant. We wish to choose this reference position as the equilibrium position of the system and consider small oscillations about this equilibrium position. Hence we define our coordinates $q_1, q_2, \ldots, q_n$ in such a way that they are zero in the equilibrium position and, in addition, we choose the arbitrary constant so as to render V zero in that position. Under these circumstances the Taylor series expansion of V about the equilibrium position is

$$V(q_1, q_2, \ldots, q_n) = \frac{\partial V}{\partial q_1} q_1 + \frac{\partial V}{\partial q_2} q_2 + \cdots + \frac{\partial V}{\partial q_n} q_n$$

$$+ \frac{1}{2} \left(\frac{\partial^2 V}{\partial q_1^2} q_1^2 + \frac{\partial^2 V}{\partial q_2^2} q_2^2 + \cdots + \frac{\partial^2 V}{\partial q_n^2} q_n^2 + 2 \frac{\partial^2 V}{\partial q_1 \, \partial q_2} q_1 q_2 \right.$$

$$\left. + 2 \frac{\partial^2 V}{\partial q_2 \, \partial q_3} q_2 q_3 + \cdots + 2 \frac{\partial^2 V}{\partial q_{n-1} \, \partial q_n} q_{n-1} q_n \right) + \cdots,$$

$$(4.1)$$

where all partial derivatives of V are evaluated at the equilibrium position $q_1 = q_2 = \cdots = q_n = 0$, hence are constant. But, according to our previous discussion (Section 2–6), in the equilibrium position $\partial V/\partial q_1 = \partial V/\partial q_2 = \cdots = \partial V/\partial q_n = 0$. Furthermore, if the motion is confined to small oscillations, terms containing the generalized coordinates at powers higher than 2 can be ignored, so (4.1) reduces to

$$V \cong \frac{1}{2} \sum_{i=1}^{n} \sum_{j=1}^{n} \frac{\partial^2 V}{\partial q_i \, \partial q_j} q_i q_j = \frac{1}{2} \sum_{i=1}^{n} \sum_{j=1}^{n} k_{ij} q_i q_j, \qquad (4.2)$$

where
$$\frac{\partial^2 V}{\partial q_i \, \partial q_j} = \frac{\partial^2 V}{\partial q_j \, \partial q_i} = k_{ij} = k_{ji} \qquad (4.3)$$

are recognized as the stiffness coefficients. Equation (4.2) has a quadratic form, as pointed out previously.

We shall be interested in stable systems for which V has a minimum in the equilibrium position. But in the equilibrium position V is zero, so for any position other than the equilibrium position V must be larger than zero. It follows that, under these circumstances, the potential energy is a *positive definite quadratic function* of the generalized coordinates.

We have previously shown that the kinetic energy is a positive definite quadratic function of the generalized velocities and is of the form

$$T = \tfrac{1}{2} \sum_{i=1}^{n} \sum_{j=1}^{n} m_{ij} \dot{q}_i \dot{q}_j,$$ (4.4)

where the inertia coefficients m_{ij} are symmetric.

For a conservative system there are no nonconservative forces, so using Lagrange's equations we can obtain the equations of motion of a linear system in the form

$$[m]\{\ddot{q}\} + [k]\{q\} = \{0\}.$$ (4.5)

Equation (4.5) results from (3.37) by setting the generalized nonconservative forces, Q_r, equal to zero. The motion corresponding to this case is known as the *undamped free vibration*. If a conservative system is imparted some energy in the form of initial displacements, initial velocities, or both, the system will vibrate indefinitely, because there is no energy dissipation.

A matrix whose elements are the coefficients of a positive definite quadratic form is said to be a *positive definite matrix*. Owing to the manner in which we defined the generalized coordinates and the arbitrary constant in the potential energy expression, $[k]$ is positive definite, and by the definition of the kinetic energy $[m]$ is positive definite. A system for which the matrices $[m]$ and $[k]$ are positive definite is called a *positive definite system*.

There are systems for which the potential energy is zero without the coordinates $q_1, q_2, \ldots, q_n$ being zero. In such a case the potential energy is a *positive quadratic function* rather than positive definite. Correspondingly, the matrix $[k]$ is said to be *positive* and not positive definite. By definition, however, $[m]$ is positive definite. A system possessing one positive matrix and one positive definite matrix is said to be *semidefinite* (see Section 4–9).

4–3 EIGENVALUE PROBLEM. NATURAL MODES OF VIBRATION

We shall be interested in the solution of the free vibration problem

$$[m]\{\ddot{q}\} + [k]\{q\} = \{0\},$$ (4.6)

where $[m]$ and $[k]$ are positive definite inertia and stiffness matrices, respectively. In particular, we wish to explore the possibility of having solutions separable in time and of the form

$$q_i(t) = u_i f(t), \qquad i = 1, 2, \ldots, n$$ (4.7)

which implies that the amplitude ratio of any two coordinates during motion does not depend on time. Physically this means that all coordinates perform synchronous motions and that the system configuration does not change its shape during motion but only its amplitude.

Introducing (4.7) in (4.6) we obtain

$$[m]\{u\}\ddot{f}(t) + [k]\{u\}f(t) = \{0\},$$ (4.8)

which implies n equations of the type

$$\sum_{j=1}^{n} m_{ij}u_j\ddot{f}(t) + \sum_{j=1}^{n} k_{ij}u_jf(t) = 0, \qquad i = 1, 2, \ldots, n.$$ (4.9)

The time dependence can be separated as follows:

$$-\frac{\ddot{f}(t)}{f(t)} = \frac{\sum_{j=1}^{n} k_{ij}u_j}{\sum_{j=1}^{n} m_{ij}u_j}, \qquad i = 1, 2, \ldots, n.$$ (4.10)

Following the standard procedure used in the separation of variables method, we recognize that the right side of (4.10) is independent of time, and, because the left side is independent of the index i, both sides must be equal to a constant. Let the constant be a positive constant ω^2, so that (4.10) leads to the relations

$$\ddot{f}(t) + \omega^2 f(t) = 0,$$ (4.11)

$$\sum_{j=1}^{n} (k_{ij} - \omega^2 m_{ij})u_j = 0, \qquad i = 1, 2, \ldots, n.$$ (4.12)

The choice of the sign of the constant is dictated by physical considerations. Our system is conservative, so the displacements must remain finite as t increases. By choosing a positive constant we obtain harmonic rather than exponential solutions, which is in agreement with the physical limitations of finite total energy.

The solution of (4.11) is

$$f(t) = A\cos(\omega t - \varphi),$$ (4.13)

where A is an arbitrary constant, and we conclude that a motion in which all the coordinates perform a harmonic motion with identical frequency ω and identical phase angle φ is possible. The question remains to determine if the frequency can take any value. The answer to this question lies in equations (4.12), which constitute a set of n linear homogeneous equations in the unknowns u_j. The problem of determining the constant ω^2 for which a set of homogeneous equations has a nontrivial solution is known as the *character-istic-value* or *eigenvalue problem*. The trivial case in which all u_j are zero must be ignored, because it represents the static equilibrium case. The eigen-value problem as given by (4.12) can be written in the matrix form

$$[k]\{u\} = \omega^2[m]\{u\}. \tag{4.14}$$

A nontrivial solution of the set of equations (4.14) is possible only if the determinant of the coefficients vanishes,†

$$\Delta = |[k] - \omega^2[m]| = 0, \tag{4.15}$$

where Δ is called the *characteristic determinant*.

Expanding the above determinant we obtain an algebraic equation of the nth order in ω^2 which is known as the *characteristic equation* or the *frequency equation*. The n roots of the characteristic equation are called *characteristic values* or *eigenvalues*.

Since $[k]$ and $[m]$ are symmetric and positive definite, it can be shown that the roots of the characteristic equation are *real and positive*.‡ The roots are denoted $\omega_1^2, \omega_2^2, \ldots, \omega_n^2$. The positive square root of these values are the *natural frequencies* ω_i of the system. The lowest frequency, ω_1, is called the *fundamental frequency*. The frequencies are arranged in order of increasing magnitude: $\omega_1 \leq \omega_2 \leq \omega_3 \leq \ldots \leq \omega_n$. In general all ω_i are distinct, although it is quite possible that two natural frequencies possess the same value. It follows that there are at most n frequencies in which harmonic motion as given by (4.13) is possible.

It remains now to determine the amplitudes u_i of these harmonic motions. Introducing the characteristic roots in (4.14) we obtain n equations of the type

$$[k]\{u\} = \omega_r^2[m]\{u\}, \qquad r = 1, 2, \ldots, n. \tag{4.16}$$

For each ω_r^2 the above equation has a nontrivial vector solution, $\{u^{(r)}\}$, called the *characteristic vector* or *eigenvector*. The column matrices $\{u^{(r)}\}$ consist of elements $u_i^{(r)}$ which are real numbers determined within a multipli-cative arbitrary constant, because $\alpha_r\{u^{(r)}\}$ is also a solution of the homogene-ous equation (4.16). It follows that although we cannot determine the amplitudes $u_i^{(r)}$ uniquely, we can determine the ratio between the elements of any vector $\{u^{(r)}\}$. Hence, for a given natural frequency ω_r, (4.16) will furnish a

---

† See F. B. Hildebrand, *Methods of Applied Mathematics*, Prentice-Hall, Englewood Cliffs, N.J., 1960, p. 30.

† See F. B. Hildebrand, *Methods of Applied Mathematics*, Prentice-Hall, Englewood Cliffs, N.J., 1960, p. 30.

‡ See F. B. Hildebrand, op. cit., p. 76.

vector $\{u^{(r)}\}$ which has a unique shape but arbitrary amplitude.† The elements $u_i^{(r)}$ are the components of the n-dimensional vector $\{u^{(r)}\}$. If one of the elements is assigned a definite value, the remaining $n - 1$ elements are uniquely determined. The characteristic vectors $\{u^{(r)}\}$ are called *modal vectors* and represent what are known as *natural modes of vibration*. It is customary to specify the value of one of the elements of $\{u^{(r)}\}$ and adjust the remaining $n - 1$ elements accordingly, thus determining uniquely the modal vectors. The process is called *normalization* and the resulting vectors are called *normal modes*. It is often convenient to normalize the vectors with respect to $[m]$ (or $[k]$ if so desired) by setting

$$\{u^{(r)}\}^T [m]\{u^{(r)}\} = 1. \tag{4.17}$$

Another frequently encountered normalization procedure is to assign the value unity to the largest element of the modal vector $\{u^{(r)}\}$ which may be convenient for plotting the mode shapes. The normalization process is just a convenience and has no physical significance. We shall use, for the most part, (4.17) to normalize the modes.

The normal modes are a characteristic of the system just as the natural frequencies are. They depend on the inertia and stiffness properties as reflected by the coefficients m_{ij} and k_{ij}. Each normal mode can be excited independently of the other. If the initial conditions or the forcing functions are such that the mode $\{u^{(r)}\}$ exclusively is excited, the motion will resemble entirely that mode shape and the system will perform a synchronous harmonic oscillation of frequency ω_r, which is the natural frequency associated with that particular mode.

It turns out that the normal modes possess an interesting and very useful property called *orthogonality*, so that in effect we have an orthonormal set of n vectors. If $\{u^{(s)}\}$ is a column matrix representing a normal mode other than the mode $\{u^{(r)}\}$, the orthogonality property is given by

$$\{u^{(r)}\}^T [m]\{u^{(s)}\} = \delta_{rs}, \tag{4.18}$$

where δ_{rs} is the Kronecker delta.

The set of vectors $\{u^{(1)}\}$, $\{u^{(2)}\}$, ..., $\{u^{(n)}\}$ is *complete* in the sense that any n-dimensional vector representing a possible motion of the system can be constructed as a linear combination of the above set of vectors. It follows that the most general motion of an n-degree-of-freedom linear system can be represented as a superposition of the normal modes of the system multiplied by some constants. The constants indicate the degree of participation of

---

† When two eigenvalues are equal, any linear combination of the corresponding eigenvectors is also an eigenvector. This case is known as degenerate.

each mode in the overall motion and are determined by the initial conditions in the case of free vibration and by the excitation forces in the case of forced vibration. The normal mode representation of a motion closely resembles the Fourier series representation of a function.

The modal vectors can be arranged in a square matrix of order n called the *modal matrix*,

$$[u] = [\{u^{(1)}\}\{u^{(2)}\}\cdots\{u^{(n)}\}], \tag{4.19}$$

so, in effect, the eigenvalue problem can be written

$$[m][u] = [k][u]\lceil\omega^{-2}\rfloor. \tag{4.20}$$

Because of the orthogonality property, the normal modes can be used to uncouple the equations of motion of a system. Indeed, if we use the linear transformation

$$\{q(t)\} = [u]\{\eta(t)\}, \tag{4.21}$$

the kinetic energy expression becomes

$$T = \tfrac{1}{2}\{\dot{q}\}^T[m]\{\dot{q}\} = \tfrac{1}{2}\{\dot{\eta}\}^T[u]^T[m][u]\{\dot{\eta}\}$$
$$= \tfrac{1}{2}\{\dot{\eta}\}^T\lceil M\rfloor\{\dot{\eta}\} = \tfrac{1}{2}\{\dot{\eta}\}^T\{\dot{\eta}\} \tag{4.22}$$

if the modes are normalized according to the scheme (4.17). In addition, the potential energy expression becomes

$$V = \tfrac{1}{2}\{q\}^T[k]\{q\} = \tfrac{1}{2}\{\eta\}^T[u]^T[k][u]\{\eta\}$$
$$= \tfrac{1}{2}\{\eta\}^T\lceil K\rfloor\{\eta\} = \tfrac{1}{2}\{\eta\}^T\lceil\omega^2\rfloor\{\eta\}, \tag{4.23}$$

so that the equations of motion (4.6) reduce to a set of n uncoupled equations of the type

$$\ddot{\eta}_r(t) + \omega_r^2\eta_r(t) = 0, \qquad r = 1, 2, \ldots, n, \tag{4.24}$$

where the variables $\eta_r(t)$ are the *normal coordinates* of the system.

Note that in (4.24) ω_r are the natural frequencies that satisfy the relation

$$\omega_r^2 = \frac{K_{rr}}{M_{rr}} = \frac{\{u^{(r)}\}^T[k]\{u^{(r)}\}}{\{u^{(r)}\}^T[m]\{u^{(r)}\}} = \{u^{(r)}\}^T[k]\{u^{(r)}\}, \qquad r = 1, 2, \ldots, n \tag{4.25}$$

when the eigenvectors $\{u^{(r)}\}$ are normalized according to (4.17).

The solution of (4.24) can be written

$$\eta_r(t) = c_r \cos(\omega_r t - \varphi_r), \qquad r = 1, 2, \ldots, n, \tag{4.26}$$

where c_r and φ_r $(r = 1, 2, \ldots, n)$ are constants of integration which can be identified as amplitudes and phase angles. Hence the free vibration of a system can be regarded as the superposition of n harmonic motions with frequencies equal to the natural frequencies of the system and with different amplitudes and phase angles,

$$\{q(t)\} = [u]\{\eta(t)\} = \sum_{r=1}^{n} c_r\{u^{(r)}\} \cos(\omega_r t - \varphi_r), \qquad (4.27)$$

where the amplitudes c_r and the corresponding phase angles φ_r are determined by the initial conditions. There are n amplitudes c_r and n phase angles φ_r, a total of $2n$ unknowns. If the n initial displacements $q_i(0)$, which are the components of the n-dimensional vector $\{q(0)\}$, and the n initial velocities $\dot{q}_i(0)$, which are the components of the n-dimensional vector $\{\dot{q}(0)\}$, are given, the determination of the amplitudes c_r and phase angles φ_r reduces to the solution of $2n$ algebraic equations in the unknowns c_r and φ_r $(r = 1, 2, \ldots, n)$.

The response of a system to initial conditions and external excitations will be discussed in great detail in Chapter 7.

We see, therefore, that the normal modes provide a good description of the dynamical properties of a system and its response. In many practical problems it is imperative to know the natural frequencies and the mode shapes. In the case of harmonic excitation, if the frequency of excitation coincides with one of the natural frequencies, a resonance condition is created in which the response tends to increase with time, perhaps causing undesirable effects. It is obvious that as the response amplitude increases, it is bound to exceed the linear range, at which point the present theory ceases to be valid.

The solution of the eigenvalue problem is of major importance in the study of vibrations and a great deal of attention will be devoted to it.

4–4 SOLUTION OF THE EIGENVALUE PROBLEM. CHARACTERISTIC DETERMINANT

There are a number of methods by which one can obtain the solution of the eigenvalue problem. The first to be presented is based on the solution of the characteristic equation. Denote

$$\frac{1}{\omega^2} = \lambda \qquad (4.28)$$

and write the equation representing the eigenvalue problem

$$\lambda[k]\{u\} - [m]\{u\} = \{0\}. \qquad (4.29)$$

Premultiply (4.29) by $[k]^{-1} = [a]$ and introduce the notation

$$[k]^{-1}[m] = [a][m] = [D], \tag{4.30}$$

where $[a]$ is the flexibility matrix and $[D]$ is called the *dynamical matrix*. Hence (4.29) can be written†

$$[\lambda \, \delta_{ij} - D_{ij}]\{u\} = \{0\}, \tag{4.31}$$

where δ_{ij} is the Kronecker delta.

The set of equations represented by (4.31) has a solution other than the trivial one if

$$\Delta(\lambda) = |\lambda \, \delta_{ij} - D_{ij}| = 0, \tag{4.32}$$

where $\Delta(\lambda)$ is an expression of the nth degree in λ. As pointed out previously, it has n real and positive roots λ_r, which are called characteristic values or eigenvalues‡ of $[D]$ related to the natural frequencies of the system by (4.28).

Suppose now that the characteristic values have been found and consider one particular root, $\lambda = \lambda_r$, to which corresponds the eigenvector $\{u^{(r)}\}$. The eigenvalue λ_r and the eigenvector $\{u^{(r)}\}$ are such that they satisfy

$$(\lambda_r[I] - [D])\{u^{(r)}\} = \{0\}. \tag{4.33}$$

Next introduce the notation

$$\lambda_r[I] - [D] = [f^{(r)}]. \tag{4.34}$$

But the determinant of the matrix $[f^{(r)}]$ satisfies the equation

$$\Delta(\lambda_r) = 0, \tag{4.35}$$

because λ_r is a root of the characteristic equation. Denoting by $[F^{(r)}]$ the adjoint of the matrix $[f^{(r)}]$, it follows that

† The same result can be obtained by writing

$$u_i = \sum_{j=1}^{n} a_{ij}F_j, \tag{a}$$

where for free vibration

$$F_j = -m_j \ddot{u}_j = \omega^2 m_j u_j, \tag{b}$$

so that

$$u_i = \omega^2 \sum_{j=1}^{n} a_{ij}m_j u_j, \tag{c}$$

in which case $[m]$ is a diagonal matrix.

‡ Sometimes also called *latent roots* of $[D]$.

$$[f^{(r)}][F^{(r)}] = \Delta(\lambda_r)[I] = [0], \tag{4.36}$$

where $[0]$ is a square null matrix. Equation (4.33) can be written

$$[f^{(r)}]\{u^{(r)}\} = \{0\}. \tag{4.37}$$

Comparing (4.36) and (4.37), we conclude that every column in the adjoint matrix $[F^{(r)}]$ must be proportional to the eigenvector $\{u^{(r)}\}$; hence they are proportional to each other.

In conclusion, to solve the eigenvalue problem of a system one must find the roots λ_r of (4.32). For each of these roots one constructs the matrix $[f^{(r)}]$ and calculates the adjoint $[F^{(r)}]$ of this matrix. The eigenvector $\{u^{(r)}\}$ is obtained by taking any of the nonvanishing columns of the adjoint matrix $[F^{(r)}]$. It is not important which column one takes, because they all differ by a multiplicative constant. Of course, the shape of the mode is unique, and every column of $[F^{(r)}]$ will yield the same mode shape after normalization.

The most difficult aspect of this method is to obtain the roots λ_r. If the degree of freedom of the system is very large, the solution of the characteristic or frequency equation can become quite laborious. There are various numerical schemes for obtaining these roots. The methods of Horner and Newton for determining the real roots of algebraic equation are presented in many texts.[†] A method known as *Graeffe's root-squaring method* will be presented here.

The frequency equation of an n-degree-of-freedom system can be written

$$\Delta(\lambda) = (\lambda - \lambda_1)(\lambda - \lambda_2) \cdots (\lambda - \lambda_n) = 0. \tag{4.38}$$

Replacing λ by $-\lambda$ in (4.38), we obtain $\Delta(-\lambda)$, so we can write

$$\Delta(\lambda)\,\Delta(-\lambda) = (-1)^n(\lambda^2 - \lambda_1^2)(\lambda^2 - \lambda_2^2) \cdots (\lambda^2 - \lambda_n^2) = 0. \tag{4.39}$$

Let us define a new frequency equation in the form

$$\Delta_2(\lambda) = (\lambda - \lambda_1^2)(\lambda - \lambda_2^2) \cdots (\lambda - \lambda_n^2) = 0, \tag{4.40}$$

where (4.40) has roots $\lambda_1^2, \lambda_2^2, \ldots, \lambda_n^2$ which are the squares of the roots of (4.38). Repeating this procedure, we obtain a sequence of polynomials $\Delta_4(\lambda), \Delta_8(\lambda), \Delta_{16}(\lambda), \ldots$. The result can be generalized by writing

$$\Delta_{2p}(\lambda) = (\lambda - \lambda_1^{2p})(\lambda - \lambda_2^{2p}) \cdots (\lambda - \lambda_n^{2p}), \qquad p = 1, 2, 4, 8, \ldots. \tag{4.41}$$

[†] See I. S. Sokolnikoff and E. S. Sokolnikoff, *Higher Mathematics for Engineers and Physicists*, 2nd ed., McGraw-Hill, New York, 1941, p. 95.

Note that p takes the values 2^q ($q = 0, 1, 2, \ldots$).

If the roots λ_r are such that

$$\lambda_1 > \lambda_2 > \lambda_3 > \cdots > \lambda_n \tag{4.42}$$

as p increases indefinitely, we have

$$\lambda_1^{2p} \gg \lambda_2^{2p} \gg \lambda_3^{2p} \gg \cdots \gg \lambda_n^{2p}. \tag{4.43}$$

We recall that all roots λ_r are real and positive. Equation (4.41) expands into

$$\Delta_{2p}(\lambda) = \lambda^n + a_1\lambda^{n-1} + a_2\lambda^{n-2} + \cdots + a_n = 0, \tag{4.44}$$

where

$$a_1 = -(\lambda_1^{2p} + \lambda_2^{2p} + \cdots + \lambda_n^{2p}),$$

$$a_2 = (\lambda_1^{2p}\lambda_2^{2p} + \lambda_1^{2p}\lambda_3^{2p} + \cdots + \lambda_{n-1}^{2p}\lambda_n^{2p}), \tag{4.45}$$

$$\vdots$$

$$a_n = (-1)^n\lambda_1^{2p}\lambda_2^{2p}\lambda_3^{2p}\cdots\lambda_n^{2p}.$$

But for large p the first term in each of the expressions in (4.45) will be very large compared with the remaining ones, and because the coefficients $a_1, a_2, \ldots, a_n$ are known, the roots $\lambda_1, \lambda_2, \ldots, \lambda_n$ can be approximated by

$$\lambda_1^{2p} = -a_1,$$

$$\lambda_2^{2p} = \frac{a_2}{\lambda_1^{2p}} = -\frac{a_2}{a_1},$$

$$\lambda_3^{2p} = -\frac{a_3}{\lambda_1^{2p}\lambda_2^{2p}} = -\frac{a_3}{a_2}, \tag{4.46}$$

$$\vdots$$

$$\lambda_n^{2p} = \frac{(-1)^n a_n}{\lambda_1^{2p}\lambda_2^{2p}\cdots\lambda_n^{2p}} = -\frac{a_n}{a_{n-1}}.$$

This method enables us to determine the magnitudes of the roots but not the sign. Owing to the nature of the present eigenvalue problem the roots must always be taken real and positive, because, by definition, $\lambda = 1/\omega^2$, where ω is real.

To obtain the roots of a polynomial we select a value for p and calculate the corresponding roots. Next we take the value $p + 1$ and check the improvement in accuracy. The procedure is repeated until the desired level of accuracy is achieved. The method is suitable for electronic computation.

An advantage of this method is that it furnishes all the roots simultaneously. On the other hand, it requires, in general, making a decision as to the sign of the roots. Furthermore, errors introduced at any stage of the iteration do not merely delay the convergence but also affect the correctness of the roots. The method is quite adequate when there is advance knowledge that all the roots are real and positive.

Example 4.1

Set up the eigenvalue problem corresponding to the system of Example 3.1 and solve the eigenvalue problem by means of the method using the characteristic determinant. Let

$$I_1 = I_2 = I_D, \qquad I_3 = 2I_D,$$

$$k_1 = k_2 = k_3 = \frac{GJ}{L}, \qquad k_4 = 2\frac{GJ}{L}. \tag{a}$$

Using results obtained in Example 3.1 we write

$$[a] = \frac{L}{7GJ} \begin{bmatrix} 5 & 3 & 1 \\ 3 & 6 & 2 \\ 1 & 2 & 3 \end{bmatrix}. \tag{b}$$

The inertia matrix of the system was given in Example 3.2,

$$[I_p] = \begin{bmatrix} I_1 & 0 & 0 \\ 0 & I_2 & 0 \\ 0 & 0 & I_3 \end{bmatrix} = I_D \begin{bmatrix} 1 & 0 & 0 \\ 0 & 1 & 0 \\ 0 & 0 & 2 \end{bmatrix}, \tag{c}$$

so the dynamical matrix assumes the form

$$[D] = [a][I_p] = \frac{LI_D}{7GJ} \begin{bmatrix} 5 & 3 & 1 \\ 3 & 6 & 2 \\ 1 & 2 & 3 \end{bmatrix} \begin{bmatrix} 1 & 0 & 0 \\ 0 & 1 & 0 \\ 0 & 0 & 2 \end{bmatrix} = \frac{LI_D}{7GJ} \begin{bmatrix} 5 & 3 & 2 \\ 3 & 6 & 4 \\ 1 & 2 & 6 \end{bmatrix}. \tag{d}$$

The characteristic determinant can be written

$$\Delta(\omega) = \left| \frac{1}{\omega^2} \delta_{ij} - D_{ij} \right| = \frac{LI_D}{7GJ} \begin{vmatrix} \lambda - 5 & -3 & -2 \\ -3 & \lambda - 6 & -4 \\ -1 & -2 & \lambda - 6 \end{vmatrix}, \tag{e}$$

where

$$\lambda = \frac{7GJ}{\omega^2 LI_D}. \tag{f}$$

Next denote

$$\Delta(\lambda) = \begin{vmatrix} \lambda - 5 & -3 & -2 \\ -3 & \lambda - 6 & -4 \\ -1 & -2 & \lambda - 6 \end{vmatrix} = \lambda^3 - 17\lambda^2 + 77\lambda - 98 = 0, \quad \text{(g)}$$

and let us obtain the roots of $\Delta(\lambda)$ by Graeffe's root-squaring method. First form

$$\Delta_2(\lambda) = \lambda^3 - 135\lambda^2 + 2{,}597\lambda - 9{,}604 = 0,$$

$$\Delta_4(\lambda) = \lambda^3 - 13{,}031\lambda^2 + 4{,}151{,}329\lambda - 92{,}236{,}816 = 0, \quad \text{(h)}$$

$$\Delta_8(\lambda) = \lambda^3 - 161{,}504{,}303\lambda^2 + 14{,}829{,}656{,}567{,}649\lambda$$

$$- 8{,}507{,}630{,}225{,}817{,}856 = 0,$$

where the latter is (4.44) with $p = 4$. At this point we wish to compute the roots λ_1, λ_2, and λ_3. Using (4.46) we obtain

$$\lambda_1^8 = -a_1 = 161{,}504{,}303, \qquad\qquad\qquad \lambda_1 = 10.6175,$$

$$\lambda_2^8 = -\frac{a_2}{a_1} = \frac{14{,}829{,}656{,}567{,}649}{161{,}504{,}303} = 91{,}822.052, \quad \lambda_2 = 4.1722, \quad \text{(i)}$$

$$\lambda_3^8 = -\frac{a_3}{a_2} = \frac{8{,}507{,}630{,}225{,}817{,}856}{14{,}829{,}656{,}567{,}649} = 573.690, \quad \lambda_3 = 2.2123.$$

Introducing λ_1 in (4.34) we have

$$[f^{(1)}] = \frac{1}{\omega_1^2}[I] - [D] = \frac{LI_D}{7GJ}\begin{bmatrix} \lambda_1 - 5 & -3 & -2 \\ -3 & \lambda_1 - 6 & -4 \\ -1 & -2 & \lambda_1 - 6 \end{bmatrix}$$

$$= \frac{LI_D}{7GJ}\begin{bmatrix} 5.6175 & -3 & -2 \\ -3 & 4.6175 & -4 \\ -1 & -2 & 4.6175 \end{bmatrix}. \quad \text{(j)}$$

The adjoint of $[f^{(1)}]$ is

$$[F^{(1)}] = \left(\frac{LI_D}{7GJ}\right)^2 \begin{bmatrix} 13.3213 & 17.8525 & 21.2350 \\ 17.8525 & 23.9388 & 28.4700 \\ 10.6175 & 14.2350 & 16.9388 \end{bmatrix}, \quad \text{(k)}$$

and it is not difficult to check that the columns of the adjoint matrix $[F^{(1)}]$ are all proportional and that any of them may be used to represent the first natural mode. Next we write

$$[f^{(2)}] = \frac{LI_D}{7GJ} \begin{bmatrix} \lambda_2 - 5 & -3 & -2 \\ -3 & \lambda_2 - 6 & -4 \\ -1 & -2 & \lambda_2 - 6 \end{bmatrix}$$

$$= -\frac{LI_D}{7GJ} \begin{bmatrix} 0.8278 & 3 & 2 \\ 3 & 1.8278 & 4 \\ 1 & 2 & 1.8278 \end{bmatrix}, \qquad (l)$$

which has the adjoint

$$F^{(2)} = \left(\frac{LI_D}{7GJ}\right)^2 \begin{bmatrix} -4.6591 & -1.4834 & 8.3444 \\ -1.4834 & -0.4869 & 2.6888 \\ 4.1722 & 1.3444 & -7.4869 \end{bmatrix}, \qquad (m)$$

and each of the columns of $[F^{(2)}]$ is proportional to the second natural mode. Similarly,

$$[f^{(3)}] = \frac{LI_D}{7GJ} \begin{bmatrix} \lambda_3 - 5 & -3 & -2 \\ -3 & \lambda_3 - 6 & -4 \\ -1 & -2 & \lambda_3 - 6 \end{bmatrix}$$

$$= -\frac{LI_D}{7GJ} \begin{bmatrix} 2.7877 & 3 & 2 \\ 3 & 3.7877 & 4 \\ 1 & 2 & 3.7877 \end{bmatrix}, \qquad (n)$$

and its adjoint is

$$[F^{(3)}] = \left(\frac{LI_D}{7GJ}\right)^2 \begin{bmatrix} 6.3467 & 7.3631 & 4.4246 \\ -7.3631 & -8.5590 & -5.1508 \\ 2.2123 & 2.5754 & 1.5590 \end{bmatrix}. \qquad (o)$$

The natural frequencies ω_1, ω_2, and ω_3 are obtained by introducing λ_1, λ_2, and λ_3 in (f). The results are summarized by taking the first column of the adjoint matrices and normalizing them according to

$$\{\theta^{(r)}\}^T [I_p]\{\theta^{(r)}\} = 1. \qquad (p)$$

The normal modes and the corresponding natural frequencies are

$$\{\theta^{(1)}\} = I_D^{-1/2} \begin{Bmatrix} 0.4959 \\ 0.6646 \\ 0.3952 \end{Bmatrix}, \qquad \omega_1 = 0.8139 \sqrt{\frac{GJ}{LI_D}},$$

$$\{\theta^{(2)}\} = I_D^{-1/2} \begin{Bmatrix} 0.6078 \\ 0.1935 \\ -0.5442 \end{Bmatrix}, \qquad \omega_2 = 1.2953 \sqrt{\frac{GJ}{LI_D}}, \qquad (q)$$

$$\{\theta^{(3)}\} = I_D^{-1/2} \begin{Bmatrix} 0.6215 \\ -0.7210 \\ 0.2166 \end{Bmatrix}, \qquad \omega_3 = 1.7788 \sqrt{\frac{GJ}{LI_D}}.$$

Note that the first mode has all the elements of the same sign. The second mode has one sign change and at some point between disks 2 and 3 the angular displacement is zero. Such a point is called a *node*. The third mode has two nodes. The presence of nodes is a typical characteristic of higher normal modes. The modes are plotted in Figure 4.1.

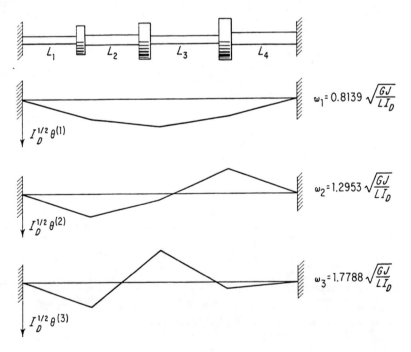

FIGURE 4.1

4-5 ORTHOGONALITY OF CHARACTERISTIC VECTORS

In Section 4-4 we have shown how to obtain the n natural frequencies ω_r and the corresponding modal vectors $\{u^{(r)}\}$ of the system. It was stated that the modal vectors possess the orthogonality property, and the corroboration of this statement follows.

The natural frequency ω_r and the characteristic vector $\{u^{(r)}\}$ satisfy the equation

$$\omega_r^2[m]\{u^{(r)}\} = [k]\{u^{(r)}\}. \tag{4.47}$$

Another solution of the eigenvalue problem, say ω_s and $\{u^{(s)}\}$, satisfies the equation

$$\omega_s^2[m]\{u^{(s)}\} = [k]\{u^{(s)}\}. \tag{4.48}$$

Premultiply both sides of (4.47) by $\{u^{(s)}\}^T$ and both sides of (4.48) by $\{u^{(r)}\}^T$ to obtain

$$\omega_r^2\{u^{(s)}\}^T[m]\{u^{(r)}\} = \{u^{(s)}\}^T[k]\{u^{(r)}\}, \tag{4.49}$$

$$\omega_s^2\{u^{(r)}\}^T[m]\{u^{(s)}\} = \{u^{(r)}\}^T[k]\{u^{(s)}\}. \tag{4.50}$$

Now transpose (4.50) recalling that $[m]$ and $[k]$ are symmetric matrices, and subtract the result from (4.49) to obtain

$$(\omega_r^2 - \omega_s^2)\{u^{(s)}\}^T[m]\{u^{(r)}\} = 0. \tag{4.51}$$

In general, any two frequencies ω_r and ω_s are not equal, so $\omega_r^2 - \omega_s^2 \neq 0$, from which it follows that

$$\{u^{(s)}\}^T[m]\{u^{(r)}\} = 0, \qquad r \neq s, \tag{4.52}$$

which is the statement of the orthogonality condition. The matrix $[m]$ plays the role of a weighting matrix. From either (4.49) or (4.50) we conclude that the orthogonality condition can also be written

$$\{u^{(s)}\}^T[k]\{u^{(r)}\} = 0, \qquad r \neq s. \tag{4.53}$$

Equations (4.52) and (4.53) hold true only if $[m]$ and $[k]$ are symmetric matrices. In many cases $[m]$ is a diagonal matrix, in which case the expansion of (4.52) has a simpler form than the expansion of (4.53).

In the case of repeated eigenvalues the associated eigenvectors are orthogonal to all the other eigenvectors but, in general, are not orthogonal to each other. One can orthogonalize them by taking them in certain linear combinations.

When $r = s$ the triple products in (4.52) and (4.53) are not zero, but yield mass and stiffness coefficients

$$M_{rr} = \{u^{(r)}\}^T[m]\{u^{(r)}\}, \tag{4.54}$$

$$K_{rr} = \{u^{(r)}\}^T[k]\{u^{(r)}\}. \tag{4.55}$$

Equations (4.52) and (4.54), on the one hand, and (4.53) and (4.55), on the other, yield

$$[u]^T[m][u] = \lceil M \rfloor, \tag{4.56}$$

$$[u]^T[k][u] = \lceil K \rfloor, \tag{4.57}$$

and if the eigenvectors $\{u^{(r)}\}$ are normalized according to (4.17), the above reduce to

$$[u]^T[m][u] = \lceil M \rfloor = [I], \tag{4.58}$$

$$[u]^T[k][u] = \lceil K \rfloor = \lceil \omega^2 \rfloor. \tag{4.59}$$

The orthogonality property has an important implication. Consider a set of n linear independent vectors in an n-dimensional space. Such a set is called a *basis* in that space. Any vector in the n-dimensional space can be generated by a linear combination of the linearly independent vectors. This fact implies solving a set of n algebraic equations, the coefficients of which are the elements of the linearly independent vectors. Hence the determinant of the coefficients cannot vanish and a nontrivial solution exists. It follows that the set of characteristic vectors $\{u^{(1)}\}, \{u^{(2)}\}, \ldots, \{u^{(n)}\}$, because it is orthogonal, hence independent, is a complete set of vectors in the sense that it can be used as a basis for the decomposition of any arbitrary n-dimensional vector $\{x\}$. This statement is known as the *expansion theorem* and can be expressed

$$\{x\} = \sum_{r=1}^{n} c_r\{u^{(r)}\}, \tag{4.60}$$

where the coefficients c_r are given by

$$c_r = \{u^{(r)}\}^T[m]\{x\}, \qquad r = 1, 2, \ldots, n. \tag{4.61}$$

The expansion theorem is of extreme importance in the field of vibrations and forms the basis of a widely used method of obtaining the system response known as *modal analysis*. Modal analysis will be used extensively throughout this text and will be discussed in detail in Chapter 7.

4-6   MATRIX ITERATION METHOD. SWEEPING TECHNIQUE

The method described in Section 4-3 involves the expansion of the determinantal equation and obtaining the roots of the frequency equation, which in the case of an n-degree-of-freedom system is an algebraic equation of nth order. A different approach uses matrix iteration procedures. We shall discuss two matrix iteration procedures: the first one iterates to one mode at a time and is known as the *matrix iteration method*, and the second one iterates to all modes simultaneously and we shall call it the *diagonalization by successive rotations method*. The first procedure will be discussed in this section.

The matrix iteration method is based on the assumption that the natural frequencies are distinct: $\omega_1 < \omega_2 < \omega_3 < \cdots < \omega_n$. The rate of convergence is deeply affected if two natural frequencies have very close values.

Equation (4.31) leads to the eigenvalue problem in the form

$$[D]\{u\} = \frac{1}{\omega^2}\{u\}, \qquad (4.62)$$

which is satisfied by every eigenvector $\{u^{(r)}\}$ with corresponding natural frequency ω_r.

The premultiplication of an arbitrary vector $\{u\}_1$ by the matrix $[D]$ represents a linear transformation that transforms the trial vector $\{u\}_1$ into another vector $\{u\}_2$. If the vector $\{u\}_1$ is one of the eigenvectors, say $\{u^{(r)}\}$, premultiplication by $[D]$ results in a vector $\{u\}_2$ which is proportional to $\{u^{(r)}\}$. If the trial vector $\{u\}_1$ is not an eigenvector, the resulting vector $\{u\}_2$ can be used as an improved trial vector. A sequence of such linear transformations eventually leads to a vector which when premultiplied by $[D]$ transforms into a vector proportional to itself. At this point convergence has been achieved, and the vector is simply the first eigenvector $\{u^{(1)}\}$ and the constant of proportionality is ω_1^{-2}. The following discussion justifies these statements.

According to the *expansion theorem* the n orthogonal eigenvectors $\{u^{(r)}\}$ of a system can be used as a basis for the decomposition of any arbitrary n-dimensional vector $\{u\}$ representing a possible motion of the system. Denote this vector $\{u\}_1$, such that

$$\{u\}_1 = c_1\{u^{(1)}\} + c_2\{u^{(2)}\} + \cdots + c_n\{u^{(n)}\} = \sum_{r=1}^{n} c_r\{u^{(r)}\}, \qquad (4.63)$$

where the constants c_r play the role of the unknowns in the set of n algebraic equations.

Any vector $\{u^{(r)}\}$ premultiplied by the dynamical matrix $[D]$ reproduces itself. Hence, if we use $\{u\}_1$, as given by (4.63), as a trial vector, premultiplication by $[D]$ yields

$$\{u\}_2 = [D]\{u\}_1 = \sum_{r=1}^{n} c_r[D]\{u^{(r)}\} = \sum_{r=1}^{n} \frac{c_r}{\omega_r^2} \{u^{(r)}\}. \qquad (4.64)$$

Next premultiply $\{u\}_2$ by $[D]$ and call the result $\{u\}_3$, to obtain

$$\{u\}_3 = [D]\{u\}_2 = \sum_{r=1}^{n} \frac{c_r}{(\omega_r^2)^2} \{u^{(r)}\}. \qquad (4.65)$$

In general, we have

$$\{u\}_p = [D]\{u\}_{p-1} = \sum_{r=1}^{n} \frac{c_r}{(\omega_r^2)^{p-1}} \{u^{(r)}\}. \qquad (4.66)$$

But the natural frequencies are such that $\omega_1 < \omega_2 < \cdots < \omega_n$. As p increases indefinitely, the first term of the series in (4.66) becomes predominantly larger than the other terms and in the limit, as $p \to \infty$, the trial vector $\{u\}_p$ will resemble in shape the first eigenvector $\{u^{(1)}\}$. Hence we have

$$\lim_{p \to \infty} \{u\}_p = \{u^{(1)}\}, \qquad (4.67)$$

and the first natural frequency is obtained from

$$\lim_{p \to \infty} \frac{u_{i,p-1}}{u_{i,p}} = \omega_1^2, \qquad (4.68)$$

where $u_{i,p-1}$ and $u_{i,p}$ are the elements in the ith row of the trial vector $\{u\}_{p-1}$ and the resulting vector $\{u\}_p$, respectively. Of course, in practice a finite number of iterations is sufficient for a good estimate of the first mode, the number of iterations depending on the accuracy desired.

For a given desired degree of accuracy there are two factors affecting the number of iterations necessary. The first one is how closely the arbitrary trial vector $\{u\}_1$ resembles the first mode. Mathematically this is equivalent to asking how large the coefficient c_1 is compared to the other $n - 1$ coefficients, if the eigenvectors $\{u^{(r)}\}$ are regarded as normalized. Obviously if c_1 is large compared to the remaining coefficients, the trial vector resembles the first mode to some degree and a smaller number of iterations will be necessary. This factor depends on the skill and experience of the analyst. The second factor depends entirely on the system, and it concerns the relative values of ω_1 and ω_2. The larger ω_2 is compared to ω_1, the faster the modes will separate and the smaller the necessary number of iterations will be. It is obvious that this method converges to the lowest mode.

One advantage of this method is that computational errors do not bring wrong results. Any error in one of the premultiplications by $[D]$ does not have a persistent damaging effect, because one can look upon the vector in error as a new trial vector. Errors, in general, will delay the convergence. Any set of numbers can be chosen for the trial vector $\{u\}_1$. Even if the numbers are such that the coefficient c_1 is almost zero, convergence will be achieved, although it may take longer. If after one iteration cycle the values of the elements of the resulting vector are too large, one can scale them down proportionately, because this does not affect the shape of the vector. Only in the unusual case in which the trial vector $\{u\}_1$ is exactly proportional to one of the modes other than the first mode, say $\{u^{(s)}\}$, does the method fail to deliver the first mode, because premultiplication by $[D]$ transforms $\{u^{(s)}\}$ into itself.

The method described above gives us the first or fundamental mode. The question remains how to obtain the higher modes. Any arbitrary trial vector premultiplied by $[D]$ would lead again to the first mode, so we must modify the procedure to obtain the second mode. To this end we must insist that the trial vector for the second mode is independent of the first eigenvector. That means the trial vector for the second mode must be orthogonal to the eigenvector $\{u^{(1)}\}$. The orthogonality relation is expressed by

$$\{u\}^T[m]\{u^{(1)}\} = 0. \tag{4.69}$$

Now introduce the notation

$$\{m^{(r)}\} = [m]\{u^{(r)}\}, \qquad r = 1, 2, \ldots, n. \tag{4.70}$$

so the orthogonality relation takes the form of the constraint equation

$$u_1 m_1^{(1)} + u_2 m_2^{(1)} + \cdots + u_n m_n^{(1)} = 0. \tag{4.71}$$

Equation (4.71) can be solved for one of the variables, say u_1, in terms of the remaining $n - 1$ variables. Letting

$$\frac{m_i^{(1)}}{m_1^{(1)}} = m_{1i}^{(1)}, \tag{4.72}$$

we obtain

$$u_1 = -m_{12}^{(1)} u_2 - m_{13}^{(1)} u_3 - \cdots - m_{1n}^{(1)} u_n, \tag{4.73}$$

and we may choose the remaining $n - 1$ variables arbitrarily,

$$u_i = u_i, \qquad i = 2, 3, \ldots, n. \tag{4.74}$$

Next construct a constraint matrix

$$[S^{(1)}] = \begin{bmatrix} 0 & -m_{12}^{(1)} & -m_{13}^{(1)} & \cdots & -m_{1n}^{(1)} \\ 0 & 1 & 0 & \cdots & 0 \\ 0 & 0 & 1 & \cdots & 0 \\ \vdots & & & & \\ 0 & 0 & 0 & \cdots & 1 \end{bmatrix}, \qquad (4.75)$$

called a *sweeping matrix*, which allows us to write (4.73) and (4.74) in the compact form

$$\{u\}_c^{(2)} = [S^{(1)}]\{u\}, \qquad (4.76)$$

where $\{u\}$ indicates an arbitrary vector and $\{u\}_c^{(2)}$ denotes a constrained vector orthogonal to the eigenvector $\{u^{(1)}\}$. At every stage of the iteration for the second mode we must be sure that the first mode is suppressed. This leads to the matrix iteration for the second mode in the form

$$[D]\{u\}_c^{(2)} = [D][S^{(1)}]\{u\}, \qquad (4.77)$$

so it is convenient to devise a new dynamical matrix,

$$[D^{(2)}] = [D][S^{(1)}], \qquad (4.78)$$

which has all the elements in the first column zero. The matrix $[D^{(2)}]$ leads to convergence to the second mode in the same way $[D]$ brings about convergence to the first. The eigenvalue problem now has the form

$$[D^{(2)}]\{u\} = \frac{1}{\omega^2}\{u\}, \qquad (4.79)$$

and the iteration to the second mode follows the same pattern as the one for the first mode. An arbitrary trial vector is selected and premultiplication by $[D^{(2)}]$ yields an improved trial vector which is in turn premultiplied by $[D^{(2)}]$. We can speed up the convergence by choosing a trial vector reasonably close to the second mode. This is not always possible nor is it necessary. Here, too, the method fails to furnish the second mode if the arbitrary trial vector happens to coincide with one of the modes other than the second one.

To obtain the third mode we must insist that both the first and second modes are suppressed from the trial vectors. Hence the trial vector must be orthogonal to both first and second modes, and the orthogonality conditions are

$$\{u\}^T[m]\{u^{(1)}\} = 0, \qquad (4.80)$$

$$\{u\}^T[m]\{u^{(2)}\} = 0, \qquad (4.81)$$

which represent two simultaneous constraint equations in the unknowns $u_1, u_2, \ldots, u_n$. Solving for u_1 and u_2 in terms of the remaining $n - 2$ unknowns and arbitrarily letting $u_i = u_i$, for $i = 3, 4, \ldots, n$, we can construct a new sweeping matrix $[S^{(2)}]$ in the form

$$[S^{(2)}] = \begin{bmatrix} 0 & 0 & m_{23}^{(2)} & m_{24}^{(2)} & \cdots & m_{2n}^{(2)} \\ 0 & 0 & -m_{13}^{(2)} & -m_{14}^{(2)} & \cdots & -m_{1n}^{(2)} \\ 0 & 0 & 1 & 0 & \cdots & 0 \\ 0 & 0 & 0 & 1 & \cdots & 0 \\ \vdots & & & & & \\ 0 & 0 & 0 & 0 & \cdots & 1 \end{bmatrix}, \tag{4.82}$$

where $$m_{ij}^{(2)} = \frac{m_i^{(2)}m_j^{(1)} - m_i^{(1)}m_j^{(2)}}{m_1^{(2)}m_2^{(1)} - m_1^{(1)}m_2^{(2)}}, \qquad i = 1, 2; j = 3, 4, \ldots, n. \tag{4.83}$$

Correspondingly we obtain a third dynamical matrix,

$$[D^{(3)}] = [D][S^{(2)}], \tag{4.84}$$

having the elements in both first and second columns zero, so the eigenvalue problem

$$[D^{(3)}]\{u\} = \frac{1}{\omega^2}\{u\} \tag{4.85}$$

yields the third mode.

The procedure follows the same pattern for the remaining modes, which are all obtained in ascending order. This is a numerical procedure, so accuracy is lost as the mode sought becomes higher. It is highly desired that at the beginning a larger number of significant figures than the one desired for the lower modes is retained, thus improving the accuracy of the higher modes.

The eigenvalue problem as given by (4.62) can also be written

$$[D]^{-1}\{u\} = \omega^2\{u\}, \tag{4.86}$$

and it follows that one can converge to the highest mode $\{u^{(n)}\}$ by premultiplying the trial vectors $\{u\}_p$ by $[D]^{-1}$ instead of $[D]$. Otherwise the method of solution is similar to the one already described. Aside from the fact that the modes are obtained in descending order, the constant of proportionality is in this case ω^2 rather than ω^{-2}.

It should be noted that to obtain the last mode it is no longer necessary to iterate, because the $n - 1$ orthogonality conditions are sufficient to define the last mode. One may wish, however, to follow the iteration procedure as a means of obtaining the last eigenvalue.

Example 4.2

Solve the eigenvalue problem of Example 4.1 by the matrix iteration method, using the sweeping technique.

In Example 4.1 we obtained the dynamical matrix

$$[D] = \frac{LI_D}{7GJ} \begin{bmatrix} 5 & 3 & 2 \\ 3 & 6 & 4 \\ 1 & 2 & 6 \end{bmatrix}, \tag{a}$$

so introducing the above matrix $[D]$ in (4.62) we obtain the following eigenvalue problem:

$$\begin{bmatrix} 5 & 3 & 2 \\ 3 & 6 & 4 \\ 1 & 2 & 6 \end{bmatrix} \begin{Bmatrix} \theta_1 \\ \theta_2 \\ \theta_3 \end{Bmatrix} = \lambda \begin{Bmatrix} \theta_1 \\ \theta_2 \\ \theta_3 \end{Bmatrix}, \qquad \lambda = \frac{7GJ}{\omega^2 LI_D}. \tag{b}$$

Let the first arbitrary vector $\{\theta\}_1$ have all its elements unity and perform the multiplication

$$\begin{bmatrix} 5 & 3 & 2 \\ 3 & 6 & 4 \\ 1 & 2 & 6 \end{bmatrix} \begin{Bmatrix} 1.0000 \\ 1.0000 \\ 1.0000 \end{Bmatrix} = \begin{Bmatrix} 10.0000 \\ 13.0000 \\ 9.0000 \end{Bmatrix} = 10.0000 \begin{Bmatrix} 1.0000 \\ 1.3000 \\ 0.9000 \end{Bmatrix}.$$

Next we use the resulting vector as an improved trial vector and perform the second iteration,

$$\begin{bmatrix} 5 & 3 & 2 \\ 3 & 6 & 4 \\ 1 & 2 & 6 \end{bmatrix} \begin{Bmatrix} 1.0000 \\ 1.3000 \\ 0.9000 \end{Bmatrix} = \begin{Bmatrix} 10.7000 \\ 14.4000 \\ 9.0000 \end{Bmatrix} = 10.7000 \begin{Bmatrix} 1.0000 \\ 1.3458 \\ 0.8411 \end{Bmatrix}.$$

The seventh iteration gives

$$\begin{bmatrix} 5 & 3 & 2 \\ 3 & 6 & 4 \\ 1 & 2 & 6 \end{bmatrix} \begin{Bmatrix} 1.0000 \\ 1.3413 \\ 0.7985 \end{Bmatrix} = \begin{Bmatrix} 10.6209 \\ 14.2418 \\ 8.4736 \end{Bmatrix} = 10.6209 \begin{Bmatrix} 1.0000 \\ 1.3409 \\ 0.7978 \end{Bmatrix}.$$

At this point it is intuitively felt that one additional iteration will not produce any appreciable change, and we take as the first eigenvector and first eigenvalue,

$$\{\theta^{(1)}\} = \begin{Bmatrix} 1.0000 \\ 1.3409 \\ 0.7978 \end{Bmatrix}, \qquad \lambda_1 = 10.6209. \tag{c}$$

The second eigenvector must be orthogonal to the first,

$$\{\theta\}^T[I_p]\{\theta^{(1)}\} = I_D \begin{Bmatrix} \theta_1 \\ \theta_2 \\ \theta_3 \end{Bmatrix}^T \begin{bmatrix} 1 & 0 & 0 \\ 0 & 1 & 0 \\ 0 & 0 & 2 \end{bmatrix} \begin{Bmatrix} 1.0000 \\ 1.3409 \\ 0.7978 \end{Bmatrix} = 0, \qquad (d)$$

which represents a constraint equation, allowing us to express θ_1 in terms of θ_2 and θ_3 as follows:

$$\theta_1 = -1.3409\theta_2 - 1.5956\theta_3, \qquad (e)$$

and in addition we take arbitrarily

$$\theta_2 = \theta_2, \qquad \theta_3 = \theta_3. \qquad (f)$$

Equations (e) and (f) are used to construct the first sweeping matrix,

$$[S^{(1)}] = \begin{bmatrix} 0 & -1.3409 & -1.5956 \\ 0 & 1 & 0 \\ 0 & 0 & 1 \end{bmatrix}. \qquad (g)$$

The iteration to the second eigenvector must use the new dynamical matrix

$$[D^{(2)}] = [D][S^{(1)}] = \frac{LI_D}{7GJ} \begin{bmatrix} 5 & 3 & 2 \\ 3 & 6 & 4 \\ 1 & 2 & 6 \end{bmatrix} \begin{bmatrix} 0 & -1.3409 & -1.5956 \\ 0 & 1 & 0 \\ 0 & 0 & 1 \end{bmatrix}$$

$$= \frac{LI_D}{7GJ} \begin{bmatrix} 0 & -3.7045 & -5.9780 \\ 0 & 1.9773 & -0.7868 \\ 0 & 0.6591 & 4.4044 \end{bmatrix}. \qquad (h)$$

We choose an arbitrary vector and write the first iteration for the second eigenvector,

$$\begin{bmatrix} 0 & -3.7045 & -5.9780 \\ 0 & 1.9773 & -0.7868 \\ 0 & 0.6591 & 4.4044 \end{bmatrix} \begin{Bmatrix} 0.0000 \\ 0.0000 \\ -1.0000 \end{Bmatrix} = \begin{Bmatrix} 5.9780 \\ 0.7868 \\ -4.4044 \end{Bmatrix} = 5.9780 \begin{Bmatrix} 1.0000 \\ 0.1316 \\ -0.7368 \end{Bmatrix}.$$

The eleventh iteration yields

$$\begin{bmatrix} 0 & -3.7045 & -5.9780 \\ 0 & 1.9773 & -0.7868 \\ 0 & 0.6591 & 4.4044 \end{bmatrix} \begin{Bmatrix} 1.0000 \\ 0.3214 \\ -0.8962 \end{Bmatrix} = \begin{Bmatrix} 4.1669 \\ 1.3406 \\ -3.7354 \end{Bmatrix} = 4.1669 \begin{Bmatrix} 1.0000 \\ 0.3217 \\ -0.8964 \end{Bmatrix},$$

at which point the iteration process was stopped. Hence we have the second eigenvector and the second eigenvalue,

$$\{\theta^{(2)}\} = \left\{ \begin{array}{c} 1.0000 \\ 0.3217 \\ -0.8964 \end{array} \right\}, \qquad \lambda_2 = 4.1669. \tag{i}$$

The third eigenvector must be orthogonal to the first two eigenvectors:

$$\{\theta\}^T[I_p]\{\theta^{(1)}\} = I_D \left\{ \begin{array}{c} \theta_1 \\ \theta_2 \\ \theta_3 \end{array} \right\}^T \left[\begin{array}{ccc} 1 & 0 & 0 \\ 0 & 1 & 0 \\ 0 & 0 & 2 \end{array} \right] \left\{ \begin{array}{c} 1.0000 \\ 1.3409 \\ 0.7978 \end{array} \right\} = 0,$$

$$\text{(j)}$$

$$\{\theta\}^T[I_p]\{\theta^{(2)}\} = I_D \left\{ \begin{array}{c} \theta_1 \\ \theta_2 \\ \theta_3 \end{array} \right\}^T \left[\begin{array}{ccc} 1 & 0 & 0 \\ 0 & 1 & 0 \\ 0 & 0 & 2 \end{array} \right] \left\{ \begin{array}{c} 1.0000 \\ 0.3217 \\ -0.8964 \end{array} \right\} = 0.$$

These two simultaneous equations, when solved for θ_1 and θ_2 in terms of θ_3, give

$$\theta_1 = 2.8623\theta_3, \qquad \theta_2 = -3.3246\theta_3, \tag{k}$$

and arbitrarily taking

$$\theta_3 = \theta_3, \tag{l}$$

we can construct a new sweeping matrix

$$[S^{(2)}] = \left[\begin{array}{ccc} 0 & 0 & 2.8623 \\ 0 & 0 & -3.3246 \\ 0 & 0 & 1 \end{array} \right]. \tag{m}$$

The dynamical matrix to be used for the iteration to the third eigenvector is

$$[D^{(3)}] = [D][S^{(2)}] = \frac{LI_D}{7GJ} \left[\begin{array}{ccc} 5 & 3 & 2 \\ 3 & 6 & 4 \\ 1 & 2 & 6 \end{array} \right] \left[\begin{array}{ccc} 0 & 0 & 2.8623 \\ 0 & 0 & -3.3246 \\ 0 & 0 & 1 \end{array} \right]$$

$$= \frac{LI_D}{7GJ} \left[\begin{array}{ccc} 0 & 0 & 6.3377 \\ 0 & 0 & -7.3607 \\ 0 & 0 & 2.2131 \end{array} \right]. \tag{n}$$

Upon arbitrarily choosing a trial vector we iterate to the third eigenvector as follows:

$$
\begin{bmatrix} 0 & 0 & 6.3377 \\ 0 & 0 & -7.3607 \\ 0 & 0 & 2.2131 \end{bmatrix} \begin{Bmatrix} 0.0000 \\ 0.0000 \\ 1.0000 \end{Bmatrix} = \begin{Bmatrix} 6.3377 \\ -7.3607 \\ 2.2131 \end{Bmatrix} = 6.3377 \begin{Bmatrix} 1.0000 \\ -1.1614 \\ 0.3492 \end{Bmatrix},
$$

$$
\begin{bmatrix} 0 & 0 & 6.3377 \\ 0 & 0 & -7.3607 \\ 0 & 0 & 2.2131 \end{bmatrix} \begin{Bmatrix} 1.0000 \\ -1.1614 \\ 0.3492 \end{Bmatrix} = \begin{Bmatrix} 2.2131 \\ -2.5704 \\ 0.7728 \end{Bmatrix} = 2.2131 \begin{Bmatrix} 1.0000 \\ -1.1614 \\ 0.3492 \end{Bmatrix}.
$$

It is easy to see that to obtain the third eigenvector there is not really any iteration process involved, because the nonzero column of the dynamical matrix $[D^{(3)}]$ is proportional to the last mode. The third eigenvector and eigenvalues are

$$
\{\theta^{(3)}\} = \begin{Bmatrix} 1.0000 \\ -1.1614 \\ 0.3492 \end{Bmatrix}, \qquad \lambda_3 = 2.2131. \tag{o}
$$

The eigenvectors can be normalized by writing

$$
\{\theta^{(r)}\}^T [I_p]\{\theta^{(r)}\} = 1, \tag{p}
$$

and the natural frequencies can be evaluated by using the second of equations (b).   The normal modes and the natural frequencies are

$$
\{\theta^{(1)}\} = I_D^{-1/2} \begin{Bmatrix} 0.4956 \\ 0.6646 \\ 0.3954 \end{Bmatrix}, \qquad \omega_1 = 0.8118 \sqrt{\frac{GJ}{LI_D}},
$$

$$
\{\theta^{(2)}\} = I_D^{-1/2} \begin{Bmatrix} 0.6075 \\ 0.1954 \\ -0.5445 \end{Bmatrix}, \qquad \omega_2 = 1.2961 \sqrt{\frac{GJ}{LI_D}}, \tag{q}
$$

$$
\{\theta^{(3)}\} = I_D^{-1/2} \begin{Bmatrix} 0.6210 \\ -0.7213 \\ 0.2169 \end{Bmatrix}, \qquad \omega_3 = 1.7784 \sqrt{\frac{GJ}{LI_D}}.
$$

The above results compare reasonably well with the results obtained by using the characteristic determinant method.

4-7 GEOMETRIC INTERPRETATION OF THE EIGENVALUE PROBLEM

Let us consider the eigenvalue problem

$$[a]\{x\} = \lambda[I]\{x\} = \lambda\{x\}, \tag{4.87}$$

where $[a]$ is an $n \times n$ symmetric matrix consisting of real elements and $\{x\}$ is a column matrix defining an n-dimensional vector. This is a special type of eigenvalue problem in the sense that one of the matrices defining the eigenvalue problem is the identity matrix. In theory there is no difficulty in obtaining the solution of the eigenvalue problem, (4.87), by means of the matrix methods discussed in preceding sections.

Perhaps we can obtain a deeper insight into the problem by exploring the geometric aspect of the eigenvalue problem. Premultiplying (4.87) by $\{x\}^T$, we obtain

$$\{x\}^T[a]\{x\} = \lambda\{x\}^T\{x\}. \tag{4.88}$$

But the vector $\{x\}$ can be normalized by writing

$$\{x\}^T\{x\} = \sum_{i=1}^{n} x_i^2 = |\mathbf{r}|^2 = r^2 = 1, \tag{4.89}$$

which is the equivalent of saying that the length r of the vector $\{x\}$ is unity. It follows that

$$\{x\}^T[a]\{x\} = \lambda, \tag{4.90}$$

which is a quadratic form in the coordinates $x_1, x_2, \ldots, x_n$.

Next let $n = 2$, so that (4.90) becomes

$$\frac{1}{\lambda}[a_{11}x_1^2 + (a_{12} + a_{21})x_1x_2 + a_{22}x_2^2] = \frac{1}{\lambda}(a_{11}x_1^2 + 2a_{12}x_1x_2 + a_{22}x_2^2) = 1, \tag{4.91}$$

which can be identified as a *conic section* with the center at the origin of the system of axes. The conic could be an ellipse, a parabola, or a hyperbola, depending on the coefficients a_{ij} (the circle can be regarded as an ellipse of zero eccentricity). Because we are dealing with small oscillations, the conic must by necessity be an ellipse. The ellipse is shown in Figure 4.2, where x_1 and x_2 represent a rectangular system of axes.

It is not difficult to show that the vector of the direction cosines of the normal $\mathbf{n}$ to the ellipse at any point (x_1, x_2), if written in matrix form, is

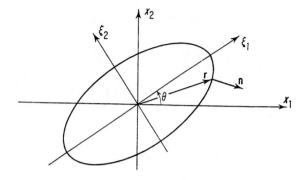

FIGURE 4.2

proportional to the vector $[a]\{x\}$. Hence the eigenvalue problem (4.87) can be interpreted as the problem of finding the directions for which the radius vector **r**, and the normal to the ellipse **n**, are parallel to each other. These directions define a set of axes known as the *principal axes* of the conic.

Let ξ_1 and ξ_2 be the directions of the principal axes and denote by θ the angle between these axes and the x_1 and x_2 directions, respectively. To express the equation of the ellipse in terms of the coordinates ξ_1 and ξ_2, we use the following linear transformation:

$$x_1 = \xi_1 \cos\theta - \xi_2 \sin\theta,$$
$$x_2 = \xi_1 \sin\theta + \xi_2 \cos\theta, \tag{4.92}$$

which can be written in the matrix form

$$\{x\} = [R]\{\xi\}, \tag{4.93}$$

where
$$[R] = \begin{bmatrix} \cos\theta & -\sin\theta \\ \sin\theta & \cos\theta \end{bmatrix} \tag{4.94}$$

is called a *rotation* matrix, because the axes ξ_1 and ξ_2 are obtained from the axes x_1 and x_2 by means of a rotation through an angle θ. Note that

$$[R]^T[R] = [R][R]^T = [I], \tag{4.95}$$

from which it follows that

$$[R]^T = [R]^{-1}, \tag{4.96}$$

which means that the transpose of the matrix $[R]$ is equal to its inverse. A linear transformation possessing the property described by (4.96) is called an *orthonormal transformation*. In this particular case it represents the relation between two rectangular sets of axes. Introducing (4.93) in (4.90), we obtain

$$\frac{1}{\lambda}\{\xi\}^T[R]^T[a][R]\{\xi\} = \frac{1}{\lambda}\{\xi\}^T[b]\{\xi\} = 1, \tag{4.97}$$

where
$$[b] = [R]^T[a][R] \tag{4.98}$$

has the elements

$$b_{11} = a_{11}\cos^2\theta + 2a_{12}\sin\theta\cos\theta + a_{22}\sin^2\theta,$$

$$b_{12} = b_{21} = -(a_{11} - a_{22})\sin\theta\cos\theta + a_{12}(\cos^2\theta - \sin^2\theta), \quad (4.99)$$

$$b_{22} = a_{11}\sin^2\theta - 2a_{12}\sin\theta\cos\theta + a_{22}\cos^2\theta.$$

The matrix $[b]$ can be reduced to a diagonal form, $b_{12} = b_{21} = 0$, if θ is such that

$$\tan 2\theta = \frac{2a_{12}}{a_{11} - a_{22}}. \tag{4.100}$$

Note that there are two values of θ satisfying (4.100). Their values differ by $\pi/2$ and they correspond to the directions ξ_1 and ξ_2, respectively, as can be seen from Figure 4.2. If θ does satisfy (4.100), (4.97) reduces to

$$\frac{b_{11}}{\lambda}\xi_1^2 + \frac{b_{22}}{\lambda}\xi_2^2 = 1, \tag{4.101}$$

which is the equation of the ellipse in its *canonical form*,† so that the principal axes of the ellipse coincide with the directions ξ_1 and ξ_2.

Introducing (4.93) in (4.87) and premultiplying by $[R]^T$, we obtain

$$[b]\{\xi\} = \lambda\{\xi\}, \tag{4.102}$$

which represents an uncoupled system of equations. Equation (4.102) gives in effect, the solution of the eigenvalue problem, which can be written‡

$$\lambda_1 = b_{11}, \quad \{\xi^{(1)}\} = \begin{Bmatrix} 1 \\ 0 \end{Bmatrix}, \quad \{x^{(1)}\} = [R]\{\xi^{(1)}\} = \begin{Bmatrix} \cos\theta \\ \sin\theta \end{Bmatrix},$$

$$\lambda_2 = b_{22}, \quad \{\xi^{(2)}\} = \begin{Bmatrix} 0 \\ 1 \end{Bmatrix}, \quad \{x^{(2)}\} = [R]\{\xi^{(2)}\} = \begin{Bmatrix} -\sin\theta \\ \cos\theta \end{Bmatrix},$$

$$\tag{4.103}$$

† The cross products are absent.
‡ Recall that for a symmetric matrix of real numbers the eigenvalues are real.

where λ_1 and λ_2 are the eigenvalues of the problem and $\{x^{(1)}\}$ and $\{x^{(2)}\}$ are the associated normal modes. We note that for the two-dimensional case the modal matrix $[x]$ coincides with the rotation matrix $[R]$.

Hence, in essence, the solution of the eigenvalue problem consists of obtaining the principal axes of a conic section, which is also consistent with reducing the quadratic expression (4.97) to its canonical form

$$\lambda_1 \xi_1^2 + \lambda_2 \xi_2^2 = \lambda. \tag{4.104}$$

Note that the eigenvalues are inversely proportional to the squares of the semimajor and semiminor axes, respectively.

For $n = 3$, (4.90) represents a *quadric surface* with the center at the origin. Because of the nature of the problem, the surface must be an ellipsoid, so the eigenvalue problem (4.87) reduces to the problem of finding the principal axes of the ellipsoid. To this end we use the linear transformation

$$\{x\} = [u]\{\xi\}, \tag{4.105}$$

where the transformation matrix $[u]$ is such that

$$[u]^T[a][u] = \lceil\lambda\rfloor, \tag{4.106}$$

where $\lceil\lambda\rfloor$ is the diagonal matrix of the eigenvalues λ_1, λ_2, and λ_3. The equation of the ellipsoid in its canonical form becomes accordingly

$$\lambda_1 \xi_1^2 + \lambda_2 \xi_2^2 + \lambda_3 \xi_3^2 = \lambda, \tag{4.107}$$

where ξ_1, ξ_2, and ξ_3 represent a mutually orthogonal system of axes coinciding with the principal axes of the ellipsoid. In this case, however, the transformation matrix $[u]$, which is the modal matrix, will have a more complicated form than the simple rotation matrix of the two-dimensional case. It can, in fact, be obtained as a sequence of rotations, as will be shown in Section 4–8.

Next let us consider the *degenerate case* in which two eigenvalues are equal, $\lambda_1 = \lambda_2$. For this case (4.107) reduces to

$$\lambda_1(\xi_1^2 + \xi_2^2) + \lambda_3 \xi_3^2 = \lambda, \tag{4.108}$$

so the ellipsoid becomes an ellipsoid of revolution with the ξ_3 axis as the symmetry axis. The intersection between the plane containing the axes ξ_1 and ξ_2 and the ellipsoid itself is a circle rather than an ellipse. The principal axes ξ_1 and ξ_2, which in the general case are the axes of symmetry of the elliptical cross section, reduce to two orthogonal diameters in the case of the circular cross section. But for a circle there is no reason to prefer one set of mutually orthogonal diameters over another, so any pair of perpendicular diameters can be regarded as a set of two principal axes. This provides a geometric interpretation of the statement made earlier (see Section 4–3)—that for equal eigenvalues any linear combination of the associated eigen-vectors is also an eigenvector.

For $n > 3$, (4.90) represents a *quadratic surface* with the center at the origin. Although we can no longer use the simple geometric representation, the concepts remain the same. The quadratic surface can be thought of as having n principal axes which are orthogonal to each other, forming a rectangular system of axes. The eigenvalue problem (4.87) can be written

$$[a][u] = [u][\lambda], \qquad (4.109)$$

where $[u]$ is a square matrix of the orthonormal modes, $[u]^T[u] = [I]$, and $[\lambda]$ is a diagonal matrix of the eigenvalues. Similarly, (4.90) assumes the form (4.106), and the quadratic surface has the canonical form

$$\lambda_1 \xi_1^2 + \lambda_2 \xi_2^2 + \lambda_3 \xi_3^2 + \cdots + \lambda_n \xi_n^2 = \lambda, \qquad (4.110)$$

where λ_r and ξ_r $(r = 1, 2, \ldots, n)$ are the eigenvalues and the associated principal coordinates of the system, respectively.

The method of solution of the eigenvalue problem employing the diagonalization of a matrix by means of successive rotations is based on the ideas presented here. The method is shown in Section 4–8.

4–8 DIAGONALIZATION BY SUCCESSIVE ROTATIONS METHOD. THE JACOBI METHOD

In contrast to the matrix iteration method employing the sweeping technique, the diagonalization by successive rotations method iterates to all eigenvectors and eigenvalues simultaneously. The diagonalization process consists of multiplications by matrices which are similar in form to coordinate transformation matrices that represent angular rotations, as shown in Section 4–7. This series of multiplications results in matrices whose diagonal elements keep increasing at the expense of the off-diagonal elements. When finally the off-diagonal elements become zero, the resulting matrix is the matrix of the eigenvalues and the continuous product of rotation matrices is the modal matrix. The method is due to Jacobi.

The general eigenvalue problem for an n-degree-of-freedom system can be written according to (4.20) in the form

$$[k][u] = [m][u][\omega^2], \qquad (4.111)$$

where $[u]$ is the modal matrix.

The method calls for the matrices $[m]^{1/2}$ and $[m]^{-1/2}$. In the case in which

$[m]$ is a diagonal matrix, the above matrices are readily obtained, because for a diagonal matrix,

$$[\lambda]^p = [\lambda^p], \tag{4.112}$$

provided the matrix is not singular. When $[m]$ is not a diagonal matrix we must solve the eigenvalue problem associated with the matrix $[m]$. To this end we write the eigenvalue problem in the form

$$[m][z] = [z][\gamma], \tag{4.113}$$

where $[z]$ is the modal matrix and $[\gamma]$ the eigenvalue matrix associated with the symmetric matrix $[m]$. The eigenvalue problem (4.113) is a special case of the eigenvalue problem (4.111) in the sense that instead of the $[k]$ matrix we have here the identity matrix $[I]$. If the eigenvectors $\{z^{(r)}\}$ are normalized so that $\{z^{(r)}\}^T\{z^{(r)}\} = 1$, we have

$$[z]^T[z] = [I], \tag{4.114}$$

and from the definition of the inverse of a matrix it follows that

$$[z]^T = [z]^{-1}, \tag{4.115}$$

or the transpose of $[z]$ is equal to its inverse. Postmultiplying (4.113) by $[z]^T$ we obtain

$$[m] = [z][\gamma][z]^T. \tag{4.116}$$

It also follows that

$$[m]^{-1} = ([z]^T)^{-1}[\gamma]^{-1}[z]^{-1} = [z][\gamma^{-1}][z]^T, \tag{4.117}$$

and

$$[m]^2 = [z][\gamma][z]^T[z][\gamma][z]^T = [z][\gamma^2][z]^T. \tag{4.118}$$

It can also be shown that

$$[m]^{1/2} = [z][\gamma^{1/2}][z]^T, \tag{4.119}$$

$$[m]^{-1/2} = [z][\gamma^{-1/2}][z]^T. \tag{4.120}$$

Equation (4.113) premultiplied by $[z]^T$ yields

$$[z]^T[m][z] = [\gamma]. \tag{4.121}$$

In view of (4.115) transformation (4.121) is a *similarity transformation*.† At present we have only the matrix $[m]$, and we must find a matrix $[z]$ such

† A transformation of the form $[b] = [Q]^{-1}[a][Q]$ is said to be a *similarity transformation* and the matrices $[a]$ and $[b]$ are said to be *similar*. One can show that the matrices $[a]$ and $[b]$ possess the same characteristic values. It follows that the sum of the diagonal elements of a matrix, known as the *trace* of the matrix, is invariant under similarity transformation. The orthonormal transformations are special types of similarity transformations.

that the triple product $[z]^T[m][z]$ yields the diagonal matrix $\lceil \gamma \rfloor$. To this end consider the matrix

$$[R] = \begin{bmatrix} & & & p & & & q & & & \\ 1 & 0 & \cdots & 0 & 0 & 0 & 0 & \cdots & \cdots & 0 \\ 0 & 1 & \cdots & 0 & 0 & 0 & 0 & \cdots & \cdots & 0 \\ \vdots & & & & & & & & & \\ 0 & 0 & \cdots & \cos\theta & 0 & 0 & -\sin\theta & \cdots & \cdots & 0 \\ 0 & 0 & \cdots & 0 & 1 & 0 & 0 & \cdots & \cdots & 0 \\ 0 & 0 & \cdots & 0 & 0 & 1 & 0 & \cdots & \cdots & 0 \\ 0 & 0 & \cdots & \sin\theta & 0 & 0 & \cos\theta & \cdots & \cdots & 0 \\ \vdots & & & & & & & & & \\ 0 & 0 & \cdots & 0 & 0 & 0 & 0 & \cdots & \cdots & 1 \end{bmatrix} \begin{matrix} \\ \\ \\ p \\ \\ \\ q \\ \\ \\ \end{matrix}, \quad (4.122)$$

which is similar in structure to a matrix representing the rotation of a rectangular system of coordinates through an angle θ. Of course, in contrast to the above matrix, the matrix representing the rotation of the rectangular coordinate system is a square matrix of order 3. Because of the similarity in structure, $[R]$ is called a rotation matrix. Now let

$$[b] = [R]^T[a][R], \qquad (4.123)$$

where $[a]$ is a symmetric matrix. It follows that the resulting matrix, $[b]$, is also symmetric. Note that (4.123) is a similarity transformation. The elements of matrix $[b]$ are given by

$$
\begin{aligned}
&b_{jk} = a_{jk}, \qquad j, k \neq p, q, \\
&\left. \begin{aligned} b_{pj} &= a_{pj}\cos\theta + a_{qj}\sin\theta \\ b_{qj} &= -a_{pj}\sin\theta + a_{qj}\cos\theta \end{aligned} \right\} \quad j \neq p, q, \\
&b_{pp} = a_{pp}\cos^2\theta + 2a_{pq}\sin\theta\cos\theta + a_{qq}\sin^2\theta, \\
&b_{pq} = a_{pq}(\cos^2\theta - \sin^2\theta) - (a_{pp} - a_{qq})\sin\theta\cos\theta, \\
&b_{qq} = a_{pp}\sin^2\theta - 2a_{pq}\sin\theta\cos\theta + a_{qq}\cos^2\theta.
\end{aligned}
\qquad (4.124)
$$

From equations (4.124) we observe that

$$\sum_{i=1}^{n}\sum_{j=1}^{n} b_{ij}^2 = \sum_{i=1}^{n}\sum_{j=1}^{n} a_{ij}^2, \qquad (4.125)$$

or the sum of the squares of *all* the elements in $[b]$ is equal to the sum of the squares of all the elements in $[a]$. We also note that

$$\sum_{j=1}^{n} b_{jj}^2 + 2b_{pq}^2 = \sum_{j=1}^{n} a_{jj}^2 + 2a_{pq}^2. \qquad (4.126)$$

In the special case in which θ is chosen such that

$$\tan 2\theta = \frac{2a_{pq}}{a_{pp} - a_{qq}}, \tag{4.127}$$

b_{pq} vanishes and it follows that

$$\sum_{j=1}^{n} b_{jj}^2 = \sum_{j=1}^{n} a_{jj}^2 + 2a_{pq}^2, \tag{4.128}$$

or the sum of the squares of the off-diagonal elements in $[b]$ is smaller by $2a_{pq}^2$ than the sum of the off-diagonal elements in $[a]$. This forms the basis of the diagonalization process. By a continuous multiplication by various rotation matrices, the off-diagonal elements of a matrix are reduced to zero, so a diagonal matrix is finally obtained. The process is expressed by

$$[m] = [m]_0,$$
$$[R]_1^T[m]_0[R]_1 = [m]_1$$
$$[R]_2^T[m]_1[R]_2 = [R]_2^T[R]_1^T[m][R]_1[R]_2 = [m]_2, \tag{4.129}$$
$$\vdots$$
$$[R]_t^T[m]_{t-1}[R]_t = [R]_t^T \cdots [R]_1^T[m][R]_1 \cdots [R]_t = [m]_t.$$

As t increases indefinitely, $[m]_t$ becomes a diagonal matrix and in the limit we have

$$\lim_{t \to \infty} [m]_t = [\gamma], \tag{4.130}$$

$$\lim_{t \to \infty} [R]_1[R]_2 \cdots [R]_{t-1}[R]_t = [z]. \tag{4.131}$$

The work is reduced by noting that the matrices $[m]_t$ are symmetric. Of course, in practice, only a finite number of multiplications are necessary to achieve diagonalization. When $[z]$ and $[\gamma]$ are obtained, $[m]^{1/2}$ and $[m]^{-1/2}$ can be easily evaluated.

Now we are in the position of solving the eigenvalue problem (4.111), which can also be written

$$[k][m]^{-1/2}[m]^{1/2}[u] = [m]^{1/2}[m]^{1/2}[u][\omega^2]. \tag{4.132}$$

Premultiply both sides of (4.132) by $[m]^{-1/2}$, so

$$[m]^{-1/2}[k][m]^{-1/2}[m]^{1/2}[u] = [m]^{1/2}[u][\omega^2] \tag{4.133}$$

and denote

$$[m]^{-1/2}[k][m]^{-1/2} = [\bar{k}], \tag{4.134}$$

$$[m]^{1/2}[u] = [\bar{u}] \tag{4.135}$$

to obtain the special form of the eigenvalue problem

$$[\bar{k}][\bar{u}] = [\bar{u}][\omega^2],\tag{4.136}$$

which is of the same form as (4.113) and, therefore, can be solved for $[\bar{u}]$ and $[\omega^2]$ in precisely the same way. The diagonal matrix $[\omega^2]$ gives the natural frequencies, whereas the modal matrix is obtained from the matrix product

$$[u] = [m]^{-1/2}[\bar{u}].\tag{4.137}$$

The Jacobi method can be used to find the eigenvalues and eigenvectors of a symmetric matrix of very large order. It is basically an iterative method, because an element reduced to zero by one rotation is, in general, no longer zero after the subsequent rotation. Furthermore, the rotation angles θ_s must be evaluated with a large degree of accuracy, because the accuracy of the results depends on these angles.

When the off-diagonal elements of the symmetric matrix $[\bar{k}]$ are large relative to its diagonal elements, the Jacobi method may require a very large number of iterations. In such cases one may wish to consider the use of a method that reduces the general symmetric matrix $[\bar{k}]$ to a symmetric tri-diagonal matrix rather than a diagonal one. A tridiagonal matrix has all its elements zero except the elements on the main diagonal and the elements directly above and below the main diagonal. This, of course, still requires the solution of the eigenvalue problem associated with the tridiagonal matrix, but this is relatively simpler to obtain than the one associated with a general symmetric matrix. The reduction of the $[\bar{k}]$ matrix to a tridiagonal form is a finite iterative process, so, in general, this procedure involves less computational work than the Jacobi method, provided the diagonal elements of $[\bar{k}]$ are not dominant and the solution of the eigenvalue problem associated with the tridiagonal matrix presents no difficulty.

We single out two methods to reduce a general symmetric matrix to a tridiagonal form. The first method is due to Givens, and, as the Jacobi method, it uses rotation matrices as transformation matrices. For a matrix of order n, the process involves $n - 2$ major steps r ($r = 1, 2, \ldots, n - 2$), and every one of these major steps renders the elements in the rth row and rth column zero, except the tridiagonal elements. A major step consists of $n - r - 1$ minor steps, in which the elements in the positions $r + 2, r + 3, \ldots, n$ in the rth row and rth column are successively reduced to zero. Every major step r does not affect the elements in the first $r - 1$ rows and columns, so the process is a finite one. Furthermore, the computation of the rotation angles θ_s in the Givens method is simpler than in the Jacobi method. The second method, due to Householder, is a variation on the method by Givens, and it is more efficient than the Givens method. It uses elementary Hermitian orthogonal matrices as transformation matrices, as opposed to rotation matrices used by the Jacobi method and the Givens method. The reduction to a tridiagonal form is done in $n - 2$ steps r $(r = 1, 2, \ldots, n - 2)$, in which all the elements in the rth row and rth column, with the exception of the tridiagonal elements, are rendered zero simultaneously. A full description of the Givens method and the Householder method can be found in the text by Ralston[†] or the text by Wilkinson.[‡]

Example 4.3

Solve the eigenvalue problem of Example 4.1 by the matrix iteration method employing the diagonalization by successive rotations technique. The eigenvalue problem of Example 4.1 may be written in the form

$$[k][\theta] = [I_p][\theta][\ulcorner\omega^2\urcorner], \tag{a}$$

where
$$[k] = \frac{GJ}{L} \begin{bmatrix} 2 & -1 & 0 \\ -1 & 2 & -1 \\ 0 & -1 & 3 \end{bmatrix}, \tag{b}$$

$$[I_p] = I_D \begin{bmatrix} 1 & 0 & 0 \\ 0 & 1 & 0 \\ 0 & 0 & 2 \end{bmatrix} \tag{c}$$

are symmetric stiffness and inertia matrices, respectively. The matrix $[I_p]$ is already in diagonal form, so without difficulty we can obtain

$$[I_p]^{-1/2} = I_D^{-1/2} \begin{bmatrix} 1 & 0 & 0 \\ 0 & 1 & 0 \\ 0 & 0 & 1/\sqrt{2} \end{bmatrix}. \tag{d}$$

We make the following transformation:

$$[\bar{k}] = [I_p]^{-1/2}[k][I_p]^{-1/2} = \frac{GJ}{LI_D} \begin{bmatrix} 2.0000 & -1.0000 & 0 \\ -1.0000 & 2.0000 & -0.7071 \\ 0 & -0.7071 & 1.5000 \end{bmatrix}, \tag{e}$$

† A. Ralston, *A First Course in Numerical Analysis*, McGraw-Hill, New York, 1965, pp. 492–499.

‡ J. H. Wilkinson, *The Algebraic Eigenvalue Problem*, Oxford Univ. Press, New York, 1965, pp. 282 and 290.

and if we let

$$[\dot{\theta}] = [I_p]^{1/2}[\theta],$$ (f)

the eigenvalue problem (a) reduces to the special form

$$[\bar{k}][\dot{\theta}] = [\dot{\theta}][\omega^2],$$ (g)

which can also be written

$$[\dot{\theta}]^T[\bar{k}][\dot{\theta}] = [\omega^2].$$ (h)

Now we shall proceed to diagonalize the matrix $[\bar{k}]$ by the method outlined in Section 4–8. The symbol θ in the rotation matrix $[R]$ will be replaced by φ, because θ here denotes the angular motion of the disks. First, let us eliminate the off-diagonal term $\bar{k}_{12}$, so that $p = 1$, $q = 2$. The angle φ_1 associated with the first rotation matrix, $[R]_1$, is given by†

$$\tan 2\varphi_1 = \frac{2\bar{k}_{12}}{\bar{k}_{11} - \bar{k}_{22}} = \frac{2(-1.0000)}{2.0000 - 2.0000} = \mp \infty \rightarrow \varphi_1 = \frac{3\pi}{4},$$ (i)

so the first rotation matrix is

$$[R]_1 = \begin{bmatrix} -0.7071 & -0.7071 & 0 \\ 0.7071 & -0.7071 & 0 \\ 0 & 0 & 1 \end{bmatrix},$$ (j)

and $$[\bar{k}]_1 = [R]_1^T[\bar{k}][R]_1 = \frac{GJ}{LI_D} \begin{bmatrix} 3.0000 & 0 & -0.5000 \\ 0 & 1.0000 & 0.5000 \\ -0.5000 & 0.5000 & 1.5000 \end{bmatrix}.$$ (k)

Next, letting $p = 1$, $q = 3$, we obtain $\varphi_2 = 163° \; 09'$, the second rotation matrix,

$$[R]_2 = \begin{bmatrix} -0.9571 & 0 & -0.2899 \\ 0 & 1 & 0 \\ 0.2899 & 0 & -0.9571 \end{bmatrix},$$ (l)

and $$[\bar{k}]_2 = [R]_2^T[\bar{k}]_1[R]_2 = \frac{GJ}{LI_D} \begin{bmatrix} 3.1516 & 0.1450 & 0 \\ 0.1450 & 1.0000 & -0.4786 \\ 0 & -0.4786 & 1.3485 \end{bmatrix}.$$ (m)

† The decision in which quadrant the angle lies will be based on the sign of the numerator, which corresponds to the value of $\sin 2\varphi_i$ ($i = 1, 2, \cdots$).

The subsequent rotation matrices are

$$[R]_3 = \begin{bmatrix} 1 & 0 & 0 \\ 0 & -0.5736 & -0.8192 \\ 0 & 0.8192 & -0.5736 \end{bmatrix}, \quad [R]_4 = \begin{bmatrix} -0.9989 & 0 & -0.0477 \\ 0 & 1 & 0 \\ 0.0477 & 0 & -0.9989 \end{bmatrix},$$

(n)

$$[R]_5 = \begin{bmatrix} 0.9984 & -0.0561 & 0 \\ 0.0561 & 0.9984 & 0 \\ 0 & 0 & 1 \end{bmatrix}, \quad [R]_6 = \begin{bmatrix} 1 & 0 & 0 \\ 0 & 1.0000 & -0.0039 \\ 0 & 0.0039 & 1.0000 \end{bmatrix},$$

and we obtain

$$[\omega^2] \cong [R]_6^T[\bar{k}]_5[R]_6 = \frac{GJ}{LI_D} \begin{bmatrix} 3.1619 & 0 & 0 \\ 0 & 1.6789 & 0 \\ 0 & 0 & 0.6594 \end{bmatrix}$$

(o)

$$[\hat{\theta}] = \prod_{i=1}^{6} [R]_i = \begin{bmatrix} -0.6209 & 0.6074 & -0.4959 \\ 0.7215 & 0.1949 & -0.6646 \\ -0.3070 & -0.7702 & -0.5592 \end{bmatrix}.$$

(p)

From (f) we obtain the modal matrix

$$[\theta] = [I_p]^{-1/2}[\hat{\theta}] = I_D^{-1/2} \begin{bmatrix} -0.6209 & 0.6074 & -0.4959 \\ 0.7215 & 0.1949 & -0.6646 \\ -0.2171 & -0.5446 & -0.3954 \end{bmatrix}.$$

(q)

For comparison purposes we list the normal modes and the natural frequencies

$$\{\theta^{(1)}\} = I_D^{-1/2} \begin{Bmatrix} 0.4959 \\ 0.6646 \\ 0.3954 \end{Bmatrix}, \qquad \omega_1 = 0.8120\sqrt{\frac{GJ}{LI_D}},$$

$$\{\theta^{(2)}\} = I_D^{-1/2} \begin{Bmatrix} 0.6074 \\ 0.1949 \\ -0.5446 \end{Bmatrix}, \qquad \omega_2 = 1.2957\sqrt{\frac{GJ}{LI_D}},$$

(r)

$$\{\theta^{(3)}\} = I_D^{-1/2} \begin{Bmatrix} 0.6209 \\ -0.7215 \\ 0.2171 \end{Bmatrix}, \qquad \omega_3 = 1.7782\sqrt{\frac{GJ}{LI_D}}.$$

We see that the solutions of the eigenvalue problem obtained by three different methods are in pretty close agreement. The slight discrepancies are due to the degree of accuracy to which the iteration process has been carried out.

4-9   SEMIDEFINITE SYSTEMS. UNRESTRAINED SYSTEMS

Consider a conservative system with the kinetic and potential energy expressions given by

$$T = \tfrac{1}{2}\{\dot{q}\}^T [m]\{\dot{q}\}, \tag{4.138}$$

$$V = \tfrac{1}{2}\{q\}^T [k]\{q\}, \tag{4.139}$$

where $[m]$ and $[k]$ are symmetric matrices. The kinetic energy, by definition, is always positive; hence $[m]$ is a positive definite matrix. We can conceive

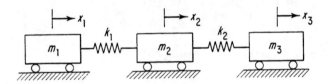

FIGURE 4.3

of a system, however, for which the potential energy can be zero without the motion being identically zero. For such a system the potential energy is larger than zero, unless it takes the minimum value, which is zero. The corresponding matrix $[k]$ is said to be *positive* rather than positive definite. A system for which $[m]$ is positive definite and $[k]$ is positive only is called a *semidefinite system*. For semidefinite systems $V = 0$ does not correspond to an equilibrium position. Unrestrained systems, or systems with no supports, are typical examples of semidefinite systems.

We can let $\{u\}$ be the amplitude of the free vibration and write the corresponding eigenvalue problem in the form

$$\omega^2 [m]\{u\} = [k]\{u\}. \tag{4.140}$$

Premultiplying both sides of (4.140) by $\{u\}^T$, we obtain

$$\omega^2 \{u\}^T [m]\{u\} = \{u\}^T [k]\{u\}, \tag{4.141}$$

and if we denote by $\{u^{(0)}\}$ the column matrix that renders the potential energy zero, we have

$$\omega_0^2 \{u^{(0)}\}^T [m] \{u^{(0)}\} = 0, \tag{4.142}$$

and as $\{u^{(0)}\}^T [m] \{u^{(0)}\} \neq 0$, it follows by necessity that $\omega_0 = 0$, or in the case of a semidefinite system there is at least one zero eigenvalue. The mode $\{u^{(0)}\}$ is called a *zero* mode and can be identified as a *rigid-body mode*.

As an illustration consider the system of Fig. 4.3, for which the kinetic energy expression is

$$T = \tfrac{1}{2}(m_1 \dot{x}_1^2 + m_2 \dot{x}_2^2 + m_3 \dot{x}_3^2)$$

$$= \tfrac{1}{2} \begin{Bmatrix} \dot{x}_1 \\ \dot{x}_2 \\ \dot{x}_3 \end{Bmatrix}^T \begin{bmatrix} m_1 & 0 & 0 \\ 0 & m_2 & 0 \\ 0 & 0 & m_3 \end{bmatrix} \begin{Bmatrix} \dot{x}_1 \\ \dot{x}_2 \\ \dot{x}_3 \end{Bmatrix} = \tfrac{1}{2}\{\dot{x}\}^T [m]\{\dot{x}\}, \tag{4.143}$$

and the potential energy expression has the form

$$V = \tfrac{1}{2}[k_1(x_2 - x_1)^2 + k_2(x_3 - x_2)^2]$$

$$= \tfrac{1}{2} \begin{Bmatrix} x_1 \\ x_2 \\ x_3 \end{Bmatrix}^T \begin{bmatrix} k_1 & -k_1 & 0 \\ -k_1 & k_1 + k_2 & -k_2 \\ 0 & -k_2 & k_2 \end{bmatrix} \begin{Bmatrix} x_1 \\ x_2 \\ x_3 \end{Bmatrix} = \tfrac{1}{2}\{x\}^T [k]\{x\}. \tag{4.144}$$

It is easy to see that the kinetic energy is always positive. We can imagine, however, the situation in which all the masses undergo the same motion,

$$x_1 = x_2 = x_3 = x_0, \tag{4.145}$$

which renders $V = 0$. Obviously, this is not an equilibrium position, and the motion may be regarded as the motion of a rigid body, from which the name rigid-body mode is derived. It can be easily seen from (4.144) that any rigid-body displacement added to the motion does not affect the potential energy.

From (4.141) we conclude that the rigid-body mode with zero natural frequency is a solution of the eigenvalue problem. Any other natural mode must be orthogonal to the rigid-body mode,

$$\{x^{(0)}\}^T [m]\{x\} = 0, \tag{4.146}$$

and because $\{x^{(0)}\}^T$ is a row matrix of equal elements, the above leads to

$$[m]\{x\} = m_1 x_1 + m_2 x_2 + m_3 x_3 = 0 \tag{4.147}$$

as the orthogonality condition. But this can be readily identified as the condition that the linear momentum for an unrestrained system in longitudinal elastic motion, with no external forces applied, must be zero,

$$m_1 \dot{x}_1 + m_2 \dot{x}_2 + m_3 \dot{x}_3 = 0. \tag{4.148}$$

Integrating (4.148) to obtain (4.147) one must let $m_1 x_1(0) + m_2 x_2(0) + m_3 x_3(0) = 0$, which means that the position of the center of mass must not be disturbed if only elastic motion is to occur.

It can be easily shown that the matrix $[k]$ has no inverse; hence it is singular. This is consistent with the fact that for an unrestrained system we cannot define flexibility influence coefficients.

Equation (4.147) can be looked upon as a constraint equation† and may be used to reduce the degree of freedom of the system by 1. To this end write

$$x_3 = -\frac{m_1}{m_3} x_1 - \frac{m_2}{m_3} x_2; \tag{4.149}$$

in addition, take arbitrarily

$$x_1 = x_1, \qquad x_2 = x_2 \tag{4.150}$$

and obtain the matrix transformation

$$\begin{Bmatrix} x_1 \\ x_2 \\ x_3 \end{Bmatrix}_c = \begin{bmatrix} 1 & 0 & 0 \\ 0 & 1 & 0 \\ -\dfrac{m_1}{m_3} & -\dfrac{m_2}{m_3} & 0 \end{bmatrix} \begin{Bmatrix} x_1 \\ x_2 \\ x_3 \end{Bmatrix} = \begin{bmatrix} 1 & 0 \\ 0 & 1 \\ -\dfrac{m_1}{m_3} & -\dfrac{m_2}{m_3} \end{bmatrix} \begin{Bmatrix} x_1 \\ x_2 \end{Bmatrix}. \tag{4.151}$$

Now denote

$$[c] = \begin{bmatrix} 1 & 0 \\ 0 & 1 \\ -\dfrac{m_1}{m_3} & -\dfrac{m_2}{m_3} \end{bmatrix}, \tag{4.152}$$

where the 3×2 matrix $[c]$ is referred to as a *constraint matrix*. This allows us to write (4.151) in a more compact form,

$$\{x\}_c = [c]\{x\}, \qquad \{\dot{x}\}_c = [c]\{\dot{x}\} \tag{4.153}$$

and, furthermore,

$$\{x\}_c^T = \{x\}^T [c]^T, \qquad \{\dot{x}\}_c^T = \{\dot{x}\}^T [c]^T. \tag{4.154}$$

† Note that this is a holonomic constraint according to the definition of Chapter 2.

The vectors $\{x\}$ and $\{\dot{x}\}$ in (4.143) and (4.144) may be regarded as constrained, because they are subject to the constraint equation (4.149). The kinetic and potential energy expressions can be written in terms of independent coordinates,

$$T = \tfrac{1}{2}\{\dot{x}\}_c^T [m]\{\dot{x}\}_c = \tfrac{1}{2}\{\dot{x}\}[c]^T [m][c]\{\dot{x}\} = \tfrac{1}{2}\{\dot{x}\}^T [m']\{\dot{x}\}, \quad (4.155)$$

$$V = \tfrac{1}{2}\{x\}_c^T [k]\{x\}_c = \tfrac{1}{2}\{x\}^T [c]^T [k][c]\{x\} = \tfrac{1}{2}\{x\}^T [k']\{x\}, \quad (4.156)$$

where

$$[m'] = [c]^T [m][c], \quad (4.157)$$

$$[k'] = [c]^T [k][c], \quad (4.158)$$

are 2×2 symmetric matrices.

The eigenvalue problem becomes

$$\omega^2 [m']\{x\} = [k']\{x\}, \quad (4.159)$$

which is a 2×2 symmetric eigenvalue problem whose solution consists of two modes $\{x^{(1)}\}$ and $\{x^{(2)}\}$ with corresponding natural frequencies ω_1 and ω_2. The modes $\{x^{(1)}\}$ and $\{x^{(2)}\}$ give only the excursion of the masses m_1 and m_2 in the corresponding modes. To obtain a complete description of the modes we must write

$$\{x^{(1)}\}_c = [c]\{x^{(1)}\}, \qquad \{x^{(2)}\}_c = [c]\{x^{(2)}\}, \quad (4.160)$$

where the satisfaction of the linear momentum equation is provided for. Of course, one should not forget the rigid-body mode, $\{x^{(0)}\}$, with the natural frequency $\omega_0 = 0$.

4–10   ENCLOSURE THEOREM

Consider the eigenvalue problem

$$[k]\{u\} = \omega^2 [m]\{u\} = \lambda [m]\{u\}, \quad (4.161)$$

which is a special eigenvalue problem in the sense that $[m]$ is a diagonal matrix of positive elements. The matrix $[k]$ is a symmetric matrix.

Let us form the following matrix products:

$$[k]\{u\} = \{v\}, \qquad [m]\{u\} = \{w\} \quad (4.162)$$

and introduce the ratios

$$\lambda_j^* = \frac{v_j}{w_j}, \quad j = 1, 2, \ldots, n, \quad (4.163)$$

where v_j and w_j are the jth elements of the column matrices $\{v\}$ and $\{w\}$, respectively. Evidently, if $\{u\}$ coincides with an eigenvector, say $\{u^{(r)}\}$, then all the ratios λ_j^* would be identically equal to the eigenvalue λ_r.

Let $\{u\}$ be an arbitrary vector with all its elements different than zero and denote the largest of the ratios λ_j^* by λ_{max}^* and the smallest by λ_{min}^*. The *enclosure theorem*† states that if all the ratios λ_j^* are positive, there is at least one eigenvalue λ of the problem, (4.161), enclosed by λ_{min}^* and λ_{max}^*,

$$\lambda_{min}^* \leq \lambda \leq \lambda_{max}^*. \tag{4.164}$$

The enclosure theorem can be used to advantage in numerical work. It provides a rough estimate of the interval within which an eigenvalue of the system under consideration lies. This interval can be narrowed by small corrections in the values of the elements of the trial vector $\{u\}$.

As an illustration consider the problem of Example 4.3, for which we have

$$[k] = \frac{GJ}{L} \begin{bmatrix} 2 & -1 & 0 \\ -1 & 2 & -1 \\ 0 & -1 & 3 \end{bmatrix},$$

$$[m] = I_D \begin{bmatrix} 1 & 0 & 0 \\ 0 & 1 & 0 \\ 0 & 0 & 2 \end{bmatrix}, \tag{4.165}$$

and in addition $\lambda = \omega^2$. Using an arbitrary vector $u_1 = 2$, $u_2 = 3$, and $u_3 = 2$, we obtain

$$\begin{Bmatrix} v_1 \\ v_2 \\ v_3 \end{Bmatrix} = \frac{GJ}{L} \begin{bmatrix} 2 & -1 & 0 \\ -1 & 2 & -1 \\ 0 & -1 & 3 \end{bmatrix} \begin{Bmatrix} 2 \\ 3 \\ 2 \end{Bmatrix} = \frac{GJ}{L} \begin{Bmatrix} 1 \\ 2 \\ 3 \end{Bmatrix},$$

$$\begin{Bmatrix} w_1 \\ w_2 \\ w_3 \end{Bmatrix} = I_D \begin{bmatrix} 1 & 0 & 0 \\ 0 & 1 & 0 \\ 0 & 0 & 2 \end{bmatrix} \begin{Bmatrix} 2 \\ 3 \\ 2 \end{Bmatrix} = I_D \begin{Bmatrix} 2 \\ 3 \\ 4 \end{Bmatrix}, \tag{4.166}$$

from which it follows that

$$(\omega^2)_1^* = \frac{v_1}{w_1} = 0.5000 \frac{GJ}{LI_D}, \qquad (\omega^2)_2^* = \frac{v_2}{w_2} = 0.6667 \frac{GJ}{LI_D},$$

$$(\omega^2)^* = \frac{v_3}{w_3} = 0.7500 \frac{GJ}{LJ_D}, \tag{4.167}$$

† For a proof see L. Collatz, *Eigenwertaufgaben mit technischen Anwendungen*, Akademische Verlagsgesellschaft, Leipzig, 1963, p. 304.

so

$$(\omega^2)^*_{min} = 0.5000 \, \frac{GJ}{LI_D}, \qquad (\omega^2)^*_{max} = 0.7500 \, \frac{GJ}{LI_D}. \qquad (4.168)$$

Referring to Example 4.3 we find that $\omega_1^2 = 0.6594(GJ/LI_D)$, so $(\omega^2)^*_{min}$ and and $(\omega^2)^*_{max}$ enclose the first eigenvalue,

$$0.5000 \, \frac{GJ}{LI_D} = (\omega^2)^*_{min} \le \omega_1^2 = 0.6594 \, \frac{GJ}{LI_D} \le (\omega^2)^*_{max} = 0.7500 \, \frac{GJ}{LI_D}. \qquad (4.169)$$

Next let us try the corrected values $u_1 = 2.4$, $u_2 = 3.2$, and $u_3 = 2.0$. The new trial vector leads to the bracketing values

$$(\omega^2)^*_{min} = \frac{v_2}{w_2} = 0.6250 \, \frac{GJ}{LI_0}, \qquad (\omega^2)^*_{max} = \frac{v_3}{w_3} = 0.7000 \, \frac{GJ}{LI_D}, \qquad (4.170)$$

and it is easy to see that the new trial vector did in fact narrow the interval within which the first eigenvalue lies. The procedure can be repeated and the results arranged in a table form.

4–11 RAYLEIGH'S QUOTIENT. PROPERTIES OF RAYLEIGH'S QUOTIENT

Consider a positive definite system and write the eigenvalue problem in the form

$$[k]\{u\} = \lambda[m]\{u\}, \qquad (4.171)$$

where $[k]$ and $[m]$ are symmetric matrices. The above equation is satisfied by any eigenvalue λ_r with the associated eigenvector $\{u^{(r)}\}$

$$[k]\{u^{(r)}\} = \lambda_r[m]\{u^{(r)}\}, \qquad r = 1, 2, \ldots, n. \qquad (4.172)$$

Premultiplying both sides of (4.172) by $\{u^{(r)}\}^T$ and dividing through by $\{u^{(r)}\}^T[m]\{u^{(r)}\}$, we obtain

$$\lambda_r = \frac{\{u^{(r)}\}^T[k]\{u^{(r)}\}}{\{u^{(r)}\}^T[m]\{u^{(r)}\}}, \qquad r = 1, 2, \ldots, n, \qquad (4.173)$$

which indicates that the eigenvalues λ_r may be obtained as the ratio of two quadratic forms as shown. This result is actually not new and was encountered before [see (4.25)].

Consider now any arbitrary vector $\{u\}$ and form the expression corresponding to (4.173),

$$\omega^2 = R(u) = \frac{\{u\}^T[k]\{u\}}{\{u\}^T[m]\{u\}}, \qquad (4.174)$$

where the scalar $R(u)$ obviously depends on the vector $\{u\}$ and is known as *Rayleigh's quotient*. When the vector $\{u\}$ coincides with one of the eigenvectors of the system, Rayleigh's quotient reduces to the associated eigenvalue.

The Rayleigh quotient has a stationary value when the arbitrary vector $\{u\}$ is in the neighborhood of an eigenvector. To show this property of Rayleigh's quotient, let us make use of the expansion theorem and assume that the vector $\{u\}$ is a superposition of the eigenvectors of the system

$$\{u\} = \sum_{r=1}^{n} c_r\{u^{(r)}\} = [u]\{c\}, \tag{4.175}$$

where $[u]$ is a square matrix of the eigenvectors and $\{c\}$ is a column matrix of the coefficients c_r. If the eigenvectors $\{u^{(r)}\}$ are normalized so that

$$[u]^T[m][u] = [I], \tag{4.176}$$

it follows that

$$[u]^T[k][u] = [\lambda], \tag{4.177}$$

where $[\lambda]$ is a diagonal matrix of the eigenvalues λ_r. Introducing (4.175) in Rayleigh's quotient expression, (4.174), and using (4.176) and (4.177), we obtain†

$$R(u) = \frac{\{c\}^T[u]^T[k][u]\{c\}}{\{c\}^T[u]^T[m][u]\{c\}} = \frac{\{c\}^T[\lambda]\{c\}}{\{c\}^T[I]\{c\}} = \frac{\sum_{i=1}^{n} c_i^2 \lambda_i}{\sum_{i=1}^{n} c_i^2}. \tag{4.178}$$

If $\{u\}$ differs little from the eigenvector $\{u^{(r)}\}$, the coefficient c_r is much larger than the remaining coefficients c_i $(i \neq r)$. Hence we can write

$$\left|\frac{c_i}{c_r}\right| = \epsilon_i \ll 1, \qquad i \neq r, \tag{4.179}$$

where ϵ_i is a small number. Dividing the numerator and denominator of (4.178) by c_r^2, we obtain

$$R(u) = \frac{\lambda_r + \sum_{i=1, i\neq r}^{n} \epsilon_i^2 \lambda_i}{1 + \sum_{i=1, i\neq r}^{n} \epsilon_i^2} = \lambda_r[1 + O(\epsilon^2)], \tag{4.180}$$

where $O(\epsilon^2)$ denotes an expression in ϵ of the second order and higher. Equation (4.180) indicates that if the arbitrary vector $\{u\}$ differs from the eigenvector $\{u^{(r)}\}$ by a small quantity of the first order, Rayleigh's quotient $R(u)$ differs from the eigenvalue λ_r by a small quantity of the second order.

† Using an energy aproach, a similar expression was derived by Lord Rayleigh in *The Theory of Sound*, Vol. 1, Dover, New York, 1945, p. 110.

Hence Rayleigh's quotient has a stationary value in the neighborhood of an eigenvector. The stationary value is actually a minimum in the neighborhood of the fundamental mode. To show this let $r = 1$ in (4.180) and write

$$R(u) = \frac{\lambda_1 + \sum_{i=2}^n \epsilon_i^2 \lambda_i}{1 + \sum_{i=2}^n \epsilon_i^2} \simeq \lambda_1 + \sum_{i=2}^n \epsilon_i^2 \lambda_i - \lambda_1 \sum_{i=2}^n \epsilon_i^2$$

$$= \lambda_1 + \sum_{i=2}^n (\lambda_i - \lambda_1)\epsilon_i^2. \tag{4.181}$$

But, in general, $\lambda_i > \lambda_1$ ($i = 2, 3, \ldots, n$), so

$$R(u) \geq \lambda_1, \tag{4.182}$$

which shows that Rayleigh's quotient is never lower than the first eigenvalue. On the other hand, if we let $r = n$ in (4.180), we obtain

$$R(u) \simeq \lambda_n - \sum_{i=1}^{n-1} (\lambda_n - \lambda_i)\epsilon_i^2, \tag{4.183}$$

and because, in general, $\lambda_n > \lambda_i$, ($i = 1, 2, \ldots, n - 1$), it follows that

$$R(u) \leq \lambda_n, \tag{4.184}$$

which implies that Rayleigh's quotient is never higher than the highest eigenvalue. Hence Rayleigh's quotient provides an *upper bound* for λ_1 and a *lower bound* for λ_n.

Consider now a restricted class of vectors $\{u\}$ which consists only of vectors that are all orthogonal to the first $s - 1$ eigenvectors:

$$\{u\}^T[m]\{u^{(i)}\} = 0, \qquad i = 1, 2, \ldots, s - 1. \tag{4.185}$$

In terms of the expansion (4.175) this means that

$$c_1 = c_2 = c_3 = \cdots = c_{s-1} = 0, \tag{4.186}$$

and Rayleigh's quotient expression becomes

$$R(u) = \frac{\sum_{i=s}^n c_i^2 \lambda_i}{\sum_{i=s}^n c_i^2}. \tag{4.187}$$

Following the same argument as before, it is not difficult to show that

$$R(u) \geq \lambda_s, \qquad s = 2, 3, \ldots, n - 1, \tag{4.188}$$

where the equality is achieved only if $c_i = 0$ for $i \neq s$. Hence Rayleigh's quotient provides an upper bound for the eigenvalue λ_s with respect to all the vectors $\{u\}$ belonging to the subclass of vectors orthogonal to the first $s - 1$ eigenvectors. The minimum value that Rayleigh's quotient can take for such trial vectors is the eigenvalue λ_s.

Let us investigate now the behavior of Rayleigh's quotient as the stiffness and mass coefficients change. Contemplating (4.174) we conclude that if the stiffness coefficients increase in value while the mass coefficients do not change, Rayleigh's quotient and, hence the natural frequencies, increase in value. This is equivalent to increasing the potential energy while the kinetic energy remains the same for the same vector $\{u\}$. If, on the other hand, the mass coefficients increase in value and the stiffness coefficients remain the same for the same vector $\{u\}$, Rayleigh's quotient and, hence the natural frequencies, decrease in value. Any constraints imposed on the system tend to reduce the number of degrees of freedom and render the system stiffer, thus increasing the values of the natural frequencies.

4-12 THE GENERAL EIGENVALUE PROBLEM

In the preceding sections a considerable amount of time was devoted to the discussion of the eigenvalue problem for discrete systems. The eigenvalue problems considered were defined exclusively by symmetric matrices. In many applications, however, it is possible to obtain an eigenvalue problem defined by the matrix equation

$$[k]\{u\} = \lambda[m]\{u\}, \tag{4.189}$$

where the square matrices $[k]$ and $[m]$ are not symmetric.† In such a case we no longer have the orthogonality relations (4.52) and (4.53), obtained for symmetric matrices $[m]$ and $[k]$, respectively. Furthermore, the expansion theorem, (4.60) and (4.61), cannot be used to decompose any arbitrary vector $\{x\}$ in terms of a set of eigenvectors $\{u^{(i)}\}$ $(i = 1, 2, \ldots, n)$, associated with the eigenvalue problem (4.189). It is possible to obtain the first eigenvalue and the associated eigenvector by using the matrix iteration method, but the difficulty arises when we want to obtain the higher modes. We shall show here a method of obtaining the solution of the general eigenvalue problem in which the matrices $[k]$ and $[m]$ are not symmetric.

Consider the general eigenvalue problem

$$[a]\{u\} = \lambda\{u\}, \tag{4.190}$$

where $[a]$ is an $n \times n$ nonsymmetric matrix of real numbers. The right side of (4.190) can be thought of as being premultiplied by the identity matrix.

† As an example see Section 6–7.

In spite of that the eigenvalue problem (4.190) is quite general, because any eigenvalue problem consisting of two nonsymmetric matrices can be brought to the form (4.190). We shall assume that all the eigenvalues λ_i ($i = 1, 2, \ldots, n$) are real and distinct. The cases in which the eigenvalues are not distinct are very rare indeed. If all the eigenvalues are distinct, the corresponding eigenvectors are linearly independent.[†] Associated with the eigenvalue problem, (4.190), we can write the *adjoint eigenvalue problem*.

$$[a]^T\{v\} = \lambda\{v\}, \tag{4.191}$$

where $[a]^T$ is the transpose of $[a]$. As $[a]$ and $[a]^T$ lead to the same characteristic determinant it follows that the eigenvalues corresponding to the two eigenvalue problems (4.190) and (4.191) are identical. Next consider a solution of (4.190) associated with the eigenvalue λ_i and a solution of (4.191) associated with the eigenvalue $\lambda_j \neq \lambda_i$ such that

$$[a]\{u^{(i)}\} = \lambda_i\{u^{(i)}\}, \tag{4.192}$$

$$[a]^T\{v^{(j)}\} = \lambda_j\{v^{(j)}\}. \tag{4.193}$$

Equation (4.192) can be transposed to yield

$$\{u^{(i)}\}^T[a]^T = \lambda_i\{u^{(i)}\}^T. \tag{4.194}$$

Now premultiply both sides of (4.193) by $\{u^{(i)}\}^T$ and postmultiply both sides of (4.194) by $\{v^{(j)}\}$ to obtain

$$\{u^{(i)}\}^T[a]^T\{v^{(j)}\} = \lambda_j\{u^{(i)}\}^T\{v^{(j)}\},$$
$$\{u^{(i)}\}^T[a]^T\{v^{(j)}\} = \lambda_i\{u^{(i)}\}^T\{v^{(j)}\}. \tag{4.195}$$

Subtracting the first of equations (4.195) from the second one, we have

$$(\lambda_i - \lambda_j)\{u^{(i)}\}^T\{v^{(j)}\} = 0, \tag{4.196}$$

and because the eigenvalues were assumed distinct, the matrix product must be zero, which gives the orthgonality relation

$$\{u^{(i)}\}^T\{v^{(j)}\} = 0, \qquad \lambda_i \neq \lambda_j, \tag{4.197}$$

which means that an eigenvector of the eigenvalue problem (4.190) associated with the eigenvalue λ_i is orthogonal, in an ordinary sense, to any eigenvector of the eigenvalue problem (4.191), corresponding to the eigenvalue $\lambda_j \neq \lambda_i$.

[†] See F. B. Hildebrand, *Methods of Applied Mathematics*, Prentice-Hall, Englewood Cliffs, N.J., 1960, p. 81.

The method of solution is similar to the matrix iteration method using the sweeping technique. In fact, the first eigenvalue and eigenvector are obtained precisely as in that method, so we can write

$$[a]\{u^{(1)}\} = \lambda_1\{u^{(1)}\}. \tag{4.198}$$

At this point the similarity ends, because any mode higher than the first is orthogonal not to the vector $\{u^{(1)}\}$ but to the vector $\{v^{(1)}\}$ which satisfies (4.191),

$$[a]^T\{v^{(1)}\} = \lambda_1\{v^{(1)}\}. \tag{4.199}$$

Hence, we must also obtain the first eigenvector associated with the transposed matrix, as indicated by (4.199). Subsequently we can write the orthogonality relation

$$\{u\}^T\{v^{(1)}\} = 0, \tag{4.200}$$

which enables us to construct a sweeping matrix of the form

$$[S^{(1)}] = \begin{bmatrix} 0 & -\dfrac{v_2^{(1)}}{v_1^{(1)}} & -\dfrac{v_3^{(1)}}{v_1^{(1)}} & \cdots & -\dfrac{v_n^{(1)}}{v_1^{(1)}} \\ 0 & 1 & 0 & \cdots & 0 \\ 0 & 0 & 1 & \cdots & 0 \\ \vdots & \vdots & \vdots & \cdots & \vdots \\ 0 & 0 & 0 & \cdots & 1 \end{bmatrix}. \tag{4.201}$$

The second mode, $\{u^{(2)}\}$, is obtained by using for the iteration a new matrix, $[a^{(2)}]$, defined by

$$[a^{(2)}] = [a][S^{(1)}]. \tag{4.202}$$

To compute the third mode, $\{u^{(3)}\}$, it is necessary to obtain first the second eigenvector of the transposed problem. To this end we must use the orthogonality relation

$$\{v\}^T\{u^{(1)}\} = 0, \tag{4.203}$$

and construct a corresponding sweeping matrix $[S_T^{(1)}]$ for the transposed problem. The second eigenvector, $\{v^{(2)}\}$, of the transposed problem is obtained by using the matrix

$$[a_T^{(2)}] = [a]^T[S_T^{(1)}]. \tag{4.204}$$

Having the vector $\{v^{(2)}\}$ we can write two orthogonality relations

$$\{u\}^T\{v^{(1)}\} = 0, \qquad \{u\}^T\{v^{(2)}\} = 0, \tag{4.205}$$

which are used to construct a second sweeping matrix $[S^{(2)}]$ and subsequently a third dynamical matrix, which is given by

$$[a^{(3)}] = [a][S^{(2)}] \tag{4.206}$$

and enables us to obtain the third mode, $\{u^{(3)}\}$.

At this point, the method of obtaining the remaining modes appears obvious. Hence we conclude that to obtain the solution of the eigenvalue problem (4.190), associated with the matrix $[a]$, by the matrix iteration method, we must also solve the eigenvalue problem (4.191), associated with the matrix $[a]^T$. The two problems are solved in a sequence of alternating steps, because one step associated with the matrix $[a]$ depends on the preceding step, associated with the matrix $[a]^T$, and vice versa.

The solution of the eigenvalue problem consists of two sets of vectors $\{u^{(i)}\}$ $(i = 1, 2, \ldots, n)$, and $\{v^{(j)}\}$ $(j = 1, 2, \cdots, n)$. These vectors are mutually orthogonal in the sense that one set of vectors is orthogonal to the other set. The vectors corresponding to identical eigenvalues, $\lambda_i = \lambda_j$, are not orthogonal and the two sets of vectors can be normalized by letting

$$\{u^{(i)}\}^T\{v^{(j)}\} = \{v^{(j)}\}^T\{u^{(i)}\} = \delta_{ij}, \qquad i,j = 1, 2, \ldots, n. \tag{4.207}$$

Equations (4.207) can be written in the compact form

$$[u]^T[v] = [v]^T[u] = [I], \tag{4.208}$$

where the $n \times n$ square matrices $[u]$ and $[v]$ have the vectors $\{u^{(i)}\}$ and $\{v^{(j)}\}$ as columns, respectively. Equation (4.207) or (4.208) can be regarded as a *biorthogonality relation*.

The set of vectors $\{v^{(j)}\}$ $(j = 1, 2, \ldots, n)$ is called the *adjoint* of the set of vectors $\{u^{(i)}\}$ $(i = 1, 2, \ldots, n)$. The vectors $\{u^{(i)}\}$ and $\{v^{(i)}\}$ $(i = 1, 2, \ldots, n)$, vectors with equal indices, are called *conjugates* of each other. For a symmetric matrix $[a]$, the conjugate vectors are equal, $\{v^{(j)}\}=\{u^{(j)}\}$ $(j=1, 2, \ldots, n)$. Such a set of vectors is called *self-adjoint*.

To use the modal analysis it is necessary to obtain both sets of vectors, $\{u^{(i)}\}$ and $\{v^{(j)}\}$. The vectors $\{u^{(i)}\}$ are linearly independent, so the set of vectors can be regarded as a basis in an n-dimensional space. The same is true for the set of vectors $\{v^{(j)}\}$. Any vector in the n-dimensional space can be decomposed in either the set $\{u^{(i)}\}$ or $\{v^{(j)}\}$ provided one has at his disposal both sets of vectors. In effect, one has a *dual expansion theorem*, so an arbitrary vector $\{x\}$ can be expressed in terms of the set $\{u^{(i)}\}$ by

$$\{x\} = \sum_{r=1}^{n} \alpha_r\{u^{(r)}\}, \tag{4.209}$$

where the coefficients α_r are given by

$$\alpha_r = \{v^{(r)}\}^T\{x\}, \qquad r = 1, 2, \ldots, n, \tag{4.210}$$

and, similarly, the same vector $\{x\}$ can be represented in terms of the set $\{v^{(s)}\}$ by

$$\{x\} = \sum_{s=1}^{n} \beta_s\{v^{(s)}\}, \tag{4.211}$$

where the coefficients β_s are obtained from

$$\beta_s = \{u^{(s)}\}^T\{x\}, \qquad s = 1, 2, \ldots, n. \tag{4.212}$$

The coefficients α_r and β_s can be regarded as the components of the same vector $\{x\}$ in the two dual reference systems.

The nonsymmetric eigenvalue problem (4.190) can also be given a geometric interpretation. To this end it will be necessary to use *skew-angular* reference systems instead of rectangular reference systems. Although the geometric interpretation of the nonsymmetric eigenvalue problem is an interesting subject, it does not serve our purpose in any way, and we shall not pursue this subject further. The interested reader is referred to the text by Lanczos.†

Problems

4.1 Given the system shown in Figure 4.4,

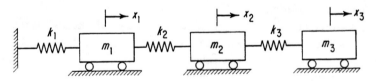

FIGURE 4.4

(a) Set up the eigenvalue problem. Indicate whether the system is positive definite or not.
(b) Let $k_1 = 2k$, $k_2 = k$, $k_3 = k$, $m_1 = 4m$, $m_2 = 2m$, and $m_3 = m$, and solve the eigenvalue problem by means of the method that employs the characteristic determinant. Solve the frequency equation by Graeffe's root-squaring method and one more method of your choice.
(c) Calculate the natural frequencies of the system and plot the corresponding normal modes. Note the number of nodes (points of zero displacement) for each mode.

† C. Lanczos, *Applied Analysis*, Prentice-Hall, Englewood Cliffs, N.J., 1964, p. 95.

4.2 Solve the eigenvalue problem of Problem 4.1 by the matrix iteration method using the sweeping procedure.

4.3 Prove (4.125) and (4.128).

4.4 Solve the eigenvalue problem of Problem 4.1 by the matrix iteration method using the matrix diagonalization by successive rotations procedure.

4.5 Set up the eigenvalue problem of Problem 4.1 in terms of the coordinates $z_1 = x_1$, $z_2 = x_2 - x_1$, and $z_3 = x_3 - x_2$, and solve the eigenvalue problem by a method of your choice. Compare the results with the results obtained in Problem 4.1 and draw conclusions.

4.6 Obtain the solution of the eigenvalue problem corresponding to the system shown in Figure 4.3. Let $k_1 = 2k$, $k_2 = k$, $m_1 = 4m$, $m_2 = 2m$, and $m_3 = m$, and plot the eigenvectors.

4.7 Apply the enclosure theorem to the system of Problem 4.1 for three distinct arbitrary vectors $\{u\}$ progressively narrowing the interval.

4.8 Write Rayleigh's quotient expression corresponding to the system of Problem 4.1. Evaluate Rayleigh's quotient for three distinct arbitrary vectors $\{u\}$: one vector nearly identical with the first true eigenvector, a second vector nearly identical with the third true eigenvector, and a third eigenvector entirely arbitrary. Draw conclusions as to the relation between Rayleigh's quotient and the eigenvalues of the system.

4.9 Repeat Problem 4.8 with the exception that k_2 is changed to $k_2 = 3k$. Use the same arbitrary vectors $\{u\}$ as in Problem 4.8 and evaluate Rayleigh's quotients. Draw conclusions as to how the change affected the values of Rayleigh's quotients.

4.10 Repeat Problem 4.8 with the exception that m_3 is changed to $m_3 = 3m$. Use the same arbitrary vectors $\{u\}$ as in Problem 4.8 and evaluate the corresponding Rayleigh's quotients. Draw conclusions as to how the change affected the values of Rayleigh's quotients.

Selected Readings

Bisplinghoff, R. L., H. Ashley, and R. L. Halfman, *Aeroelasticity*, Addison-Wesley, Reading, Mass., 1957.

Collatz, L., *Eigenwertaufgaben mit technishen Anwendungen*, Akademische Verlagsgesellschaft, Leipzig, 1963.

Courant, R., and D. Hilbert, *Methods of Mathematical Physics*, Vol. 1, Interscience, New York, 1961.

Crandall, S. H., *Engineering Analysis*, McGraw-Hill, New York, 1956.

Frazer, R. A., W. J. Duncan, and A. R. Collar, *Elementary Matrices*, Cambridge Univ. Press, New York, 1957.

Hildebrand, F. B., *Methods of Applied Mathematics*, Prentice-Hall, Englewood Cliffs, N.J., 1960.

Lanczos, C., *Applied Analysis*, Prentice-Hall, Englewood Cliffs, N.J., 1964.

Rayleigh, Lord, *The Theory of Sound*, Vol. 1, Dover, New York, 1945.

von Karman, T., and M. A. Biot, *Mathematical Methods in Engineering*, McGraw-Hill, New York, 1940.

CHAPTER 5

NATURAL
MODES OF
VIBRATION.
CONTINUOUS
SYSTEMS

5–1 GENERAL CONSIDERATIONS

In the study of discrete systems we considered elements such as springs, dampers, and masses. The only property ascribed to springs was compliance, which is the property of deforming under loads. Under our simplifying assumptions the springs were massless. On the other hand, the masses were assumed to have no compliance, so, in effect, they were assumed to be rigid. These assumptions allowed us to represent a system consisting of n discrete masses by an n-degree-of-freedom mathematical model resulting in a set of n coupled ordinary differential equations describing the motion of the masses.

In reality, however, there are many cases in which it is not possible to identify discrete masses or springs. Furthermore, there may not be any valid reason to assume that the masses cannot deform and that the springs have no mass. In fact, every material portion of the system may possess both mass and elasticity and, moreover, these properties may vary from point to point. In this case one must refine the assumptions and conceive a continuous mathematical model. To formulate the equations of motion we make use of techniques employed in the theory of mechanics of deformable bodies which leads to a set of three equations of motion, in terms of the displacements $u(x, y, z, t)$, $v(x, y, z, t)$, and $w(x, y, z, t)$, which must be satisfied at every point of the system and which are subject to initial conditions and conditions at the boundaries of the system. Here the displacements u, v, and w play the role of coordinates and x, y, and z are space variables. Each of the variables x, y, and z can take an infinity of values within the region occupied by the system, so the system possesses an infinite number of degrees of freedom.

The equations of motion in terms of displacements are associated with the name of Navier and can be found in any text on elasticity.† Their solution is complicated considerably by the boundary conditions. The discussion of Navier's equations lies outside the scope of this text.

In many instances, because of the particular configuration of the continuous body, it is possible to formulate the equations of motion in terms of one or two displacement components, depending on one or two spatial coordinates and time. Such cases, too, can be plagued by certain types of boundary conditions, rendering a solution in closed form impossible.

The classification of systems as discrete or continuous is quite often arbitrary. The same system can be regarded at times as discrete and described by ordinary differential equations and at other times as continuous and described by partial differential equations. The choice between the two models depends on many factors. It is safe to say that, in general, a discrete system can be solved with more ease than a continuous one. At the same time the information obtained by using a discrete model may not be as accurate as the information obtained by representing the system by a continuous model. Hence, based on his individual needs, one must choose between the expediency and the accuracy of the solution.

In view of the above discussion, we must conclude that the discrete and continuous systems are intimately related. In fact, if the discrete model is made finer and finer, by decreasing the size of the masses and increasing the number of degrees of freedom, one obtains, in the limit, a continuous system (as an example see Problem 5.1). Therefore, it is not surprising that many of the concepts encountered in discrete systems have counterparts in continuous systems. Again one can formulate an eigenvalue problem, which in the case of continuous systems takes the form of differential equations and less frequently the form of integral equations, as opposed to algebraic equations for discrete systems. The solution of the eigenvalue problem for a continuous system yields an infinite set of eigenvalues and corresponding eigenfunctions, in contrast to discrete systems, for which we obtain a finite set of eigenvalues and eigenvectors. If the body is finite, an infinite set of discrete natural frequencies will be obtained. As the boundaries recede to infinity the discrete spectrum of natural frequencies becomes a continuous one. The concept of orthogonality discussed in conjunction with eigenvectors applies to eigenfunctions as well. The counterpart of a discrete system with symmetric mass

---

† See I. S. Sokolnikoff, *Mathematical Theory of Elasticity*, 2nd ed., McGraw-Hill, New York, 1956.

and stiffness matrices is the self-adjoint continuous system. The question of boundary conditions does not arise in the case of discrete systems except in an indirect way, because the influence coefficients depend on the manner in which the system is supported. In the case of continuous systems, the boundary-condition problem assumes a large degree of subtlety and more elaborate discussion is necessary.

The differential equations of motion together with the associated boundary conditions constitute a boundary-value problem. The boundary-value problem becomes an eigenvalue problem when the differential equation of motion and the boundary conditions are homogeneous and depend on a parameter λ, and, moreover, a nontrivial solution is obtained only for certain values of the parameter λ.

In the present chapter the problem of the transverse vibration of a bar is used to illustrate a typical boundary-value problem and to emphasize the significance of the geometric and natural boundary conditions. The same problem, simplified somewhat, is used to derive a typical eigenvalue problem. The transition between the boundary-value problem and the eigenvalue problem is effected by means of the separation of variables method. After discussing the essential features of an eigenvalue problem, the eigenvalue problem is generalized by means of the operator notation. The general formulation enables one to present a unified approach, which has the advantage of revealing the common features of vibrating continuous systems. In some cases it is possible to reach conclusions about the dynamical properties of a certain system without actually carrying out the solution of the eigenvalue problem. To this end, concepts such as admissible functions, comparison functions, eigenfunctions, self-adjoint systems, and positive definite systems are discussed. Some of these concepts are particularly useful in the case in which one seeks an approximate solution of the eigenvalue problem as shown in Chapter 6. A discussion of the orthogonality property of the eigenfunctions and the expansion theorem is presented. The usefulness of the orthogonality property and the expansion theorem is brought into sharper focus in Chapter 7, where we discuss the response of continuous systems by means of modal analysis. A large number of eigenvalue problems are solved.

5-2 BOUNDARY-VALUE PROBLEM FORMULATION

As an example of boundary-value problems consider a nonuniform bar in transverse vibration.

At any point x the bar has a mass per unit length $m(x)$, a cross-sectional area $A(x)$, and an area moment of inertia $I(x)$ about the neutral axis (Figure 5.1). We have only one space variable, x, so this is a one-dimensional problem.

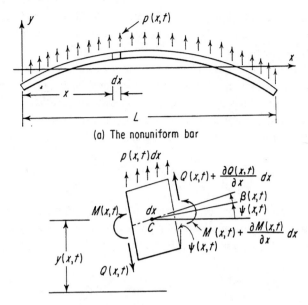

(a) The nonuniform bar

(b) Free-body diagram of a beam element

FIGURE 5.1

The total deflection $y(x, t)$ of the bar at a point x consists of two parts, one caused by bending and one by shear, so the slope of the deflection curve at the point x can be written

$$\frac{\partial y(x, t)}{\partial x} = \psi(x, t) + \beta(x, t), \tag{5.1}$$

where $\psi(x, t)$ is the angle of rotation due to bending and $\beta(x, t)$ is the angle of distortion due to shear. As usual, the linear deflection and angular deflection are assumed small.

The relation between the bending moment and the bending deformation is

$$M(x, t) = EI(x)\frac{\partial \psi(x, t)}{\partial x}, \tag{5.2}$$

and the relation between the shearing force and shearing deformation is given by†

$$Q(x, t) = k'GA(x)\beta(x, t), \tag{5.3}$$

† See S. Timoshenko, *Vibration Problems in Engineering*, 3rd ed., Van Nostrand, Princeton, N.J., 1955.

where G is the shear modulus and k' is a numerical factor depending on the shape of the cross section. Because of shear alone, the element undergoes distortion but no rotation.

To formulate the boundary-value problem we shall make use of the extended Hamilton principle, as given by (2.61). The power of this approach becomes evident when we observe that it furnishes automatically the correct number of boundary conditions and their correct expressions.

The kinetic energy is due to translation and rotation and is expressed

$$T(t) = \tfrac{1}{2} \int_0^L \left[\frac{\partial y(x, t)}{\partial t} \right]^2 m(x) \, dx + \tfrac{1}{2} \int_0^L \left[\frac{\partial \psi(x, t)}{\partial t} \right]^2 J(x) \, dx, \qquad (5.4)$$

where L is the length of the bar and $J(x)$ is the mass moment of inertia per unit length about the neutral axis which passes through the center C as shown in Figure 5.1(b). But $J(x)$ is related to $I(x)$ by

$$J(x) = \rho I(x) = \frac{m(x)}{A(x)} I(x) = k^2(x) m(x), \qquad (5.5)$$

where ρ is the mass density and $k(x)$ is the radius of gyration about the neutral axis. The variation of T can be readily written

$$\delta T = \int_0^L m \frac{\partial y}{\partial t} \, \delta\!\left(\frac{\partial y}{\partial t} \right) dx + \int_0^L k^2 m \frac{\partial \psi}{\partial t} \, \delta\!\left(\frac{\partial \psi}{\partial t} \right) dx. \qquad (5.6)$$

The virtual work consists of conservative and nonconservative work. Since the external load is in the direction of the displacement, the virtual work for the whole bar is

$$\delta W(t) = \delta W_c(t) + \delta W_{nc}(t) = -\delta V(t) + \int_0^L p(x, t) \, \delta y(x, t) \, dx, \qquad (5.7)$$

where $V(t)$ is the potential energy whose expression is

$$V(t) = \tfrac{1}{2} \int_0^L M(x, t) \frac{\partial \psi(x, t)}{\partial x} \, dx + \tfrac{1}{2} \int_0^L Q(x, t) \beta(x, t) \, dx$$

$$= \tfrac{1}{2} \int_0^L EI(x) \left[\frac{\partial \psi(x, t)}{\partial x} \right]^2 dx + \tfrac{1}{2} \int_0^L k' G A(x) \beta^2(x, t) \, dx. \qquad (5.8)$$

Hence the variation of potential energy has the form

$$\delta V = \int_0^L EI \frac{\partial \psi}{\partial x} \, \delta\!\left(\frac{\partial \psi}{\partial x} \right) dx + \int_0^L k' G A \beta \, \delta\beta \, dx$$

$$= \int_0^L EI \frac{\partial \psi}{\partial x} \, \delta\!\left(\frac{\partial \psi}{\partial x} \right) dx + \int_0^L k' G A \left(\frac{\partial y}{\partial x} - \psi \right) \delta\!\left(\frac{\partial y}{\partial x} - \psi \right) dx. \qquad (5.9)$$

Introduction of (5.6), (5.7), and (5.9) in the variational principle (2.61) leads to

$$\int_{t_1}^{t_2} (\delta T + \delta W)\, dt$$

$$= \int_{t_1}^{t_2} \left[\int_0^L m \frac{\partial y}{\partial t} \delta\!\left(\frac{\partial y}{\partial t}\right) dx + \int_0^L k^2 m \frac{\partial \psi}{\partial t} \delta\!\left(\frac{\partial \psi}{\partial t}\right) dx \right.$$

$$- \int_0^L EI \frac{\partial \psi}{\partial x} \delta\!\left(\frac{\partial \psi}{\partial x}\right) dx - \int_0^L k'GA\!\left(\frac{\partial y}{\partial x} - \psi\right) \delta\!\left(\frac{\partial y}{\partial x} - \psi\right) dx$$

$$\left. + \int_0^L p\, \delta y\, dx \right] dt = 0. \tag{5.10}$$

The order of integrations with respect to x and t is interchangeable and the variation and differentiation operators are commutative, so we can perform the following integrations by parts:

$$\int_{t_1}^{t_2} m \frac{\partial y}{\partial t} \delta\!\left(\frac{\partial y}{\partial t}\right) dt = \int_{t_1}^{t_2} m \frac{\partial y}{\partial t} \frac{\partial}{\partial t} (\delta y)\, dt = m \frac{\partial y}{\partial t} \delta y \Big|_{t_1}^{t_2}$$

$$- \int_{t_1}^{t_2} \frac{\partial}{\partial t}\!\left(m \frac{\partial y}{\partial t}\right) \delta y\, dt = -\int_{t_1}^{t_2} m \frac{\partial^2 y}{\partial t^2} \delta y\, dt,$$

because δy vanishes at $t = t_1$ and $t = t_2$. In a similar fashion we obtain

$$\int_{t_1}^{t_2} k^2 m \frac{\partial \psi}{\partial t} \delta\!\left(\frac{\partial \psi}{\partial t}\right) dt = -\int_{t_1}^{t_2} k^2 m \frac{\partial^2 \psi}{\partial t^2} \delta \psi\, dt.$$

On the other hand, integration over the spatial variable yields

$$\int_0^L EI \frac{\partial \psi}{\partial x} \delta\!\left(\frac{\partial \psi}{\partial x}\right) dx = \int_0^L EI \frac{\partial \psi}{\partial x} \frac{\partial}{\partial x} (\delta \psi)\, dx$$

$$= \left(EI \frac{\partial \psi}{\partial x}\right) \delta \psi \Big|_0^L - \int_0^L \frac{\partial}{\partial x}\!\left(EI \frac{\partial \psi}{\partial x}\right) \delta \psi\, dx,$$

$$\int_0^L k'GA\!\left(\frac{\partial y}{\partial x} - \psi\right) \delta\!\left(\frac{\partial y}{\partial x} - \psi\right) dx = \int_0^L k'GA\!\left(\frac{\partial y}{\partial x} - \psi\right) \frac{\partial}{\partial x} (\delta y)\, dx$$

$$- \int_0^L k'GA\!\left(\frac{\partial y}{\partial x} - \psi\right) \delta \psi\, dx$$

$$= \left[k'GA\!\left(\frac{\partial y}{\partial x} - \psi\right)\right] \delta y \Big|_0^L$$

$$- \int_0^L \frac{\partial}{\partial x} \left[k'GA\!\left(\frac{\partial y}{\partial x} - \psi\right)\right] \delta y\, dx$$

$$- \int_0^L k'GA\!\left(\frac{\partial y}{\partial x} - \psi\right) \delta \psi\, dx.$$

Using the above expressions in (5.10) produces

$$\int_{t_1}^{t_2} \left\{ -\int_0^L m \frac{\partial^2 y}{\partial t^2} \, \delta y \, dx - \int_0^L k^2 m \frac{\partial^2 \psi}{\partial t^2} \, \delta \psi \, dx - \left(EI \frac{\partial \psi}{\partial x} \right) \delta \psi \Big|_0^L \right.$$

$$+ \int_0^L \frac{\partial}{\partial x} \left(EI \frac{\partial \psi}{\partial x} \right) \delta \psi \, dx - \left[k'GA \left(\frac{\partial y}{\partial x} - \psi \right) \right] \delta y \Big|_0^L$$

$$+ \int_0^L \frac{\partial}{\partial x} \left[k'GA \left(\frac{\partial y}{\partial x} - \psi \right) \right] \delta y \, dx$$

$$\left. + \int_0^L k'GA \left(\frac{\partial y}{\partial x} - \psi \right) \delta \psi \, dx + \int_0^L p \, \delta y \, dx \right\} dt$$

$$= \int_{t_1}^{t_2} \left[\int_0^L \left\{ \frac{\partial}{\partial x} \left[k'GA \left(\frac{\partial y}{\partial x} - \psi \right) \right] - m \frac{\partial^2 y}{\partial t^2} + p \right\} \delta y \, dx \right.$$

$$+ \int_0^L \left\{ \left[\frac{\partial}{\partial x} \left(EI \frac{\partial \psi}{\partial x} \right) + k'GA \left(\frac{\partial y}{\partial x} - \psi \right) \right] - k^2 m \frac{\partial^2 \psi}{\partial t^2} \right\} \delta \psi \, dx$$

$$\left. - \left(EI \frac{\partial \psi}{\partial x} \right) \delta \psi \Big|_0^L - \left[k'GA \left(\frac{\partial y}{\partial x} - \psi \right) \right] \delta y \Big|_0^L \right] dt = 0. \qquad (5.11)$$

The virtual displacements $\delta \psi$ and δy are arbitrary and independent, so they can be taken equal to zero at $x = 0$ and $x = L$ and arbitrary for $0 < x < L$; therefore, we must have

$$\frac{\partial}{\partial x} \left[k'GA \left(\frac{\partial y}{\partial x} - \psi \right) \right] - m \frac{\partial^2 y}{\partial t^2} + p = 0, \qquad (5.12)$$

$$\frac{\partial}{\partial x} \left(EI \frac{\partial \psi}{\partial x} \right) + k'GA \left(\frac{\partial y}{\partial x} - \psi \right) - k^2 m \frac{\partial^2 \psi}{\partial t^2} = 0 \qquad (5.13)$$

throughout the domain. In addition, if we write

$$\left(EI \frac{\partial \psi}{\partial x} \right) \delta \psi \Big|_0^L = 0, \qquad (5.14)$$

$$\left[k'GA \left(\frac{\partial y}{\partial x} - \psi \right) \right] \delta y \Big|_0^L = 0 \qquad (5.15)$$

we take into account the possibility that either $EI(\partial \psi / \partial x)$ or $\delta \psi$, on the one hand, and either $k'GA[(\partial y / \partial x) - \psi]$ or δy, on the other, vanishes at any of the ends $x = 0$ and $x = L$. Equations (5.12) and (5.13) are the differential equations of motion that must be satisfied over the length of the bar and (5.14) and (5.15) represent the boundary conditions. The four equations together constitute the boundary-value problem. Equation (5.14) requires that either the bending moment or the bar rotation variation vanish at each end and (5.15) requires that either the shearing force or the deflection variation be zero at each end. It is the satisfaction of these boundary conditions that renders the solution of the differential equations unique.

The geometry of the system is not always able to provide the necessary number of boundary conditions. As an example, in the case of a bar clamped at both ends, the geometry requires that the rotation of the bar cross section and the deflection be zero at both ends for a correct total of four boundary conditions. But in the case of a bar hinged at both ends, the geometry requires that only the deflection be zero at both ends, so two boundary conditions are missing. The beauty of the variational principle is that it provides the correct number of boundary conditions, which are also physically correct, as indicated by (5.14) and (5.15).

The boundary conditions resulting from pure geometric compatibility are called *geometric* (or *essential* or *imposed*) *boundary conditions*. In the particular case of a bar in bending, the geometric boundary conditions are concerned with the deflection or rotation at the boundary. In the cases in which the geometry does not provide a sufficient number of boundary conditions, the remaining conditions are supplied by the moment or shearing force balance, as indicated by the expression in parentheses in (5.14) and the expression in brackets in (5.15), respectively. The boundary conditions resulting from moment or shearing force balance are called *natural* (or *additional* or *dynamical*) *boundary conditions*. The geometric and natural boundary conditions supplement each other and add up to the right number of boundary conditions, and their satisfaction ensures that the solution of the differential equations is unique.

Let us examine the various possible boundary conditions and convince ourselves that (5.14) and (5.15) are indeed satisfied in each case.

(a) Bar Clamped at Both Ends

The deflection and the rotation are zero at both ends:

$$y(0, t) = 0, \qquad y(L, t) = 0,$$
$$\psi(0, t) = 0, \qquad \psi(L, t) = 0. \tag{5.16}$$

All boundary conditions are geometric (imposed) boundary conditions. The deflection and the rotation are zero at both ends, hence constant, so their variations are zero and (5.14) and (5.15) are satisfied.

(b) Bar Clamped at One End and Hinged at the Other

At the clamped end the deflection and the rotation are zero. Let the clamped end be at $x = 0$, so that

$$y(0, t) = 0, \qquad \psi(0, t) = 0. \tag{5.17}$$

At the hinged end, $x = L$, the geometry requires only that the deflection be zero,

$$y(L, t) = 0, \tag{5.18}$$

and because the rotation there is not zero, (5.14) is satisfied only if

$$M(L, t) = EI \frac{\partial \psi}{\partial x}\bigg|_{x=L} = 0, \tag{5.19}$$

which is indeed the case for a hinged end. Equation (5.19) represents a natural (additional) boundary condition, so in this case we have three geometric boundary conditions and one natural boundary condition.

(c) Bar Hinged at Both Ends

In this case the boundary conditions are

$$y(0, t) = 0, \qquad y(L, t) = 0,$$

$$M(0, t) = EI \frac{\partial \psi}{\partial x}\bigg|_{x=0} = 0, \qquad M(L, t) = EI \frac{\partial \psi}{\partial x}\bigg|_{x=L} = 0, \tag{5.20}$$

so at each end there is one geometric and one natural boundary condition.

(d) Bar Clamped at One End and Free at the Other

If the end $x = 0$ is clamped, then

$$y(0, t) = 0, \qquad \psi(0, t) = 0. \tag{5.21}$$

At the free end neither the deflection nor the rotation is zero, so we must have

$$M(L, t) = EI \frac{\partial \psi}{\partial x}\bigg|_{x=L} = 0,$$

$$Q(L, t) = \left[k'GA\left(\frac{\partial y}{\partial x} - \psi\right)\right]_{x=L} = 0, \tag{5.22}$$

which reflects the fact that both the bending moment and the shearing force vanish at a free end. Hence at the clamped end we have two imposed boundary conditions and at the free end we have two natural boundary conditions.

(e) Bar Free at Both Ends

In this case all boundary conditions are natural and there are no imposed boundary conditions:

$$M(0, t) = EI \frac{\partial \psi}{\partial x}\bigg|_{x=0} = 0, \qquad M(L, t) = EI \frac{\partial \psi}{\partial x}\bigg|_{x=L} = 0,$$

$$Q(0, t) = \left[k'GA\left(\frac{\partial y}{\partial x} - \psi \right) \right]_{x=0} = 0, \tag{5.23}$$

$$Q(L, t) = \left[k'GA\left(\frac{\partial y}{\partial x} - \psi \right) \right]_{x=L} = 0.$$

In conclusion we see that the variational principle always supplies the natural boundary conditions needed to supplement the geometric boundary conditions.

5-3   THE EIGENVALUE PROBLEM

The formulation of Section 5–2 included the shear deformation effect and the rotatory inertia effect, the latter caused by the angular acceleration of a bar element. The shear deformation effect is reflected in the second integral in the potential energy expression, (5.8), and the rotatory inertia effect is represented by the second integral in the kinetic energy expression, (5.4). When the cross-sectional dimensions are small compared with the length of the bar, both the shear and rotatory inertia effects can be neglected.† Let us assume that this is the case and if the external load is zero, $p(x, t) = 0$, (5.12) and (5.13) can be combined into a single equation,

$$\frac{\partial^2}{\partial x^2} \left[EI(x) \frac{\partial^2 y(x, t)}{\partial x^2} \right] = -m(x) \frac{\partial^2 y(x, t)}{\partial t^2}, \tag{5.24}$$

which is the free vibration differential equation.

As a result of neglecting the shear and rotatory inertia effects, we obtain

$$\psi(x, t) = \frac{\partial y(x, t)}{\partial x}, \tag{5.25}$$

† See S. Timoshenko, *Vibration Problems in Engineering*, 3rd ed., Van Nostrand, Princeton, N.J., 1955, pp. 341–342.

and it follows from (5.2) that

$$M(x, t) = EI(x) \frac{\partial^2 y(x, t)}{\partial x^2} \tag{5.26}$$

and from (5.2), (5.3), and (5.13) that

$$Q(x, t) = -\frac{\partial M(x, t)}{\partial x} = -\frac{\partial}{\partial x} \left[EI(x) \frac{\partial^2 y(x, t)}{\partial x^2} \right]. \tag{5.27}$$

Assume that the solution of (5.24) is separable in time and space and of the form

$$y(x, t) = Y(x)f(t). \tag{5.28}$$

Introducing expression (5.28) in (5.24), and dividing through by $m(x) Y(x)f(t)$ we obtain

$$\frac{1}{m(x) Y(x)} \frac{d^2}{dx^2} \left[EI(x) \frac{d^2 Y(x)}{dx^2} \right] = -\frac{1}{f(t)} \frac{d^2 f(t)}{dt^2}. \tag{5.29}$$

The left side of (5.29) depends on x only, and the right side depends on t only. Both x and t are independent variables, so (5.29) has a solution only if both sides are constant. As in the case of discrete systems, on the basis of physical considerations we must choose a positive value, ω^2, for this constant. Equation (5.29) leads to two ordinary differential equations,

$$\frac{d^2}{dx^2} \left[EI(x) \frac{d^2 Y(x)}{dx^2} \right] - \omega^2 m(x) Y(x) = 0, \tag{5.30}$$

$$\frac{d^2 f(t)}{dt^2} + \omega^2 f(t) = 0. \tag{5.31}$$

Equation (5.31) gives a harmonic rather than an exponential solution, which is consistent with the fact that a conservative system has constant total energy.

Of particular significance is (5.30), which is a fourth order homogeneous ordinary differential equation and, as such, must be supplemented by four boundary conditions, two boundary conditions for each end. The problem of determining the values of the parameter ω^2 for which a homogeneous linear differential equation of the type (5.30) has a nontrivial solution, $Y(x)$, satisfying homogeneous boundary conditions, is called the *characteristic-value* or *eigenvalue problem*. Such values of the parameter ω^2 are called *characteristic values* or *eigenvalues* and the associated nontrivial solutions $Y(x)$ are called *characteristic functions* or *eigenfunctions*. Equation (5.30) is a homogeneous differential equation, so its solution cannot be determined uniquely except for the shape, because a solution multiplied by any arbitrary constant is also a solution. The four boundary conditions determine uniquely the shape of the solution, leaving the amplitude arbitrary, and also yield a *characteristic equation* or *frequency equation*, which, when solved, gives the characteristic values of the problem. If the bar is finite in length, the solution of the frequency equation consists of an infinite sequence of discrete eigenvalues related to the natural frequencies ω_r ($r = 1, 2, 3, \ldots$). To each of the natural frequencies corresponds a characteristic function or eigenfunction $A_r Y_r(x)$ ($r = 1, 2, 3, \ldots$), where A_r can be looked upon as the arbitrary amplitude and $Y_r(x)$ is the unique mode shape associated with the frequency ω_r. The functions $A_r Y_r(x)$ are the natural modes, which upon normalization become the normal modes of vibration. The normal modes $Y_r(x)$ and the associated natural frequencies ω_r evidently depend on the stiffness $EI(x)$ and mass $m(x)$ distributions and, of course, the four boundary conditions. In short, they are a characteristic of the system.

The most common types of boundary conditions encountered in the bending vibration of bars are the following.

(a) Clamped End at $x = 0$ or $x = L$

The deflection and the slope of the deflection curve must be zero at the clamped end:

$$y(x, t) = Y(x)f(t) = 0 \qquad \text{at } x = 0 \quad \text{or} \quad x = L.$$

Therefore, $\qquad\qquad Y(0) = 0 \qquad \text{or} \qquad Y(L) = 0.$ (5.32)

$$\frac{\partial y(x, t)}{\partial x} = \frac{dY(x)}{dx} f(t) = 0 \qquad \text{at } x = 0 \quad \text{or} \quad x = L.$$

Therefore, $\qquad \dfrac{dY(x)}{dx}\bigg|_{x=0} = 0 \qquad \text{or} \qquad \dfrac{dY(x)}{dx}\bigg|_{x=L} = 0.$ (5.33)

(b) Hinged End at $x = 0$ or $x = L$

The deflection and the bending moment are zero at a hinged end. The condition that the deflection be zero again yields equations (5.32). The condition concerning the bending moment gives

$$M(x, t) = EI(x)\frac{\partial^2 y(x, t)}{\partial x^2} = EI(x)\frac{d^2 Y(x)}{dx^2} f(t) = 0 \quad \text{at } x = 0 \text{ or } x = L.$$

Therefore, $\quad EI(x)\dfrac{d^2 Y(x)}{dx^2}\bigg|_{x=0} = 0 \qquad \text{or} \qquad EI(x)\dfrac{d^2 Y(x)}{dx^2}\bigg|_{x=L} = 0.$ (5.34)

(c) Free End at $x = 0$ or $x = L$

The bending moment and the shearing force are zero at a free end. The vanishing of the bending moment is ensured by the satisfaction of either of the expressions (5.34). The vanishing of the shearing force leads to

$$Q(x, t) = -\frac{\partial}{\partial x}\left[EI(x)\frac{\partial^2 y(x, t)}{\partial x^2}\right] = -\frac{d}{dx}\left[EI(x)\frac{d^2 Y(x)}{dx^2}\right]f(t) = 0$$

$$\text{at } x = 0 \quad \text{or} \quad x = L.$$

Therefore,
$$\frac{d}{dx}\left[EI(x)\frac{d^2 Y(x)}{dx^2}\right]_{x=0} = 0$$

(5.35)

or
$$\frac{d}{dx}\left[EI(x)\frac{d^2 Y(x)}{dx^2}\right]_{x=L} = 0.$$

These are not the only possible boundary conditions by any means; we shall encounter other types later.

5–4 GENERAL FORMULATION OF THE EIGENVALUE PROBLEM

The eigenvalue problem as defined by (5.30) is only one example of a large class of problems lending themselves to the same general formulation.

Let w be a function of one or two independent spatial variables, so that in effect we confine ourselves to one- or two-dimensional problems. The expression

$$L[w] = A_1 w + A_2 \frac{\partial w}{\partial x} + A_3 \frac{\partial w}{\partial y} + A_4 \frac{\partial^2 w}{\partial x^2} + A_5 \frac{\partial^2 w}{\partial x\, \partial y} + \cdots \qquad (5.36)$$

in which the coefficients $A_1, A_2, A_3, \ldots$ are known functions of the spatial variables x and y, represents a linear homogeneous differential expression for the function w. The expression $L[w]$ is a linear homogeneous combination of the function w and its derivatives through order $2p$, so the differential expression is said to be of order $2p$. We can define a linear differential operator of the form

$$L = A_1 + A_2 \frac{\partial}{\partial x} + A_3 \frac{\partial}{\partial y} + A_4 \frac{\partial^2}{\partial x^2} + A_5 \frac{\partial^2}{\partial x\, \partial y} + \cdots, \qquad (5.37)$$

where the linearity of the operator L is expressed by the relation

$$L[C_1 w_1 + C_2 w_2] = C_1 L[w_1] + C_2 L[w_2], \qquad (5.38)$$

in which C_1 and C_2 are constants.

Equation (5.30) represents the eigenvalue problem for a bar in bending vibration and can be looked upon as a special case of the equation†

$$L[w] = \lambda M[w], \tag{5.39}$$

where λ is a parameter and L and M are linear homogeneous differential operators of orders $2p$ and $2q$, respectively. The operators are of the type (5.37), and their order is such that $p > q$. Equation (5.39) must be satisfied by w at every point of a domain D. Associated with the differential equation (5.39) there are p boundary conditions that the function w must satisfy at every point of the boundary S of the domain D. The boundary conditions are of the type

$$B_i[w] = \lambda C_i[w], \qquad i = 1, 2, \dots, p, \tag{5.40}$$

where B_i and C_i are linear homogeneous differential operators involving derivatives normal to the boundary and along the boundary through order $2p - 1$. In the one-dimensional case domain D is a continuum depending upon one spatial variable only and the boundaries of the domain are two end points. In the two-dimensional case domain D is a continuum defined by two spatial variables and boundary S consists of one or more closed curves. Equation (5.40) implies that at each point of the boundary there are p independent equations that w must satisfy.

The eigenvalue problem consists of seeking the values of the parameter λ for which there are nonvanishing functions w satisfying the differential equation and the boundary conditions. Such parameters are called eigenvalues and the corresponding functions are called eigenfunctions.

The problem defined by (5.39) and (5.40) is called a *general* eigenvalue problem. In the special case in which (5.39) is of the form

$$L[w] = \lambda M \cdot w, \tag{5.41}$$

in which the operator M is only a function of the spatial variables, we have a *special* eigenvalue problem.‡

Let us consider the case in which the *eigenvalue λ does not appear in the boundary conditions*, so that at every point of the boundary S we have

$$B_i[w] = 0, \qquad i = 1, 2, \dots, p, \tag{5.42}$$

---

† Note that (5.39) represents the eigenvalue problem for continuous systems and is the counterpart of (4.14) for discrete systems.

‡ Note that the case in which the operator M is only a function of the spatial variables is analogous to the case of a diagonal mass matrix, $[m]$, for a discrete system.

and for this case we wish to distinguish among the following classes of functions†:

1. *Admissible functions*, which are any arbitrary functions satisfying all the *geometric* boundary conditions of the eigenvalue problem and are p times differentiable over domain D.‡

2. *Comparison functions*, which are any arbitrary functions satisfying all the boundary conditions (geometric and natural) of the eigenvalue problem and are $2p$ times differentiable over domain D.

3. *Eigenfunctions*, which, of course, satisfy all the boundary conditions and the differential equation of the eigenvalue problem.

The eigenvalue problem as defined by (5.39) and (5.42) is said to be *self-adjoint* if, for any two arbitrary comparison functions u and v, the statements

$$\int_D uL[v]\,dD = \int_D vL[u]\,dD, \tag{5.43}$$

$$\int_D uM[v]\,dD = \int_D vM[u]\,dD \tag{5.44}$$

hold true. Whether the system is self-adjoint or not can be established by means of integrations by parts.

If for any such comparison function u,

$$\int_D uL[u]\,dD \geq 0, \tag{5.45}$$

the operator L is said to be *positive*. The operator L is said to be *positive definite* if the integral is zero only if u is identically zero. There is a similar definition in regard to the operator M. If both L and M are positive definite, the eigenvalue problem is said to be positive definite. We shall refer to systems for which L is only positive and M is positive definite as *semidefinite systems*. Again one can use integrations by parts to establish whether the operators are positive definite. As it turns out, most problems treated here are self-adjoint in the sense indicated by (5.43) and (5.44). In many cases the operator M is just a function of the spatial coordinates, which can be readily recognized as the distributed mass. In such cases it becomes obvious that M is positive definite without having to perform integrations by parts. Equation (5.30) serves as such an example.

---

† See L. Collatz, *Eigenwertprobleme*, Chelsea, New York, 1948, p. 59.

‡ Admissible functions will be discussed in Chapter 6. Both admissible and comparison functions are used in approximate methods and, for this reason, the two classes of functions are defined together, although the admissible functions are not used before Chapter 6.

The solution of the homogeneous eigenvalue problem, (5.39) and (5.42), consists of a denumerably† infinite sequence of eigenvalues $\lambda_1, \lambda_2, \ldots$ and a corresponding sequence of eigenfunctions $w_1, w_2, \ldots$. The problem is homogeneous, so the amplitudes of the eigenfunctions w_r $(r = 1, 2, \ldots)$ are arbitrary and only the shapes of the eigenfunctions can be determined uniquely. *If the system is positive definite all eigenvalues λ_r are positive. In the case of a semidefinite system in which L is only positive, $\lambda = 0$ is also an eigenvalue.* If two eigenvalues have the same numerical value, any linear combination of the corresponding eigenfunctions is also an eigenfunction.

As in the case of the eigenvectors of discrete systems with symmetric $[m]$ and $[k]$ matrices, the eigenfunctions w_r of a self-adjoint eigenvalue problem possess the orthogonality property. The analogy between discrete and continuous systems comes into sharper focus when we realize that there is an expansion theorem concerning the eigenfunctions w_r.

5-5 GENERALIZED ORTHOGONALITY. EXPANSION THEOREM

Let λ_r and λ_s be two distinct eigenvalues and w_r and w_s be the corresponding eigenfunctions resulting from the solution of a self-adjoint eigenvalue problem, so that we can write

$$L[w_r] = \lambda_r M[w_r], \tag{5.46}$$

$$L[w_s] = \lambda_s M[w_s]. \tag{5.47}$$

If we multiply (5.46) by w_s and (5.47) by w_r, subtract one result from the other, and integrate both sides of the equation over domain D, we obtain

$$\int_D (w_s L[w_r] - w_r L[w_s]) \, dD = \int_D (\lambda_r w_s M[w_r] - \lambda_s w_r M[w_s]) \, dD. \tag{5.48}$$

The eigenfunctions w_r and w_s are solutions of a self-adjoint problem, so we have

$$\int_D w_s L[w_r] \, dD = \int_D w_r L[w_s] \, dD, \tag{5.49}$$

$$\int_D w_s M[w_r] \, dD = \int_D w_r M[w_s] \, dD; \tag{5.50}$$

† In many texts the word *countable* is used instead of *denumerable*.

thus (5.48) reduces to

$$(\lambda_r - \lambda_s) \int_D w_r M[w_s] \, dD = 0. \tag{5.51}$$

But in general $\lambda_r \neq \lambda_s$, so we obtain

$$\int_D w_r M[w_s] \, dD = 0 \qquad \text{for } \lambda_r \neq \lambda_s, \tag{5.52}$$

which is known as the *generalized orthogonality condition*, and the functions w_r and w_s are said to be orthogonal in a generalized sense. It follows immediately from either (5.46) or (5.47) that

$$\int_D w_r L[w_s] \, dD = 0 \qquad \text{for } \lambda_r \neq \lambda_s. \tag{5.53}$$

Hence the knowledge that a system is self-adjoint allows us to make a positive statement about the orthogonality of the eigenfunctions, provided the corresponding eigenvalues are distinct, $\lambda_r \neq \lambda_s$. In the case of multiple eigenvalues the corresponding eigenfunctions are not orthogonal to each other, although they are orthogonal to the remaining eigenfunctions of the set. They can be orthogonalized, however, by choosing as eigenfunctions proper linear combinations† of them.

The eigenfunctions w_r can be normalized with respect to M by writing

$$\int_D w_r M[w_r] \, dD = 1, \qquad r = 1, 2, \ldots, \tag{5.54}$$

which in effect determines the amplitude of the eigenfunctions w_r, which otherwise is arbitrary. Equations (5.52) and (5.54) can be combined into

$$\int_D w_r M[w_s] \, dD = \delta_{rs}, \tag{5.55}$$

where δ_{rs} is the Kronecker delta.

In the special case in which M is not a differential operator but just a function of the spatial coordinates, the orthogonality condition becomes

$$\int_D M w_r w_s \, dD = 0 \qquad \text{for } \lambda_r \neq \lambda_s, \tag{5.56}$$

in which case the functions $\sqrt{M} w_r$ and $\sqrt{M} w_s$ are said to be orthogonal in the ordinary sense.

---

† See C. Lanczos, *Applied Analysis*, Prentice-Hall, Englewood Cliffs, N.J., 1964; see also Section 5–10 of this book.

The family of natural modes w_r satisfying (5.55) constitutes a complete set of orthonormal modes.[†] The corresponding *expansion theorem* may be stated as follows: *Any function w satisfying the homogeneous boundary conditions of the system and for which L[w] is continuous can be represented by an absolutely and uniformly convergent series in the eigenfunctions in the form*

$$w = \sum_{r=1}^{\infty} c_r w_r, \tag{5.57}$$

where the coefficients c_r are given by

$$c_r = \int_D w M[w_r] \, dD, \qquad r = 1, 2, \dots . \tag{5.58}$$

Equations (5.57) and (5.58) constitute the expansion theorem for continuous systems and are the counterpart of the expansion theorem (4.60) and (4.61) for discrete systems. The expansion theorem plays a major role in the field of vibrations and will be frequently used to obtain the system response by *modal analysis*. This underscores the importance of solving the eigenvalue problem and obtaining a set of orthogonal modes. Unfortunately the solution of the eigenvalue problems for continuous systems is not as straightforward as for discrete systems, and quite frequently it is not possible to obtain a closed form solution. Cases for which exact solutions have been found generally relate to uniform systems with relatively simple boundary conditions. In a large number of cases only approximate solutions can be obtained.

The modal analysis for obtaining the response of continuous systems will be presented in Chapter 7.

5–6 VIBRATION OF STRINGS

One of the simplest, yet most interesting, examples of a continuous system is the flexible string. In general, a continuous system vibrating transversely may derive its restoring forces from two sources: axial tension and bending stiffness. In Section 5–2 we examined the case of a bar vibrating transversely for which the resultant axial force was zero. At the other extreme we find the string, for which it is commonly assumed that the bending stiffness is negligible and consequently that the restoring forces are entirely due to the axial tension.

Consider an element of string as shown in Figure 5.2 and let $T(x)$ be the tension, $p(x, t)$ the external transverse force per unit length, and $\rho(x)$ the mass

[†] See R. Courant and D. Hilbert, *Methods of Mathematical Physics*, Interscience, New York, Vol. I, 1961, p. 360.

per unit length at point x. Let us assume that the displacement $y(x, t)$ is small and write Newton's equation of motion in the transverse direction for the element of string

$$\left[T(x) + \frac{\partial T(x)}{\partial x} dx\right]\left[\frac{\partial y(x, t)}{\partial x} + \frac{\partial^2 y(x, t)}{\partial x^2} dx\right] + p(x, t) dx - T(x) \frac{\partial y(x, t)}{\partial x}$$

$$= \rho(x) dx \frac{\partial^2 y(x, t)}{\partial t^2},$$

which reduces to

$$\frac{\partial}{\partial x}\left[T(x) \frac{\partial y(x, t)}{\partial x}\right] + p(x, t) = \rho(x) \frac{\partial^2 y(x, t)}{\partial t^2}. \tag{5.59}$$

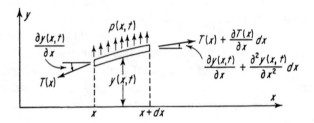

FIGURE 5.2

Consider the free vibration case, $p(x, t) = 0$, for which (5.59) becomes

$$\frac{\partial}{\partial x}\left[T(x) \frac{\partial y(x, t)}{\partial x}\right] = \rho(x) \frac{\partial^2 y(x, t)}{\partial t^2}. \tag{5.60}$$

Equation (5.60) must be satisfied over the entire length of the string. The string cannot have any displacement at a fixed end, so, if the string is fixed at $x = 0$, we must have

$$y(0, t) = 0. \tag{5.61}$$

If on the other hand, the end $x = L$ is free to move transversely, there cannot be any force component in the transverse direction, so the corresponding boundary condition is

$$T(x) \frac{\partial y(x, t)}{\partial x}\bigg|_{x=L} = 0. \tag{5.62}$$

The boundary condition (5.61) is of the geometric type and condition (5.62) is a natural boundary condition. We note that both boundary conditions are homogeneous.

Next let us find the natural modes of vibration. To this end we try the separation of variables method and write the displacement $y(x, t)$ in the form

$$y(x, t) = Y(x)f(t), \tag{5.63}$$

where $f(t)$ is a harmonic function of time. If ω denotes the frequency of $f(t)$, then

$$\frac{\partial^2 y(x, t)}{\partial t^2} = -\omega^2 Y(x)f(t) \tag{5.64}$$

and the eigenvalue problem reduces to the differential equation

$$-\frac{d}{dx}\left[T(x)\frac{dY(x)}{dx}\right] = \omega^2 \rho(x) Y(x), \tag{5.65}$$

which must be satisfied throughout the domain $0 < x < L$ and which is subject to the homogeneous boundary conditions

$$Y(0) = 0, \tag{5.66}$$

$$T(x)\frac{dY(x)}{dx}\bigg|_{x=L} = 0 \tag{5.67}$$

to be satisfied at the end points.

An eigenvalue problem such as described by (5.65) to (5.67) is called a *Sturm-Liouville problem*, after its first investigators.†

The parallel between the eigenvalue problem formulated above and the general formulation discussed in Sections 5–4 and 5–5 becomes evident if we denote

$$L = -\frac{d}{dx}\left(T\frac{d}{dx}\right), \qquad M = \rho, \qquad \lambda = \omega^2,$$

$$B_1 = 1 \quad \text{at } x = 0, \qquad B_1 = T\frac{d}{dx} \quad \text{at } x = L, \tag{5.68}$$

$$p = 1, \qquad q = 0.$$

Although the same notation is used, the reader should be able to distinguish between the operator L and the length of the string L.

---

† Note that the Sturm-Liouville eigenvalue problem is of second order only and, consequently, it is a special case of the general formulation of Sections 5–4 and 5–5.

Let $Y_r(x)$ and $Y_s(x)$ be two arbitrary functions satisfying the boundary conditions (5.66) and (5.67) and perform the following integration by parts:

$$\int_0^L Y_r L[Y_s]\, dx = -\int_0^L Y_r \frac{d}{dx}\left(T\frac{dY_s}{dx}\right) dx = -Y_r\left(T\frac{dY_s}{dx}\right)\Big|_0^L$$

$$+ \int_0^L \frac{dY_r}{dx}\left(T\frac{dY_s}{dx}\right) dx = \int_0^L T\frac{dY_r}{dx}\frac{dY_s}{dx}\, dx \quad (5.69)$$

where boundary conditions (5.66) and (5.67) have been accounted for. Similarly, we arrive at

$$\int_0^L Y_s L[Y_r]\, dx = \int_0^L T\frac{dY_r}{dx}\frac{dY_s}{dx}\, dx, \qquad (5.70)$$

and it is obvious that

$$\int_0^L Y_r L[Y_s]\, dx = \int_0^L Y_s L[Y_r]\, dx, \qquad (5.71)$$

which proves that the operator L is self-adjoint.† In addition,

$$\int_0^L Y_r M[Y_s]\, dx = \int_0^L Y_s M[Y_r]\, dx = \int_0^L \rho Y_r Y_s\, dx, \qquad (5.72)$$

so M is also self-adjoint. It follows that the eigenvalue problem is self-adjoint.
  Now consider the integral

$$\int_0^L Y_r L[Y_r]\, dx = -\int_0^L Y_r \frac{d}{dx}\left(T\frac{dY_r}{dx}\right) dx = -Y_r\left(T\frac{dY_r}{dx}\right)\Big|_0^L$$

$$+ \int_0^L \frac{dY_r}{dx}\left(T\frac{dY_r}{dx}\right) dx = \int_0^L T\left(\frac{dY_r}{dx}\right)^2 dx \geq 0. \quad (5.73)$$

The above integral is always positive, because by definition $T(x) > 0$. It is equal to zero only if $Y_r(x)$ is constant throughout the domain. Because of boundary condition (5.66), however, this constant must be zero, which would imply a trivial solution. It follows that the operator L is positive definite. From

$$\int_0^L Y_r M[Y_r]\, dx = \int_0^L \rho Y_r^2\, dx \geq 0, \qquad (5.74)$$

we conclude that the operator M is also positive definite, so the eigenvalue problem is positive definite.

---

  † The same conclusion is reached in the more general type of boundary condition, $a(x)y(x, t) + b(x)[\partial y(x, t)/\partial x] = 0$ at each of the end points.

Consider the case of a *uniform string*, ρ = constant, subjected to a constant tension T and *fixed at both ends*. Under these circumstances the eigenvalue problem reduces to the solution of the differential equation

$$\frac{d^2 Y(x)}{dx^2} + \beta^2 Y(x) = 0, \qquad \beta^2 = \omega^2 \frac{\rho}{T}, \tag{5.75}$$

which is subject to the boundary conditions

$$Y(0) = 0, \qquad Y(L) = 0. \tag{5.76}$$

The solution of (5.75) is simply

$$Y(x) = C_1 \sin \beta x + C_2 \cos \beta x. \tag{5.77}$$

The first of the boundary conditions of (5.76) indicates that $C_2 = 0$, and the second condition leads to the characteristic equation or frequency equation

$$\sin \beta L = 0, \tag{5.78}$$

which is the counterpart of the characteristic determinant obtained in the case of discrete systems. Equation (5.78) has an infinite number of solutions

$$\beta_r = r \frac{\pi}{L}, \qquad r = 1, 2, \ldots, \tag{5.79}$$

where the eigenvalues† β_r are related to the natural frequencies ω_r by

$$\omega_r = \beta_r \sqrt{\frac{T}{\rho}} = r\pi \sqrt{\frac{T}{\rho L^2}}, \qquad r = 1, 2, \ldots. \tag{5.80}$$

The frequency ω_1 is called the *fundamental frequency* and the higher frequencies are known as *overtones*. Overtones that are *integral multiples* of the fundamental frequency are called *higher harmonics*, and the fundamental frequency in such a case is called the *fundamental harmonic*. There are only a small number of vibrating systems having harmonic overtones and most of them are used as musical instruments because they produce pleasant sounds.

The corresponding eigenfunctions are

$$Y_r(x) = A_r \sin r\pi \frac{x}{L}, \tag{5.81}$$

† The name is used somewhat loosely here, because β_r is related to the eigenvalue λ_r.

and it is easy to see that they are orthogonal. If they are also normalized, so that

$$\int_0^L \rho Y_r^2(x)\,dx = 1, \qquad r = 1, 2, \ldots, \tag{5.82}$$

we obtain an orthonormal set of modes

$$Y_r(x) = \sqrt{\frac{2}{\rho L}} \sin r\pi \frac{x}{L}, \qquad r = 1, 2, \ldots. \tag{5.83}$$

The first three modes are shown in Figure 5.3. We note that there are points at which the displacements are zero. These points are called *nodes*. The mode Y_r has $r - 1$ nodes (excluding the end points) which occur at points $x = n(L/r)$ $(n = 1, 2, \ldots, r - 1)$.

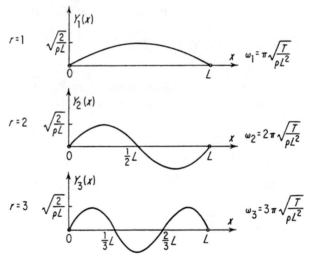

FIGURE 5.3

The natural frequencies ω_r constitute an infinite set of discrete values, so the string in question is said to have a *discrete frequency spectrum*. Whereas the shape of the modes is independent of the tension T or the mass per unit length ρ, the natural frequencies increase as T increases and decrease as ρ increases. For given values of T and ρ the natural frequencies decrease as the length L of the string increases. In fact, as the length L increases more and more, the values of the natural frequencies draw closer and closer and, as L approaches infinity, we obtain a *continuous frequency spectrum*. At this point it is no longer meaningful to speak of natural frequencies and natural modes, and a different point of view must be adopted. In the case of bodies of infinite dimensions, we regard the motion as consisting of *traveling waves*. In fact, we must not confine ourselves to infinite bodies, and we can obtain a solution in terms of waves in the case of finite bodies also. This approach will be discussed later.

5–7 LONGITUDINAL VIBRATION OF RODS

The differential equation for the longitudinal motion of a thin rod was derived in Chapter 2. Referring to these results, if $u(x, t)$ is the axial displacement, the differential equation of motion is

$$\frac{\partial}{\partial x}\left[EA(x)\frac{\partial u(x, t)}{\partial x}\right] = m(x)\frac{\partial^2 u(x, t)}{\partial t^2}, \tag{5.84}$$

which must be satisfied over the domain $0 < x < L$. In addition, u must be such that at the end points we have

$$\left(EA\frac{\partial u}{\partial x}\right)\delta u\Big|_0^L = 0. \tag{5.85}$$

If the bar is clamped at the end $x = 0$, the boundary condition is

$$u(0, t) = 0, \tag{5.86}$$

and if the end $x = L$ is free, we have

$$EA(x)\frac{\partial u(x, t)}{\partial x}\Big|_{x=L} = 0. \tag{5.87}$$

Comparing (5.84), (5.86), and (5.87) with (5.60) through (5.62,) we conclude that they possess the same structures. In fact, the conclusions of Section 5–6 concerning the self-adjointness and positive definiteness of the system apply equally well here if we replace $y(x, t)$ by $u(x, t)$, $\rho(x)$ by $m(x)$, and $T(x)$ by $EA(x)$. In Section 5–6 we solved the eigenvalue problem for a string fixed at both ends. By using the analogy pointed out above, we can use these results for the case of a rod clamped at both ends. Of course, one must remember that $y(x, t)$ is a transverse displacement and $u(x, t)$ is a longitudinal displacement.

Let us pursue further the case of a *clamped-free* rod, for which the eigenvalue problem reduces to the differential equation

$$-\frac{d}{dx}\left[EA(x)\frac{dU(x)}{dx}\right] = \omega^2 m(x)U(x), \tag{5.88}$$

which must be satisfied throughout the domain $0 \leq x \leq L$, and the homogeneous boundary conditions

$$U(0) = 0, \tag{5.89}$$

$$EA(x) \frac{dU(x)}{dx}\bigg|_{x=L} = 0, \tag{5.90}$$

to be satisfied at the end point. There is no difficulty in demonstrating that the problem is self-adjoint and positive definite.

Let us consider the case in which the rod is *uniform* and calculate the natural modes and the corresponding natural frequencies. For a uniform rod the eigenvalue problem reduces to the solution of the differential equation

$$\frac{d^2U(x)}{dx^2} + \beta^2 U(x) = 0, \qquad \beta^2 = \omega^2 \frac{m}{EA}, \tag{5.91}$$

which is subject to boundary conditions (5.89) and (5.90). The solution of (5.91) is

$$U(x) = C_1 \sin \beta x + C_2 \cos \beta x. \tag{5.92}$$

Boundary condition (5.89) yields $C_2 = 0$, and from boundary condition (5.90) we obtain the frequency equation

$$\cos \beta L = 0, \tag{5.93}$$

which yields the eigenvalues

$$\beta_r = (2r - 1)\frac{\pi}{2L}, \qquad r = 1, 2, \ldots, \tag{5.94}$$

so that the natural frequencies ω_r are

$$\omega_r = \beta_r \sqrt{\frac{EA}{m}} = (2r - 1)\frac{\pi}{2} \sqrt{\frac{EA}{mL^2}}, \qquad r = 1, 2, \ldots. \tag{5.95}$$

The corresponding eigenfunctions have the form

$$U_r(x) = A_r \sin (2r - 1)\frac{\pi}{2}\frac{x}{L}, \qquad r = 1, 2, \ldots \tag{5.96}$$

and they are orthogonal. Let us normalize them and adjust the coefficients A_r such that

$$\int_0^L mU_r^2(x)\, dx = 1, \qquad r = 1, 2, \ldots, \tag{5.97}$$

from which we obtain the orthonormal set

$$U_r(x) = \sqrt{\frac{2}{mL}} \sin (2r - 1) \frac{\pi}{2} \frac{x}{L}, \qquad r = 1, 2, \dots \tag{5.98}$$

The first three modes are shown in Figure 5.4. The rth mode, $U_r(x)$, has nodes (other than $x = 0$) at points $x = n\,[2L/(2r - 1)]\,(n = 1, 2, \dots, r - 1)$.

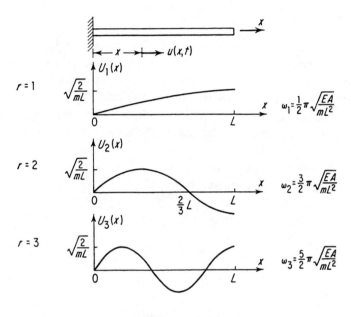

FIGURE 5.4

It will prove instructive to investigate a different kind of boundary condition—namely the case in which *both ends are free*. The formulation follows the same pattern and once again we find that (5.88) must be satisfied throughout the domain $0 < x < L$, but in contrast to the previous case the boundary conditions in this case are

$$EA(x) \frac{dU(x)}{dx}\bigg|_{x=0} = 0, \tag{5.99}$$

$$EA(x) \frac{dU(x)}{dx}\bigg|_{x=L} = 0. \tag{5.100}$$

This problem, too, is self-adjoint. Let us now check to see whether the system is positive definite or not. To this end consider the integral

$$\int_0^L U_r L[U_r] \, dx = -\int_0^L U_r \frac{d}{dx} \left(EA \frac{dU_r}{dx} \right) dx$$

$$= -U_r \left(EA \frac{dU_r}{dx} \right)\Big|_0^L + \int_0^L \frac{dU_r}{dx} \left(EA \frac{dU_r}{dx} \right) dx$$

$$= \int_0^L EA \left(\frac{dU_r}{dx} \right)^2 dx \geq 0, \qquad (5.101)$$

where boundary conditions (5.99) and (5.100) have been accounted for. When we examine (5.101), however, we note that it is zero when $U_r(x)$ is a constant, but this constant *does not* necessarily have to be zero. It follows that for these boundary conditions the operator L is only positive, not positive definite as in the previous case. In all cases encountered here the operator M is positive definite. Hence the free-free rod is a semidefinite system.

Letting the rod be uniform we must satisfy (5.91) again, and the solution of this differential equation remains in the form (5.92). But in contrast with the previous case, boundary condition (5.99) yields $C_1 = 0$, whereas boundary condition (5.100) gives the frequency equation

$$\sin \beta L = 0, \qquad (5.102)$$

which leads to the eigenvalues

$$\beta_r = r \frac{\pi}{L}, \qquad r = 0, 1, 2, \ldots, \qquad (5.103)$$

so $\beta_0 = 0$ is also an eigenvalue, corroborating an earlier statement made in connection with semidefinite systems. Corresponding to the eigenvalues other than β_0 we have the eigenfunctions

$$U_r(x) = A_r \cos r\pi \frac{x}{L}, \qquad r = 1, 2, \ldots. \qquad (5.104)$$

For $\beta = \beta_0 = 0$, (5.91) becomes

$$\frac{d^2 U(x)}{dx^2} = 0, \qquad (5.105)$$

which has the solution

$$U_0(x) = A_0 + A_0' x. \qquad (5.106)$$

Upon consideration of the boundary conditions, it reduces to

$$U_0(x) = A_0. \tag{5.107}$$

Hence to the eigenvalue $\beta = \beta_0 = 0$ there corresponds a mode that is interpreted as the displacement of the rod as a whole. This is known as a rigid-body mode and is typical of unrestrained systems (semidefinite systems) for which there are no forces or moment exerted by the supports. In this particular case we are concerned with forces in the longitudinal direction only and not with moments. Denoting the resultant force in the longitudinal direction $F(t)$ we have

$$F(t) = \int_0^L m(x) \frac{\partial^2 u(x, t)}{\partial t^2} \, dx = \ddot{f}(t) \int_0^L m(x)U(x) \, dx = 0, \tag{5.108}$$

because there is no external force present. The above leads to the equations

$$\int_0^L m(x)U_r(x) \, dx = 0, \qquad r = 1, 2, \ldots, \tag{5.109}$$

which can be merely interpreted as the statement of the fact that the rigid-body mode is orthogonal to the elastic modes. The orthogonality relation

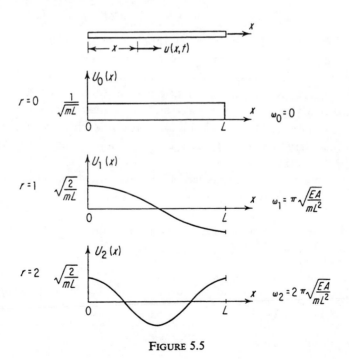

FIGURE 5.5

and the normalization statement may be combined into

$$\int_0^L m(x)U_r(x)U_s(x)\,dx = \delta_{rs}, \qquad r, s = 0, 1, 2, \ldots \tag{5.110}$$

The first three modes are plotted in Figure 5.5. In this case the nodes of $U_r(x)$ are at the points $x = (L/2r)(2n-1)$ $(n = 1, 2, \ldots, r)$.

5-8 BOUNDARY CONDITIONS DEPENDING ON THE EIGENVALUE λ

Let us consider the eigenvalue problem which consists of the differential equation

$$L[w] = \lambda M[w] \tag{5.111}$$

and the associated boundary conditions. Equation (5.111) must be satisfied throughout a domain D, where L and M are linear homogeneous differential operators of orders $2p$ and $2q$, respectively, such that $p > q$. In addition, at every point of the boundaries S of domain D, w must satisfy the boundary conditions

$$B_i[w] = 0, \qquad\qquad i = 1, 2, \ldots, k, \tag{5.112}$$

$$B_j[w] = \lambda C_j[w], \qquad j = 1, 2, \ldots, l; \, l = p - k, \tag{5.113}$$

where B_i, B_j, and C_j are linear homogeneous differential operators containing derivatives normal to the boundaries and along the boundaries through order $2p - 1$. Although the boundary conditions as given by (5.113) are homogeneous, they are different from the conditions given by (5.112) because they contain λ, which takes different values for different eigenfunctions. Equations (5.112) and (5.113) imply that at each boundary point there are p boundary conditions to be satisfied, of which l boundary conditions contain λ and the remaining $k = p - l$ do not contain λ.

If for any two arbitrary functions u and v satisfying all the boundary conditions (5.112), the following statements hold true:

$$\int_D uL[v]\,dD + \sum_{j=1}^{l} \int_S uB_j[v]\,dS = \int_D vL[u]\,dD + \sum_{j=1}^{l} \int_S vB_j[u]\,dS, \tag{5.114}$$

$$\int_D uM[v]\,dD + \sum_{j=1}^{l} \int_S uC_j[v]\,dS = \int_D vM[u]\,dD + \sum_{j=1}^{l} \int_S vC_j[u]\,dS, \tag{5.115}$$

the eigenvalue problem is said to be *self-adjoint*.

If for any function u satisfying all the boundary conditions (5.112),

$$\int_D uL[u]\,dD + \sum_{j=1}^{l} \int_S uB_j[u]\,dS \geq 0, \tag{5.116}$$

the operator L is said to be positive and it is said to be positive definite if the expression vanishes only if u is identically zero. Correspondingly, if for such a function u,

$$\int_D uM[u]\,dD + \sum_{j=1}^{l} \int_S uC_j[u]\,dS \geq 0, \tag{5.117}$$

the operator M is said to be positive, and it is said to be positive definite if the equality holds true only if u vanishes identically. If both L and M are positive definite, the eigenvalue problem is said to be positive definite. Recall that in a special case treated previously, M was not an operator but a prescribed positive function of the spatial variables.

The eigenvalue problem defined by (5.111) to (5.113) yields a denumerably infinite sequence of eigenvalues λ_r and corresponding eigenfunctions w_r of arbitrary amplitudes. Corresponding to the definition of a self-adjoint problem as outlined by (5.114) and (5.115), there is a modified orthogonality relation for the eigenfunctions w_r, as we are about to show.

Let λ_r and λ_s be two distinct eigenvalues and let w_r and w_s be the corresponding eigenfunctions satisfying the self-adjoint eigenvalue problem described above, such that we have

$$L[w_r] = \lambda_r M[w_r], \tag{5.118}$$

$$L[w_s] = \lambda_s M[w_s]. \tag{5.119}$$

Multiply (5.118) by w_s and (5.119) by w_r, subtract one result from the other, integrate over domain D, and obtain

$$\int_D (w_s L[w_r] - w_r L[w_s])\,dD = \int_D (\lambda_r w_s M[w_r] - \lambda_s w_r M[w_s])\,dD. \tag{5.120}$$

But the eigenfunctions w_r and w_s are solutions of the self-adjoint problem, so we can write

$$\int_D (w_s L[w_r] - w_r L[w_s])\,dD = \sum_{j=1}^{l} \int_S (w_r B_j[w_s] - w_s B_j[w_r])\,dS, \tag{5.121}$$

$$\int_D w_s M[w_r]\,dD = \int_D w_r M[w_s]\,dD$$

$$+ \sum_{j=1}^{l} \int_S (w_r C_j[w_s] - w_s C_j[w_r])\,dS. \tag{5.122}$$

Equations (5.121) and (5.122) introduced in (5.120) yield

$$(\lambda_r - \lambda_s)\left[\int_D w_r M[w_s]\, dD + \sum_{j=1}^{l} \int_S w_r C_j[w_s]\, dS\right] = 0, \qquad (5.123)$$

and if the eigenvalues λ_r and λ_s are distinct, we obtain the orthogonality relation

$$\int_D w_r M[w_s]\, dD + \sum_{j=1}^{l} \int_S w_r C_j[w_s]\, dS = 0. \qquad (5.124)$$

Here, too, the orthogonal functions w_r can be normalized to give an ortho-normal set of eigenfunctions for which the expansion theorem applies.

In Section 5–9 we shall present an example of the case discussed above.

5-9 TORSIONAL VIBRATION OF BARS

Figure 5.6(a) represents a nonuniform bar such that the x axis coincides with the neutral axis, that is, the axis along which there is no strain. Figure 5.6(b) shows a free-body diagram for an element dx of the bar.

In general, if the cross section is not circular, there is some warping of the cross-sectional plane associated with torsional motion. Furthermore, if the cross-sectional area is circular but the bar is nonuniform, although there is no

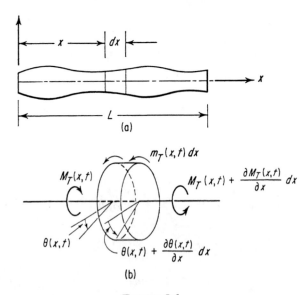

FIGURE 5.6

warping of the cross-sectional planes, the displacements are not proportional to the radial distance from the axis of twist. These subjects are treated in texts on elasticity.† We shall assume here that the shape of the cross-sectional area and the nonuniformity of the bar are such that the motion can be regarded as rotation of the cross-sectional plane as a whole and without warping. This, of course, is precisely what happens in the case of a uniform circular shaft in torsion.

Let G be the shear modulus and let $J(x)$ be a geometric property of the cross section, which in the case of a circular cross section is the polar moment of inertia.

If $\theta(x, t)$ represents the angle of twist of a cross section at the point x and at time t, the relation between the deformation and the twisting moment is

$$M_T(x, t) = GJ(x) \frac{\partial \theta(x, t)}{\partial x}, \tag{5.125}$$

where the product $GJ(x)$ is called *torsional stiffness*.

Denoting by $I(x)$ the mass polar moment of inertia per unit length of bar and by $m_T(x, t)$ the external twisting moment per unit length of bar, we can write the equation for the rotational motion of the bar element in the form

$$\left[M_T(x, t) + \frac{\partial M_T(x, t)}{\partial x} \, dx \right] + m_T(x, t) \, dx - M_T(x, t) = I(x) \, dx \, \frac{\partial^2 \theta(x, t)}{\partial t^2}, \tag{5.126}$$

which reduces to

$$\frac{\partial M_T(x, t)}{\partial x} + m_T(x, t) = I(x) \frac{\partial^2 \theta(x, t)}{\partial t^2}. \tag{5.127}$$

In view of (5.125), we can write (5.127) as

$$\frac{\partial}{\partial x} \left[GJ(x) \frac{\partial \theta(x, t)}{\partial x} \right] + m_T(x, t) = I(x) \frac{\partial^2 \theta(x, t)}{\partial t^2}, \tag{5.128}$$

which is the equation of motion in torsion.

In the case in which $m_T(x, t) = 0$, (5.128) reduces to the equation for the free torsional vibration of a bar,

$$\frac{\partial}{\partial x} \left[GJ(x) \frac{\partial \theta(x, t)}{\partial x} \right] = I(x) \frac{\partial^2 \theta(x, t)}{\partial t^2}, \tag{5.129}$$

† See S. P. Timoshenko and J. N. Goodier, *Theory of Elasticity*, 2nd ed., McGraw-Hill, New York, 1951.

which is of precisely the same form as (5.60) for the transverse vibration of strings or (5.84) for longitudinal vibration of rods. Furthermore, for a clamped end at $x = 0$, we obtain the boundary condition

$$\theta(0, t) = 0, \tag{5.130}$$

and for a free end at $x = L$ the boundary condition is

$$GJ(x) \left. \frac{\partial \theta(x, t)}{\partial x} \right|_{x=L} = 0, \tag{5.131}$$

which expresses the fact that the twisting moment vanishes at that point. Hence the analogy between lateral vibration of strings, longitudinal vibration of rods, and torsional vibration of bars is complete, and the results obtained for one system apply equally well to any of the other systems.

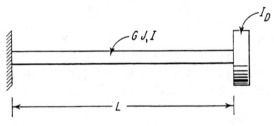

FIGURE 5.7

At this moment we wish to explore one example which presents a certain degree of novelty. Consider a circular bar with a *rigid disk attached at one end*. The torsional rigidity of the bar is $GJ(x)$, its mass polar moment of inertia per unit length is $I(x)$, and the mass polar moment of inertia of the disk is I_D (Figure 5.7). This case is interesting to investigate because of the concentrated inertia at the end $x = L$. Whereas at the clamped end, $x = 0$, the boundary condition remains as given by (5.130), at the end $x = L$ we have

$$GJ(x) \left. \frac{\partial \theta(x, t)}{\partial x} \right|_{x=L} = -I_D \left. \frac{\partial^2 \theta(x, t)}{\partial t^2} \right|_{x=L}, \tag{5.132}$$

where the right side of (5.132) indicates the presence of an inertia moment caused by the disk. Letting

$$\theta(x, t) = \Theta(x)f(t) \tag{5.133}$$

and recalling that $f(t)$ is harmonic, the eigenvalue problem reduces to the differential equation

$$-\frac{d}{dx}\left[GJ(x) \frac{d\Theta(x)}{dx} \right] = \omega^2 I(x)\Theta(x), \tag{5.134}$$

to be satisfied throughout the domain $0 < x < L$, and the boundary conditions

$$\Theta(0) = 0, \tag{5.135}$$

$$GJ(x) \frac{d\Theta(x)}{dx}\bigg|_{x=L} = \omega^2 I_D \Theta(x)\bigg|_{x=L} \tag{5.136}$$

at the ends.

Contemplating (5.134) through (5.136), we conclude that the eigenvalue problem is of the form described in Section 5–8. The analogous quantities are

$$L = -\frac{d}{dx}\left(GJ\frac{d}{dx}\right), \qquad M = I, \qquad \lambda = \omega^2,$$

$$B_1 = 1, \quad k = 1, \quad l = 0 \qquad \text{at } x = 0,$$

$$B_1 = GJ\frac{d}{dx}, \quad C_1 = I_D, \quad k = 0, \quad l = 1 \qquad \text{at } x = L, \tag{5.137}$$

$$p = 1, \qquad q = 0.$$

The domain D is the interval $0 < x < L$ and the boundaries S are just the two end points $x = 0$ and $x = L$.

Let the bar be uniform and denote,

$$\frac{\omega^2 I}{GJ} = \beta^2, \tag{5.138}$$

so that (5.134) reduces to

$$\frac{d^2\Theta(x)}{dx^2} + \beta^2\Theta(x) = 0, \tag{5.139}$$

which has the solution

$$\Theta(x) = C_1 \sin \beta x + C_2 \cos \beta x. \tag{5.140}$$

From boundary condition (5.135) it follows that $C_2 = 0$. Boundary condition (5.136) gives

$$GJC_1\beta \cos \beta L = \omega^2 I_D C_1 \sin \beta L,$$

yielding the frequency equation

$$\tan \beta L = \frac{IL}{I_D}\frac{1}{\beta L}, \tag{5.141}$$

which is a transcendental equation in βL. It has an infinity of solutions $\beta_r L$ which must be obtained numerically and are related to the natural frequencies ω_r by

$$\omega_r = \beta_r L \sqrt{\frac{GJ}{IL^2}}, \qquad r = 1, 2, \ldots. \tag{5.142}$$

Note that the natural frequencies ω_r are no longer integral multiples of the fundamental frequency ω_1.

The natural modes are given by

$$\Theta_r(x) = A_r \sin \beta_r x, \qquad r = 1, 2, \ldots. \tag{5.143}$$

and they are orthogonal. The orthogonality condition follows from (5.124) and has the form

$$\int_0^L I\Theta_r(x)\Theta_s(x)\, dx + I_D\Theta_r(L)\Theta_s(L) = 0, \quad r, s = 1, 2, \ldots; r \neq s. \tag{5.144}$$

The first three modes for a ratio $IL/I_D = 1$ are plotted in Figure 5.8.

From (5.141) we deduce that as $\beta L \to \infty$ the solutions of the frequency equation become integral multiples of π. Specifically it can be easily shown that $\beta_r L \simeq (r-1)\pi$ for large values of $\beta_r L$. (The reader is urged to verify this statement.) From (5.141) and (5.143) we reach the conclusion that the very high order modes have nodes at $x = L$, so for these modes the rigid disk is at rest.

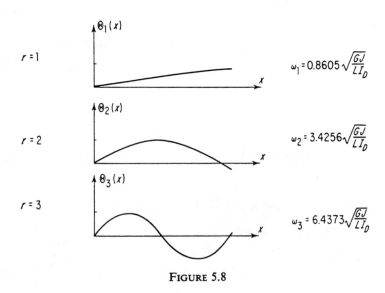

$$r = 1 \qquad\qquad \omega_1 = 0.8605 \sqrt{\frac{GJ}{LI_D}}$$

$$r = 2 \qquad\qquad \omega_2 = 3.4256 \sqrt{\frac{GJ}{LI_D}}$$

$$r = 3 \qquad\qquad \omega_3 = 6.4373 \sqrt{\frac{GJ}{LI_D}}$$

FIGURE 5.8

The system of Figure 5.7 may be approached differently. One can assume that the mass of the disk is uniformly distributed over a short distance ϵ and take the limit by letting $\epsilon \to 0$ (see Problem 5.8).

5-10 BENDING VIBRATIONS OF BARS

In Section 5–3 we used the case of a bar in bending vibration to illustrate an eigenvalue problem. It was shown there that the problem reduces to finding the values of ω and $Y(x)$ so that the equation

$$\frac{d^2}{dx^2}\left[EI(x)\frac{d^2 Y(x)}{dx^2}\right] = \omega^2 m(x) Y(x) \qquad (5.145)$$

has a nontrivial solution over the domain $0 < x < L$, where $Y(x)$ satisfies four boundary conditions, two at each end. If the displacement vanishes at either end, we have

$$Y(0) = 0 \qquad \text{or} \qquad Y(L) = 0. \qquad (5.146)$$

The vanishing of the slope at either end leads to

$$\left.\frac{dY(x)}{dx}\right|_{x=0} = 0 \qquad \text{or} \qquad \left.\frac{dY(x)}{dx}\right|_{x=L} = 0. \qquad (5.147)$$

If the bending moment is zero at either end,

$$\left.EI(x)\frac{d^2 Y(x)}{dx^2}\right|_{x=0} = 0 \qquad \text{or} \qquad \left.EI(x)\frac{d^2 Y(x)}{dx^2}\right|_{x=L} = 0, \qquad (5.148)$$

and, finally, if the shearing force vanishes at either end, we have

$$\left.\frac{d}{dx}\left[EI(x)\frac{d^2 Y(x)}{dx^2}\right]\right|_{x=0} = 0 \quad \text{or} \quad \left.\frac{d}{dx}\left[EI(x)\frac{d^2 Y(x)}{dx^2}\right]\right|_{x=L} = 0. \qquad (5.149)$$

Boundary conditions (5.146) and (5.147) are of the imposed (or geometric) type and boundary conditions (5.148) and (5.149) are of the natural (or additional) type. They are all homogeneous boundary conditions. These are not the only possible boundary conditions. In Section 5–9 we examined a case in which the eigenvalues entered into the boundary conditions, and it turns out that such a case is possible here, too.

The above formulation fits quite nicely into the general formulation of the special type of eigenvalue problem of Section 5–4. To show this we consider the case of a bar clamped at the end $x = 0$ and free at the end $x = L$. The analogy is expressed by

$$L = \frac{d^2}{dx^2}\left(EI\frac{d^2}{dx^2}\right), \qquad M = m, \qquad \lambda = \omega^2,$$

$$B_1 = 1, \quad B_2 = \frac{d}{dx} \qquad \text{at } x = 0,$$

$$B_1 = EI\frac{d^2}{dx^2}, \quad B_2 = \frac{d}{dx}\left(EI\frac{d^2}{dx^2}\right) \qquad \text{at } x = L, \qquad (5.150)$$

$$p = 2, \qquad q = 0.$$

It can be easily shown that the problem is self-adjoint and positive definite.

Let us find now the natural frequencies and normal modes of the uniform *clamped-free* beam. The beam is *uniform*, so the differential equation to be satisfied over the length of the beam reduces to

$$\frac{d^4 Y(x)}{dx^4} - \beta^4 Y(x) = 0, \qquad \beta^4 = \frac{\omega^2 m}{EI}. \qquad (5.151)$$

At the clamped end, $x = 0$, we have the imposed boundary conditions

$$Y(0) = 0, \qquad (5.152)$$

$$\left.\frac{dY(x)}{dx}\right|_{x=0} = 0. \qquad (5.153)$$

At the free end, because the bar is uniform, we have the boundary conditions

$$\left.\frac{d^2 Y(x)}{dx^2}\right|_{x=L} = 0, \qquad (5.154)$$

$$\left.\frac{d^3 Y(x)}{dx^3}\right|_{x=L} = 0, \qquad (5.155)$$

and they are natural boundary conditions. The general solution of (5.151) is

$$Y(x) = C_1 \sin \beta x + C_2 \cos \beta x + C_3 \sinh \beta x + C_4 \cosh \beta x. \quad (5.156)$$

Using boundary conditions (5.152) to (5.155), we obtain the frequency equation

$$\cos \beta L \cosh \beta L = -1, \qquad (5.157)$$

which must be solved numerically and yields an infinity of solutions β_r.

Corresponding to the eigenvalues β_r we obtain the natural modes

$$Y_r(x) = A_r[(\sin \beta_r L - \sinh \beta_r L)(\sin \beta_r x - \sinh \beta_r x)$$
$$+ (\cos \beta_r L + \cosh \beta_r L)(\cos \beta_r x - \cosh \beta_r x)],$$

$$r = 1, 2, \ldots, \quad (5.158)$$

which are not yet normalized. It can be easily shown that the modes $Y_r(x)$ are orthogonal. They constitute a complete set of orthogonal eigenfunctions. The first three natural modes are plotted in Figure 5.9. Note that the mode $Y_r(x)$ has $r - 1$ nodes.

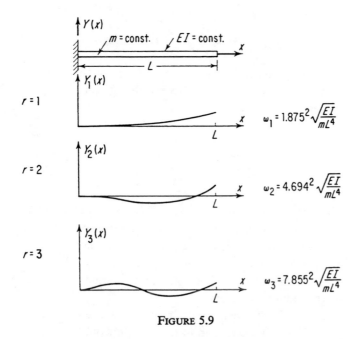

FIGURE 5.9

Another case of interest is the *free-free uniform* bar. As the differential equation (5.151) still holds true, we obtain the same general solution, (5.156). In this case, however, all boundary conditions are natural. Specifically, at the end $x = 0$ we have

$$\left.\frac{d^2 Y(x)}{dx^2}\right|_{x=0} = 0, \qquad \left.\frac{d^3 Y(x)}{dx^3}\right|_{x=0} = 0, \qquad (5.159)$$

whereas at the end $x = L$ the boundary conditions are

$$\left.\frac{d^2 Y(x)}{dx^2}\right|_{x=L} = 0, \qquad \left.\frac{d^3 Y(x)}{dx^3}\right|_{x=L} = 0. \qquad (5.160)$$

Consequently, we obtain the frequency equation

$$\cos \beta L \cosh \beta L = 1, \tag{5.161}$$

which again is a transcendental equation and must be solved numerically for the eigenvalues β_r. In contrast with the clamped-free case, however, we note that (5.161) yields a zero double root, $\beta_0 = \beta_1 = 0$.†

For $\beta = 0$ (5.151) reduces to

$$\frac{d^4 Y(x)}{dx^4} = 0, \tag{5.162}$$

which has the general solution

$$Y(x) = D_1 + D_2 x + D_3 x^2 + D_4 x^3, \tag{5.163}$$

and it is easy to see that to satisfy the boundary conditions we must have $D_3 = D_4 = 0$. Hence, corresponding to the double root $\beta = \beta_0 = \beta_1 = 0$, there are two modes

$$Y_0(x) = A_0, \tag{5.164}$$

$$Y_1(x) = A_1\left(x - \frac{L}{2}\right), \tag{5.165}$$

where $Y_0(x)$ is selected as a symmetric and $Y_1(x)$ as an antisymmetric mode.‡ They correspondingly represent rigid-body transverse translation and rigid-body rotation about the center of mass of the bar. Since $\beta = 0$ is a double root, any combination of $Y_0(x)$ and $Y_1(x)$ is also a natural mode. This explains why it is possible to choose $Y_1(x)$ to represent rotation about the center of mass rather than any other point.

Equation (5.161) further yields an infinite sequence of discrete eigenvalues β_r $(r = 2, 3, \ldots)$, which correspond to the eigenfunctions

$$Y_r(x) = A_r[(\cos \beta_r L - \cosh \beta_r L)(\sin \beta_r x + \sinh \beta_r x)$$

$$- (\sin \beta_r L - \sinh \beta_r L)(\cos \beta_r x + \cosh \beta_r x)],$$

$$r = 2, 3, \ldots. \tag{5.166}$$

The natural modes as given by (5.164) through (5.166) are not normalized. Again the natural modes are orthogonal.

---

† Such a system is a semidefinite system, as pointed out in Section 5–4.
‡ Note that this particular linear combination renders the eigenfunctions $Y_0(x)$ and $Y_1(x)$ orthogonal.

The orthogonality condition

$$\int_0^L m(x)\,Y_0(x)\,Y_r(x)\,dx = 0, \qquad r = 2, 3, \ldots \tag{5.167}$$

reduces to

$$\int_0^L m(x)\,Y_r(x)\,dx = 0, \qquad r = 2, 3, \ldots, \tag{5.168}$$

which can be interpreted as a statement of the fact that for free vibration there are no external transverse forces on the bar. On the other hand, the orthogonality condition

$$\int_0^L m(x)\,Y_1(x)\,Y_r(x)\,dx = 0, \qquad r = 2, 3, \ldots \tag{5.169}$$

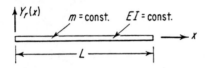

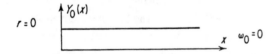

FIGURE 5.10

leads to

$$\int_0^L m(x)\left(x - \frac{L}{2}\right) Y_r(x)\, dx = 0, \qquad r = 2, 3, \ldots, \tag{5.170}$$

which can be regarded as a statement that for free vibration the external moment about the center of mass is zero. Both conditions (5.168) and (5.170) are typical of a completely unrestrained bar (semidefinite system) in bending vibration.

The natural modes $Y_r(x)$ ($r = 0, 1, 2, 3, \ldots$) constitute a complete set of orthogonal modes for which the expansion theorem applies. The first four modes are plotted in Figure 5.10.

As βL increases $\cosh \beta L$ increases very rapidly, so *for large βL* the roots of the frequency equation (5.161) may be approximated by $\beta_r L \cong (2r - 1)(\pi/2)$. Of course, one should always bear in mind that the elementary theory used is valid for the lower modes only, because as the mode number increases the effects of shearing deformation and rotatory inertia become significant.[†]

5-11 VIBRATION OF MEMBRANES

Now we wish to consider two-dimensional systems which, in the equilibrium position, lie in a plane. We shall be concerned in particular with membranes and plates. The membranes have no bending resistance and the restoring forces are due exclusively to tension, just as in the case of strings in the one-dimensional systems. A drumhead is an example. On the other hand, plates do have bending stiffness, which is responsible for the restoring forces in a manner similar to bars in bending in the one-dimensional case.

The differential equation for the displacement of a thin uniform membrane subjected to uniform tension is obtained in a way similar to the equation for strings. It is assumed that in the equilibrium position the membrane lies entirely in one plane under a uniform tension T. The displacement w of any point on the membrane is in the direction normal to the reference plane and must be sufficiently small that the tensile force per unit length of membrane, T, remains constant. Under these conditions the differential equation for the displacement w of a thin membrane of uniform thickness can be expressed

$$T\mathbf{V}^2 w + p = \rho\, \frac{\partial^2 w}{\partial t^2}, \tag{5.171}$$

where $\mathbf{V}^2$ is the Laplace operator, which in this particular case is a two-

† See S. Timoshenko, *Vibration Problems in Engineering*, Van Nostrand, Princeton, N.J., 1937, p. 341.

dimensional operator; p the external pressure, and ρ the mass per unit area of membrane.

If the external force is zero, we obtain the free vibration differential equation

$$T\mathbf{V}^2 w = \rho \frac{\partial^2 w}{\partial t^2}, \tag{5.172}$$

which must be satisfied over a two-dimensional domain D bounded by one or more curves S. If the membrane is fixed at points (a_1, b_1) on a segment of the boundary, we have the boundary condition

$$w = 0 \qquad \text{at } (a_1, b_1); \tag{5.173}$$

if, on the other hand, the membrane is free to deflect transversely at some other points (a_2, b_2) on a different segment of the boundary, then there cannot be any force component in the transverse direction. Consequently, the boundary condition at these points is

$$T\frac{\partial w}{\partial n} = 0 \qquad \text{at } (a_2, b_2), \tag{5.174}$$

where the derivative is taken normal to the boundary and in the reference plane of the membrane. Other types of boundary conditions are possible [see (5.189)]. Whereas for a string we must specify conditions at two end points, for a membrane we must specify not only the conditions at the boundaries but also the shape of the boundary curves.

Now write the displacement w as the product of a function W of the spatial variables only and a function f depending on time only,

$$w = Wf, \tag{5.175}$$

and, because for free vibration f is harmonic and of frequency ω, (5.172) to (5.174) reduce to the differential equation

$$-T\mathbf{V}^2 W = \omega^2 \rho W, \tag{5.176}$$

to be satisfied throughout domain D and boundary conditions

$$W = 0 \qquad \text{at } (a_1, b_1), \tag{5.177}$$

$$T\frac{\partial W}{\partial n} = 0 \qquad \text{at } (a_2, b_2). \tag{5.178}$$

The curves S contain all points (a_1, b_1) and (a_2, b_2), and only one of the boundary conditions (5.177) and (5.178) must be satisfied at any boundary point.

The eigenvalue problem is recognized as an eigenvalue problem of the special type, because

$$L = -T\nabla^2, \quad M = \rho. \tag{5.179}$$

The boundary condition operators become

$$B_1 = 1 \quad \text{or} \quad B_1 = T\frac{\partial}{\partial n} \quad \text{on } S. \tag{5.180}$$

Now let us check whether the problem is self-adjoint. To this end consider two arbitrary comparison functions u and v satisfying either one of boundary conditions (5.177) and (5.178) at any point of boundaries S. Recall first from vector analysis that the Laplacian ∇^2 is the divergence of a gradient, $\nabla^2 = \nabla \cdot \nabla$, and, because

$$\nabla \cdot (\varphi \nabla \psi) = \varphi \nabla^2 \psi + \nabla \varphi \cdot \nabla \psi, \tag{5.181}$$

we can let $\varphi = u$ and $\psi = v$ and obtain

$$u\nabla^2 v = u\nabla \cdot \nabla v = \nabla \cdot (u\nabla v) - \nabla u \cdot \nabla v. \tag{5.182}$$

The divergence theorem can be written

$$\int_D \nabla \cdot A \, dD = \int_S A_n \, dS, \tag{5.183}$$

where A_n is the component of the vector A along the exterior normal to boundary S of domain D. Letting

$$A = u\nabla v, \tag{5.184}$$

it follows that

$$A_n = u\frac{\partial v}{\partial n}. \tag{5.185}$$

Equations (5.182) to (5.185) can be used to write

$$\int_D uL[v] \, dD = -T\int_D u\nabla^2 v \, dD = -T\int_D [\nabla \cdot (u\nabla v) - \nabla u \cdot \nabla v] \, dD$$

$$= -T\left[\int_S u\frac{\partial v}{\partial n} \, dS - \int_D \nabla u \cdot \nabla v \, dD\right]. \tag{5.186}$$

Similarly,

$$\int_D vL[u] \, dD = -T\left[\int_S v\frac{\partial u}{\partial n} \, dS - \int_D \nabla u \cdot \nabla v \, dD\right], \tag{5.187}$$

and it follows that

$$\int_D (uL[v] - vL[u])\, dD = -T \int_S \left(u \frac{\partial v}{\partial n} - v \frac{\partial u}{\partial n} \right) dS. \qquad (5.188)$$

If u and v satisfy the boundary condition†

$$aW + b \frac{\partial W}{\partial n} = 0 \qquad \text{on } S, \qquad (5.189)$$

where a and b may depend on the spatial variables but not on the eigen-values of the problem, the line integral in (5.188) reduces to zero and the problem is self-adjoint, because, obviously, M is self-adjoint.

Similarly, it can be shown that if the boundary condition (5.177) is satisfied over part of curve S, the system is positive definite, and if the boundary condition (5.178) is satisfied over the entire length of S, the system is only positive.

If W_r and W_s are two eigenfunctions of the self-adjoint problem corresponding to two distinct eigenvalues, $\lambda_r \neq \lambda_s$, then they satisfy the orthogonality relation

$$\int_D \rho W_r W_s\, dD = 0, \qquad r \neq s. \qquad (5.190)$$

In the above discussion we carefully avoided reference to any particular set of coordinates. The deflection w can be given in terms of rectangular coordinates and time or curvilinear coordinates and time. Accordingly, the Laplacian ∇^2 can be expressed in terms of rectangular or curvilinear coordinates. In effect, the shape of the boundary curves dictates the choice of coordinates, because the only way we can deal effectively with boundary conditions is to formulate the problem in the coordinates corresponding to the shape of the boundaries. In fact, there are only a few shapes of boundaries that lend themselves to closed form solutions. We shall confine ourselves to a discussion of rectangular and circular membranes.

(a) Rectangular Membranes

Consider a rectangular membrane extending over a domain D defined by $0 < x < a$ and $0 < y < b$. The boundaries of the domain are the straight lines $x = 0, a$ and $y = 0, b$. Let

$$\frac{T}{\rho} = c^2, \qquad (5.191)$$

---

† Note that the boundary condition (5.189) contains the boundary conditions (5.177) and (5.178) as limiting cases as $b \to 0$ and $a \to 0$, respectively.

where c is known as the wave propagation velocity (discussed in a later chapter), and write (5.176)

$$\nabla^2 W(x, y) + \beta^2 W(x, y) = 0, \qquad \beta^2 = \left(\frac{\omega}{c}\right)^2, \tag{5.192}$$

where the Laplacian, in rectangular coordinates, has the form

$$\nabla^2 = \frac{\partial^2}{\partial x^2} + \frac{\partial^2}{\partial y^2}. \tag{5.193}$$

If the membrane is *clamped* at the boundaries, the boundary conditions are all of the geometric type, and we have the expressions

$$W(x, y) = 0 \qquad \text{along } x = 0,a, \tag{5.194}$$

$$W(x, y) = 0 \qquad \text{along } y = 0,b. \tag{5.195}$$

Equation (5.192), which must be satisfied through domain D, together with boundary conditions (5.194) and (5.195), constitute the eigenvalue problem.

Equation (5.192) can be solved by the method of separation of variables. To this end, let the solution have the form

$$W(x, y) = X(x) Y(y), \tag{5.196}$$

and upon substitution in (5.192) we obtain

$$\frac{d^2 X(x)}{dx^2} Y(y) + X(x) \frac{d^2 Y(y)}{dy^2} + \beta^2 X(x) Y(y) = 0,$$

which can be divided through by $X(x) Y(y)$ to yield

$$\frac{1}{X(x)} \frac{d^2 X(x)}{dx^2} + \frac{1}{Y(y)} \frac{d^2 Y(y)}{dy^2} + \beta^2 = 0. \tag{5.197}$$

This leads to the equations

$$\frac{d^2 X(x)}{dx^2} + \alpha^2 X(x) = 0, \tag{5.198}$$

$$\frac{d^2 Y(y)}{dy^2} + \gamma^2 Y(y) = 0, \tag{5.199}$$

where

$$\alpha^2 + \gamma^2 = \beta^2. \tag{5.200}$$

The solution of (5.198) is

$$X(x) = C_1 \sin \alpha x + C_2 \cos \alpha x, \tag{5.201}$$

and the solution of (5.199) is

$$Y(y) = C_3 \sin \gamma y + C_4 \cos \gamma y, \tag{5.202}$$

so that

$$W(x, y) = A_1 \sin \alpha x \sin \gamma y + A_2 \sin \alpha x \cos \gamma y$$
$$+ A_3 \cos \alpha x \sin \gamma y + A_4 \cos \alpha x \cos \gamma y, \tag{5.203}$$

where A_1, A_2, A_3, and A_4, as well as α and γ, must be determined by means of the boundary conditions.

The first of boundary conditions (5.194) gives

$$A_3 \sin \gamma y + A_4 \cos \gamma y = 0, \tag{5.204}$$

which can hold true for any y, assuming $\gamma \neq 0$, only if A_3 and A_4 are zero. Assuming A_3 and A_4 are zero, the second of boundary conditions (5.194) requires that

$$A_1 \sin \alpha a \sin \gamma y + A_2 \sin \alpha a \cos \gamma y = 0, \tag{5.205}$$

which can be satisfied if A_1 and A_2 are zero. This would give us the trivial solution $W(x, y) = 0$, however, so that we must consider the other possibility,

$$\sin \alpha a = 0. \tag{5.206}$$

Similarly, the first of boundary conditions (5.195) leads us to the conclusion that $A_2 = A_4 = 0$, whereas the second of boundary conditions (5.195) gives a second condition,

$$\sin \gamma b = 0. \tag{5.207}$$

Equations (5.206) and (5.207) play the role of characteristic or frequency equations, because together they define the eigenvalues of the system. Indeed, (5.206) yields an infinite sequence of discrete roots

$$\alpha_m a = m\pi, \qquad m = 1, 2, \ldots, \tag{5.208}$$

and (5.207) gives another infinite sequence of roots,

$$\gamma_n b = n\pi, \qquad n = 1, 2, \ldots. \tag{5.209}$$

It follows that the solution of the eigenvalue problem consists of the eigenvalues

$$\beta_{mn} = \sqrt{\alpha_m^2 + \gamma_n^2} = \pi \sqrt{\left(\frac{m}{a}\right)^2 + \left(\frac{n}{b}\right)^2}, \qquad m, n = 1, 2, \ldots \tag{5.210}$$

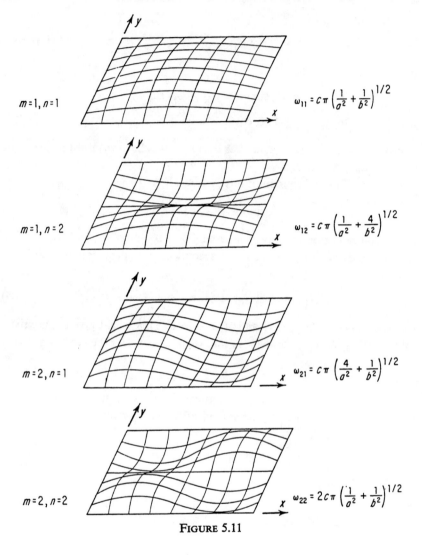

$$\omega_{11} = c\pi \left(\frac{1}{a^2} + \frac{1}{b^2}\right)^{1/2}$$

$$\omega_{12} = c\pi \left(\frac{1}{a^2} + \frac{4}{b^2}\right)^{1/2}$$

$$\omega_{21} = c\pi \left(\frac{4}{a^2} + \frac{1}{b^2}\right)^{1/2}$$

$$\omega_{22} = 2c\pi \left(\frac{1}{a^2} + \frac{1}{b^2}\right)^{1/2}$$

FIGURE 5.11

and the corresponding eigenfunctions

$$W_{mn} = A_{mn} \sin\frac{m\pi x}{a} \sin\frac{n\pi y}{b}, \qquad m, n = 1, 2, \ldots. \qquad (5.211)$$

On the basis of earlier results one concludes that the problem, as formulated here, is self-adjoint, and it can be readily demonstrated that it is also positive definite. The orthogonality relation follows from (5.190),

$$\int_0^a \int_0^b \rho W_{mn}(x, y) W_{rs}(x, y)\, dx\, dy = 0, \qquad m \neq r \text{ and/or } n \neq s. \qquad (5.212)$$

The first four eigenfunctions are plotted in Figure 5.11. The nodes are straight lines; the number of nodes parallel to the x axis is $n - 1$ and the number parallel to the y axis is $m - 1$.

We note that some of the natural frequencies are integral multiples of the fundamental frequency, $\omega_{mm} = m\omega_{11}$, but not all higher frequencies are integral multiples of the fundamental frequency. For example, ω_{12} is not an integral multiple of ω_{11}. Hence the sounds produced by vibrating membranes are not as pleasant as the sounds produced by strings or any other system with harmonic overtones.

In the special case in which the ratio $R = a/b$ is a rational number, we have repeated eigenvalues $\omega_{mn} = \omega_{rs}$ if

$$m^2 + R^2 n^2 = r^2 + R^2 s^2. \tag{5.213}$$

For a ratio $R = 4/3$, we note that $\omega_{35} = \omega_{54}$, $\omega_{83} = \omega_{46}$, etc.

For a square membrane, $a = b$, (5.213) reduces to

$$m^2 + n^2 = r^2 + s^2, \tag{5.214}$$

in which case we obtain repeated frequencies $\omega_{mn} = \omega_{nm}$. Hence two distinct eigenfunctions W_{mn} and W_{nm} correspond to the same frequency, so there are fewer frequencies than modes. Such a case is called a *degenerate* case. Any-time we obtain repeated eigenvalues any linear combination of the corresponding natural modes is also a natural mode. For these cases a large variety of nodal patterns occur. For the square membrane the nodal lines are no longer straight lines except in special cases. The reader who wishes to pursue this subject further is referred to the text by Courant and Hilbert.†

(b) Circular Membranes

Let us now consider a circular membrane extending over a domain D defined by $0 < r < a$. The boundary of the domain is the circle S given by the equation $r = a$.

Let r and θ be polar coordinates, so that the differential equation to be satisfied throughout domain D is

$$\nabla^2 W(r, \theta) + \beta^2 W(r, \theta) = 0, \qquad \beta^2 = \left(\frac{\omega}{c}\right)^2, \tag{5.215}$$

† R. Courant and D. Hilbert, *Methods of Mathematical Physics*, Vol. 1, Interscience, New York, 1961, p. 302.

where the Laplacian, in polar coordinates, is given by

$$\nabla^2 = \frac{\partial^2}{\partial r^2} + \frac{1}{r}\frac{\partial}{\partial r} + \frac{1}{r^2}\frac{\partial^2}{\partial \theta^2}. \qquad (5.216)$$

We can obtain a general solution of (5.215) without reference to boundary conditions, and we recall that to render this solution unique we must specify and satisfy the boundary conditions of the problem. Assuming a solution of the form

$$W(r, \theta) = R(r)\Theta(\theta), \qquad (5.217)$$

(5.125) reduces to

$$\left(\frac{d^2R}{dr^2} + \frac{1}{r}\frac{dR}{dr}\right)\Theta + \frac{R}{r^2}\frac{d^2\Theta}{d\theta^2} + \beta^2 R\Theta = 0,$$

which can be separated into two equations,

$$\frac{d^2\Theta}{d\theta^2} + m^2\Theta = 0, \qquad (5.218)$$

$$\frac{d^2R}{dr^2} + \frac{1}{r}\frac{dR}{dr} + \left(\beta^2 - \frac{m^2}{r^2}\right)R = 0, \qquad (5.219)$$

where the constant m^2 has been chosen to obtain a harmonic equation in θ. Furthermore, because the solution of (5.218) must be continuous, implying that the solution for $\theta = \theta_0$ must be identical to the solution for $\theta = \theta_0 + j2\pi$ ($j = 1, 2, \ldots$), for any value θ_0, m must be an integer. Equation (5.218) therefore yields trigonometric functions as solutions:

$$\Theta_m(\theta) = C_{1m}\sin m\theta + C_{2m}\cos m\theta, \qquad m = 0, 1, 2, \ldots. \qquad (5.220)$$

Equation (5.219), on the other hand, is a Bessel equation and its solution is

$$R_m(r) = C_{3m}J_m(\beta r) + C_{4m}Y_m(\beta r), \qquad m = 0, 1, 2, \ldots, \qquad (5.221)$$

where $J_m(\beta r)$ and $Y_m(\beta r)$ are Bessel functions of order m and of the first and second kind, respectively. The general solution can be written

$$W_m(r, \theta) = A_{1m}J_m(\beta r)\sin m\theta + A_{2m}J_m(\beta r)\cos m\theta$$

$$+ A_{3m}Y_m(\beta r)\sin m\theta + A_{4m}Y_m(\beta r)\cos m\theta,$$

$$m = 0, 1, 2, \ldots. \qquad (5.222)$$

Now consider a continuous membrane *fixed* at the boundary $r = a$, which implies that the deflection at the rim is zero. Hence the boundary condition at $r = a$ is

$$W_m(a, \theta) = 0, \qquad m = 0, 1, 2, \ldots. \tag{5.223}$$

At every interior point of the membrane the displacement must be finite. But Bessel functions of the second kind tend to infinity as the argument approaches zero. It follows that $A_{3m} = A_{4m} = 0$, so (5.222) reduces to

$$W_m(r, \theta) = A_{1m}J_m(\beta r) \sin m\theta + A_{2m}J_m(\beta r) \cos m\theta,$$

$$m = 0, 1, 2, \ldots. \tag{5.224}$$

At $r = a$, however, we have

$$W_m(a, \theta) = A_{1m}J_m(\beta a) \sin m\theta + A_{2m}J_m(\beta a) \cos m\theta = 0,$$

$$m = 0, 1, 2, \ldots,$$

regardless of the value of θ. This can be satisfied only if

$$J_m(\beta a) = 0, \qquad m = 0, 1, 2, \ldots. \tag{5.225}$$

Equation (5.225) represents an infinite set of characteristic or frequency equations, and for each m there is an infinite number of discrete solutions β_{mn} which correspond to the zeros of the Bessel functions J_m. As an illustration the Bessel functions of zero and first order are plotted in Figure 5.12. The intersections with the x axis provide the roots $\beta_{mn}a$, from which we obtain the natural frequencies $\omega_{mn} = c\beta_{mn}$. For each frequency ω_{mn} there are two modes,

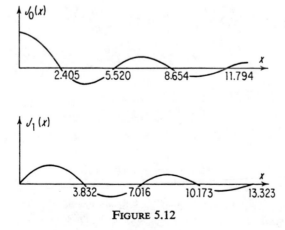

FIGURE 5.12

except when $m = 0$, for which we obtain only one mode. It follows that for $m \neq 0$ the natural modes are degenerate. The modes can be written

$$W_{0n}(r, \theta) = A_{0n} J_0\left(\frac{\omega_{0n}}{c} r\right), \qquad n = 1, 2, \ldots, \tag{5.226}$$

$$\left.\begin{array}{l} W_{mnc}(r, \theta) = A_{mnc} J_m\left(\frac{\omega_{mn}}{c} r\right) \cos m\theta \\[3mm] W_{mns}(r, \theta) = A_{mns} J_m\left(\frac{\omega_{mn}}{c} r\right) \sin m\theta \end{array}\right\} \quad m, n = 1, 2, \ldots. \tag{5.227}$$

The problem is self-adjoint and positive definite. The natural modes are orthogonal, and from (5.190) we can write the orthogonality relation

$$\int_D \rho W_{mn} W_{pq} \, dD = \int_0^{2\pi} \int_0^a \rho W_{mn} W_{pq} r \, dr \, d\theta = 0, \quad m \neq p \text{ and/or } n \neq q. \tag{5.228}$$

The natural modes can be normalized by writing†

$$\int_D \rho W_{0n}^2 \, dD = \int_0^{2\pi} \int_0^a \rho A_{0n}^2 J_0^2\left(\frac{\omega_{0n}}{c} r\right) r \, dr \, d\theta$$

$$= \pi \rho a^2 A_{0n}^2 J_1^2\left(\frac{\omega_{0n}}{c} a\right) = 1,$$

so that

$$A_{0n}^2 = \frac{1}{\pi \rho a^2 J_1^2[(\omega_{0n}/c)a]}. \tag{5.229}$$

Also

$$\int_D \rho W_{mnc}^2 \, dD = \int_0^{2\pi} \int_0^a \rho A_{mnc}^2 J_m^2\left(\frac{\omega_{mn}}{c} r\right) \cos^2 m\theta r \, dr \, d\theta$$

$$= \frac{\pi}{2} \rho A_{mnc}^2 a^2 J_{m+1}^2\left(\frac{\omega_{mn}}{c} a\right) = 1$$

or

$$A_{mnc}^2 = \frac{2}{\pi \rho a^2 J_{m+1}^2[(\omega_{mn}/c)a]}. \tag{5.230}$$

Similarly,

$$A_{mns}^2 = \frac{2}{\pi \rho a^2 J_{m+1}^2[(\omega_{mn}/c)a]}. \tag{5.231}$$

† For integrals and formulas concerning Bessel functions see N. W. McLachlan, *Bessel Functions for Engineers*, 2nd ed., Oxford Univ. Press, New York, 1961, p. 190.

The normal modes take the form

$$W_{0n}(r,\,\theta) = \frac{1}{\sqrt{\pi\rho}\,aJ_1[(\omega_{0n}/c)a]}\,J_0\!\left(\frac{\omega_{0n}}{c}\,r\right), \qquad n = 1, 2, \ldots, \tag{5.232}$$

$$\left.\begin{aligned}W_{mnc}(r,\,\theta)\\ W_{mns}(r,\,\theta)\end{aligned}\right\} = \frac{\sqrt{2}}{\sqrt{\pi\rho}\,aJ_{m+1}[(\omega_{mn}/c)a]}\,J_m\!\left(\frac{\omega_{mn}}{c}\,r\right)\left.\begin{aligned}\cos m\theta\\ \sin m\theta\end{aligned}\right\} \quad m, n = 1, 2, \ldots. \tag{5.233}$$

The nodal lines are circles $r = $ const and straight diametrical lines $\theta = $ const. For $m = 0$ there are no diametrical nodes and there are $n - 1$ circular

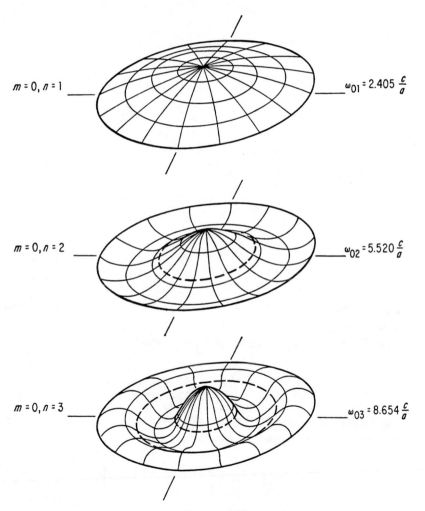

$m = 0, n = 1$ $\omega_{01} = 2.405\,\frac{c}{a}$

$m = 0, n = 2$ $\omega_{02} = 5.520\,\frac{c}{a}$

$m = 0, n = 3$ $\omega_{03} = 8.654\,\frac{c}{a}$

FIGURE 5.13

nodes. The first three modes W_{0n} are plotted in Figure 5.13. For $m = 1$ there is just one diametrical node and $n - 1$ circular nodes. The first two modes W_{1nc} are plotted in Figure 5.14. In general the mode W_{mn} has m equally

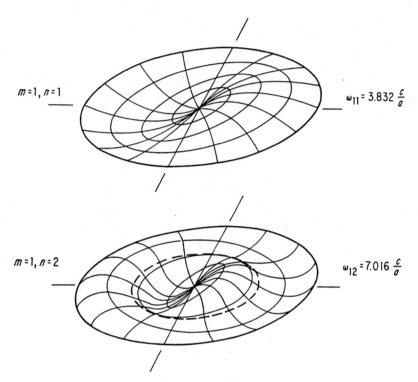

$m = 1, n = 1$ $\omega_{11} = 3.832 \frac{c}{a}$

$m = 1, n = 2$ $\omega_{12} = 7.016 \frac{c}{a}$

FIGURE 5.14

spaced diametrical nodes and $n - 1$ circular nodes (the boundary is excluded) of radius $r_i = (\beta_{mi}/\beta_{mn})a$ $(i = 1, 2, \ldots, n - 1)$.

Note that for very large argument we have the relation

$$\lim_{z \to \infty} J_m(z) = \sqrt{\frac{2}{\pi z}} \cos\left(z - \frac{2m + 1}{4}\pi\right), \tag{5.234}$$

so the frequency equation (5.225) leads us to the conclusion that the very large natural frequencies can be approximated by

$$\omega_{mn} = \left(\frac{m}{2} + n - \frac{1}{4}\right)\frac{c\pi}{a} \qquad \text{for large } n, \tag{5.235}$$

where both m and n are integers.

5–12 VIBRATION OF PLATES

In contrast to membranes, plates have bending stiffness in the same manner as bars in bending. There is one distinction to be made between bars and plates in bending, however. The bar is essentially a one-dimensional system. When it bends a portion of the cross section undergoes tension and the remaining portion undergoes compression, with the neutral axis the dividing line between the two regions. The part in tension will tend to contract laterally and the part in compression will tend to expand laterally. As long as the width of the bar is small, this lateral contraction and expansion is free to take place and there are no lateral stresses. In the elementary theory of bars the width is assumed small and the lateral deformation effect is ignored. As the width of the bar increases, this effect tends to bend the cross section, so that a curvature is produced in the plane of the cross section in addition to the curvature in the plane of bending. Let us now consider a plate and imagine for the moment that the plate is made up of individual parallel bars obtained by dividing the plate by means of vertical planes. We can imagine two such adjacent bars which in the undeformed position have a common lateral surface that is part of the dividing vertical plane. When the plate begins to bend these two adjacent elements, if allowed to behave like bars in bending, would expand and contract laterally, so that in expanding each element would cross the dividing surface and occupy space belonging to the adjacent element; in contracting each element would pull away from the dividing surface, which would result in the creation of a void in the material. This situation is not possible, so internal lateral stresses must arise to prevent it from happening. Furthermore, in the case of plates one can think of two planes of bending producing, in general, two distinct curvatures. In addition to bending there is also a twist present, because an element of area of a plate can be thought as belonging to two orthogonal strips and, as such, bending in one plane can be looked upon as rotation about the normal to the plane of bending. For an adjacent strip this rotation may be different, causing the orthogonal strip to twist. The elementary theory of plates is based on the following assumptions:

1. The deflection, w, of a plate is small when compared with the plate thickness h.

2. The normal stresses in the direction transverse to the plate can be ignored.

3. There is no force resultant on the cross-sectional area of a plate element. The middle plane of a plate does not undergo deformation during bending and can be regarded as a neutral plane.

4. Any straight line normal to the middle plane before deformation remains a straight line normal to the neutral plane during deformation.

These assumptions are reasonable for a relatively thin plate with no forces acting in the middle plane.

The differential equation for plate vibration can be found in many texts on the theory of elasticity. The equation for a plate of uniform thickness subjected to the external transverse load per unit area, p, has the form

$$-D_E \nabla^4 w + p = \rho \frac{\partial^2 w}{\partial t^2}, \qquad D_E = \frac{Eh^3}{12(1 - \nu^2)}, \qquad (5.236)$$

where the operator $\nabla^4 = \nabla^2 \nabla^2$ is known as the *biharmonic operator* and D_E denotes the plate flexural rigidity. The symbol ν denotes Poisson's ratio and is equal to the ratio between the lateral tensile (or compressive) strain and the longitudinal compressive (or tensile) strain in a simple tension test. Poisson's ratio depends on the elastic properties of the material and enters into the equation because of the fact that the system does not act as individual bars, independent of each other, but as a plate, as discussed previously.

In the case of free vibration, $p = 0$, (5.236) reduces to

$$-D_E \nabla^4 w = \rho \frac{\partial^2 w}{\partial t^2}, \qquad (5.237)$$

which must be satisfied throughout the area of the plate. In addition, there are two boundary conditions to be satisfied at any point on the boundary. Denote by n and s the coordinates in the directions normal and tangential to the boundary, so that for a *clamped edge* we have

$$w = 0 \quad \text{and} \quad \frac{\partial w}{\partial n} = 0, \qquad (5.238)$$

which are both geometric boundary conditions. At a *simply supported edge* the boundary conditions are

$$w = 0 \quad \text{and} \quad M_n = 0, \qquad (5.239)$$

where M_n represents the bending moment per unit length associated with the cross section whose normal is n. The first of boundary conditions (5.239) is geometric and the second is a natural boundary condition. In the case of a *free edge* the boundary conditions are

$$M_n = 0 \quad \text{and} \quad V_n = Q_n - \frac{\partial M_{ns}}{\partial s} = 0, \qquad (5.240)$$

which are both natural boundary conditions. In (5.240) V_n denotes the vertical force, Q_n is the shearing force, and M_{ns} is the twisting moment about the direction n. All three quantities V_n, Q_n, and M_{ns} are for one unit length of boundary and are associated with the cross section whose normal is n. The second of boundary conditions (5.240) is associated with the name of Kirchhoff and is of some historical interest. Poisson believed that M_n, Q_n, and M_{ns} must be independently zero at a free boundary, yielding a total number of three boundary conditions, one too many for a fourth order differential equation. Later, however, Kirchhoff cleared up the problem by pointing out that Q_n and M_{ns} are related as indicated in the second of equations (5.240). If variational principles are used to formulate the problem, the right number of boundary conditions emerge in their correct form.

The relations between the moments and shearing forces and deformations, in terms of normal and tangential coordinates, are

$$M_n = -D_E \nabla^2 w + (1 - \nu)D_E\left(\frac{1}{R}\frac{\partial w}{\partial n} + \frac{\partial^2 w}{\partial s^2}\right), \qquad (5.241)$$

$$M_{ns} = (1 - \nu)D_E\left(\frac{\partial^2 w}{\partial n\, \partial s} - \frac{1}{R}\frac{\partial w}{\partial s}\right), \qquad (5.242)$$

$$Q_n = -D_E \frac{\partial}{\partial n}\nabla^2 w, \qquad (5.243)$$

where the Laplacian has the form

$$\nabla^2 = \frac{\partial^2}{\partial n^2} + \frac{1}{R}\frac{\partial}{\partial n} + \frac{\partial^2}{\partial s^2} \qquad (5.244)$$

and R denotes the radius of curvature of the boundary curve.

To formulate the eigenvalue problem we let w be given by

$$w = Wf, \qquad (5.245)$$

where W depends on the spatial coordinates only and f is a time-dependent harmonic function of frequency ω. With this in mind, (5.237) reduces to

$$D_E \nabla^4 W = \omega^2 \rho W. \qquad (5.246)$$

The boundary conditions for a clamped edge reduce to

$$W = 0 \quad \text{and} \quad \frac{\partial W}{\partial n} = 0. \qquad (5.247)$$

The boundary conditions for a simply supported edge become

$$W = 0 \quad \text{and} \quad \nabla^2 W - (1 - \nu)\left(\frac{1}{R}\frac{\partial W}{\partial n} + \frac{\partial^2 W}{\partial s^2}\right) = 0,$$

and since W is constant along the edge, the above assumes the simplified form

$$W = 0 \quad \text{and} \quad \frac{\partial^2 W}{\partial n^2} + \frac{\nu}{R}\frac{\partial W}{\partial n} = 0. \tag{5.248}$$

Along a free edge the boundary conditions are

$$\nabla^2 W - (1 - \nu)\left(\frac{1}{R}\frac{\partial W}{\partial n} + \frac{\partial^2 W}{\partial s^2}\right) = 0,$$

$$\frac{\partial}{\partial n}\nabla^2 W + (1 - \nu)\frac{\partial}{\partial s}\left(\frac{\partial^2 W}{\partial n\,\partial s} - \frac{1}{R}\frac{\partial W}{\partial s}\right) = 0. \tag{5.249}$$

Again one must recognize that the eigenvalue problem for the transverse vibration of a uniform plate follows the pattern of the general formulation of Section 5–4. In this case the linear operators are

$$L = D_E \nabla^4, \qquad M = \rho, \tag{5.250}$$

and it follows that the eigenvalue problem is of the special type.

From vector analysis we recall the following results:

$$\nabla \cdot (u\mathbf{F}) = \nabla u \cdot \mathbf{F} + u\nabla \cdot \mathbf{F},$$

$$\nabla \cdot (\Phi \nabla u) = \Phi \nabla^2 u + \nabla \Phi \cdot \nabla u, \tag{5.251}$$

where u and Φ are scalar functions and $\mathbf{F}$ is a vector function. Letting $\mathbf{F} = \nabla \nabla^2 v$, we can write

$$u\nabla^4 v = u\nabla^2 \nabla^2 v = u\nabla \cdot \nabla \nabla^2 v$$

$$= \nabla \cdot (u\nabla \nabla^2 v) - \nabla u \cdot \nabla \nabla^2 v, \tag{5.252}$$

and further letting $\Phi = \nabla^2 v$, we obtain

$$\nabla u \cdot \nabla \nabla^2 v = \nabla \cdot (\nabla^2 v \nabla u) - \nabla^2 u \nabla^2 v. \tag{5.253}$$

It follows that

$$u\nabla^4 v = \nabla \cdot (u\nabla \nabla^2 v) - \nabla \cdot (\nabla^2 v \nabla u) + \nabla^2 u \nabla^2 v, \tag{5.254}$$

and, similarly,

$$v\nabla^4 u = \nabla \cdot (v\nabla \nabla^2 u) - \nabla \cdot (\nabla^2 u \nabla v) + \nabla^2 v \nabla^2 u. \tag{5.255}$$

We now form the expression

$$\int_D (uL[v] - vL[u]) \, dD = D_E \int_D (u\nabla^4 v - v\nabla^4 u) \, dD$$

$$= D_E \int_D [\nabla \cdot (u\nabla\nabla^2 v) - \nabla \cdot (v\nabla\nabla^2 u) - \nabla \cdot (\nabla^2 v \nabla u)$$

$$+ \nabla \cdot (\nabla^2 u \nabla v)] \, dD. \qquad (5.256)$$

Again using the divergence theorem, (5.183), the integral over domain D can be transformed into an integral over boundary S,

$$\int_D (uL[v] - vL[u]) \, dD = D_E \int_S \left[u \frac{\partial}{\partial n} (\nabla^2 v) - v \frac{\partial}{\partial n} (\nabla^2 u) \right.$$

$$\left. - \nabla^2 v \frac{\partial u}{\partial n} + \nabla^2 u \frac{\partial v}{\partial n} \right] dS. \quad (5.257)$$

The eigenvalue problem is self-adjoint if for any two arbitrary comparison functions u and v the integral over the boundary in (5.257) vanishes.

As in the case of vibration of thin membranes, the shape of the boundary dictates the type of coordinates to be used. For plate vibrations, however, the satisfaction of the boundary conditions turns out to be a much more formidable obstacle than for the membrane.

Only rectangular and circular plates will be discussed here.

(a) Rectangular Plates

Consider a uniform rectangular plate extending over a domain D defined by $0 < x < a$ and $0 < y < b$. The boundaries of the domain are the straight lines $x = 0, a$ and $y = 0, b$. Equation (5.246), in rectangular coordinates, takes the form

$$\nabla^4 W(x, y) - \beta^4 W(x, y) = 0, \qquad \beta^4 = \frac{\omega^2 \rho}{D_E}, \qquad (5.258)$$

where the biharmonic operator, in rectangular coordinates, is given by

$$\nabla^4 = \nabla^2 \nabla^2 = \frac{\partial^4}{\partial x^4} + 2 \frac{\partial^4}{\partial x^2 \, \partial y^2} + \frac{\partial^4}{\partial y^4}. \qquad (5.259)$$

Equation (5.258) can be written in the operator form

$$(\nabla^4 - \beta^4) W(x, y) = (\nabla^2 + \beta^2)(\nabla^2 - \beta^2) W(x, y) = 0, \qquad (5.260)$$

which leads to

$$(\nabla^2 - \beta^2)W = W_1, \qquad (\nabla^2 + \beta^2)W_1 = 0. \tag{5.261}$$

Because β^2 is constant, the solution of the first of (5.261) is†

$$W = W_1 + W_2, \tag{5.262}$$

where W_2 is the solution of the homogeneous equation

$$(\nabla^2 - \beta^2)W_2 = [\nabla^2 + (i\beta)^2]W_2 = 0. \tag{5.263}$$

The second of equations (5.261) resembles the equation for the vibration of a thin uniform membrane whose general solution was obtained in Section 5–11. The solution of (5.263) has the same form as the solution of the second of equations (5.261), except that β is replaced by $i\beta$. Hence the general solution of (5.258) is

$$W(x, y) = A_1 \sin \alpha x \sin \gamma y + A_2 \sin \alpha x \cos \gamma y + A_3 \cos \alpha x \sin \gamma y$$

$$+ A_4 \cos \alpha x \cos \gamma y + A_5 \sinh \alpha_1 x \sinh \gamma_1 y$$

$$+ A_6 \sinh \alpha_1 x \cosh \gamma_1 y + A_7 \cosh \alpha_1 x \sinh \gamma_1 y$$

$$+ A_8 \cosh \alpha_1 x \cosh \gamma_1 y, \qquad \alpha^2 + \gamma^2 = \alpha_1^2 + \gamma_1^2 = \beta^2.$$

$$\tag{5.264}$$

Now consider a *simply supported* plate, for which the general boundary conditions (5.248) assume the form

$$W = 0 \qquad \text{and} \qquad \frac{\partial^2 W}{\partial n^2} = 0, \tag{5.265}$$

because for a straight boundary the radius of curvature of the boundary curve is infinite, $R \to \infty$. It follows that the boundary conditions are

$$W = 0 \quad \text{and} \quad \frac{\partial^2 W}{\partial x^2} = 0 \qquad \text{along } x = 0,a, \tag{5.266}$$

$$W = 0 \quad \text{and} \quad \frac{\partial^2 W}{\partial y^2} = 0 \qquad \text{along } y = 0,b. \tag{5.267}$$

---

† Note that a proportionality factor multiplying W_1 was discarded. Since W_1 is obtained by solving a homogeneous equation, the second of (5.261), such a factor is unnecessary.

When the above boundary conditions are used, we conclude that all the coefficients A_i, with the exception of A_1, vanish, and in addition we obtain two equations which α and γ must satisfy,

$$\sin \alpha a = 0, \qquad \sin \gamma b = 0; \tag{5.268}$$

these can be regarded as characteristic equations, because together they define the characteristic values of the system. Their solutions are

$$\alpha_m a = m\pi, \qquad m = 1, 2, \ldots,$$
$$\gamma_n b = n\pi, \qquad n = 1, 2, \ldots, \tag{5.269}$$

so the natural frequencies of the system become

$$\omega_{mn} = \beta_{mn}^2 \sqrt{\frac{D_E}{\rho}} = \pi^2 \left[\left(\frac{m}{a} \right)^2 + \left(\frac{n}{b} \right)^2 \right] \sqrt{\frac{D_E}{\rho}}, \qquad m, n = 1, 2, \ldots, \tag{5.270}$$

and the corresponding natural modes are

$$W_{mn}(x, y) = A_{mn} \sin \frac{m\pi x}{a} \sin \frac{n\pi y}{b}, \qquad m, n = 1, 2, \ldots, \tag{5.271}$$

which resemble the modes of the clamped rectangular membrane. The mode shapes are similar to the mode shapes of a clamped rectangular membrane but the natural frequencies are different.

It is easy to see that boundary conditions (5.266) and (5.267) render the line integral in (5.257) zero, so the eigenvalue problem is self-adjoint. It follows immediately that the eigenfunctions are orthogonal, as can readily be seen from (5.271).

(b) Circular Plates

Now consider a uniform circular plate extending over a domain D given by $0 < r < a$, where the boundary of the domain is the circle S defined by $r = a$. Because of the shape of the boundary we use the polar coordinates r and θ, so the differential equation to be satisfied over domain D is

$$\nabla^4 W(r, \theta) - \beta^4 W(r, \theta) = 0, \qquad \beta^4 = \frac{\omega^2 \rho}{D_E}, \tag{5.272}$$

where the biharmonic operator, in polar coordinates, has the form

$$\nabla^4 = \nabla^2 \nabla^2 = \left(\frac{\partial^2}{\partial r^2} + \frac{1}{r} \frac{\partial}{\partial r} + \frac{1}{r^2} \frac{\partial^2}{\partial \theta^2} \right) \left(\frac{\partial^2}{\partial r^2} + \frac{1}{r} \frac{\partial}{\partial r} + \frac{1}{r^2} \frac{\partial^2}{\partial \theta^2} \right). \tag{5.273}$$

Equation (5.272) can be written in operator form, which leads to

$$(\nabla^2 + \beta^2) W_1(r, \theta) = 0 \tag{5.274}$$

$$[\nabla^2 + (i\beta)^2] W_2(r, \theta) = 0, \tag{5.275}$$

which must be satisfied over the domain D. Equation (5.274) has precisely the form of the equation for the vibration of a thin uniform membrane (the definition of β is not the same as for the membrane), so its solution is given by (5.222). The solution of (5.275) is obtained from (5.222) by replacing β by $i\beta$. The Bessel functions of imaginary argument, $J_m(ix)$ and $Y_m(ix)$, are called *modified or hyperbolic Bessel functions* and denoted $I_m(x)$ and $K_m(x)$, respectively. The hyperbolic Bessel functions are not equal to but are proportional to the ordinary Bessel functions of imaginary argument.† Hence the solution of (5.272) is the sum of the solutions of (5.274) and (5.275) and has the form

$$W_{mn}(r, \theta) = [A_{1m}J_m(\beta r) + A_{3m}Y_m(\beta r) + B_{1m}I_m(\beta r) + B_{3m}K_m(\beta r)] \sin m\theta$$
$$+ [A_{2m}J_m(\beta r) + A_{4m}Y_m(\beta r) + B_{2m}I_m(\beta r) + B_{4m}K_m(\beta r)] \cos m\theta,$$
$$m = 0, 1, 2, \ldots, \tag{5.276}$$

which is subject to given boundary conditions.

As an example consider the case of a *clamped plate*, for which the boundary conditions are

$$W(a, \theta) = 0, \tag{5.277}$$

$$\left. \frac{\partial W(r, \theta)}{\partial r} \right|_{r=a} = 0. \tag{5.278}$$

In addition, at every interior point the solution must be finite. This immediately eliminates Bessel functions of the second kind, Y_m and K_m, which become infinite at $r = 0$. Hence the solution (5.276) reduces to

$$W_{mn}(r, \theta) = [A_{1m}J_m(\beta r) + B_{1m}I_m(\beta r)] \sin m\theta$$
$$+ [A_{2m}J_m(\beta r) + B_{2m}I_m(\beta r)] \cos m\theta,$$
$$m = 0, 1, 2, \ldots. \tag{5.279}$$

Boundary condition (5.277) yields

$$B_{1m} = -\frac{J_m(\beta a)}{I_m(\beta a)} A_{1m}, \qquad B_{2m} = -\frac{J_m(\beta a)}{I_m(\beta a)} A_{2m},$$
$$m = 0, 1, 2, \ldots, \tag{5.280}$$

† See N. W. McLachlan, *Bessel Functions for Engineers*, 2nd ed., Oxford Univ. Press, New York, 1961, p. 113.

so that

$$W_m(r, \theta) = \left[J_m(\beta r) - \frac{J_m(\beta a)}{I_m(\beta a)} I_m(\beta r) \right] (A_{1m} \sin m\theta + A_{2m} \cos m\theta),$$

$$m = 0, 1, 2, \ldots. \quad (5.281)$$

Boundary condition (5.278) leads to the set of characteristic equations

$$\left[\frac{d}{dr} J_m(\beta r) - \frac{J_m(\beta a)}{I_m(\beta a)} \frac{d}{dr} I_m(\beta r) \right]_{r=a} = 0, \qquad m = 0, 1, 2, \ldots. \quad (5.282)$$

But

$$\frac{d}{dr} J_m(\beta r) = \beta \left[J_{m-1}(\beta r) - \frac{m}{\beta r} J_m(\beta r) \right],$$

$$(5.283)$$

$$\frac{d}{dr} I_m(\beta r) = \beta \left[I_{m-1}(\beta r) - \frac{m}{\beta r} I_m(\beta r) \right],$$

so the set of characteristic equations reduces to

$$I_m(\beta a) J_{m-1}(\beta a) - J_m(\beta a) I_{m-1}(\beta a) = 0, \qquad m = 0, 1, 2, \ldots. \quad (5.284)$$

For a given m one must solve (5.284) numerically and obtain the eigen-values β_{mn} and subsequently the natural frequencies

$$\omega_{mn} = \beta_{mn}^2 \sqrt{\frac{D_E}{\rho}}. \quad (5.285)$$

For each frequency ω_{mn} there are two corresponding natural modes, except for $m = 0$, for which just one mode is obtained. Hence, as for the membrane, all modes for which $m \neq 0$ are degenerate. The natural modes can be written

$$W_{0n}(r, \theta) = A_{0n}[I_0(\beta_{0n}a)J_0(\beta_{0n}r) - J_0(\beta_{0n}a)I_0(\beta_{0n}r)],$$

$$n = 1, 2, \ldots, \quad (5.286)$$

$$\left. \begin{aligned} W_{mnc}(r, \theta) &= A_{mnc}[I_m(\beta_{mn}a)J_m(\beta_{mn}r) \\ &\quad - J_m(\beta_{mn}a)I_m(\beta_{mn}r)] \cos m\theta \\ W_{mns}(r, \theta) &= A_{mns}[I_m(\beta_{mn}a)J_m(\beta_{mn}r) \\ &\quad - J_m(\beta_{mn}a)I_m(\beta_{mn}r)] \sin m\theta \end{aligned} \right\} \quad m, n = 1, 2, \ldots. \quad (5.287)$$

For $m = 0$ there are no diametrical nodes and there are $n - 1$ circular nodes. The modes W_{01} and W_{02} are plotted in Figure 5.15. For $m = 1$ there

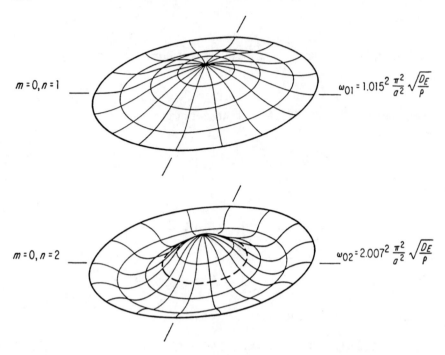

$m = 0, n = 1$ _____ $\omega_{01} = 1.015^2 \dfrac{\pi^2}{a^2} \sqrt{\dfrac{D_E}{\rho}}$

$m = 0, n = 2$ _____ $\omega_{02} = 2.007^2 \dfrac{\pi^2}{a^2} \sqrt{\dfrac{D_E}{\rho}}$

FIGURE 5.15

is one diametrical node and $n - 1$ circular nodes. The mode W_{11c} is plotted in Figure 5.16. Note that the overtones are not harmonic.

It is easy to see that boundary conditions (5.277) and (5.278) render the contour integral in (5.257) zero, so the problem is self-adjoint. Consequently the natural modes are orthogonal.

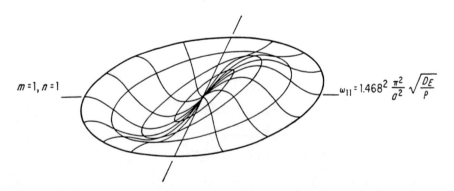

$m = 1, n = 1$ _____ $\omega_{11} = 1.468^2 \dfrac{\pi^2}{a^2} \sqrt{\dfrac{D_E}{\rho}}$

FIGURE 5.16

Defining the modified Bessel functions of the first kind by the relation $I_m(x) = i^{-m}J_m(ix)$ and using (5.234), it can be shown that, for large argument, the solutions of the characteristic equation (5.284) tend to

$$\beta_{mn} = \left(\frac{m}{2} + n\right)\frac{\pi}{2} \qquad \text{for large } n, \qquad (5.288)$$

and, consequently, the natural frequencies tend to

$$\omega_{mn} = \left(\frac{m}{2} + n\right)^2 \frac{\pi^2}{a^2}\sqrt{\frac{D_E}{\rho}} \qquad \text{for large } n. \qquad (5.289)$$

Because of the limitations of the elementary plate theory, however, this result is only of academic value.

5-13   ENCLOSURE THEOREM

Consider the special self-adjoint eigenvalue problem consisting of the differential equation

$$L[w] = \lambda M w, \qquad (5.290)$$

which must be satisfied by w throughout domain D. In (5.290) L is a linear homogeneous differential operator of order $2p$ and M is a function of the spatial variables. In addition, w must satisfy the conditions

$$B_i[w] = 0, \qquad i = 1, 2, \ldots, p \qquad (5.291)$$

at every point of the boundary, where B_i are linear homogeneous differential operators of order $2p - 1$, as defined previously. Now define

$$\frac{L[u]}{Mu} = \lambda^* \qquad (5.292)$$

such that if u is an eigenfunction, say w_r, then λ^* is the corresponding eigenvalue λ_r, hence a constant. If u is a comparison function satisfying all the boundary conditions of the problem, however, then λ^* will vary throughout domain D. Let λ^*_{max} and λ^*_{min} be the maximum and minimum values assumed by λ^* throughout domain D. Then, according to the *enclosure theorem*,† there is at least one eigenvalue λ, of the special eigenvalue problem, (5.290) and (5.291), such that

$$\lambda^*_{min} \le \lambda \le \lambda^*_{max}. \qquad (5.293)$$

† See L. Collatz, *Eigenwertprobleme*, Chelsea, New York, 1948, p. 135.

The enclosure theorem can be used as a quick check to obtain an estimate of the interval within which an eigenvalue lies. It is very similar to the theorem shown for discrete systems.

As an illustration consider the case of a uniform rod clamped at one end and free at the other end and vibrating longitudinally. In this case

$$L = -EA \frac{d^2}{dx^2}, \qquad M = \rho, \tag{5.294}$$

and the boundary conditions are

$$w(0) = \left.\frac{dw(x)}{dx}\right|_{x=L} = 0. \tag{5.295}$$

A comparison function satisfying the above boundary conditions can be taken in the form

$$w = C\left[\left(\frac{x}{L}\right)^3 - 3\frac{x}{L}\right], \tag{5.296}$$

so that

$$L[w] = -EA \frac{6C}{L^2} \frac{x}{L}, \tag{5.297}$$

and subsequently

$$\lambda^* = \frac{L[w]}{\rho w} = \frac{EA}{\rho L^2} \frac{6}{3 - (x/L)^2}. \tag{5.298}$$

It is easy to check that over the range $0 \leq x \leq L$,

$$\lambda^*_{min} = 2 \frac{EA}{\rho L^2}, \qquad \lambda^*_{max} = 3 \frac{EA}{\rho L^2}. \tag{5.299}$$

The first eigenvalue associated with the problem is

$$\lambda_1 = \left(\frac{\pi}{2}\right)^2 \frac{EA}{\rho L^2} = 2.4674 \frac{EA}{\rho L^2}, \tag{5.300}$$

and it is enclosed by the above two values.

5–14 RAYLEIGH'S QUOTIENT AND ITS PROPERTIES

The general eigenvalue problem for a continuous system was presented in Section 5–4, and it had the form

$$L[w] = \lambda M[w], \tag{5.301}$$

where L and M are linear homogeneous differential operators of orders $2p$ and $2q$, respectively, such that $p > q$. Any eigenvalue λ_r with the associated eigenfunction w_r must satisfy (5.301) throughout domain D and, of course, must satisfy the associated boundary conditions of the problem. Hence we can write

$$L[w_r] = \lambda_r M[w_r], \qquad r = 1, 2, \ldots. \tag{5.302}$$

Multiplying both sides of (5.302) by w_r, integrating over domain D, and dividing through by $\int_D w_r M[w_r]\, dD$, we obtain

$$\lambda_r = \frac{\displaystyle\int_D w_r L[w_r]\, dD}{\displaystyle\int_D w_r M[w_r]\, dD}, \qquad r = 1, 2, \ldots. \tag{5.303}$$

Consider a self-adjoint positive definite continuous system and let u be a comparison function, satisfying all the boundary conditions of the problem. Assume that the boundary conditions do not depend on the eigenvalue λ, and write

$$\omega^2 = R(u) = \frac{\displaystyle\int_D u L[u]\, dD}{\displaystyle\int_D u M[u]\, dD}, \tag{5.304}$$

where $R(u)$ is the expression of Rayleigh's quotient for continuous systems and is always positive for positive definite system. When the system is positive only, rather than positive definite, Rayleigh's quotient can be zero. Since $R(u)$ depends on the function u, it is a functional. Obviously any eigenfunction w_r can be used as a comparison function, in which case Rayleigh's quotient reduces to the corresponding eigenvalue, λ_r.

As in the case of discrete systems, for which Rayleigh's quotient has a stationary value in the neighborhood of an eigenvector, in the case of continuous systems Rayleigh's quotient has a stationary value in the neighborhood of an eigenfunction. To show this we call upon the expansion theorem and represent the comparison function u in the form

$$u = \sum_{i=1}^{\infty} c_i w_i, \tag{5.305}$$

where w_i are the eigenfunctions of the system, satisfying differential equation (5.301) throughout domain D and all the boundary conditions of the problem, and c_i are constants. Let the eigenfunctions be normalized according to the relation

$$\int_D w_i M[w_i] \, dD = 1, \qquad i = 1, 2, \ldots, \tag{5.306}$$

from which it follows immediately that

$$\int_D w_i L[w_i] \, dD = \lambda_i, \qquad i = 1, 2, \ldots. \tag{5.307}$$

Introducing (5.305) through (5.307) in (5.304) and recalling that L and M are linear operators, Rayleigh's quotient becomes

$$R(u) = \frac{\displaystyle\int_D \sum_{i=1}^{\infty} c_i w_i L\left[\sum_{j=1}^{\infty} c_j w_j\right] dD}{\displaystyle\int_D \sum_{i=1}^{\infty} c_i w_i M\left[\sum_{j=1}^{\infty} c_j w_j\right] dD}$$

$$= \frac{\displaystyle\sum_{i=1}^{\infty} \sum_{j=1}^{\infty} c_i c_j \int_D w_i L[w_j] \, dD}{\displaystyle\sum_{i=1}^{\infty} \sum_{j=1}^{\infty} c_i c_j \int_D w_i M[w_j] \, dD} = \frac{\displaystyle\sum_{i=1}^{\infty} c_i^2 \lambda_i}{\displaystyle\sum_{i=1}^{\infty} c_i^2}, \tag{5.308}$$

where the orthogonality of the eigenfunctions w_i has been accounted for.

Using a method similar to the method used for discrete systems, one can readily show that $R(u)$ has a stationary value in the neighborhood of any eigenfunction w_i. The stationary value is a minimum in the neighborhood of the first eigenfunction,

$$R(u) \geq \lambda_1, \tag{5.309}$$

so Rayleigh's quotient provides an *upper bound* for λ_1 with respect to all comparison functions u. When the comparison function coincides with the first eigenfunction, Rayleigh's quotient is identically equal to the first eigenvalue. For a continuous system there is no "highest" eigenvalue, so that the quotient cannot provide a lower bound.

It can be shown that the sth eigenvalue, λ_s, is the minimum value that Rayleigh's quotient can assume with respect to the subclass of comparison functions u satisfying the orthogonality condition

$$\int_D u M[w_i] \, dD = 0, \qquad i = 1, 2, \ldots, s - 1, \tag{5.310}$$

which is equivalent to saying that in the series expansion (5.305),

$$c_1 = c_2 = \cdots = c_{s-1} = 0. \tag{5.311}$$

Hence for this subclass of functions

$$R(u) \geq \lambda_s, \qquad s = 2, 3, \ldots, \tag{5.312}$$

or Rayleigh's quotient provides an upper bound for λ_s with respect to all comparison functions which are orthogonal to the first $s - 1$ eigenfunctions.

As an illustration let us consider again the case of a uniform rod clamped at $x = 0$ and free at the end $x = L$. The rod is vibrating longitudinally. As in Section 5–13 we assume a comparison function in the form

$$u(x) = C\left[\left(\frac{x}{L}\right)^3 - 3\frac{x}{L}\right], \tag{5.313}$$

so that

$$L[u] = -EA\frac{d^2u}{dx^2} = -\frac{6EAC}{L^2}\frac{x}{L},$$
$$\tag{5.314}$$
$$M[u] = \rho u = C\rho\left[\left(\frac{x}{L}\right)^3 - 3\frac{x}{L}\right].$$

Next we calculate

$$\int_D uL[u]\,dD = -\frac{6EAC^2}{L^2}\int_0^L \frac{x}{L}\left[\left(\frac{x}{L}\right)^3 - 3\frac{x}{L}\right]dx = \frac{24}{5}\frac{EAC^2}{L}, \tag{5.315}$$

$$\int_D uM[u]\,dD = \rho C^2\int_0^L \left[\left(\frac{x}{L}\right)^3 - 3\frac{x}{L}\right]^2 dx = \frac{68}{35}C^2\rho L. \tag{5.316}$$

Introducing (5.315) and (5.316) in (5.304) we obtain

$$R(u) = \frac{\displaystyle\int_D uL[u]\,dD}{\displaystyle\int_D uM[u]\,dD} = \frac{168}{68}\frac{AE}{\rho L^2} = 2.4706\,\frac{EA}{\rho L^2}. \tag{5.317}$$

Recalling that the first eigenvalue of the system considered is

$$\lambda_1 = \left(\frac{\pi}{2}\right)^2\frac{EA}{\rho L^2} = 2.4674\,\frac{EA}{\rho L^2}, \tag{5.318}$$

we see that the inequality (5.309) is verified. We also must note that the value of Rayleigh's quotient as given by (5.317) is remarkably close to the true first eigenvalue.

The first eigenvalue, λ_1, is the minimum value $R(u)$ can take, so one can construct a minimizing sequence of comparison functions to minimize Rayleigh's quotient. That is precisely the procedure used in cases in which an exact solution of the eigenvalue problem is not easy to obtain and an approximate method is desired. The Rayleigh-Ritz method employs a linear combination of comparison functions multiplied by some adjustable coefficients. The coefficients are so adjusted as to render Rayleigh's quotient stationary. We shall discuss the Rayleigh-Ritz method in detail in Chapter 6.

5–15   EIGENVALUE PROBLEM. INTEGRAL FORMULATION

The displacement $w(x, t)$ at a point x in a one-dimensional continuous system can be given in the integral form

$$w(x, t) = \int_0^L a(x, \zeta)p(\zeta, t) \, d\zeta, \qquad (5.319)$$

where $a(x, \zeta)$ is the flexibility influence function and $p(\zeta, t)$ is the distributed load at point ζ (see Section 3–2).

In the case of free vibration, the load is due entirely to the inertia force. The free vibration is harmonic, so we can write

$$p(x, t) = -m(x) \frac{\partial^2 w(x, t)}{\partial t^2} = \omega^2 m(x)w(x, t). \qquad (5.320)$$

Denoting by $w(x)$ the amplitude of the harmonic motion, we introduce (5.320) in (5.319), cancel the time dependence in both sides, and obtain

$$w(x) = \omega^2 \int_0^L a(x, \zeta)m(\zeta)w(\zeta) \, d\zeta, \qquad (5.321)$$

which is known as a homogeneous linear integral equation of the second kind.[†] At this point we must recall that the influence function $a(x, \zeta)$ satisfies the boundary conditions of the problem. Equation (5.321) represents an eigenvalue problem in integral form, which is the equivalent[‡] of the special

---

† Also called a Fredholm homogeneous integral equation of the second kind.
‡ See F. B. Hildebrand, *Methods of Applied Mathematics*, Prentice-Hall, Englewood Cliffs, N.J., 1960, p. 388.

eigenvalue problem

$$L[w] = \omega^2 m w, \tag{5.322}$$

for which the eigenvalue ω^2 does not enter into the boundary conditions.

Equation (5.321) may be generalized in the form

$$w(P) = \lambda \int_D G(P, Q) m(Q) w(Q) \, dD(Q), \tag{5.323}$$

where P denotes the position of the point at which the displacement $w(P)$ is observed. This position may be defined by one or two spatial coordinates according to the nature of the problem. The function $G(P, Q)$ is a more general type of influence function called a *Green's function*. For a self-adjoint system Green's function is symmetric† in P and Q, $G(P, Q) = G(Q, P)$. The kernel $G(P, Q) m(Q)$ of the integral transformation (5.323) is not symmetric unless $m(Q)$ is constant. It can be symmetrized, however. To this end we introduce the function

$$v(P) = m^{1/2}(P) w(P), \tag{5.324}$$

and multiplying both sides of (5.323) by $m^{1/2}(P)$ we obtain

$$v(P) = \lambda \int_D K(P, Q) v(Q) \, dD(Q), \tag{5.325}$$

where the kernel $K(P, Q)$ is symmetric,

$$K(P, Q) = G(P, Q) m^{1/2}(P) m^{1/2}(Q) = K(Q, P). \tag{5.326}$$

For certain values λ_i, (5.325) has nontrivial solutions $v^{(i)}(P)$, which are related to the solutions $w^{(i)}(P)$ of (5.323) by (5.324). The values λ_i are the eigenvalues and $w^{(i)}(P)$ are the associated eigenfunctions of the system. Whereas the functions $v^{(i)}(P)$ are orthogonal in the ordinary sense, the functions $w^{(i)}(P)$ are orthogonal with respect to the function $m(P)$. To show this we consider two distinct solutions of (5.325),

$$v^{(i)}(P) = \lambda_i \int_D K(P, Q) v^{(i)}(Q) \, dD(Q), \tag{5.327}$$

$$v^{(j)}(P) = \lambda_j \int_D K(P, Q) v^{(j)}(Q) \, dD(Q). \tag{5.328}$$

† See R. Courant and D. Hilbert, *Methods of Mathematical Physics*, Vol. 1, Interscience, New York, 1961, p. 354.

Multiplying (5.327) by $v^{(j)}(P)$, integrating over domain D, and using (5.328), we obtain

$$\int_D v^{(i)}(P)v^{(j)}(P)\,dD(P) = \lambda_i \int_D v^{(j)}(P)\left[\int_D K(P,\,Q)v^{(i)}(Q)\,dD(Q)\right]dD(P)$$

$$= \lambda_i \int_D v^{(i)}(Q)\left[\int_D K(Q,\,P)v^{(j)}(P)\,dD(P)\right]dD(Q)$$

$$= \frac{\lambda_i}{\lambda_j}\int_D v^{(i)}(Q)v^{(j)}(Q)\,dD(Q), \tag{5.329}$$

from which we obtain

$$(\lambda_i - \lambda_j)\int_D v^{(i)}(P)v^{(j)}(P)\,dD(P) = 0, \tag{5.330}$$

and for two distinct eigenvalues the orthogonality relation is

$$\int_D v^{(i)}(P)v^{(j)}(P)\,dD(P) = 0 \qquad \text{for } \lambda_i \neq \lambda_j. \tag{5.331}$$

Introducing (5.324) in (5.331) and normalizing the eigenfunctions, we write

$$\int_D m(P)w^{(i)}(P)w^{(j)}(P)\,dD(P) = \delta_{ij}, \tag{5.332}$$

where δ_{ij} is the Kronecker delta.

As for the differential equation formulation, there is an expansion theorem concerning the eigenfunctions $w^{(i)}(P)$ according to which we may represent a function satisfying the boundary conditions and possessing a continuous $L[w]$ by an infinite series

$$w(P) = \sum_{i=1}^{\infty} c_i w^{(i)}(P), \tag{5.333}$$

where the coefficients c_i are given by

$$c_i = \int_D m(P)w(P)w^{(i)}(P)\,dD(P), \qquad i = 1, 2, \ldots . \tag{5.334}$$

There are a number of methods for solving (5.325). The description of these methods is beyond the scope of this text. The interested reader will find good treatments in the books by Courant and Hilbert, Hildebrand, Lovitt, Tricomi, etc. (see Selected Readings at the end of this chapter). We shall concentrate here on the iteration method, which is similar in principle to the matrix iteration method.

Denote by $w_1^{(1)}(P)$ the first trial function to be used for the first iteration to the first mode. Introduce $w_1^{(1)}(P)$ in the integral in (5.323), perform the integration, and denote the result by $w_2^{(1)}(P)$, so that

$$w_2^{(1)}(P) = \int_D G(P, Q)m(Q)w_1^{(1)}(Q)\, dD(Q)$$

$$= \sum_{i=1}^{\infty} c_i \int_D G(P, Q)m(Q)w^{(i)}(Q)\, dD(Q) = \sum_{i=1}^{\infty} c_i \frac{w^{(i)}(P)}{\lambda_i}. \quad (5.335)$$

Using $w_2^{(1)}(P)$ as an improved trial function, we obtain

$$w_3^{(1)}(P) = \int_D G(P, Q)m(Q)w_2^{(1)}(Q)\, dD(Q) = \sum_{i=1}^{\infty} c_i \frac{w^{(i)}(P)}{\lambda_i^2}. \quad (5.336)$$

In general, we have

$$w_n^{(1)}(P) = \sum_{i=1}^{\infty} c_i \frac{w^{(i)}(P)}{\lambda_i^{n-1}}. \quad (5.337)$$

If the eigenvalues are such that $\lambda_1 < \lambda_2 < \lambda_3 < \cdots$, the first term of the series in (5.337) becomes increasingly large in comparison with the remaining ones, and as $n \to \infty$, $w_n^{(1)}(P)$ becomes proportional to the first eigenfunction,

$$\lim_{n \to \infty} w_n^{(1)}(P) = w^{(1)}(P), \quad (5.338)$$

and at this point λ_1 is obtained as the ratio of two subsequent trial functions,

$$\lambda_1 = \lim_{n \to \infty} \frac{w_n^{(1)}(P)}{w_{n+1}^{(1)}(P)}. \quad (5.339)$$

To obtain the second mode we must insist that the trial function $w_1^{(2)}(P)$, used for the first iteration to the second mode, be entirely free of the first mode. To this end we write it in the form

$$w_1^{(2)}(P) = \varphi_1^{(2)}(P) - a_1 w^{(1)}(P), \quad (5.340)$$

where $\varphi_1^{(2)}(P)$ is any arbitrary function and a_1 is a coefficient which is determined from the orthogonality requirement

$$\int_D m(P)w_1^{(2)}(P)w^{(1)}(P)\, dD(P) = \int_D m(P)\varphi_1^{(2)}(P)w^{(1)}(P)\, dD(P)$$

$$- a_1 \int_D m(P)[w^{(1)}(P)]^2\, dD(P) = 0, \quad (5.341)$$

which yields

$$a_1 = \frac{\int_D m(P)\varphi_1^{(2)}(P)w^{(1)}(P)\,dD(P)}{\int_D m(P)[w^{(1)}(P)]^2\,dD(P)},\tag{5.342}$$

and if $w^{(1)}(P)$ is normalized,

$$a_1 = \int_D m(P)\varphi_1^{(2)}(P)w^{(1)}(P)\,dD(P).\tag{5.343}$$

Introduce (5.340) in the integral in (5.323), perform the integration, and denote the result by $\varphi_2^{(2)}(P)$,

$$\varphi_2^{(2)}(P) = \int_D G(P,\,Q)m(Q)w_1^{(2)}(Q)\,dD(Q).\tag{5.344}$$

For the next iteration we use

$$w_2^{(2)}(P) = \varphi_2^{(2)}(P) - a_2 w^{(1)}(P),\tag{5.345}$$

where for normalized $w^{(1)}(P)$ we have

$$a_2 = \int_D m(P)\varphi_2^{(2)}(P)w^{(1)}(P)\,dD(P).\tag{5.346}$$

In general, for the nth iteration we use

$$w_n^{(2)}(P) = \varphi_n^{(2)}(P) - a_n w^{(1)}(P).\tag{5.347}$$

Convergence is achieved when as $n \to \infty$, $a_n \to 0$ and

$$\lim_{n \to \infty} w_n^{(2)}(P) = w^{(2)}(P),\tag{5.348}$$

$$\lambda_2 = \lim_{n \to \infty} \frac{w_n^{(2)}(P)}{w_{n+1}^{(2)}(P)}.\tag{5.349}$$

Similarly, for the third mode we use a trial function

$$w_1^{(3)}(P) = \varphi_1^{(3)}(P) - a_1 w^{(1)}(P) - b_1 w^{(2)}(P),\tag{5.350}$$

where a_1 and b_1 are obtained by insisting that $w_1^{(3)}(P)$ be orthogonal to both $w^{(1)}(P)$ and $w^{(2)}(P)$. The same procedure as previously described is used to iterate to the third mode and, subsequently, to higher modes. In practice, a finite number n of iterations is needed at each step.

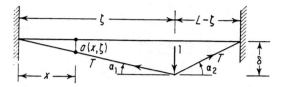

FIGURE 5.17

As an illustration consider the free vibration of a string of uniformly distributed mass m, clamped at both ends and subjected to a constant tension T. The flexibility influence function $a(x, \zeta)$ is obtained by applying a unit load at point ζ and calculating the deflection at point x (see Figure 5.17). For small angles α_1 and α_2, the equilibrium condition at the point of load application is

$$T\frac{\delta}{\zeta} + T\frac{\delta}{L - \zeta} = 1,$$

from which we obtain the influence function,

$$a(x, \zeta) = \delta\frac{x}{\zeta} = \frac{x(L - \zeta)}{TL}, \qquad \zeta > x, \qquad (5.351)$$

and it can be readily shown that

$$a(x, \zeta) = \frac{\zeta(L - x)}{TL}, \qquad \zeta < x. \qquad (5.352)$$

As expected, Green's function $G(x, \zeta) = a(x, \zeta)$ is symmetric in x and ζ, because, obviously, the system is self-adjoint.

Let us use the iteration method to solve the eigenvalue problem and to this end assume first

$$w_1^{(1)} = \frac{x}{L}, \qquad (5.353)$$

so that

$$w_2^{(1)}(x) = \int_0^L G(x, \zeta)m(\zeta)w_1^{(1)}(\zeta)\,d\zeta$$

$$= \frac{m}{TL^2}\left[\int_0^x \zeta(L - x)\zeta\,d\zeta + \int_x^L x(L - \zeta)\zeta\,d\zeta\right]$$

$$= \frac{mL^2}{6T}\left(\frac{x}{L} - \frac{x^3}{L^3}\right). \qquad (5.354)$$

Next we obtain in sequence

$$w_3^{(1)}(x) = \int_0^L G(x, \zeta) m(\zeta) w_2^{(1)}(\zeta)\, d\zeta$$

$$= \frac{mL^2}{6T} \frac{7mL^2}{60T} \left(\frac{x}{L} - \frac{10}{7} \frac{x^3}{L^3} + \frac{3}{7} \frac{x^5}{L^5} \right), \tag{5.355}$$

$$w_4^{(1)}(x) = \int_0^L G(x, \zeta) m(\zeta) w_3^{(1)}(\zeta)\, d\zeta$$

$$= \frac{mL^2}{6T} \frac{7mL^2}{60T} \frac{31mL^2}{294T} \left(\frac{x}{L} - \frac{49}{31} \frac{x^3}{L^3} + \frac{21}{31} \frac{x^5}{L^5} - \frac{3}{31} \frac{x^7}{L^7} \right), \tag{5.356}$$

$$w_5^{(1)}(x) = \int_0^L G(x, \zeta) m(\zeta) w_4^{(1)}(\zeta)\, d\zeta,$$

$$= \frac{mL^2}{6T} \frac{7mL^2}{60T} \frac{31mL^2}{294T} \frac{2667mL^2}{26040T}$$

$$\times \left(\frac{x}{L} - \frac{4340}{2667} \frac{x^3}{L^3} + \frac{2058}{2667} \frac{x^5}{L^5} - \frac{420}{2667} \frac{x^7}{L^7} + \frac{35}{2667} \frac{x^9}{L^9} \right). \tag{5.357}$$

At this point we may wish to pause and check the convergence. Letting $n = 4$ we write

$$\omega_1^2 \simeq \frac{w_4^{(1)}(x)}{w_5^{(1)}(x)} = \frac{26040T}{2667mL^2}, \qquad \omega_1 \simeq 3.12 \sqrt{\frac{T}{mL^2}}. \tag{5.358}$$

The approximation is quite good, because the exact value of the first natural frequency is $\omega_1 = \pi \sqrt{T/mL^2}$. It can also be easily checked that $w_5^{(1)}(x)$, as given by (5.357) is almost proportional to $\sin(\pi x/L)$, which is the first natural mode. Hence (5.357) and (5.358) can be accepted as representing the first natural mode and the first natural frequency, respectively. To obtain the second mode we must use a trial function orthogonal to $w^{(1)}(x)$. This is left as an exercise to the reader.

Quite often Green's function has a more complicated form than the one encountered here and the integration may have to be performed numerically or graphically. This is bound to introduce errors. In such cases it may prove more satisfactory to adopt an approximate method to obtain the first few modes. Approximate methods in conjunction with the integral formulation will be discussed in Chapter 6.

Problems

5.1 Consider a system consisting of discrete masses connected by springs in series as shown in Figure 5.18. Write the equation of motion for mass m_i using Newton's second law, arrange the equation in incremental form, take the limit, and derive the differential equation of motion for the longitudinal vibration of a thin bar.

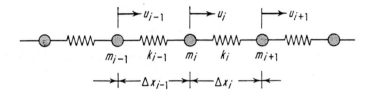

FIGURE 5.18

5.2 Derive the boundary-value problem of Section 5–2 by writing the force and moment equations of motion. Explain the boundary conditions for a clamped rod in transverse vibration.

5.3 Formulate the eigenvalue problem for a uniform cantilever beam in transverse vibration. Neglect the shear deformation but consider the rotatory inertia effect. Write the problem to fit the general formulation of Section 5–4 and identify the corresponding differential operators.

5.4 Derive the boundary-value problem of a string in transverse vibration by means of Hamilton's principle. Set up the eigenvalue problem for a string fixed at one end and free at the other end and check whether the system is self-adjoint and positive definite. Assume that the string is uniform and plot the first three normal modes and compute the corresponding natural frequencies.

5.5 Formulate the eigenvalue problem for the nonuniform bar clamped at one end and with a linear spring attached at the other end as shown in Figure 5.19. Check whether the system is self-adjoint and positive definite.

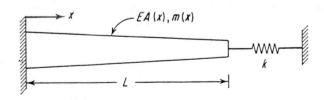

FIGURE 5.19

5.6 Let the bar in Problem 5.5 be uniform and determine the first three natural frequencies and normal modes for $EA = 10kL$. Plot the first three modes.

5.7   Consider the system of Problem 5.5, let

$$EA(x) = 2EA\left(1 - \frac{x}{L}\right), \qquad m(x) = 2m\left(1 - \frac{x}{L}\right), \qquad k = 0,$$

and solve the corresponding eigenvalue problem. Plot the first three modes and indicate the first three natural frequencies.

5.8   Consider a shaft consisting of two uniform segments as shown in Figure 5.20. Set up the eigenvalue problem for the torsional vibration of the system and derive the frequency equation. Show that as $L_2 \to 0$ such that

$$\lim_{L_2 \to 0} I_2 L_2 = I_D,$$

the frequency equation reduces to (5.141) and the orthogonality condition reduces to (5.144).

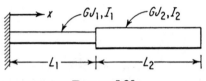

FIGURE 5.20

5.9   Formulate the eigenvalue problem for the bending vibration of a uniform bar hinged at one end and free at the other end, as shown in Figure 5.21. Obtain the first three natural frequencies and normal modes and plot the modes. Check whether the system is self-adjoint and positive definite. Indicate the orthogonality relation.

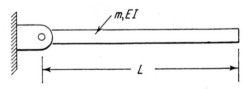

FIGURE 5.21

5.10   Formulate the eigenvalue problem for the bending vibration of a uniform cantilever beam with a concentrated mass, prevented from rotating, at the end, as shown in Figure 5.22. Derive the frequency equation and obtain the first two modes for $M = 0.5mL$.

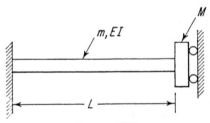

FIGURE 5.22

5.11 Formulate the eigenvalue problem for the transverse vibration of a bar with each end hinged to a spring as shown in Figure 5.23. The springs can deflect vertically only, and in the equilibrium position the bar is horizontal. Assume small motions and derive the frequency equation for the uniform bar. Check whether the system is self-adjoint and positive definite.

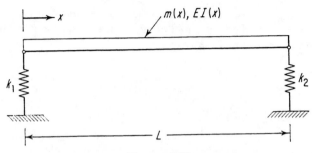

FIGURE 5.23

5.12 Derive the differential equation of motion and the associated boundary conditions of a thin rectangular membrane by means of Hamilton's principle. Give the physical interpretation of the boundary conditions.

5.13 Show that for a uniform rectangular membrane with the side ratio $R = a/b = 4/3$, we obtain the pairs of equal natural frequencies $\omega_{35} = \omega_{54}$ and $\omega_{83} = \omega_{46}$. Find another pair of equal frequencies.

5.14 A uniform square membrane has repeated natural frequencies $\omega_{mn} = \omega_{nm}$. Any linear combination of modes W_{mn} and W_{nm} is also a mode. Plot the nodal lines for the mode

$$f(x, y, c) = W_{13}(x, y) + cW_{31}(x, y)$$

for the values $c = 0, \frac{1}{2}, 1$.

5.15 Derive the frequency equation for a uniform annular membrane defined over the domain $b \le r \le a$ and fixed at the boundaries $r = b$ and $r = a$.

5.16 Derive the differential equation of motion for a nonuniform plate.

5.17 Check whether a uniform rectangular plate simply supported along the boundaries is self-adjoint. Repeat the problem for a uniform circular plate clamped at the boundary $r = a$.

5.18 Consider a uniform string fixed at both ends, assume a comparison function in the form

$$u(x) = 2x(x - L) - \left(\frac{x}{L}\right)^2 (x - L)^2,$$

and plot

$$\lambda^*(x) = \frac{L[u(x)]}{Mu(x)}$$

as a function of x over the range $0 \le x \le L$. Use the enclosure theorem and predict the range within which there is at least one eigenvalue present.

5.19   Consider a uniform string fixed at both ends and calculate the value of Rayleigh's quotient for the comparison function

$$u(x) = \frac{x}{L} - \left(\frac{x}{L}\right)^3.$$

Draw conclusions.

5.20   Formulate the eigenvalue problem in the integral form for the longitudinal vibration of the uniform rod shown in Figure 5.24. Use the iteration method and obtain the first eigenvalue of the system.

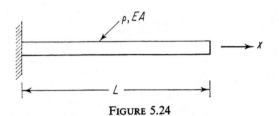

FIGURE 5.24

Selected Readings

Bisplinghoff, R. L., H. Ashley, and R. L. Halfman, *Aeroelasticity*, Addison-Wesley, Reading, Mass., 1957.

Collatz, L., *Eigenwertaufgaben mit technishen Anwendungen*, Akademische Verlagsgesellschaft, Leipzig, Germany, 1963.

Collatz, L., *Eigenwertprobleme*, Chelsea, New York, 1948.

Courant R., and D. Hilbert, *Methods of Mathematical Physics*, Vol. 1, Inter-science, New York, 1961.

Crandall, S. H., *Engineering Analysis*, McGraw-Hill, New York, 1956.

Hildebrand, F. B., *Methods of Applied Mathematics*, Prentice-Hall, Englewood Cliffs, N.J., 1960.

Lanczos, C., *The Variational Principles of Mechanics*, Univ. of Toronto Press, Toronto, Canada, 1964.

Lovitt, W. V., *Linear Integral Equations*, McGraw-Hill, New York, 1924.

McLachlan, N. W., *Bessel Functions for Engineers*, Oxford Univ. Press, New York, 1961.

Mikhlin, S. G., *Integral Equations*, Pergamon Press, New York, 1957.

Morse, P. M., *Vibration and Sound*, 2nd ed., McGraw-Hill, New York, 1948.

Rayleigh, Lord, *The Theory of Sound*, Vol. 1, Dover, New York, 1945.

Sokolnikoff, I. S., *Mathematical Theory of Elasticity*, 2nd ed., McGraw-Hill, New York, 1956.

Timoshenko, S., and S. Woinowsky-Krieger, *Theory of Plates and Shells*, 2nd ed., McGraw-Hill, New York, 1959.

Timoshenko, S., *Vibration Problems in Engineering*, Van Nostrand, New York, 1937.

Tong, K. N., *Theory of Mechanical Vibration*, Wiley, New York, 1960.

Tricomi, F. G., *Integral Equations*, Interscience, New York, 1957.

Weinstock, R., *Calculus of Variations*, McGraw-Hill, New York, 1952.

CHAPTER 6

NATURAL MODES OF VIBRATION. APPROXIMATE METHODS

6–1 GENERAL DISCUSSION

In Chapter 5 we discussed vibration problems associated with various continuous systems by placing the emphasis on the similarities between various types of continuous systems, as well as the common features of continuous and discrete systems. First we presented an example of a boundary-value problem and then used the separation of variables method to derive the corresponding eigenvalue problem. A large number of eigenvalue problems were derived and "exact" solutions, consisting of infinite sets of eigenfunctions and associated eigenvalues, obtained. For the most part the systems were relatively simple, in the sense that their mass and stiffness distributions were uniform and the boundary conditions relatively easy to account for. This may have left us with a feeling of false security, however, because in the physical world there are many instances when the formulation of the boundary-value problem may pose insurmountable difficulties. In some cases, when the boundary-value problem can be formulated, it may not be possible to solve the differential equation. In other cases, although it is possible to solve the differential equation, it is not possible to satisfy the associated boundary conditions. In such cases one may be content with an approximate solution of the boundary-value problem or of the eigenvalue problem, if a solution by means of modal analysis is sought.

There are various schemes for obtaining approximate solutions of the eigenvalue problem. We distinguish two major classes according to the method of approach, although all the methods treated here have one thing in common: they all approximate a continuous system by an n-degree-of-freedom-system, where n is a finite number depending on the accuracy desired. The first class of methods assumes a solution in the form of a finite series consisting of known functions multiplied by unknown coefficients. By adopting various points of view, which will be discussed in detail in this chapter, the eigenvalue problem for a continuous system is converted into an eigenvalue problem similar, in structure, to the eigenvalue problem of a discrete system. Depending on the method used, one can select as assumed functions comparison functions, admissible functions, or functions satisfying the differential equation but not the boundary conditions. If the series consists of n functions, the corresponding eigenvalue problem yields n eigenvalues and n associated eigenvectors. The components of every one of the resulting n-dimensional eigenvectors are multiplied by the respective assumed functions to obtain the desired eigenfunctions. A second class of methods represents the continuous system by a lumped model. Within this second class one can further distinguish between two subclasses: step-by-step methods and methods employing influence coefficients. The step-by-step methods assume that the mass of the system is concentrated at chosen points, which will be called *stations*, whereas the segments between two stations, called *fields*, are massless and possess uniform stiffness properties. The natural frequencies and subsequently the natural modes can be obtained by using either recurrence formulas or transfer matrices leading to a characteristic equation. We shall adopt the latter approach. The method employing influence coefficients also assumes that the mass of the system is concentrated at chosen points but, in contrast with the step-by-step methods, it makes no assumptions as to the stiffness properties between any two adjacent points and, from this standpoint, it describes better the stiffness properties of the system. At the same time, the evaluation of the influence coefficients may be extremely difficult. The method employing influence coefficients is broader in scope than the step-by-step methods because it lends itself to a general formulation and, in principle, can be applied to any system. The influence coefficients method leads to an eigenvalue problem similar to the one for discrete systems. In fact, the discrete models of Chapter 4 may be regarded as very crude approximations of continuous systems.

In the present chapter a large variety of approximate methods are presented. In the discussion of the methods assuming solutions in the form of finite series, the conditions under which the use of admissible functions rather than comparison functions is warranted are brought out. Differential as well as integral formulations are presented. The basic idea behind the step-by-step

methods, as applied to continuous systems, consists of replacing the differential expressions for the equations of motion and the force-deformation and
moment-deformation relations by finite difference expressions. The difference
expressions can be conveniently arranged in matrix form, and to this end the
concepts of station and field transfer matrices are introduced. Finally, the
lumped-parameter method employing influence coefficients is presented and
applied to obtain the natural modes of a semidefinite system.

All the methods presented in this chapter are formulated in a way that
allows expedient numerical solutions by means of high-speed electronic
computer.

6-2 RAYLEIGH'S ENERGY METHOD

Rayleigh's method is generally used to obtain an approximation for the
first natural frequency, or fundamental frequency, of a system without having
to solve the differential equations of motion. It can be used for a discrete or
continuous system. The method is based on *Rayleigh's principle*, which may
be enunciated as follows: The frequency of vibration of a conservative system
vibrating about an equilibrium position has a stationary value in the neighborhood of a natural mode. This stationary value is actually a minimum in the
neighborhood of the fundamental mode. The method is based on the fact
that natural modes execute harmonic motions.

Consider a discrete system and let the time-dependent displacement of
mass m_i be of the form

$$q_i(t) = u_i f(t), \tag{6.1}$$

where $f(t)$ is a harmonic function. The above can be written in the matrix
form

$$\{q(t)\} = \{u\} f(t). \tag{6.2}$$

The kinetic energy of the system can be written

$$T(t) = \tfrac{1}{2}\{u\}^T[m]\{u\}\dot{f}^2(t), \tag{6.3}$$

and the potential energy has the form

$$V(t) = \tfrac{1}{2}\{u\}^T[k]\{u\}f^2(t). \tag{6.4}$$

For a conservative system the total energy E is constant. When the system
passes through the equilibrium position, the potential energy is zero and the
kinetic energy has a maximum equal to the total energy. On the other hand,
when the system reaches the maximum displacement, the kinetic energy is

zero and the potential energy has a maximum equal to the total energy of the system. This can be expressed

$$E = 0 + T_{\max} = V_{\max} + 0, \tag{6.5}$$

from which it follows that

$$T_{\max} = V_{\max}. \tag{6.6}$$

If the harmonic function $f(t)$ has the frequency ω, then

$$T_{\max} = \tfrac{1}{2}\{u\}^T[m]\{u\}\omega^2 = T^*\omega^2, \tag{6.7}$$

where

$$T^* = \tfrac{1}{2}\{u\}^T[m]\{u\} \tag{6.8}$$

is called *reference kinetic energy*. In addition we have

$$V_{\max} = \tfrac{1}{2}\{u\}^T[k]\{u\}. \tag{6.9}$$

Equations (6.7) and (6.9) introduced in (6.6) give

$$\omega^2 = R(u) = \frac{\{u\}^T[k]\{u\}}{\{u\}^T[m]\{u\}} = \frac{V_{\max}}{T^*}, \tag{6.10}$$

which is easily recognized as Rayleigh's quotient expression obtained in Chapter 4 directly from the eigenvalue problem. There we also proved the stationarity of Rayleigh's quotient, which justifies the statement of Rayleigh's principle.

Equation (6.10) takes the minimum value when the vector $\{u\}$ is identical to the first eigenvector $\{u^{(1)}\}$. For any vector $\{u\}$, other than the first eigenvector, the value of Rayleigh's quotient will be larger. Equation (6.10) may be used to obtain an estimate of the first natural frequency. The procedure consists of arbitrarily selecting a trial vector $\{u\}$ representing the first natural mode. Substituting the trial vector in (6.10) we obtain an estimate of the first natural frequency. Owing to the stationarity of Rayleigh's quotient, remarkably good estimates of the fundamental frequency can be obtained even if the trial vector does not resemble the true first eigenvector too closely. Evidently the closer the trial vector resembles the first eigenvector, the better the estimate of the fundamental frequency. Hence the quality of the estimate obtained depends on the experience and skill in selecting a trial vector.

Of course, one can derive a similar expression for continuous systems. If we wish to approach the problem from an energy point of view, it is more convenient to discuss specific systems than a general case. As an illustration let us consider the longitudinal vibration of a thin rod. To this end let the longitudinal displacement be given by

$$u(x, t) = U(x)f(t), \tag{6.11}$$

where $U(x)$ plays the role of the maximum displacement at point x and $f(t)$ expresses the harmonic time dependence. The kinetic energy takes the form

$$T(t) = \tfrac{1}{2} \int_0^L m(x) \left[\frac{\partial u(x, t)}{\partial t} \right]^2 dx = \tfrac{1}{2} f^2(t) \int_0^L m(x) U^2(x) \, dx \tag{6.12}$$

and the potential energy becomes

$$V(t) = \tfrac{1}{2} \int_0^L EA(x) \left[\frac{\partial u(x, t)}{\partial x} \right]^2 dx = \tfrac{1}{2} f^2(t) \int_0^L EA(x) \left[\frac{dU(x)}{dx} \right]^2 dx. \tag{6.13}$$

Equations (6.12) and (6.13) lead to

$$\omega^2 = R(U(x)) = \frac{\displaystyle\int_0^L EA(x)[dU(x)/dx]^2 \, dx}{\displaystyle\int_0^L m(x) U^2(x) \, dx}. \tag{6.14}$$

Equation (6.14) can be obtained directly from the general expression of Rayleigh's quotient, (5.304) through an integration by parts. Here, too, Rayleigh's quotient, (6.14), takes the minimum value when $U(x)$ is the first normal mode. To obtain an estimate of the fundamental frequency one must assume an expression $U(x)$ for the first natural mode satisfying all the boundary conditions of the problem.

This method is quite useful when the system stiffness and mass are distributed nonuniformly and an exact solution of the eigenvalue problem is impossible to obtain.

Example 6.1

As an illustration of Rayleigh's energy method let us estimate the fundamental frequency of a nonuniform clamped-free bar vibrating longitudinally (Figure 6.1). The mass per unit length is given by

$$m(x) = 2m\left(1 - \frac{x}{L}\right) \tag{a}$$

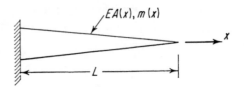

FIGURE 6.1

and the stiffness has the expression

$$EA(x) = 2EA\left(1 - \frac{x}{L}\right). \tag{b}$$

Although the eigenvalue problem can be solved in closed form, we shall use Rayleigh's energy method for comparison purposes. Let us assume as the fundamental mode the first eigenfunction of a uniform clamped-free bar

$$U(x) = \sin\frac{\pi x}{2L}, \tag{c}$$

because it obviously satisfies the boundary conditions of the problem. Next form the integrals

$$\int_0^L EA(x)\left[\frac{dU(x)}{dx}\right]^2 dx = 2EA\left(\frac{\pi}{2L}\right)^2 \int_0^L \left(1 - \frac{x}{L}\right)\cos^2\frac{\pi x}{2L}\, dx$$

$$= \frac{EA}{2L}\left(1 + \frac{\pi^2}{4}\right), \tag{d}$$

$$\int_0^L m(x)U^2(x)\, dx = 2m\int_0^L \left(1 - \frac{x}{L}\right)\sin^2\frac{\pi x}{2L}\, dx = \frac{mL}{2}\left(1 - \frac{4}{\pi^2}\right). \tag{e}$$

Introducing (d) and (e) in (6.14) we obtain

$$\omega^2 = R(U) = \frac{\int_0^L EA(x)[dU(x)/dx]^2\, dx}{\int_0^L m(x)U^2(x)\, dx} = \frac{1 + (\pi^2/4)}{1 - (4/\pi^2)}\frac{EA}{mL^2} = 5.8305\frac{EA}{mL^2},$$

from which we obtain the estimated fundamental frequency

$$\omega = 2.4146\sqrt{\frac{EA}{mL^2}}. \tag{f}$$

The eigenvalue problem associated with the system of Figure 6.1 can actually be solved in closed form (see Problem 5.7). The true fundamental frequency is

$$\omega_1 = 2.4048\sqrt{\frac{EA}{mL^2}}, \tag{g}$$

so Rayleigh's energy method furnishes an estimate about 0.4 per cent higher than the true fundamental frequency. Normally one cannot expect such good estimates. The reason the estimated value is so close to the true value of the fundamental frequency is that the comparison function chosen closely resembles the fundamental mode.

6–3 RAYLEIGH–RITZ METHOD

Quite often continuous systems lead to eigenvalue problems which, for all practical purposes, are impossible to solve. This is frequently the case when the stiffness or mass distribution of the system is nonuniform or the shape of the boundary curves cannot be described in terms of known functions. Yet it may be imperative to obtain information about the physical system and in particular about natural frequencies. Quite often it is sufficient to know the value of only a limited number of lowest frequencies rather than all the frequencies. The higher frequencies cannot be taken too seriously anyway, even if an exact solution of the eigenvalue problem is obtained, because the nature of the assumptions employed in defining the models in most elementary theories restricts the validity of the solutions to the lower modes only.

In Section 6–2 it was shown how to obtain an estimate of the fundamental frequency without solving the eigenvalue problem. Specifically it was shown that Rayleigh's quotient provides an upper bound for the first eigenvalue λ_1,

$$R(u) \geq \lambda_1, \tag{6.15}$$

where the equality sign holds true if and only if the comparison function u is actually the first eigenfunction of the system. This implies that the true fundamental frequency is always smaller than the estimated one. The natural thing to do is to try to minimize the estimate. This is the essence of Rayleigh-Ritz method.

The method consists of selecting a trial family of comparison functions u_i, satisfying *all*† the boundary conditions of the problem, and constructing a linear combination

$$w_n = \sum_{i=1}^{n} a_i u_i, \tag{6.16}$$

† This requirement will be re-examined later (see Section 6–4).

where the u_i are known functions of the spatial coordinates linearly independent over domain D and the a_i are unknown coefficients to be determined. The set of comparison functions u_i is called a *generating set*. The above expression for w_n is substituted in Rayleigh's quotient and the coefficients a_i determined so as to render the quotient stationary. Essentially, in doing that, one approximates an infinitely-many-degrees-of-freedom system by an n-degrees-of-freedom system, so the constraints

$$a_{n+1} = a_{n+2} = \cdots = 0 \qquad (6.17)$$

are imposed on the system. Constraints have a tendency to raise the stiffness of the system, so the estimated frequency will be higher than the true fundamental frequency. By increasing the number n of comparison functions, constraints are eliminated, resulting in an estimate which is not higher and may be lower than the previous one.

If we substitute w_n in Rayleigh's quotient we obtain

$$R(w_n) = \frac{\displaystyle\int_D w_n L[w_n]\, dD}{\displaystyle\int_D w_n M[w_n]\, dD} = \frac{N(w_n)}{D(w_n)}, \qquad (6.18)$$

where

$$N(w_n) = \int_D w_n L[w_n]\, dD, \qquad (6.19)$$

$$D(w_n) = \int_D w_n M[w_n]\, dD \qquad (6.20)$$

are the numerator and denominator of Rayleigh's quotient, respectively. Of course, both the numerator and denominator depend on the function w_n and are both positive for a positive definite system.

The necessary conditions for the minimum† of Rayleigh's quotient are

$$\frac{\partial R(w_n)}{\partial a_j} = \frac{D(w_n)[\partial N(w_n)/\partial a_j] - N(w_n)[\partial D(w_n)/\partial a_j]}{D^2(w_n)} = 0,$$

$$j = 1, 2, \ldots, n. \quad (6.21)$$

Denote the minimum estimated value of Rayleigh's quotient

$$\min R(w_n) = {}^n\Lambda, \qquad (6.22)$$

† Recall that for a continuous system Rayleigh's quotient has no maximum.

so that conditions (6.21) become

$$\frac{\partial N(w_n)}{\partial a_j} - {}^n\Lambda \frac{\partial D(w_n)}{\partial a_j} = 0, \qquad j = 1, 2, \ldots, n. \tag{6.23}$$

Now introduce the notation

$$\left.\begin{aligned} k_{ij} &= \int_D u_i L[u_j] \, dD \\ m_{ij} &= \int_D u_i M[u_j] \, dD \end{aligned}\right\} \qquad i, j = 1, 2, \ldots, n, \tag{6.24}$$

and if the system is self-adjoint we have

$$k_{ij} = k_{ji}, \qquad m_{ij} = m_{ji}. \tag{6.25}$$

The operators L and M are linear, so we can write

$$N = \int_D \sum_{i=1}^n a_i u_i L\left[\sum_{j=1}^n a_j u_j\right] dD$$

$$= \sum_{i=1}^n \sum_{j=1}^n a_i a_j \int_D u_i L[u_j] \, dD = \sum_{i=1}^n \sum_{j=1}^n k_{ij} a_i a_j, \tag{6.26}$$

and, similarly,

$$D = \int_D \sum_{i=1}^n a_i u_i M\left[\sum_{j=1}^n a_j u_j\right] dD = \sum_{i=1}^n \sum_{j=1}^n m_{ij} a_i a_j. \tag{6.27}$$

Taking the partial derivatives with respect to a_r and recalling the symmetry of the coefficients k_{ij}, we write

$$\frac{\partial N}{\partial a_r} = \sum_{i=1}^n \sum_{j=1}^n \left(k_{ij} \frac{\partial a_i}{\partial a_r} a_j + k_{ij} a_i \frac{\partial a_j}{\partial a_r}\right)$$

$$= \sum_{i=1}^n \sum_{j=1}^n (k_{ij} \delta_{ir} a_j + k_{ij} a_i \delta_{jr})$$

$$= \sum_{j=1}^n k_{rj} a_j + \sum_{i=1}^n k_{ir} a_i = 2 \sum_{j=1}^n k_{rj} a_j, \qquad r = 1, 2, \ldots, n. \tag{6.28}$$

Similarly,

$$\frac{\partial D}{\partial a_r} = 2 \sum_{j=1}^n m_{rj} a_j, \qquad r = 1, 2, \ldots, n. \tag{6.29}$$

Introducing (6.28) and (6.29) in (6.23) we obtain

$$\sum_{j=1}^{n} (k_{rj} - {}^{n}\Lambda m_{rj})a_j = 0, \qquad r = 1, 2, \ldots, n, \tag{6.30}$$

which represents a set of n homogeneous algebraic equations in the unknowns a_j known as *Galerkin's equations*. They represent the eigenvalue problem for an n-degree-of-freedom system and can be written in the matrix form

$$[k]\{a\} = {}^{n}\Lambda[m]\{a\}, \tag{6.31}$$

where $[k]$ and $[m]$ are $n \times n$ symmetric matrices. The solution of the eigenvalue problem (6.31) can be obtained by any of the methods discussed previously and, if the system is positive definite, it consists of n positive eigenvalues ${}^{n}\Lambda_r$ and n corresponding eigenvectors $\{a^{(r)}\}$. The eigenvalues ${}^{n}\Lambda_r$ of the n-degree-of-freedom system, when arranged in increasing order of magnitude, provide *upper bounds*† for the true eigenvalues λ_r,

$$ {}^{n}\Lambda_r \geq \lambda_r, \qquad r = 1, 2, \ldots, n. \tag{6.32}$$

The eigenfunctions associated with the estimated eigenvalues ${}^{n}\Lambda_r$ are obtained by introducing the coefficients $a_i^{(r)}$ in (6.16),

$$w_n^{(r)} = \sum_{i=1}^{n} a_i^{(r)} u_i. \tag{6.33}$$

It can be easily shown that the eigenfunctions are orthogonal in a generalized sense because the eigenvectors $\{a^{(r)}\}$ are orthogonal with respect to the matrices $[m]$ and $[k]$.

The Rayleigh-Ritz method calls for construction of a minimizing sequence consisting of series of comparison functions

$$
\begin{aligned}
w_1 &= a_1 u_1, \\
w_2 &= a_1 u_1 + a_2 u_2 = \sum_{i=1}^{2} a_i u_i, \\
&\;\;\vdots \\
w_n &= a_1 u_1 + a_2 u_2 + \cdots + a_n u_n = \sum_{i=1}^{n} a_i u_i,
\end{aligned}
\tag{6.34}
$$

and the calculation of the corresponding frequencies ${}^{n}\Lambda_r$ and the associated modes $w_n^{(r)}$, where n indicates the number of comparison functions used in

† This statement can be proved by means of Courant's maximum-minimum principle. See L. Collatz, *Eigenwertprobleme*, Chelsea, New York, 1948, p. 237, or R. Weinstock. *Calculus of Variations*, McGraw-Hill, New York, 1952, p. 171.

the series. Because every new term added to the series reduces the number of constraints by one, the estimated values, $^n\Lambda_r$, of Rayleigh's quotient should approach the eigenvalues from above,

$$\lim_{n \to \infty} R(w_n^{(r)}) = \lim_{n \to \infty} {}^n\Lambda_r = \lambda_r. \qquad (6.35)$$

One has yet to prove that the minimizing sequence converges to the solution. In general, because of the stationarity of Rayleigh's quotient, one can expect that the estimated eigenvalues will provide a better approximation for the true eigenvalues than the estimated eigenfunctions will provide for the true eigenfunctions.

The quality of the estimates depends to a large extent on the choice of the functions u_i. The need to select a set of functions u_i is the main criticism of the Rayleigh-Ritz method, as one is hesitant about deciding upon a set of functions u_i. In the beginning the method seems surrounded by vagueness but, with experience, one becomes more and more confident using it. In many cases it is possible to base the decision as to the choice of functions u_i on the knowledge of the eigenfunctions of a system with a slightly different configuration.

In selecting the functions u_i it may be possible to reduce the computational work by choosing orthogonal functions such as trigonometric functions, Bessel functions, Legendre polynomials, etc. Furthermore, it may prove advantageous to use a set of functions that is orthonormal in a generalized sense. To this end we start with the given set of normalized, but not necessarily orthogonal, comparison functions u_i $(i = 1, 2, \ldots, n)$, and consider a new set of comparison functions v_i $(i = 1, 2, \ldots, n)$, obtained from the first set by means of the relations

$$v_1 = u_1,$$

$$v_2 = a_{12}v_1 + u_2,$$

$$v_3 = a_{13}v_1 + a_{23}v_2 + u_3, \qquad (6.36)$$

$$\vdots$$

$$v_n = a_{1n}v_1 + a_{2n}v_2 + a_{3n}v_3 + \cdots + u_n,$$

where the coefficients a_{rs} are chosen so as to render the functions v_i orthonormal in a generalized sense. For example, the coefficient a_{12} can be obtained by writing

$$\int_D v_1 M[v_2] \, dD = a_{12} \int_D v_1 M[v_1] \, dD + \int_D v_1 M[u_2] \, dD = 0,$$

from which we obtain

$$a_{12} = -\frac{\int_D v_1 M[u_2]\, dD}{\int_D v_1 M[v_1]\, dD} = -\int_D v_1 M[u_2]\, dD, \qquad (6.37)$$

because $v_1 = u_1$ was already normalized. The coefficients a_{13} and a_{23} can be obtained by writing

$$\int_D v_1 M[v_3]\, dD = a_{13} \int_D v_1 M[v_1]\, dD + a_{23} \int_D v_1 M[v_2]\, dD$$

$$+ \int_D v_1 M[u_3]\, dD = 0,$$

$$\int_D v_2 M[v_3]\, dD = a_{13} \int_D v_2 M[v_1]\, dD + a_{23} \int_D v_2 M[v_2]\, dD$$

$$+ \int_D v_2 M[u_3]\, dD = 0,$$

which lead to

$$a_{13} = -\int_D v_1 M[u_3]\, dD, \qquad a_{23} = -\int_D v_2 M[u_3]\, dD, \qquad (6.38)$$

and, in general, it can be shown that

$$a_{rs} = -\int_D v_r M[u_s]\, dD, \qquad \begin{cases} r = 1, 2, \ldots, n-1, \\ s = 2, 3, \ldots, n. \end{cases} \qquad (6.39)$$

This procedure is called the *Schmidt orthogonalization process*.† If such orthonormal comparison functions, v_i, are used, the eigenvalue problem reduces to the special type

$$[k]\{a\} = {}^n\Lambda\{a\}. \qquad (6.40)$$

When the system possesses certain symmetry it is possible to reduce the computational work. To this end consider a one-dimensional system, $-L/2 \leq x \leq L/2$, possessing the following properties:

$$L[u_i(x)] = L[u_i(-x)], \qquad M[u_i(x)] = M[u_i(-x)] \qquad (6.41)$$

† It is also referred to in the literature as the *Gram-Schmidt orthogonalization process*.

if the comparison function $u_i(x)$ is an even function of x,

$$u_i(x) = u_i(-x), \tag{6.42}$$

and

$$L[u_i(x)] = -L[u_i(-x)], \qquad M[u_i(x)] = -M[u_i(-x)] \tag{6.43}$$

if the comparison function $u_i(x)$ is an odd function of x,

$$u_i(x) = -u_i(-x). \tag{6.44}$$

Equations (6.41) to (6.44) imply that the system possesses symmetric stiffness and mass distributions where the center of symmetry is the origin $x = 0$. It follows that (see Problem 6.5)

$$k_{ij} = 2 \int_0^{L/2} u_i L[u_j] \, dx, \qquad m_{ij} = 2 \int_0^{L/2} u_i M[u_j] \, dx \tag{6.45}$$

if the comparison functions u_i and u_j are such that

$$u_i(x) = u_i(-x) \qquad \text{and} \qquad u_j(x) = u_j(-x) \tag{6.46}$$

or $\qquad u_i(x) = -u_i(-x) \qquad \text{and} \qquad u_j(x) = -u_j(-x), \tag{6.47}$

which means that they are either both even or both odd functions of x. On the other hand, if

$$u_i(x) = u_i(-x) \qquad \text{and} \qquad u_j(x) = -u_j(-x) \tag{6.48}$$

or $\qquad u_i(x) = -u_i(-x) \qquad \text{and} \qquad u_j(x) = u_j(-x), \tag{6.49}$

which means that one of the comparison functions is an even function of x and the other an odd function of x, we obtain

$$k_{ij} = 0, \qquad m_{ij} = 0. \tag{6.50}$$

If the system is self-adjoint and the comparison functions are chosen such that some of them are even functions of x and the remaining ones odd functions of x,

$$u_i(x) = u_i(-x), \qquad 1 \le i \le r,$$

$$u_i(x) = -u_i(-x), \qquad r < i \le n, \tag{6.51}$$

the eigenvalue problem (6.31) can be written in terms of partitioned matrices as

$$\begin{bmatrix} [k]_e & | & [0] \\ \hline [0] & | & [k]_o \end{bmatrix} \begin{Bmatrix} \{a\}_e \\ \{a\}_o \end{Bmatrix} = {}^{n}\Lambda \begin{bmatrix} [m]_e & | & [0] \\ \hline [0] & | & [m]_o \end{bmatrix} \begin{Bmatrix} \{a\}_e \\ \{a\}_o \end{Bmatrix}, \tag{6.52}$$

which reduces to

$$[k]_e\{a\}_e = {}^{r}\Lambda[m]_e\{a\}_e, \tag{6.53}$$

$$[k]_o\{a\}_o = {}^{n-r}\Lambda[m]_o\{a\}_o, \tag{6.54}$$

which are two independent eigenvalue problems of order r and $n - r$ for the even and odd modes, respectively. The matrices $[k]_e$ and $[m]_e$ are symmetric matrices of order r and $[k]_o$ and $[m]_o$ are symmetric matrices of order $n - r$.

Thus, for a system with properties as described by (6.41) through (6.44), it is advantageous to select a set of comparison functions $u_i(x)$ consisting only of even and odd functions of x because it is simpler to solve two eigenvalue problems of order r and $n - r$, respectively, than to solve one eigenvalue problem of order n.

Example 6.2

Let us consider the system of Example 6.1, and although the eigenvalue problem can be solved in closed form we shall use the Rayleigh-Ritz method to evaluate the natural frequencies and natural modes of the system shown in Figure 6.1. We shall use as the minimizing sequence

$$w_1 = a_1 u_1 = a_1 \sin \frac{\pi x}{2L},$$

$$w_2 = a_1 u_1 + a_2 u_2 = a_1 \sin \frac{\pi x}{2L} + a_2 \sin \frac{3\pi x}{2L}, \tag{a}$$

$$w_3 = a_1 u_1 + a_2 u_2 + a_3 u_3 = a_1 \sin \frac{\pi x}{2L} + a_2 \sin \frac{3\pi x}{2L} + a_3 \sin \frac{5\pi x}{2L}.$$

The mass per unit length and the stiffness are

$$m(x) = 2m\left(1 - \frac{x}{L}\right), \qquad EA(x) = 2EA\left(1 - \frac{x}{L}\right), \tag{b}$$

so the operators L and M are

$$L = 2EA\left[\frac{1}{L}\frac{d}{dx} - \left(1 - \frac{x}{L}\right)\frac{d^2}{dx^2}\right], \qquad M = 2m\left(1 - \frac{x}{L}\right). \qquad (c)$$

Obviously the latter is not really a differential operator.

Using equations (6.24) we write

$$k_{11} = \int_0^L u_1 L[u_1]\,dx$$

$$= \int_0^L \sin\frac{\pi x}{2L}\left\{2EA\left[\frac{1}{L}\frac{d}{dx} - \left(1 - \frac{x}{L}\right)\frac{d^2}{dx^2}\right]\sin\frac{\pi x}{2L}\right\}dx$$

$$= 2EA\int_0^L \sin\frac{\pi x}{2L}\left[\frac{\pi}{2L^2}\cos\frac{\pi x}{2L} + \left(1 - \frac{x}{L}\right)\frac{\pi^2}{4L^2}\sin\frac{\pi x}{2L}\right]dx$$

$$= \frac{EA}{2L}\left(1 + \frac{\pi^2}{4}\right), \qquad (d)$$

$$m_{11} = \int_0^L u_1 M[u_1]\,dx = 2m\int_0^L \left(1 - \frac{x}{L}\right)\sin^2\frac{\pi x}{2L}\,dx = \frac{mL}{2}\left(1 - \frac{4}{\pi^2}\right), \quad (e)$$

and if we limit the sequence to only one term, we obtain from (6.30),

$$^1\Lambda_1 = \frac{k_{11}}{m_{11}} = \frac{EA}{mL^2}\frac{1 + (\pi^2/4)}{1 - (4/\pi^2)} = 5.8305\frac{EA}{mL^2}, \qquad (f)$$

which gives us the estimated fundamental frequency

$$^1\omega_1 = 2.4146\sqrt{\frac{EA}{mL^2}}. \qquad (g)$$

The corresponding mode is as given by the first of equations (a). We note that the result is exactly as given by Rayleigh's energy method.

Next we wish to use the second expression in the minimizing sequence (a). The system is self-adjoint, so we have

$$k_{12} = k_{21} = \int_0^L u_1 L[u_2]\,dx$$

$$= \int_0^L \sin\frac{\pi x}{2L}\left\{2EA\left[\frac{1}{L}\frac{d}{dx} - \left(1 - \frac{x}{L}\right)\frac{d^2}{dx^2}\right]\sin\frac{3\pi x}{2L}\right\}dx$$

$$= 2EA\int_0^L \sin\frac{\pi x}{2L}\left[\frac{3\pi}{2L^2}\cos\frac{3\pi x}{2L} + \left(1 - \frac{x}{L}\right)\frac{9\pi^2}{4L^2}\sin\frac{3\pi x}{2L}\right]dx = \frac{3EA}{2L}.$$

$$(h)$$

Similarly, we obtain

$$k_{22} = \int_0^L u_2 L[u_2] \, dx = \frac{EA}{2L} \left(1 + \frac{9\pi^2}{4} \right).$$ (i)

In addition,

$$m_{12} = m_{21} = \int_0^L u_1 M[u_2] \, dx = \frac{2mL}{\pi^2},$$ (j)

$$m_{22} = \int_0^L u_2 M[u_2] \, dx = \frac{mL}{2} \left(1 - \frac{4}{9\pi^2} \right).$$ (k)

The eigenvalue problem (6.31) takes the form

$$\frac{EA}{2L} \begin{bmatrix} 1 + \dfrac{\pi^2}{4} & 3 \\ 3 & 1 + \dfrac{9\pi^2}{4} \end{bmatrix} \begin{Bmatrix} a_1 \\ a_2 \end{Bmatrix} = {}^2\Lambda \frac{mL}{2} \begin{bmatrix} 1 - \dfrac{4}{\pi^2} & \dfrac{4}{\pi^2} \\ \dfrac{4}{\pi^2} & 1 - \dfrac{4}{9\pi^2} \end{bmatrix} \begin{Bmatrix} a_1 \\ a_2 \end{Bmatrix},$$ (l)

which yields the eigenvalues and eigenvectors

$$\begin{aligned}
{}^2\Lambda_1 &= 5.7897 \frac{EA}{mL^2}, & \{a^{(1)}\} &= \begin{Bmatrix} 1.0000 \\ -0.0369 \end{Bmatrix}, \\
\end{aligned}$$ (m)

$$\begin{aligned}
{}^2\Lambda_2 &= 30.5717 \frac{EA}{mL^2}, & \{a^{(2)}\} &= \begin{Bmatrix} 1.0000 \\ -1.5651 \end{Bmatrix}.
\end{aligned}$$

The above results lead to the estimated natural frequencies and natural modes

$$\begin{aligned}
{}^2\omega_1 &= 2.4062 \sqrt{\frac{EA}{mL^2}}, & w_2^{(1)} &= \sin \frac{\pi x}{2L} - 0.0369 \sin \frac{3\pi x}{2L}, \\
\end{aligned}$$ (n)

$$\begin{aligned}
{}^2\omega_2 &= 5.5292 \sqrt{\frac{EA}{mL^2}}, & w_2^{(2)} &= \sin \frac{\pi x}{2L} - 1.5651 \sin \frac{3\pi x}{2L},
\end{aligned}$$

and we note that using two comparison functions instead of one yields a better estimate,

$$ {}^2\omega_1 < {}^1\omega_1. $$ (o)

The true first and second natural frequencies are (see Problem 5.7)

$$\omega_1 = 2.4048\sqrt{\frac{EA}{ML^2}} < {}^2\omega_1 = 2.4062\sqrt{\frac{EA}{ML^2}} < {}^1\omega_1,$$

$$\omega_2 = 5.5201\sqrt{\frac{EA}{ML^2}} < {}^2\omega_2 = 5.5292\sqrt{\frac{EA}{ML^2}},$$ (p)

and we note that the estimates are very good indeed.

The third expression in the minimizing sequence allows us to evaluate the additional stiffness coefficients

$$k_{13} = k_{31} = \int_0^L u_1 L[u_3]\, dx = \frac{5EA}{18L},$$

$$k_{23} = k_{32} = \int_0^L u_2 L[u_3]\, dx = \frac{15EA}{2L},$$ (q)

$$k_{33} = \int_0^L u_3 L[u_3]\, dx = \frac{EA}{2L}\left(1 + \frac{25\pi^2}{4}\right),$$

and the additional mass coefficients

$$m_{13} = m_{31} = \int_0^L u_1 M[u_3]\, dx = -\frac{2mL}{9\pi^2},$$

$$m_{23} = m_{32} = \int_0^L u_2 M[u_3]\, dx = \frac{2mL}{\pi^2},$$ (r)

$$m_{33} = \int_0^L u_3 M[u_3]\, dx = \frac{mL}{2}\left(1 - \frac{4}{25\pi^2}\right).$$

The eigenvalue problem (6.31) can be written

$$\frac{EA}{2L}\begin{bmatrix} 1 + \dfrac{\pi^2}{4} & 3 & \dfrac{5}{9} \\[2mm] 3 & 1 + \dfrac{9\pi^2}{4} & 15 \\[2mm] \dfrac{5}{9} & 15 & 1 + \dfrac{25\pi^2}{4} \end{bmatrix}\begin{Bmatrix} a_1 \\ a_2 \\ a_3 \end{Bmatrix}$$

$$= {}^3\Lambda \frac{mL}{2}\begin{bmatrix} 1 - \dfrac{4}{\pi^2} & \dfrac{4}{\pi^2} & -\dfrac{4}{9\pi^2} \\[2mm] \dfrac{4}{\pi^2} & 1 - \dfrac{4}{9\pi^2} & \dfrac{4}{\pi^2} \\[2mm] -\dfrac{4}{9\pi^2} & \dfrac{4}{\pi^2} & 1 - \dfrac{4}{25\pi^2} \end{bmatrix}\begin{Bmatrix} a_1 \\ a_2 \\ a_3 \end{Bmatrix}. \quad (s)$$

The solution of the eigenvalue problem (s) consists of the eigenvalues and eigenvectors

$$^3\Lambda_1 = 5.7837 \frac{EA}{mL^2}, \quad \{a^{(1)}\} = \begin{Bmatrix} 1.0000 \\ -0.0319 \\ -0.0072 \end{Bmatrix},$$

$$^3\Lambda_2 = 30.4878 \frac{EA}{mL^2}, \quad \{a^{(2)}\} = \begin{Bmatrix} 1.0000 \\ -1.5540 \\ 0.0666 \end{Bmatrix}, \tag{t}$$

$$^3\Lambda_3 = 75.0751 \frac{EA}{mL^2}, \quad \{a^{(3)}\} = \begin{Bmatrix} 1.0000 \\ -1.2110 \\ 2.0270 \end{Bmatrix},$$

and consequently the estimated natural frequencies and natural modes are

$$^3\omega_1 = 2.4049\sqrt{\frac{EA}{mL^2}}, \quad w_3^{(1)} = \sin\frac{\pi x}{2L} - 0.0319 \sin\frac{3\pi x}{2L} - 0.0072 \sin\frac{5\pi x}{2L},$$

$$^3\omega_2 = 5.5216\sqrt{\frac{EA}{mL^2}}, \quad w_3^{(2)} = \sin\frac{\pi x}{2L} - 1.5440 \sin\frac{3\pi x}{2L} + 0.0666 \sin\frac{5\pi x}{2L}, \tag{u}$$

$$^3\omega_3 = 8.6646\sqrt{\frac{EA}{mL^2}}, \quad w_3^{(3)} = \sin\frac{\pi x}{2L} - 1.2110 \sin\frac{3\pi x}{2L} + 2.0270 \sin\frac{5\pi x}{2L}.$$

We note that the estimates of the first two natural frequencies have improved and, in addition, we obtained an estimate of the third natural frequencies. All the estimates are slightly higher than the true natural frequencies, and the first two estimates are extremely close to the true values. The estimated natural frequencies and the true natural frequencies are

$$\omega_1 = 2.4048\sqrt{\frac{EA}{mL^2}} < {}^3\omega_1 = 2.4049\sqrt{\frac{EA}{mL^2}} < {}^2\omega_1 < {}^1\omega_1,$$

$$\omega_2 = 5.5201\sqrt{\frac{EA}{mL^2}} < {}^3\omega_2 = 5.5216\sqrt{\frac{EA}{mL^2}} < {}^2\omega_2, \tag{v}$$

$$\omega_3 = 8.6537\sqrt{\frac{EA}{mL^2}} < {}^3\omega_3 = 8.6646\sqrt{\frac{EA}{mL^2}}.$$

Example 6.3

Let us consider again the system of Example 6.1 and use the orthogonalization process to derive a minimizing sequence consisting of orthogonal functions. To this end we let the orthogonal functions be

$$v_1 = u_1 = \sin \frac{\pi x}{2L},$$

$$v_2 = a_{12}v_1 + u_2 = a_{12} \sin \frac{\pi x}{2L} + \sin \frac{3\pi x}{2L},$$

$$v_3 = a_{13}v_1 + a_{23}v_2 + u_3 \tag{a}$$

$$= a_{13} \sin \frac{\pi x}{2L} + a_{23}\left(a_{12} \sin \frac{\pi x}{2L} + \sin \frac{3\pi x}{2L}\right) + \sin \frac{5\pi x}{2L},$$

. . . .

The operator M is just the function

$$m(x) = 2m\left(1 - \frac{x}{L}\right), \tag{b}$$

so the function v_2 is orthogonal to v_1 if the coefficient a_{12} takes the value

$$a_{12} = -\frac{\int_0^L mv_1u_2\, dx}{\int_0^L mv_1^2\, dx} = -\frac{\int_0^L 2m[1 - (x/L)] \sin (\pi x/2L) \sin (3\pi x/2L)\, dx}{\int_0^L 2m[1 - (x/L)] \sin^2 (\pi x/2L)\, dx}$$

$$= -\frac{2mL/\pi^2}{(mL/2)[1 - (4/\pi^2)]} = -\frac{1}{[(\pi^2/4) - 1]}, \tag{c}$$

so

$$v_2 = -\frac{1}{[(\pi^2/4) - 1]} \sin \frac{\pi x}{2L} + \sin \frac{3\pi x}{2L}. \tag{d}$$

The condition that v_3 be orthogonal to v_1 leads to

$$a_{13} = -\frac{\int_0^L mv_1u_3\, dx}{\int_0^L mv_1^2\, dx} = -\frac{\int_0^L 2m[1 - (x/L)] \sin (\pi x/2L) \sin (5\pi x/2L)\, dx}{\int_0^L 2m[1 - (x/L)] \sin^2 (\pi x/2L)\, dx}$$

$$= -\frac{-2mL/9\pi^2}{(mL/2)[1 - (4/\pi^2)]} = \frac{1}{9[(\pi^2/4) - 1]}. \tag{e}$$

Similarly, the condition that v_3 be orthogonal to v_2 gives

$$a_{23} = -\frac{\int_0^L m v_2 u_3 \, dx}{\int_0^L m v_2^2 \, dx}$$

$$= -\frac{\int_0^L 2m\left(1 - \frac{x}{L}\right)\left[-\frac{1}{(\pi^2/4) - 1} \sin\frac{\pi x}{2L} + \sin\frac{3\pi x}{2L}\right] \sin\frac{5\pi x}{2L} \, dx}{\left\{\int_0^L 2m\left(1 - \frac{x}{L}\right)\left[\frac{1}{[(\pi^2/4) - 1]^2} \sin^2\frac{\pi x}{2L}\right.\right.}$$
$$\left.\left. -\frac{2}{(\pi^2/4) - 1} \sin\frac{\pi x}{2L} \sin\frac{3\pi x}{2L} + \sin^2\frac{3\pi x}{2L}\right] dx\right\}$$

$$= -\frac{\dfrac{2mL}{9\pi^2}\dfrac{1}{(\pi^2/4) - 1} + \dfrac{2mL}{\pi^2}}{\dfrac{1}{[(\pi^2/4) - 1]^2}\dfrac{mL}{2}\left(1 - \dfrac{4}{\pi^2}\right) - \dfrac{2}{(\pi^2/4) - 1}\dfrac{2mL}{\pi^2} + \dfrac{mL}{2}\left(1 - \dfrac{4}{9\pi^2}\right)}$$

$$= \frac{(9\pi^2/4) - 8}{9 - [(\pi^2/4) - 1][(9\pi^2/4) - 1]}, \tag{f}$$

so

$$v_3 = \frac{1}{9[(\pi^2/4) - 1]} \sin\frac{\pi x}{2L} + \frac{(9\pi^2/4) - 8}{9 - [(\pi^2/4) - 1][(9\pi^2/4) - 1]}$$

$$\times \left[-\frac{1}{(\pi^2/4) - 1} \sin\frac{\pi x}{2L} + \sin\frac{3\pi x}{2L}\right] + \sin\frac{5\pi x}{2L}$$

$$= -\frac{(9\pi^2/4) + 80}{9\{9 - [(\pi^2/4) - 1][(9\pi^2/4) - 1]\}} \sin\frac{\pi x}{2L}$$

$$+ \frac{(9\pi^2/4) - 8}{9 - [(\pi^2/4) - 1][(9\pi^2/4) - 1]} \sin\frac{3\pi x}{2L} + \sin\frac{5\pi x}{2L}. \tag{g}$$

The same process can be continued to obtain $v_4, v_5, \ldots, v_n$.

When the Rayleigh-Ritz method is used in conjunction with the minimizing sequence

$$w_1 = a_1 v_1,$$
$$w_2 = a_1 v_1 + a_2 v_2,$$
$$w_3 = a_1 v_1 + a_2 v_2 + a_3 v_3, \tag{h}$$

$$\cdot \quad \cdot \quad \cdot \quad \cdot \quad \cdot \quad \cdot \quad \cdot$$

the resulting eigenvalue problem is of the special type in the sense that the inertia matrix is diagonal.

Although an eigenvalue problem of the special type may be easier to solve, the orthogonalization process may become sufficiently involved to cancel this advantage. We note, however, that the same integrals which must be evaluated for the orthogonalization process must also be evaluated to obtain the inertia coefficients m_{ij}. On the other hand, evaluation of the coefficients k_{ij} may become more laborious. Hence one must weigh the advantage of using the orthogonalization process and solving a special eigenvalue problem as opposed to solving a relatively more complex eigenvalue problem.

6–4 RAYLEIGH–RITZ METHOD. RE-EXAMINATION OF BOUNDARY CONDITION REQUIREMENTS

In Section 6–3 we discussed the Rayleigh-Ritz method, which consisted of assuming a solution of the eigenvalue problem in the form of a finite series of comparison functions. Thus we avoided the difficult task of solving the differential equation defining the eigenvalue problem. One is still faced with the problem of choosing a set of comparison functions which, according to the definition, must satisfy both the geometric and natural boundary conditions of the problem and must be $2p$ times differentiable over the domain D such that the operators L and M are defined throughout that domain. These requirements are quite extensive and we wish to examine the possibility of relaxing some of them. Indeed, we will show that the chosen functions need satisfy only the geometric boundary conditions and be only p times differentiable over the domain D, so that in effect we can use for the minimizing sequence series consisting of *admissible functions* instead of comparison functions (see Section 5–4 for the definition of the various classes of functions). However, this would require the writing of Rayleigh's quotient in terms of the maximum potential energy, V_{max}, and reference kinetic energy, T^*.

In writing the energy expressions it is more convenient to refer to a specific system, so let us consider a bar in longitudinal vibration with one end

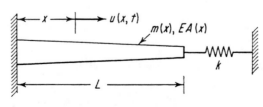

FIGURE 6.2

clamped and the other one attached to a spring (Figure 6.2). The eigenvalue
problem for such a system is defined by the differential equation

$$-\frac{d}{dx}\left[EA(x)\frac{dU(x)}{dx}\right] = \omega^2 m(x)U(x) \qquad (6.55)$$

to be satisfied over the domain $0 < x < L$, and by the boundary conditions

$$U(0) = 0, \qquad (6.56)$$

$$EA(L)\frac{dU(x)}{dx}\bigg|_{x=L} = -kU(L). \qquad (6.57)$$

It is easy to see that boundary condition (6.56) is of the geometric type and
that (6.57) represents a natural boundary condition because it reflects the
axial force balance at the end $x = L$. The eigenvalue problem is of the
special type, the operator L being defined by

$$L = -\frac{d}{dx}\left[EA(x)\frac{d}{dx}\right], \qquad (6.58)$$

which is of second order ($p = 1$), whereas M is just a function of the spatial
coordinate x,

$$M = m(x), \qquad (6.59)$$

and represents the mass distribution of the system. It can easily be shown
that the system is self-adjoint and positive definite.

The numerator of Rayleigh's quotient, after an integration by parts, takes
the form

$$\int_D uL[u]\,dD = -\int_0^L U(x)\frac{d}{dx}\left[EA(x)\frac{dU(x)}{dx}\right]\,dx$$

$$= -U(x)EA(x)\frac{dU(x)}{dx}\bigg|_0^L + \int_0^L EA(x)\left[\frac{dU(x)}{dx}\right]^2\,dx,$$

and if boundary conditions (6.56) and (6.57) are used, the above reduces to

$$\int_D uL[u]\,dD = kU^2(L) + \int_0^L EA(x)\left[\frac{dU(x)}{dx}\right]^2\,dx = 2V_{max}. \qquad (6.60)$$

The denominator of Rayleigh's quotient is simply

$$\int_D uM[u]\,dD = \int_0^L m(x)U^2(x)\,dx = 2T^*, \qquad (6.61)$$

so Rayleigh's quotient assumes the form

$$\omega^2 = R(U(x)) = \frac{kU^2(L) + \int_0^L EA(x)[dU(x)/dx]^2 \, dx}{\int_0^L m(x)U^2(x) \, dx} = \frac{V_{max}}{T^*}, \qquad (6.62)$$

and we note that by writing Rayleigh's quotient as

$$\omega^2 = R(u) = \frac{V_{max}}{T^*} \qquad (6.63)$$

instead of the form given in (5.304), we automatically take into account the natural boundary condition (6.57). In addition, the quotient in the form (6.63) is defined for functions that are only p times rather than $2p$ times differentiable. Hence if we use the Rayleigh-Ritz method with the quotient in the form (6.63) we can relax the requirements and demand that the functions used in the minimizing sequence satisfy only the geometric boundary conditions and be only p times differentiable, which is precisely the definition of *admissible functions*. The requirement on the differentiability of the functions has no particular consequence, because the chosen functions would, most likely, be $2p$ times differentiable anyway. On the other hand, the fact that we can use admissible rather than comparison functions is of great importance, because the class of comparison functions is merely a subclass of the admissible functions class; hence there is a considerably larger class of functions to choose from.

In conclusion, if we use the Rayleigh's quotient in the form (6.63), the minimizing sequence of functions can be written

$$w_n = \sum_{i=1}^{n} a_i \phi_i, \qquad (6.64)$$

where a_i are adjustable coefficients and ϕ_i denotes a set of admissible functions. The procedure for the minimization of Rayleigh's quotient remains the same as the one described in Section 6–3.

It may be worth mentioning that when the quotient is in the form (6.63), it is possible to treat systems with both distributed and discrete properties. In the example of Figure 6.2 one can regard $EA(x)$ as distributed stiffness and the spring k as a discrete stiffness. The same applies to distributed and concentrated masses, in which case the kinetic energy expression contains additional terms, accounting for the kinetic energy associated with the discrete masses.

Example 6.4   Admissible vs. Comparison Functions

Consider the system of Figure 6.2 and use the minimizing sequence

$$w_n = \sum_{i=1}^{n} a_i \phi_i(x) \tag{a}$$

to obtain an approximate solution of the eigenvalue problem associated with the system. For comparison purposes we shall solve the problem twice, the first time by selecting as the functions ϕ_i ($i = 1, 2, \ldots, n$) a set of admissible functions, and the second time using a set of comparison functions. Let the stiffness and mass distributions be

$$EA(x) = \frac{6}{5} EA\left(1 - \frac{1}{3}\frac{x}{L}\right), \qquad m(x) = \frac{6}{5} m\left(1 - \frac{1}{3}\frac{x}{L}\right). \tag{b}$$

Rayleigh's quotient can be written

$$R(w_n) = \frac{kw_n^2(L) + \int_0^L EA(x)(dw_n/dx)^2 \, dx}{\int_0^L m(x)w_n^2 \, dx}$$

$$= \frac{\left\{ k\left[\sum_{i=1}^{n} a_i\phi_i(L)\right]\left[\sum_{j=1}^{n} a_j\phi_j(L)\right] + \frac{6EA}{5} \int_0^L \left(1 - \frac{1}{3}\frac{x}{L}\right)\left[\sum_{i=1}^{n} a_i\phi_i'(x)\right] \times \left[\sum_{j=1}^{n} a_j\phi_j'(x)\right] dx \right\}}{\frac{6m}{5} \int_0^L \left(1 - \frac{1}{3}\frac{x}{L}\right)\left[\sum_{i=1}^{n} a_i\phi_i(x)\right]\left[\sum_{j=1}^{n} a_j\phi_j(x)\right] dx}$$

$$= \frac{\sum_{i=1}^{n}\sum_{j=1}^{n} k_{ij}a_i a_j}{\sum_{i=1}^{n}\sum_{j=1}^{n} m_{ij}a_i a_j}, \tag{c}$$

where the stiffness and mass coefficients are given by

$$k_{ij} = k\phi_i(L)\phi_j(L) + \frac{6EA}{5} \int_0^L \left(1 - \frac{1}{3}\frac{x}{L}\right)\phi_i'(x)\phi_j'(x) \, dx, \tag{d}$$

$$m_{ij} = \frac{6m}{5} \int_0^L \left(1 - \frac{1}{3}\frac{x}{L}\right)\phi_i(x)\phi_j(x) \, dx.$$

The problem possesses one geometric boundary condition, at $x = 0$, and a natural one, at $x = L$. As admissible functions we can choose the eigenfunctions of a uniform bar fixed at $x = 0$ and free at $x = L$, whereas the comparison functions can be taken as the eigenfunctions of a uniform bar fixed at $x = 0$ and with a spring attached at the end $x = L$ (see Problem 5.6). Both sets of functions can be written

$$\phi_i(x) = \sin \beta_i x, \qquad i = 1, 2, \ldots, n. \tag{e}$$

Of course the coefficients β_i will be different in each case.

The stiffness coefficients for $i = j$ are

$$k_{ii} = k \sin^2 \beta_i L + \frac{6EA}{5} \beta_i^2 \int_0^L \left(1 - \frac{1}{3}\frac{x}{L}\right) \cos^2 \beta_i x \, dx$$

$$= \left(k + \frac{EA}{10L}\right) \sin^2 \beta_i L + \frac{EA}{2L} (\beta_i L)^2 + \frac{EA}{5L} (\beta_i L) \sin \beta_i L, \tag{f}$$

and, similarly, for $i \neq j$ they are

$$k_{ij} = k \sin \beta_i L \sin \beta_j L + \frac{6EA}{5} \beta_i \beta_j \int_0^L \left(1 - \frac{1}{3}\frac{x}{L}\right) \cos \beta_i x \cos \beta_j x \, dx$$

$$= k \sin \beta_i L \sin \beta_j L + \frac{EA}{5L} (\beta_i L)(\beta_j L)\left\{2 \frac{\sin (\beta_i - \beta_j)L}{(\beta_i - \beta_j)L} + 2 \frac{\sin (\beta_i + \beta_j)L}{(\beta_i + \beta_j)L}\right.$$

$$\left. + \frac{1 - \cos (\beta_i - \beta_j)L}{[(\beta_i - \beta_j)L]^2} + \frac{1 - \cos (\beta_i + \beta_j)L}{[(\beta_i + \beta_j)L]^2}\right\}. \tag{g}$$

The mass coefficients for $i = j$ are

$$m_{ii} = \frac{6m}{5} \int_0^L \left(1 - \frac{1}{3}\frac{x}{L}\right) \sin^2 \beta_i x \, dx$$

$$= \frac{mL}{10} \left[5 - 2 \frac{\sin 2\beta_i L}{\beta_i L} - \frac{\sin^2 \beta_i L}{(\beta_i L)^2}\right], \tag{h}$$

and for $i \neq j$ we have

$$m_{ij} = \frac{6m}{5} \int_0^L \left(1 - \frac{1}{3}\frac{x}{L}\right) \sin \beta_i x \sin \beta_j x \, dx$$

$$= \frac{mL}{5} \left\{2 \frac{\sin (\beta_i - \beta_j)L}{(\beta_i - \beta_j)L} - 2 \frac{\sin (\beta_i + \beta_j)L}{(\beta_i + \beta_j)L}\right.$$

$$\left. + \frac{1 - \cos (\beta_i - \beta_j)L}{[(\beta_i - \beta_j)L]^2} - \frac{1 - \cos (\beta_i + \beta_j)L}{[(\beta_i + \beta_j)L]^2}\right\}. \tag{i}$$

We are primarily interested in an evaluation of the results obtained by using admissible functions as opposed to using comparison functions, so we shall not be concerned with a sequence and we shall choose instead only one value for n, say $n = 4$. The two cases evaluated for

$$k = \frac{EA}{6.25L} \tag{j}$$

are presented below.

(a) Admissible Functions

The eigenfunctions of a uniform rod fixed at $x = 0$ and free at $x = L$ are $\sin (2i - 1)(\pi x/2L)$ $(i = 1, 2, 3, 4)$, so

$$\beta_i = (2i - 1)\frac{\pi}{2L}, \qquad i = 1, 2, 3, 4, \tag{k}$$

for which we obtain the stiffness matrix

$$[k] = \frac{EA}{50L}
\begin{bmatrix}
13 + \dfrac{25\pi^2}{4} & 7 & \frac{97}{9} & -\frac{37}{9} \\[2ex]
7 & 13 + \dfrac{225\pi^2}{4} & 67 & \frac{61}{5} \\[2ex]
\frac{97}{9} & 67 & 13 + \dfrac{625\pi^2}{4} & 167 \\[2ex]
-\frac{37}{9} & \frac{61}{5} & 167 & 13 + \dfrac{1225\pi^2}{4}
\end{bmatrix} \tag{l}$$

and the mass matrix

$$[m] = \frac{mL}{10}
\begin{bmatrix}
5 - \dfrac{4}{\pi^2} & \dfrac{4}{\pi^2} & -\dfrac{4}{9\pi^2} & \dfrac{4}{9\pi^2} \\[2ex]
\dfrac{4}{\pi^2} & 5 - \dfrac{4}{9\pi^2} & \dfrac{4}{\pi^2} & -\dfrac{4}{25\pi^2} \\[2ex]
-\dfrac{4}{9\pi^2} & \dfrac{4}{\pi^2} & 5 - \dfrac{4}{25\pi^2} & \dfrac{4}{\pi^2} \\[2ex]
\dfrac{4}{9\pi^2} & -\dfrac{4}{25\pi^2} & \dfrac{4}{\pi^2} & 5 - \dfrac{4}{49\pi^2}
\end{bmatrix} . \tag{m}$$

The eigenvalue problem assumes the form

$$[k]\{a\} = {}^4\Lambda[m]\{a\}, \tag{n}$$

where the matrices $[k]$ and $[m]$ are given by (1) and (m) and the eigenvalues $^4\Lambda_r$ are equal to the corresponding natural frequencies squared. The solution of the eigenvalue problem, obtained by means of an electronic computer, is given in the form of the matrix of the eigenvalues and the matrix of the eigenvectors

$$[\Lambda] = \frac{EA}{mL^2} \begin{bmatrix} 3.2465 & 0 & 0 & 0 \\ 0 & 23.0558 & 0 & 0 \\ 0 & 0 & 62.5432 & 0 \\ 0 & 0 & 0 & 121.8333 \end{bmatrix} \tag{o}$$

$$[a] = \begin{bmatrix} 1.0000 & 1.0000 & 1.0000 & 1.0000 \\ 0.0000 & -11.3831 & -2.6524 & -1.0603 \\ -0.0078 & 0.2119 & 41.2533 & 4.1102 \\ 0.0021 & 0.0585 & -1.0830 & -73.3248 \end{bmatrix}. \tag{p}$$

Hence the first four estimated natural frequencies are

$$\omega_1 = 1.8018\sqrt{\frac{EA}{mL^2}}, \qquad \omega_2 = 4.8016\sqrt{\frac{EA}{mL^2}},$$

$$\omega_3 = 7.9084\sqrt{\frac{EA}{mL^2}}, \qquad \omega_4 = 11.0378\sqrt{\frac{EA}{mL^2}}, \tag{q}$$

and the corresponding four natural modes are

$$w^{(1)}(x) = \sin\frac{\pi x}{2L} + 0.0000\sin\frac{3\pi x}{2L} - 0.0078\sin\frac{5\pi x}{2L} + 0.0021\sin\frac{7\pi x}{2L},$$

$$w^{(2)}(x) = \sin\frac{\pi x}{2L} - 11.3831\sin\frac{3\pi x}{2L} + 0.2119\sin\frac{5\pi x}{2L} + 0.0585\sin\frac{7\pi x}{2L},$$

$$w^{(3)}(x) = \sin\frac{\pi x}{2L} - 2.6524\sin\frac{3\pi x}{2L} + 41.2533\sin\frac{5\pi x}{2L} - 1.0830\sin\frac{7\pi x}{2L}, \tag{r}$$

$$w^{(4)}(x) = \sin\frac{\pi x}{2L} - 1.0603\sin\frac{3\pi x}{2L} + 4.1102\sin\frac{5\pi x}{2L} - 73.3248\sin\frac{7\pi x}{2L}.$$

Note that the natural modes are not normalized.

(b) Comparison Functions

The eigenfunctions of a uniform rod fixed at $x = 0$ and with a spring k at $x = L$ are also $\sin \beta_i x$ $(i = 1, 2, \ldots)$; however, in contrast with the previous case, the constants β_i are not odd multiples of $\pi/2$ but must satisfy the characteristic equation

$$EA(L)(\beta_i L) \cos \beta_i L = -kL \sin \beta_i L,$$

and because $EA(L) = \frac{4}{5}EA$, we obtain the characteristic equation

$$\tan \beta_i L = -\frac{4}{5}\frac{EA}{kL}(\beta_i L) = -5(\beta_i L). \tag{s}$$

The first four roots of (s) are

$$\beta_1 L = 1.6887, \quad \beta_2 L = 4.7544, \quad \beta_3 L = 7.8794, \quad \beta_4 L = 11.0137 \tag{t}$$

and note that the higher characteristic values approach the values $(2i - 1)\pi/2$. The solution of (s), as well as the balance of the results, were obtained by means of an electronic computer. Introducing the values given by (t) in (f) to (i) we obtain the stiffness and mass matrices,

$$[k] = \frac{EA}{L}\begin{bmatrix} 1.6033 & 0.3417 & 0.0581 & 0.0856 \\ 0.3417 & 11.4820 & 1.5345 & 0.0842 \\ 0.0581 & 1.5345 & 31.2220 & 3.5333 \\ 0.0856 & 0.0842 & 3.5333 & 60.8311 \end{bmatrix}, \tag{u}$$

$$[m] = mL\begin{bmatrix} 0.4931 & 0.0424 & -0.0043 & 0.0046 \\ 0.0424 & 0.4991 & 0.0410 & -0.0016 \\ -0.0043 & 0.0410 & 0.4997 & 0.0407 \\ 0.0046 & -0.0016 & 0.0407 & 0.4998 \end{bmatrix}. \tag{v}$$

The solution of the eigenvalue problem (n) consists of the eigenvalues

$$[\lambda] = \frac{EA}{mL^2}\begin{bmatrix} 3.2428 & 0 & 0 & 0 \\ 0 & 23.0534 & 0 & 0 \\ 0 & 0 & 62.5410 & 0 \\ 0 & 0 & 0 & 121.8582 \end{bmatrix} \tag{w}$$

and the eigenvectors

$$[a] = \begin{bmatrix} 1.0000 & 1.0000 & 1.0000 & 1.0000 \\ -0.0202 & -15.2182 & -3.4380 & -1.0703 \\ -0.0015 & 0.0446 & 63.4411 & 5.2650 \\ -0.0011 & 0.0144 & -2.0893 & -106.3806 \end{bmatrix}, \tag{x}$$

from which we obtain the estimated natural frequencies

$$\omega_1 = 1.8008\sqrt{\frac{EA}{mL^2}}, \qquad \omega_2 = 4.8014\sqrt{\frac{EA}{mL^2}},$$

$$\omega_3 = 7.9083\sqrt{\frac{EA}{mL^2}}, \qquad \omega_4 = 11.0389\sqrt{\frac{EA}{mL^2}}, \tag{y}$$

and the natural modes

$$w^{(1)}(x) = \sin 1.6887\frac{x}{L} - 0.0202 \sin 4.7544\frac{x}{L} - 0.0015 \sin 7.8794\frac{x}{L}$$
$$- 0.0011 \sin 11.0137\frac{x}{L},$$

$$w^{(2)}(x) = \sin 1.6887\frac{x}{L} - 15.2182 \sin 4.7544\frac{x}{L} + 0.0446 \sin 7.8794\frac{x}{L}$$
$$+ 0.0144 \sin 11.0137\frac{x}{L},$$

$$w^{(3)}(x) = \sin 1.6887\frac{x}{L} - 3.4380 \sin 4.7544\frac{x}{L} + 63.4411 \sin 7.8794\frac{x}{L}$$
$$- 2.0893 \sin 11.0137\frac{x}{L},$$

$$w^{(4)}(x) = \sin 1.6887\frac{x}{L} - 1.0703 \sin 4.7544\frac{x}{L} + 5.2650 \sin 7.8794\frac{x}{L}$$
$$- 106.3806 \sin 11.0137\frac{x}{L}, \tag{z}$$

where the natural modes are not normalized.

We conclude that, in this particular case, there is no significant difference between the results obtained by using admissible functions and the ones obtained by using comparison functions. The reason is that the admissible functions closely resemble the corresponding comparison functions. In other cases the difference in results may be more pronounced.

6–5 ASSUMED-MODES METHOD. LAGRANGE'S EQUATIONS

We shall now consider a method which is different in approach but gives the same results as the Rayleigh-Ritz method. It will be referred to as the *assumed-modes method.*† The method consists of assuming a solution of the free vibration problem in the form of a series composed of a linear combination of admissible functions ϕ_i, which are functions of the spatial coordinates, multiplied by time-dependent generalized coordinates $q_i(t)$,

$$w_n = \sum_{i=1}^{n} \phi_i q_i(t),$$ (6.65)

which in essence treats a continuous system as an n-degree-of-freedom system.

The kinetic energy expression can be written

$$T(t) = \tfrac{1}{2} \sum_{i=1}^{n} \sum_{j=1}^{n} m_{ij} \dot{q}_i(t) \dot{q}_j(t),$$ (6.66)

where the mass coefficients m_{ij} depend on the mass distribution of the system and the chosen admissible functions ϕ_i. Similarly, the potential energy is

$$V(t) = \tfrac{1}{2} \sum_{i=1}^{n} \sum_{j=1}^{n} k_{ij} q_i(t) q_j(t),$$ (6.67)

where the stiffness coefficients k_{ij} depend on the stiffness properties of the system and the admissible functions ϕ_i and its derivatives. Discrete masses or discrete springs, as encountered previously, are accounted for in the kinetic and potential energy expressions, respectively.

Lagrange's equations for a conservative discrete system have the form

$$\frac{d}{dt}\left(\frac{\partial T}{\partial \dot{q}_r}\right) - \frac{\partial T}{\partial q_r} + \frac{\partial V}{\partial q_r} = 0, \qquad r = 1, 2, \ldots, n,$$ (6.68)

which, upon using (6.66) and (6.67) lead to the equations of motion

$$\sum_{j=1}^{n} m_{rj} \ddot{q}_j + \sum_{j=1}^{n} k_{rj} q_j = 0, \qquad r = 1, 2, \ldots, n,$$ (6.69)

and assuming that the time dependence of q_j is harmonic, they reduce to a set of homogeneous equations representing the eigenvalue problem

$$\sum_{j=1}^{n} (k_{rj} - {}^n\Lambda m_{rj}) q_j = 0, \qquad {}^n\Lambda = \omega^2, \qquad r = 1, 2, \ldots, n,$$ (6.70)

which have precisely the same form as Galerkin's equations, (6.30).

The matrix form of the eigenvalue problem (6.70) is

$$[k]\{q\} = {}^n\Lambda[m]\{q\},$$ (6.71)

† In some texts on the subject this method is called the *Rayleigh-Ritz method.*

which can be solved by one of the previously discussed methods. The solution consists of n eigenvalues ${}^n\Lambda_r$, which are related to the estimated natural frequencies of the system, and n associated eigenvectors $\{q^{(r)}\}$, which can be looked upon as the amplitude of the time-dependent harmonic motion and can be used to obtain the n eigenfunctions

$$W_n^{(r)}(x) = \sum_{i=1}^{n} \phi_i(x)q_i^{(r)}, \tag{6.72}$$

where the subscript n indicates that the continuous system has been approximated by an n-degree-of-freedom system. As the number of degrees of freedom is increased, the estimated natural frequencies of the system approach the true natural frequencies from above.

The assumed-modes method can be used successfully in the case of systems with both distributed and discrete properties.

6–6 GALERKIN'S METHOD†

Galerkin's method assumes a solution of the eigenvalue problem in the form of a series of n comparison functions satisfying all the boundary conditions and being $2p$ times differentiable. In general the series solution will not satisfy the differential equation defining the eigenvalue problem unless by some coincidence the series is composed of eigenfunctions. Substituting the series of comparison functions in the differential equation an *error* will be obtained. At this point one insists that the integral of the weighted error over the domain be zero. The weighting functions are exactly the n comparison functions.

Consider the eigenvalue problem

$$L[w] = \lambda M[w], \tag{6.73}$$

where L and M are self-adjoint linear homogeneous differential operators of order $2p$ and $2q$, as defined previously. The function w is subject to boundary conditions which, in general, do not depend on the eigenvalue λ. We assume a solution of the eigenvalue problem in the form

$$w_n = \sum_{j=1}^{n} a_j u_j, \tag{6.74}$$

where a_j are coefficients to be determined and u_j are comparison functions. Introduce (6.74) in (6.73) and denote the error by ϵ, so that

$$\epsilon = L[w_n] - {}^n\Lambda M[w_n], \tag{6.75}$$

† Also known as the *method of weighting functions*.

where $^n\Lambda$ is the corresponding estimate of the eigenvalue λ. The conditions that the weighted error integrated over the domain be zero are

$$\int_D \epsilon u_r \, dD = 0, \qquad r = 1, 2, \ldots, n. \tag{6.76}$$

Consider

$$\int_D u_r L[w_n] \, dD = \sum_{j=1}^{n} a_j \int_D u_r L[u_j] \, dD$$

$$= \sum_{j=1}^{n} k_{rj} a_j, \qquad r = 1, 2, \ldots, n, \tag{6.77}$$

where the coefficients k_{rj} are symmetric:

$$k_{rj} = k_{jr} = \int_D u_r L[u_j] \, dD, \tag{6.78}$$

because L is self-adjoint. Similarly, let

$$\int_D u_r M[w_n] \, dD = \sum_{j=1}^{n} m_{rj} a_j, \qquad r = 1, 2, \ldots, n, \tag{6.79}$$

where the coefficients m_{rj} are given by

$$m_{rj} = m_{jr} = \int_D u_r M[u_j] \, dD \tag{6.80}$$

and are symmetric because M is self-adjoint. Equations (6.74) through (6.80) yield

$$\sum_{j=1}^{n} (k_{rj} - {}^n\Lambda m_{rj}) a_j = 0, \qquad r = 1, 2, \ldots, n, \tag{6.81}$$

which are called Galerkin's equations and represent an eigenvalue problem for an n-degree-of-freedom system. The above equations are similar to (6.30) obtained by using the Rayleigh-Ritz method. The equations would be identical if in using the Rayleigh-Ritz method the minimizing sequence consists of a series of comparison rather than admissible functions. The eigenvalues and eigenfunctions are obtained in precisely the same manner as in the Rayleigh-Ritz method.

As an illustration consider the problem of the bending vibration of a bar for which

$$L = \frac{d^2}{dx^2} \left(EI \frac{d^2}{dx^2} \right), \qquad M = m. \tag{6.82}$$

It follows that the stiffness coefficients have the form

$$k_{rj} = \int_D u_r L[u_j]\, dD = \int_0^L u_r \frac{d^2}{dx^2}\left(EI\frac{d^2 u_j}{dx^2}\right) dx$$

$$= u_r\left[\frac{d}{dx}\left(EI\frac{d^2 u_j}{dx^2}\right)\right]\Big|_0^L - \frac{du_r}{dx}\left(EI\frac{d^2 u_j}{dx^2}\right)\Big|_0^L + \int_0^L EI\frac{d^2 u_r}{dx^2}\frac{d^2 u_j}{dx^2}$$

$$(6.83)$$

and the mass coefficients are

$$m_{rj} = \int_D u_r M[u_j]\, dD = \int_0^L m u_r u_j\, dx, \tag{6.84}$$

and for a self-adjoint system the coefficients k_{rj} and m_{rj} are symmetric. We note that the comparison functions u_j must satisfy all the boundary conditions of the problem, so the coefficients k_{rj} and m_{rj} as calculated by using (6.83) and (6.84) will be identical with the corresponding coefficients obtained by using the same comparison functions in conjunction with the Rayleigh-Ritz method.

As soon as a set of comparison functions u_j is selected, the coefficients k_{rj} and m_{rj} can be computed according to (6.83) and (6.84), respectively. This leads to the eigenvalue problem (6.81), which can be solved for the estimated eigenvalues $^n\Lambda_r$ and the eigenvectors $\{a^{(r)}\}$; the latter are introduced in (6.74) to obtain the eigenfunctions

$$w_n^{(r)} = \sum_{j=1}^n a_j^{(r)} u_j. \tag{6.85}$$

6-7 COLLOCATION METHOD

Another method based on the error concept is the collocation method. The collocation method also assumes an approximate solution w_n as a combination of some functions v_j and associated coefficients a_j. According to the nature of the functions v_j, the problem falls into one of three classifications†:

1. *Boundary method*—the functions v_j satisfy the differential equation but not the boundary conditions.

2. *Interior method*—the functions v_j satisfy the boundary conditions but not the differential equation.

3. *Mixed method*—the functions v_j satisfy neither the differential equation nor the boundary conditions.

† See L. Collatz, *The Numerical Treatment of Differential Equations*, Springer-Verlag, Berlin, 1960.

The method consists of selecting a function w_n and a set of locations we shall call stations s_r. The stations can be on the boundaries of domain D, within domain D, or both on the boundaries and within domain D. The coefficients a_j are determined by insisting that at the selected stations w_n satisfies either the boundary conditions or the differential equation, depending upon whether one deals with the boundary, interior, or mixed problem.

We shall describe here the interior method, and to this end let us consider the eigenvalue problem (6.73), assume a solution in the form (6.74), and write the corresponding error, ϵ, in the form (6.75). At this point we require that the error ϵ vanishes at the stations s_r within domain D, so we obtain the conditions

$$L[w_n(s_r)] - {}^n\Lambda M[w_n(s_r)] = \sum_{j=1}^{n} a_j L[v_j(s_r)]$$

$$-{}^n\Lambda \sum_{j=1}^{n} a_j M[v_j(s_r)] = 0, \qquad r = 1, 2, \ldots, n. \qquad (6.86)$$

Denoting

$$k_{rj} = L[v_j(s_r)], \qquad m_{rj} = M[v_j(s_r)], \qquad (6.87)$$

we conclude that the coefficients a_j are obtained from the equations

$$\sum_{j=1}^{n} (k_{rj} - {}^n\Lambda m_{rj})a_j = 0, \qquad r = 1, 2, \ldots, n. \qquad (6.88)$$

The eigenvalue problem (6.88) can be written in the matrix form

$$[k]\{a\} = {}^n\Lambda[m]\{a\}, \qquad (6.89)$$

where, in general, the matrices $[k]$ and $[m]$ are *not symmetric*.

The advantage of the collocation method, as opposed to Galerkin's method, lies in the relative ease with which the coefficients k_{rj} and m_{rj} are evaluated. On the other hand, when the matrices $[k]$ and $[m]$ are indeed nonsymmetric, the solution of the eigenvalue problem is complicated by the fact that we have no longer at our disposal the orthogonality relations associated with symmetric $[k]$ and $[m]$ matrices. Equation (6.89) can be written

$$[D]\{u\} = \lambda\{u\}, \qquad (6.90)$$

where

$$[D] = [k]^{-1}[m] \qquad (6.91)$$

is a dynamical matrix and

$$\lambda = \frac{1}{{}^n\Lambda} \qquad (6.92)$$

is a parameter.

A method for obtaining the solution of the general eigenvalue problem, when the customary orthogonality relations are not available, was discussed in Section 4–12.

Example 6.5

Let us consider again the system shown in Figure 6.1 and formulate the eigenvalue problems by means of the collocation method. The mass per unit length and the stiffness are, as before,

$$m(x) = 2m\left(1 - \frac{x}{L}\right), \qquad EA(x) = 2EA\left(1 - \frac{x}{L}\right). \tag{a}$$

The continuous system is approximated by a three-degree-of-freedom system by using the comparison functions

$$v_1 = \sin\frac{\pi x}{2L}, \qquad v_2 = \sin\frac{3\pi x}{2L}, \qquad v_3 = \sin\frac{5\pi x}{2L}. \tag{b}$$

We chose the stations at the locations

$$s_1 = \frac{L}{4}, \qquad s_2 = \frac{L}{2}, \qquad s_3 = \frac{3L}{4}. \tag{c}$$

The operator L has the form

$$L = -\frac{d}{dx}\left[EA(x)\frac{d}{dx}\right] = 2EA\left[\frac{1}{L}\frac{d}{dx} - \left(1 - \frac{x}{L}\right)\frac{d^2}{dx^2}\right] \tag{d}$$

and the operator M is just the function

$$M = 2m\left(1 - \frac{x}{L}\right). \tag{e}$$

Hence we have

$$L[v_1(x)] = 2EA\left[\frac{\pi}{2L^2}\cos\frac{\pi x}{2L} + \left(1 - \frac{x}{L}\right)\frac{\pi^2}{4L^2}\sin\frac{\pi x}{2L}\right],$$

$$L[v_2(x)] = 2EA\left[\frac{3\pi}{2L^2}\cos\frac{3\pi x}{2L} + \left(1 - \frac{x}{L}\right)\frac{9\pi^2}{4L^2}\sin\frac{3\pi x}{2L}\right], \tag{f}$$

$$L[v_3(x)] = 2EA\left[\frac{5\pi}{2L^2}\cos\frac{5\pi x}{2L} + \left(1 - \frac{x}{L}\right)\frac{25\pi^2}{4L^2}\sin\frac{5\pi x}{2L}\right],$$

and

$$M[v_1(x)] = 2m\left(1 - \frac{x}{L}\right) \sin \frac{\pi x}{2L},$$

$$M[v_2(x)] = 2m\left(1 - \frac{x}{L}\right) \sin \frac{3\pi x}{2L}, \qquad \text{(g)}$$

$$M[v_3(x)] = 2m\left(1 - \frac{x}{L}\right) \sin \frac{5\pi x}{2L}.$$

Using the first of equations (6.87) we write

$$k_{11} = L[v_1(s_1)] = L\left[v_1\left(\frac{L}{4}\right)\right]$$

$$= \frac{\pi EA}{L^2}\left[\cos\frac{\pi}{8} + \left(1 - \frac{1}{4}\right)\frac{\pi}{2}\sin\frac{\pi}{8}\right] = 1.3748\frac{\pi EA}{L^2}. \qquad \text{(h)}$$

The remaining coefficients k_{rj} are obtained in a similar fashion. The stiffness matrix has the form

$$[k] = \frac{\pi EA}{L^2}\begin{bmatrix} 1.3748 & 3(3.6480) & 5(5.0595) \\ 1.2625 & 3(0.9590) & -5(3.4839) \\ 0.7455 & -3(1.3748) & 5(0.1725) \end{bmatrix}, \qquad \text{(i)}$$

and it is easy to see that it is not symmetric. Using the second of equations (6.87) we have

$$m_{11} = M[v_1(s_1)] = M\left[v_1\left(\frac{L}{4}\right)\right] = 2m\left(1 - \frac{1}{4}\right)\sin\frac{\pi}{8} = 0.2870 \times 2m. \quad \text{(j)}$$

The rest of the coefficients m_{rj} are obtained with the same relative ease, so we can write the inertia matrix

$$[m] = 2m\begin{bmatrix} 0.2870 & 0.6929 & 0.6929 \\ 0.3535 & 0.3535 & -0.3535 \\ 0.2310 & -0.0957 & -0.0957 \end{bmatrix}, \qquad \text{(k)}$$

and this matrix, too, is nonsymmetric.

Now compute the inverse of the $[k]$ matrix and obtain

$$[k]^{-1} = \frac{L^2}{10\pi EA}\begin{bmatrix} 1.5932 & 2.6132 & 6.0503 \\ 0.3233 & 0.4059 & -1.2836 \\ 0.1689 & -0.3176 & 0.2265 \end{bmatrix},$$

which enables us to compute the dynamical matrix

$$[D] = [k]^{-1}[m] = \frac{mL^2}{5\pi EA} \begin{bmatrix} 2.7786 & 1.4487 & -0.3989 \\ -0.0602 & 0.4903 & 0.2033 \\ -0.0115 & -0.0170 & 0.2076 \end{bmatrix}. \qquad (l)$$

The dynamical matrix (l) yields the first natural frequency and eigenvector,

$$\omega_1 = 2.3939\sqrt{\frac{EA}{mL^2}}, \qquad \{u^{(1)}\} = \left\{ \begin{matrix} 1.0000 \\ -0.0271 \\ -0.0043 \end{matrix} \right\}, \qquad (m)$$

and the transposed dynamical matrix, $[D]^T$, yields the first eigenvalue and conjugate eigenvector,

$$\omega_1 = 2.3939\sqrt{\frac{EA}{mL^2}}, \qquad \{v^{(1)}\} = \left\{ \begin{matrix} 1.0173 \\ 0.6555 \\ -0.1079 \end{matrix} \right\}. \qquad (n)$$

Having $\{v^{(1)}\}$ we write the orthogonality condition

$$\{u\}^T\{v^{(1)}\} = 1.0173u_1 + 0.6555u_2 - 0.1079u_3 = 0,$$

which enables us to construct the first sweeping matrix

$$[S^{(1)}] = \begin{bmatrix} 0 & -0.6444 & 0.1061 \\ 0 & 1 & 0 \\ 0 & 0 & 1 \end{bmatrix},$$

and subsequently the new dynamical matrix

$$[D^{(2)}] = [D][S^{(1)}] = \frac{mL^2}{5\pi EA} \begin{bmatrix} 0 & -0.3418 & -0.1041 \\ 0 & 0.5291 & 0.1969 \\ 0 & -0.0096 & 0.2064 \end{bmatrix}, \qquad (o)$$

giving the second natural frequency and second eigenvector,

$$\omega_2 = 5.4793\sqrt{\frac{EA}{mL^2}}, \qquad \{u^{(2)}\} = \left\{ \begin{matrix} 1.0000 \\ -1.5449 \\ 0.0464 \end{matrix} \right\}. \qquad (p)$$

To calculate $\{v^{(2)}\}$ we write

$$\{v^T\}\{u^{(1)}\} = 1.0000v_1 - 0.0271v_2 - 0.0043v_3 = 0,$$

so

$$[D_T^{(2)}] = [D]^T[S_T^{(2)}]$$

$$= \frac{mL^2}{5\pi EA} \begin{bmatrix} 2.7786 & -0.0602 & -0.0115 \\ 1.4487 & 0.4903 & -0.0170 \\ -0.3989 & 0.2033 & 0.2076 \end{bmatrix} \begin{bmatrix} 0 & 0.0271 & 0.0043 \\ 0 & 1 & 0 \\ 0 & 0 & 1 \end{bmatrix}$$

$$= \frac{mL^2}{5\pi EA} \begin{bmatrix} 0 & 0.0151 & 0.0004 \\ 0 & 0.5296 & -0.0108 \\ 0 & 0.1925 & 0.2093 \end{bmatrix}, \tag{q}$$

which yields

$$\omega_2 = 5.4804\sqrt{\frac{EA}{mL^2}}, \qquad \{v^{(2)}\} = \begin{Bmatrix} -0.0190 \\ -0.6476 \\ -0.3973 \end{Bmatrix}. \tag{r}$$

For the third eigenvector we write the orthogonality relations

$$\{u\}^T\{v^{(1)}\} = 1.0173u_1 + 0.6555u_2 - 0.1079u_3 = 0,$$

$$\{u\}^T\{v^{(2)}\} = -0.0190u_1 - 0.6476u_2 - 0.3973u_3 = 0,$$

which lead to the third dynamical matrix,

$$[D^{(3)}] = [D][S^{(2)}]$$

$$= \frac{mL^2}{5\pi EA} \begin{bmatrix} 2.7786 & 1.4487 & -0.3989 \\ -0.0602 & 0.4903 & 0.2033 \\ -0.0115 & -0.0170 & 0.2076 \end{bmatrix} \begin{bmatrix} 0 & 0 & 0.5110 \\ 0 & 0 & -0.6284 \\ 0 & 0 & 1.0000 \end{bmatrix}$$

$$= \frac{mL^2}{5\pi EA} \begin{bmatrix} 0 & 0 & 0.1106 \\ 0 & 0 & -0.1356 \\ 0 & 0 & 0.2124 \end{bmatrix}. \tag{s}$$

Using $[D^{(3)}]$ we obtain the third natural frequency and eigenvector,

$$\omega_3 = 8.5997\sqrt{\frac{EA}{mL^2}}, \qquad \{u^{(3)}\} = \begin{Bmatrix} 0.5207 \\ -0.6384 \\ 1.0000 \end{Bmatrix}. \tag{t}$$

For the third conjugate eigenvector we have the orthogonality relations

$$\{v\}^T\{u^{(1)}\} = v_1 - 0.0271v_2 - 0.0043v_3 = 0,$$

$$\{v\}^T\{u^{(2)}\} = v_1 - 1.5449v_2 + 0.0464v_3 = 0,$$

which lead to

$$[D_T^{(3)}] = [D]^T[S_T^{(3)}]$$

$$= \frac{mL^2}{5\pi EA}\begin{bmatrix} 2.7786 & -0.0602 & -0.0115 \\ 1.4487 & 0.4903 & -0.0170 \\ -0.3989 & 0.2033 & 0.2076 \end{bmatrix}\begin{bmatrix} 0 & 0 & 0.0052 \\ 0 & 0 & 0.0334 \\ 0 & 0 & 1.0000 \end{bmatrix}$$

$$= \frac{mL^2}{5\pi EA}\begin{bmatrix} 0 & 0 & 0.0009 \\ 0 & 0 & 0.0069 \\ 0 & 0 & 0.2123 \end{bmatrix}. \tag{u}$$

Using $[D_T^{(3)}]$ we obtain the third natural frequency and conjugate eigenvector,

$$\omega_3 = 8.6017\sqrt{\frac{EA}{mL^2}}, \qquad \{v^{(3)}\} = \begin{Bmatrix} 0.0043 \\ 0.0288 \\ 1.0162 \end{Bmatrix}. \tag{v}$$

Note that the sets $\{u^{(i)}\}$ and $\{v^{(j)}\}$ are normalized, so they satisfy (4.207).
The eigenfunctions corresponding to the set $\{u^{(i)}\}$ are

$$w_u^{(1)}(x) = \sin\frac{\pi x}{2L} - 0.0271\sin\frac{3\pi x}{2L} - 0.0043\sin\frac{5\pi x}{2L},$$

$$w_u^{(2)}(x) = \sin\frac{\pi x}{2L} - 1.5449\sin\frac{3\pi x}{2L} + 0.0464\sin\frac{5\pi x}{2L}, \tag{w}$$

$$w_u^{(3)}(x) = 0.5207\sin\frac{\pi x}{2L} - 0.6384\sin\frac{3\pi x}{2L} + \sin\frac{5\pi x}{2L},$$

and the eigenfunctions corresponding to the set $\{v^{(j)}\}$ are

$$w_v^{(1)}(x) = 1.0713\sin\frac{\pi x}{2L} + 0.6555\sin\frac{3\pi x}{2L} - 0.1079\sin\frac{5\pi x}{2L},$$

$$w_v^{(2)}(x) = -0.0190\sin\frac{\pi x}{2L} - 0.6476\sin\frac{3\pi x}{2L} - 0.3973\sin\frac{5\pi x}{2L}, \tag{x}$$

$$w_v^{(3)}(x) = 0.0043\sin\frac{\pi x}{2L} + 0.0288\sin\frac{3\pi x}{2L} + 1.0162\sin\frac{5\pi x}{2L}.$$

6–8 ASSUMED-MODES METHOD. INTEGRAL FORMULATION

The assumed-modes method as presented in Section 6–5 makes use of the potential energy expressions in terms of stiffness coefficients k_{ij}, which, in general, involve derivatives of the chosen admissible functions with respect to the spatial variables. In many cases, for various reasons, one may be interested in avoiding the differentiation of the admissible functions. This can be accomplished by writing the potential energy expression in terms of influence functions. In the one-dimensional case we can write

$$V(t) = \tfrac{1}{2} \int_0^L p(x,\, t) \left[\int_0^L a(x,\, \zeta) p(\zeta,\, t)\, d\zeta \right] dx, \tag{6.93}$$

where $a(x,\, \zeta)$ is the flexibility influence function and $p(x,\, t)$ is the distributed load at point x (see Section 3–3).

For free vibration we have the relation [see (5.320)]

$$p(x,\, t) = \omega^2 m(x) w(x,\, t), \tag{6.94}$$

where $w(x,\, t)$ is the displacement at point x, so (6.93) takes the form

$$V(t) = \tfrac{1}{2}\omega^4 \int_0^L m(x) w(x,\, t) \left[\int_0^L a(x,\, \zeta) m(\zeta) w(\zeta,\, t)\, d\zeta \right] dx. \tag{6.95}$$

The above can be generalized by writing

$$V(t) = \tfrac{1}{2}\omega^4 \int_D m(P) w(P,\, t) \left[\int_D G(P,\, Q) m(Q) w(Q,\, t)\, dD(Q) \right] dD(P), \tag{6.95}$$

where P and Q represent positions that may be defined by one or two spatial variables and $G(P,\, Q)$ is Green's function that is symmetric in P and Q for a self-adjoint system.

As in previous cases, the continuous system is approximated by an n-degree-of-freedom system by writing the displacement in the form

$$w_n(P,\, t) = \sum_{i=1}^n \phi_i(P) q_i(t), \tag{6.96}$$

where $\phi_i(P)$ are admissible functions and $q_i(t)$ are time-dependent generalized coordinates, so that (6.95) reduces to

$$V(t) = \tfrac{1}{2} {}^n\Lambda^2 \sum_{i=1}^{n} \sum_{j=1}^{n} q_i(t)q_j(t) \int_D m(P)\phi_i(P)$$

$$\times \left[\int_D G(P, Q)m(Q)\phi_j(Q) \, dD(Q) \right] dD(P)$$

$$= \tfrac{1}{2} {}^n\Lambda^2 \sum_{i=1}^{n} \sum_{j=1}^{n} G_{ij}q_i(t)q_j(t), \qquad (6.97)$$

where ${}^n\Lambda$ provides an estimate for ω^2 and

$$G_{ij} = G_{ji} = \int_D m(P)\phi_i(P) \left[\int_D G(P, Q)m(Q)\phi_j(Q) \, dD(Q) \right] dD(P) \quad (6.98)$$

are symmetric constants. The kinetic energy has the form

$$T(t) = \tfrac{1}{2} \sum_{i=1}^{n} \sum_{j=1}^{n} m_{ij}\dot{q}_i(t)\dot{q}_j(t), \qquad (6.99)$$

which is the same as (6.66), where the coefficients m_{ij} are given by

$$m_{ij} = m_{ji} = \int_D m(P)\phi_i(P)\phi_j(P) \, dD(P). \qquad (6.100)$$

Using Lagrange's equations we obtain the equations of motion

$$\sum_{j=1}^{n} m_{rj}\ddot{q}_j + {}^n\Lambda^2 \sum_{j=1}^{n} G_{rj}q_j = 0, \qquad r = 1, 2, \ldots, n, \qquad (6.101)$$

and, as $\ddot{q}_j = -{}^n\Lambda q_j$, the above equations reduce to

$$\sum_{j=1}^{n} (m_{rj} - {}^n\Lambda G_{rj})q_j = 0, \qquad r = 1, 2, \ldots, n, \qquad (6.102)$$

which represent an eigenvalue problem where the q_j play the role of the amplitudes of the harmonic motion. Note that equations (6.102) are similar to (6.70) except that ${}^n\Lambda$ multiplies G_{rj} instead of m_{rj}. Of course, the coefficients G_{rj} involve integrations, whereas the coefficients k_{rj} of Section 6–5 involve differentiations. The eigenvalue problem may be written in the matrix form

$$[m]\{q\} = {}^n\Lambda[G]\{q\}, \qquad (6.103)$$

where the matrices $[m]$ and $[G]$ are symmetric.

From this point on the method of solution resembles the procedure described in Section 6–5.

Example 6.6

Let us consider the problem of a uniform string subjected to constant tension and estimate the natural frequencies and natural modes by means of the assumed-modes method in its integral form. First, we have

$$m(x) = m = \text{const}, \qquad T(x) = T = \text{const}. \tag{a}$$

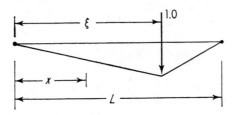

FIGURE 6.3

Green's function, or flexibility influence function, for the system shown in Figure 6.3 is

$$G(x, \xi) = a(x, \xi) = \frac{\xi(L - x)}{TL}, \qquad \xi < x,$$

$$G(x, \xi) = a(x, \xi) = \frac{x(L - \xi)}{TL}, \qquad \xi > x. \tag{b}$$

We note that $G(x, \xi)$ is symmetric in x and ξ.
As admissible functions let us choose

$$\phi_1(x) = \frac{x}{L} - \left(\frac{x}{L}\right)^2, \qquad \phi_2(x) = \frac{x}{L} - 3\left(\frac{x}{L}\right)^2 + 2\left(\frac{x}{L}\right)^3. \tag{c}$$

First evaluate the integral

$$\int_D G(P, Q)m(Q)\phi_1(Q)\, dD(Q) = \int_0^L G(x, \xi)m(\xi)\phi_1(\xi)\, d\xi$$

$$= \frac{m(L - x)}{TL} \int_0^x \xi\left[\frac{\xi}{L} - \left(\frac{\xi}{L}\right)^2\right] d\xi$$

$$+ \frac{mx}{TL} \int_x^L (L - \xi)\left[\frac{\xi}{L} - \left(\frac{\xi}{L}\right)^2\right] d\xi$$

$$= \frac{mL^2}{12T} \left[\frac{x}{L} - 2\left(\frac{x}{L}\right)^3 + \left(\frac{x}{L}\right)^4\right], \tag{d}$$

so that using (6.98) we obtain

$$G_{11} = \int_0^L m(x)\phi_1(x)\left[\int_0^L G(x,\xi)m(\xi)\phi_1(\xi)\,d\xi\right]dx$$

$$= \frac{m^2L^2}{12T}\int_0^L\left[\frac{x}{L} - \left(\frac{x}{L}\right)^2\right]\left[\frac{x}{L} - 2\left(\frac{x}{L}\right)^3 + \left(\frac{x}{L}\right)^4\right]dx = \frac{17}{5,040}\frac{m^2L^3}{T}. \quad (e)$$

Similarly, we obtain

$$G_{12} = G_{21} = 0, \qquad G_{22} = \frac{1}{8,400}\frac{m^2L^3}{T}. \quad (f)$$

Using (6.100) we write

$$m_{ij} = \int_D m(P)\phi_i(P)\phi_j(P)\,dD(P) = m\int_0^L \phi_i(x)\phi_j(x)\,dx, \quad (g)$$

from which we obtain

$$m_{11} = m\int_0^L\left[\frac{x}{L} - \left(\frac{x}{L}\right)^2\right]^2 dx = m\int_0^L\left[\left(\frac{x}{L}\right)^2 - 2\left(\frac{x}{L}\right)^3 + \left(\frac{x}{L}\right)^4\right]dx$$

$$= \tfrac{1}{30}mL, \quad (h)$$

and in a similar way,

$$m_{12} = m_{21} = 0, \qquad m_{22} = \tfrac{1}{210}mL. \quad (i)$$

The eigenvalue problem (6.103) becomes

$$\frac{mL}{30}\begin{bmatrix}1 & 0\\0 & \frac{1}{7}\end{bmatrix}\begin{Bmatrix}q_1\\q_2\end{Bmatrix} = {}^2\Lambda\frac{m^2L^3}{1,680T}\begin{bmatrix}\frac{17}{3} & 0\\0 & \frac{1}{5}\end{bmatrix}\begin{Bmatrix}q_1\\q_2\end{Bmatrix}, \quad (j)$$

which represents two eigenvalue problems of order 1. This result should surprise no one, because the two assumed modes consist of one even and one odd function of the variable $x - (L/2)$. The system possesses the properties indicated by (6.41) through (6.44), so the eigenvalue problem separates into two independent eigenvalue problems: one for even modes and one for odd modes.

The solution of (j) can be written

$$^2\Lambda_1 = \frac{168}{17}\frac{T}{mL^2}, \qquad \{q^{(1)}\} = \begin{Bmatrix}1\\0\end{Bmatrix},$$

$$^2\Lambda_2 = 40\frac{T}{mL^2}, \qquad \{q^{(2)}\} = \begin{Bmatrix}0\\1\end{Bmatrix}, \quad (k)$$

which leads to the estimated natural frequencies and natural modes

$$\omega_1 = 3.1436\sqrt{\frac{T}{mL^2}}, \qquad w^{(1)}(x) = \frac{x}{L} - \left(\frac{x}{L}\right)^2,$$

$$\omega_2 = 6.3246\sqrt{\frac{T}{mL^2}}, \qquad w^{(2)}(x) = \frac{x}{L} - 3\left(\frac{x}{L}\right)^2 + \left(\frac{x}{L}\right)^3. \tag{1}$$

The estimated natural frequencies are remarkably close to the first two true natural frequencies $\omega_1 = \pi\sqrt{T/mL^2}$ and $\omega_2 = 2\pi\sqrt{T/mL^2}$. This is due to the fact that the two assumed modes, $\phi_1(x)$ and $\phi_2(x)$, resemble very closely the first two natural modes of the system, $\sin(\pi x/L)$ and $\sin(2\pi x/L)$.

6-9 GALERKIN'S METHOD. INTEGRAL FORMULATION

By adopting a different point of view, Galerkin's method leads to the same results as the assumed-modes method. We refer to the eigenvalue problem for continuous systems in the integral form

$$w(P) = \lambda \int_D G(P, Q)m(Q)w(Q) \, dD(Q), \tag{6.104}$$

where $w(P)$ is the displacement at any point P and $G(P, Q)$ is the symmetric Green's function. Multiplying both sides of (6.104) by $m^{1/2}(P)$ and substituting

$$v(P) = m^{1/2}(P)w(P) \tag{6.105}$$

in the result, we obtain

$$v(P) = \lambda \int_D k(P, Q)v(Q) \, dD(Q), \tag{6.106}$$

where

$$k(P, Q) = k(Q, P) = G(P, Q)m^{1/2}(P)m^{1/2}(Q) \tag{6.107}$$

is the symmetric kernel of the integral transformation.

Again we approximate the continuous system by an n-degree-of-freedom system and assume the corresponding modified displacement $v_n(P)$ in the form

$$v_n(P) = \sum_{j=1}^{n} a_j \Phi_j(P), \tag{6.108}$$

where a_j are coefficients to be determined and Φ_j are comparison functions.

In doing this, an error $\epsilon(P)$ is introduced, so that

$$\epsilon(P) = v_n(P) - {}^n\Lambda \int_D k(P, Q)v_n(Q)\, dD(Q), \qquad (6.109)$$

where ${}^n\Lambda$ is an estimate of the eigenvalue λ. The coefficients a_j are determined in such a way that the weighted averages of the error over domain D are zero. The weighting functions are just the functions $\Phi_j(P)$. This is expressed by the conditions

$$\int_D \epsilon(P)\Phi_r(P)\, dD(P) = 0, \qquad r = 1, 2, \ldots, n, \qquad (6.110)$$

which lead to

$$\sum_{j=1}^n a_j \int_D \Phi_r(P)\Phi_j(P)\, dD(P)$$
$$- {}^n\Lambda \sum_{j=1}^n a_j \int_D \Phi_r(P)\left[\int_D k(P, Q)\Phi_j(Q)\, dD(Q)\right] dD(P) = 0,$$
$$r = 1, 2, \ldots, n. \quad (6.111)$$

Now let

$$\Phi_j(P) = m^{1/2}(P)\phi_j(P), \qquad j = 1, 2, \ldots, n, \qquad (6.112)$$

and denote

$$G_{rj} = \int_D \Phi_r(P)\left[\int_D k(P, Q)\Phi_j(Q)\, dD(Q)\right] dD(P)$$
$$= \int_D m(P)\phi_r(P)\left[\int_D G(P, Q)m(Q)\phi_j(Q)\, dD(Q)\right] dD(P) = G_{jr}, \qquad (6.113)$$

$$m_{rj} = \int_D \Phi_r(P)\Phi_j(P)\, dD(P) = \int_D m(P)\phi_r(P)\phi_j(P)\, dD(P) = m_{jr}. \qquad (6.114)$$

The coefficients G_{rj} and m_{rj} are symmetric and precisely of the same form as the coefficients given by (6.98) and (6.100), respectively. With this notation, equations (6.111) become

$$\sum_{j=1}^n (m_{rj} - {}^n\Lambda G_{rj})a_j = 0, \qquad r = 1, 2, \ldots, n, \qquad (6.115)$$

which are of the form (6.102), and their solution follows the procedure described in Section 6–8.

6–10 COLLOCATION METHOD. INTEGRAL FORMULATION

The collocation method, as Galerkin's method, considers the error resulting from treating a continuous system as an n-degree-of-freedom system by assuming a solution w_n. In contrast with the averaging method of Galerkin, the collocation method sets the error zero at various points called stations.

The eigenvalue problem for a continuous system has the integral form

$$w(P) = \lambda \int_D G(P, Q) m(Q) w(Q) \, dD(Q). \tag{6.116}$$

Let the solution of the n-degree-of-freedom system be

$$w_n(P) = \sum_{j=1}^{n} a_j \phi_j(P), \tag{6.117}$$

where $\phi_j(P)$ are comparison functions, and write the error resulting from this assumption as

$$\epsilon(P) = w_n(P) - {}^n\Lambda \int_D G(P, Q) m(Q) w_n(Q) \, dD(Q)$$

$$= \sum_{j=1}^{n} a_j \phi_j(P) - {}^n\Lambda \sum_{j=1}^{n} a_j \int_D G(P, Q) m(Q) \phi_j(Q) \, dD(Q). \tag{6.118}$$

Letting the error vanish at the points P_r $(r = 1, 2, \ldots, n)$, we obtain the conditions

$$\epsilon(P_r) = 0, \qquad r = 1, 2, \ldots, n, \tag{6.119}$$

and denoting

$$m_{rj} = \phi_j(P_r), \qquad G_{rj} = \int_D G(P_r, Q) m(Q) \phi_j(Q) \, dD(Q), \tag{6.120}$$

conditions (6.119) lead to the eigenvalue problem

$$\sum_{j=1}^{n} (m_{rj} - {}^n\Lambda G_{rj}) a_j = 0, \qquad r = 1, 2, \ldots, n, \tag{6.121}$$

which can be written in the matrix form

$$[m]\{a\} = {}^n\Lambda [G][a]. \tag{6.122}$$

In contrast with Galerkin's method, however, the matrices $[m]$ and $[G]$ in the collocation method, in general, are not symmetric. Hence the work saved in the evaluation of the coefficients m_{rj} and G_{rj} may be spent in solving a more laborious eigenvalue problem. On the other hand, if the solution of the eigenvalue problem is obtained by means of a high-speed computer, the collocation method is worth considering.

The method of solution of the eigenvalue problem consisting of non-symmetric matrices $[m]$ and $[G]$ was presented in Section 4–12 and an example solved in Section 6–7.

6-11 HOLZER'S METHOD FOR TORSIONAL VIBRATION

The approximate methods discussed previously treat the continuous system as an n-degree-of-freedom system, where n is a finite number. This is accomplished by assuming a solution in the form of a series of known functions. The methods invariably lead to an eigenvalue problem similar in structure to the eigenvalue problem of an n-degree-of-freedom discrete system. The solution of the eigenvalue problem consists, in general, of n distinct eigenvalues, which provide approximations for the first n natural frequencies of the system, and n corresponding eigenvectors, which are used to construct n eigenfunctions for the continuous system.

An entirely different approach was originated by Holzer in connection with the torsional vibration of shafts problem and extended by Myklestad to the bending vibration of beams problem. Although both methods can be applied to uniform systems, their real usefulness lies in their application to nonuniform systems. Holzer's method approximates a continuous system by an n-degree-of-freedom system,† as the previously discussed methods do, but in contrast with the previous methods this method regards the system as having the mass concentrated in n *lumps* at n points of the system which we call *stations*. The portion between the lumped masses is assumed massless and of uniform stiffness and will be referred to as a *field*. The differential equations of the continuous system are approximated by finite difference equations relating the deformations and forces or moments, as the case may be, between two sides of a station or two ends of a field, so that, in essence, this is a step-by-step or chain method.

For a continuous torsional system the relation between the angle of twist $\theta(x, t)$ and the twisting moment $M_T(x, t)$ is

$$\frac{\partial \theta(x, t)}{\partial x} = \frac{M_T(x, t)}{GJ(x)}, \tag{6.123}$$

† Actually the number of degrees of freedom depends on the end conditions, as we shall see later.

where $GJ(x)$ is the torsional stiffness. Using the right-hand rule, the twisting moment is positive if the moment vector is in the same direction as the normal to the cross section. For free vibration the equation of motion is

$$\frac{\partial M_T(x, t)}{\partial x} = I(x)\frac{\partial^2 \theta(x, t)}{\partial t^2}. \tag{6.124}$$

Letting both $\theta(x, t)$ and $M_T(x, t)$ be harmonic and of frequency ω, we can rewrite (6.123) and (6.124) in the form

$$\frac{d\Theta(x)}{dx} = \frac{M_T(x)}{GJ(x)}, \tag{6.125}$$

$$\frac{dM_T(x)}{dx} = -\omega^2 I(x)\Theta(x), \tag{6.126}$$

where $\Theta(x)$ and $M_T(x)$ are the amplitudes of $\theta(x, t)$ and $M_T(x, t)$, respectively. In the future the subscript of M_T will be dropped.

Now consider a nonuniform shaft and approximate it by a number of rigid disks connected by massless circular shafts of uniform stiffness as shown in Figure 6.4. The differential equations (6.125) and (6.126) can be approx-

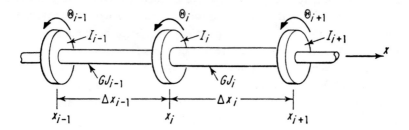

FIGURE 6.4

imated by difference equations. To this end denote by Θ_i^L and M_i^L the angular displacement and torque on the *left side of disk i* and by Θ_i^R and M_i^R the angular displacement and torque on the *right side of disk i*. Of course, disk i is rigid, so we have

$$\Theta_i^L = \Theta_i^R = \Theta_i. \tag{6.127}$$

The mass polar moment of inertia of disk i can be written

$$I_i \cong I(x_i)\,\Delta x_i. \tag{6.128}$$

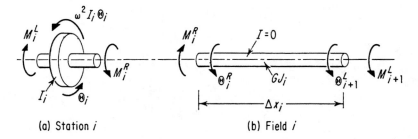

(a) Station i (b) Field i

FIGURE 6.5

At this point we wish to draw free-body diagrams for one station and an adjacent field, as shown in Figure 6.5.† Equation (6.126) in difference form, when applied to station i, leads to

$$M_i^R = M_i^L - \omega^2 I_i \Theta_i,\qquad (6.129)$$

where I_i is given by (6.128). Equation (6.125) in difference form, when applied to field i, becomes

$$\frac{\Theta_{i+1}^L - \Theta_i^R}{\Delta x_i} \cong \frac{1}{2}\frac{M_{i+1}^L + M_i^R}{GJ_i},\qquad (6.130)$$

where the right side of (6.130) represents the average moment over the field i. Since field i has no inertia, however, (6.126) in difference form reduces to

$$\frac{M_{i+1}^L - M_i^R}{\Delta x_i} = 0,\qquad (6.131)$$

from which we conclude that

$$M_{i+1}^L = M_i^R,\qquad (6.132)$$

thus enabling us to rewrite (6.130) in the form

$$\Theta_{i+1}^L = \Theta_i^R + M_{i+1}^L \frac{\Delta x_i}{GJ_i}.\qquad (6.133)$$

Now introduce the torsional flexibility influence coefficient a_i, given by

$$a_i = \frac{\Delta x_i}{GJ_i},\qquad (6.134)$$

which can be interpreted as the angular deflection at station $i + 1$ due to a unit moment $M_{i+1}^L = 1$ when the station i is prevented from rotating. With

† Note that Figure 6.5(b) uses the notation defined above for stations as far as the superscripts L and R are concerned.

this in mind and recalling (6.132), (6.133) assumes the form

$$\Theta^L_{i+1} = \Theta^R_i + a_i M^R_i. \tag{6.135}$$

Equations (6.127) and (6.129) can be written in the matrix form

$$\begin{Bmatrix} \Theta^R_i \\ M^R_i \end{Bmatrix} = \begin{bmatrix} 1 & 0 \\ -\omega^2 I_i & 1 \end{bmatrix} \begin{Bmatrix} \Theta^L_i \\ M^L_i \end{Bmatrix} \tag{6.136}$$

or, in more compact form,

$$\begin{Bmatrix} \Theta \\ M \end{Bmatrix}^R_i = [T_S]_i \begin{Bmatrix} \Theta \\ M \end{Bmatrix}^L_i, \tag{6.137}$$

where $[T_S]_i$ will be called a *station transfer matrix*, which relates the angular displacements and torques on both sides of station i. The column matrices are called *state vectors*. In a similar way, (6.132) and (6.135) can be written in the matrix form

$$\begin{Bmatrix} \Theta^L_{i+1} \\ M^L_{i+1} \end{Bmatrix} = \begin{bmatrix} 1 & a_i \\ 0 & 1 \end{bmatrix} \begin{Bmatrix} \Theta^R_i \\ M^R_i \end{Bmatrix}, \tag{6.138}$$

or

$$\begin{Bmatrix} \Theta \\ M \end{Bmatrix}^L_{i+1} = [T_F]_i \begin{Bmatrix} \Theta \\ M \end{Bmatrix}^R_i, \tag{6.139}$$

where $[T_F]_i$ is a *field transfer matrix* relating the angular displacement and torque at the left end of field i (right side of station i) with the displacement and torque at the right end of field i (left side of station $i + 1$).

Consider the system as consisting of $n + 1$ *stations* bounding n *fields*. One can go from one side of a station to the other and from one end of a field to the other end, in a step-by-step fashion, by letting i vary from 1 to $n + 1$ in (6.137) and from 1 to n in (6.139), respectively. In this manner we obtain

$$\begin{Bmatrix} \Theta \\ M \end{Bmatrix}^R_1 = [T_S]_1 \begin{Bmatrix} \Theta \\ M \end{Bmatrix}^L_1,$$

$$\begin{Bmatrix} \Theta \\ M \end{Bmatrix}^L_2 = [T_F]_1 \begin{Bmatrix} \Theta \\ M \end{Bmatrix}^R_1 = [T_F]_1[T_S]_1 \begin{Bmatrix} \Theta \\ M \end{Bmatrix}^L_1 = [T]_1 \begin{Bmatrix} \Theta \\ M \end{Bmatrix}^L_1,$$

$$\begin{Bmatrix} \Theta \\ M \end{Bmatrix}^R_2 = [T_S]_2 \begin{Bmatrix} \Theta \\ M \end{Bmatrix}^L_2 = [T_S]_2[T]_1 \begin{Bmatrix} \Theta \\ M \end{Bmatrix}^L_1, \tag{6.140}$$

$$\vdots$$

$$\begin{Bmatrix} \Theta \\ M \end{Bmatrix}^R_{n+1} = [T_S]_{n+1} \begin{Bmatrix} \Theta \\ M \end{Bmatrix}^L_{n+1} = [T_S]_{n+1}[T]_n[T]_{n-1} \cdots [T]_2[T]_1 \begin{Bmatrix} \Theta \\ M \end{Bmatrix}^L_1$$

$$= [T] \begin{Bmatrix} \Theta \\ M \end{Bmatrix}^L_1,$$

where

$$[T]_i = [T_F]_i [T_S]_i \tag{6.141}$$

is a *transfer matrix* relating the angular displacement and torque on the left side of station $i + 1$ to the ones on the left side of station i, and

$$[T] = [T_S]_{n+1}[T]_n[T]_{n-1}\cdots[T]_2[T]_1 = [T_S]_{n+1}\prod_{i=n}^{1}[T]_i \tag{6.142}$$

will be referred to as the *overall transfer matrix*, which relates the left side of station 1 with the right of station $n + 1$. Because the order is important in a matrix product, in the product notation in (6.142) the index i takes values from n to 1.

The last of equations (6.140) implies that

$$\Theta_{n+1}^R = T_{11}\Theta_1^L + T_{12}M_1^L, \qquad M_{n+1}^R = T_{21}\Theta_1^L + T_{22}M_1^L, \tag{6.143}$$

where T_{ij} ($i, j = 1, 2$) are the elements of the matrix $[T]$. At each end of the shaft we must be given one boundary condition concerning the angular displacement or the torque. There are several types of boundary conditions to be considered:

1. *Free-free shaft.* The corresponding boundary conditions are

$$M_1^L = 0, \qquad M_{n+1}^R = 0. \tag{6.144}$$

From the second of equations (6.143) we conclude that we must have

$$T_{21} = 0, \tag{6.145}$$

which is an algebraic equation of order $n + 1$ in ω^2 and is simply the *frequency equation*. In this case we must obtain one zero root, as can be easily verified, because the system is semidefinite.

2. *Free-clamped shaft.* The boundary conditions are

$$M_1^L = 0, \qquad \Theta_{n+1}^R = \Theta_{n+1}^L = 0, \tag{6.146}$$

which lead to the frequency equation

$$T_{11} = 0, \tag{6.147}$$

which is of the order n in ω^2.

3. *Clamped-free shaft.* The boundary conditions are

$$\Theta_1^L = \Theta_1^R = 0, \qquad M_{n+1}^R = 0, \tag{6.148}$$

and the frequency equation is

$$T_{22} = 0, \tag{6.149}$$

which is also of order n in ω^2.

4. *Clamped-clamped shaft*. The boundary conditions

$$\Theta_1^L = \Theta_1^R = 0, \qquad \Theta_{n+1}^R = \Theta_{n+1}^L = 0 \tag{6.150}$$

lead to the frequency equation

$$T_{12} = 0, \tag{6.151}$$

which is of order $n - 1$ in ω^2.

The reason for obtaining various orders for the frequency equation is that the nature of the boundary conditions determines the number of degrees of freedom of the system. For a clamped end the corresponding disk is prevented from rotating, which in effect reduces the number of degrees of freedom by 1. The reduction of the degree of freedom can be effected by setting the moment of inertia of the clamped disk equal to zero.† Boundary conditions other than the ones listed are possible (see Problem 6.12).

The solution of the frequency equation may be made by Graeffe's root-squaring or any other numerical method. Upon obtaining the natural frequencies ω_r we can substitute them in equations (6.140) and calculate the mode shape as well as twisting moment diagram corresponding to each frequency.

It is possible to avoid the solution of the frequency equation, which, in general, is a high order algebraic equation, by adopting a different procedure. The procedure consists of choosing arbitrarily the value of the unspecified quantity at one of the boundaries, assuming a value for the frequency ω and checking whether the boundary condition at the other end is satisfied. For example, in the case of the free-free shaft we start with $M_1^L = 0$ and with the arbitrary value $\Theta_1^L = 1$, assume a value for ω, and calculate M_{n+1}^R by following the steps indicated by (6.140). If $M_{n+1}^R = 0$, then ω is one of the natural frequencies ω_r, where the number r can be determined by observing the number of nodes. If $M_{n+1}^R \neq 0$, then ω is not one of the natural frequencies, and another value of ω must be tried. The procedure can be rendered more efficient by plotting the curve $M_{n+1}^R(\omega)$ vs. ω. The roots ω_r are obtained at the points at which $M_{n+1}^R(\omega)$ intersects the ω axis. Of course, this procedure can be carried out quite efficiently by means of an electronic computer.

---

† If station 1, or station $n + 1$, is clamped, the matrix $[T_s]_1$ or $[T_s]_{n+1}$ reduces to the identity matrix. Note that the number of masses that can move does not coincide with the degree of freedom of the system if the system is semidefinite.

After the natural frequencies are obtained, it is possible to obtain the natural modes $\{\Theta^{(r)}\}$ by introducing, in sequence, the natural frequencies ω_r in the station matrices in (6.140). To this end transfer matrices relating the state vector on either side of station i to the state vector on the left side of station 1 are needed.

This method of approach can also be used for geared and branched torsional systems.

6-12   MYKLESTAD'S METHOD FOR BENDING VIBRATION

Myklestad's method for the bending vibration of bars problem uses the same general ideas as Holzer's method. The presentation of the method in this section is different from the original presentation of Myklestad,† although the basic ideas are the same. The following presentation parallels closely the description of Holzer's method of Section 6–11.

In the absence of external forces we have the free vibration equations (see Section 5–3)

$$\frac{\partial Q(x, t)}{\partial x} = m(x) \frac{\partial^2 w(x, t)}{\partial t^2}, \tag{6.152}$$

$$Q(x, t) = -\frac{\partial M(x, t)}{\partial x}, \tag{6.153}$$

where $w(x, t)$ is the lateral displacement, $Q(x, t)$ the shearing force, and $M(x, t)$ the bending moment. Since $w(x, t)$, $Q(x, t)$, and $M(x, t)$ are harmonic we can write

$$\frac{dQ(x)}{dx} = -\omega^2 m(x) w(x), \tag{6.154}$$

$$Q(x) = -\frac{dM(x)}{dx}, \tag{6.155}$$

where $w(x)$, $Q(x)$, and $M(x)$ are the corresponding amplitudes and ω is the frequency of the harmonic time dependence.

Again we concentrate the masses in n lumps at n stations and regard the fields between the concentrated masses as massless but possessing uniform

† See N. O. Myklestad, *J. Aeron. Sci.*, **11**, 153–162 (1944).

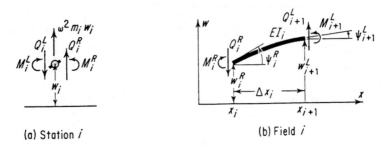

(a) Station i (b) Field i

FIGURE 6.6

bending stiffness. Figure 6.6 shows the corresponding free-body diagrams. Equation (6.154) in difference form, applied at station i, yields

$$Q_i^R = Q_i^L - \omega^2 m_i w_i, \tag{6.156}$$

where $m_i \cong m(x_i) \, \Delta x_i$. In addition, because of continuity, we can write

$$w_i^R = w_i^L = w_i, \qquad \psi_i^R = \psi_i^L = \psi_i, \tag{6.157}$$

where ψ_i is the tangent to the deflection curve at station i. If the rotatory inertia of mass m_i is neglected, it follows that

$$M_i^R = M_i^L. \tag{6.158}$$

Equation (6.155) in difference form, when applied to field i, gives

$$M_{i+1}^L = M_i^R - Q_i^R \, \Delta x_i. \tag{6.159}$$

In addition, we must have the relation between the forces and moments, on the one hand, and the linear and angular displacement, on the other. We write these relations in terms of influence coefficients defined as follows:

a_i^{wQ} linear displacement at station $i + 1$ due to a unit force, $Q_{i+1}^L = 1$, when station i is regarded as fixed (zero linear and angular deflection at station i),

a_i^{wM} linear displacement at station $i + 1$ due to a unit moment, $M_{i+1}^L = 1$, when station i is regarded as fixed,

$a_i^{\psi Q}$ angular displacement at station $i + 1$ due to a unit force, $Q_{i+1}^L = 1$, when station i is regarded as fixed, and

$a_i^{\psi M}$ angular displacement at station $i + 1$ due to a unit moment, $M_{i+1}^L = 1$, when station i is regarded as fixed.

In terms of the influence coefficients and in conjunction with Figure 6.6(*b*) we can write

$$w_{i+1}^L = w_i^R + \psi_i^R \Delta x_i + a_i^{wM} M_{i+1}^L + a_i^{wQ} Q_{i+1}^L, \tag{6.160}$$

$$\psi_{i+1}^L = \psi_i^R + a_i^{\psi M} M_{i+1}^L + a_i^{\psi Q} Q_{i+1}^L. \tag{6.161}$$

In addition, as there are no inertia forces throughout field *i*, we have

$$Q_{i+1}^L = Q_i^R. \tag{6.162}$$

Introducing (6.159) and (6.162) in (6.160) and (6.161), the latter take the form

$$w_{i+1}^L = w_i^R + \Delta x_i \psi_i^R + a_i^{wM} M_i^R + (a_i^{wQ} - \Delta x_i a_i^{wM}) Q_i^R, \tag{6.163}$$

$$\psi_{i+1}^L = \psi_i^R + a_i^{\psi M} M_i^R + (a_i^{\psi Q} - \Delta x_i a_i^{\psi M}) Q_i^R. \tag{6.164}$$

As in the case of Holzer's method, we can define a station transfer matrix and a field transfer matrix. Using (6.156) to (6.158), we write the matrix relation

$$
\begin{Bmatrix} w_i^R \\ \psi_i^R \\ M_i^R \\ Q_i^R \end{Bmatrix} =
\begin{bmatrix} 1 & 0 & 0 & 0 \\ 0 & 1 & 0 & 0 \\ 0 & 0 & 1 & 0 \\ -\omega^2 m_i & 0 & 0 & 1 \end{bmatrix}
\begin{Bmatrix} w_i^L \\ \psi_i^L \\ M_i^L \\ Q_i^L \end{Bmatrix}, \tag{6.165}
$$

or, in more compact form,

$$
\begin{Bmatrix} w \\ \psi \\ M \\ Q \end{Bmatrix}_i^R = [T_S]_i \begin{Bmatrix} w \\ \psi \\ M \\ Q \end{Bmatrix}_i^L, \tag{6.166}
$$

where $[T_S]_i$ is the station transfer matrix pertaining to station *i*. Similarly, (6.159) and (6.162) to (6.164) lead to

$$
\begin{Bmatrix} w \\ \psi \\ M \\ Q \end{Bmatrix}_{i+1}^L = [T_F]_i \begin{Bmatrix} w \\ \psi \\ M \\ Q \end{Bmatrix}_i^R, \tag{6.167}
$$

where the field transfer matrix $[T_F]_i$, associated with the field i, has the form

$$[T_F]_i = \begin{bmatrix} 1 & \Delta x_i & a_i^{wM} & a_i^{wQ} - \Delta x_i a_i^{wM} \\ 0 & 1 & a_i^{\psi M} & a_i^{\psi Q} - \Delta x_i a_i^{\psi M} \\ 0 & 0 & 1 & -\Delta x_i \\ 0 & 0 & 0 & 1 \end{bmatrix}. \tag{6.168}$$

If the shear deformation effect is neglected, the influence coefficients have the values

$$a_i^{wQ} = \frac{(\Delta x_i)^3}{3EI_i}, \qquad a_i^{wM} = \frac{(\Delta x_i)^2}{2EI_i}, \qquad a_i^{\psi Q} = \frac{(\Delta x_i)^2}{2EI_i}, \qquad a_i^{\psi M} = \frac{\Delta x_i}{EI_i}, \tag{6.169}$$

and letting $\Delta x_i/EI_i = a_i$ we have

$$[T_F]_i = \begin{bmatrix} 1 & \Delta x_i & a_i\,\Delta x_i/2 & -a_i(\Delta x_i)^2/6 \\ 0 & 1 & a_i & -a_i\,\Delta x_i/2 \\ 0 & 0 & 1 & -\Delta x_i \\ 0 & 0 & 0 & 1 \end{bmatrix}. \tag{6.170}$$

Again one can devise a transfer matrix,

$$[T]_i = [T_F]_i [T_S]_i, \tag{6.171}$$

covering the station i and the field i, and subsequently an overall transfer matrix† given by‡

$$[T] = [T_S]_{n+1} \prod_{i=n}^{1} [T]_i, \tag{6.172}$$

which allows us to write

$$\begin{Bmatrix} w \\ \psi \\ M \\ Q \end{Bmatrix}_{n+1}^{R} = [T] \begin{Bmatrix} w \\ \psi \\ M \\ Q \end{Bmatrix}_{1}^{L}. \tag{6.173}$$

† For the product notation see the comment made in Section 6–11 in connection with (6.142).

‡ The matrices $[T_S]_1$ and $[T_S]_{n+1}$ reduce to the identity matrix if the corresponding boundary stations undergo no displacements, so for every such boundary the degree of freedom of the system is reduced by 1. For semidefinite systems the number of masses free to move does not coincide with the degree of freedom of the system.

Equation (6.173) represents four algebraic equations

$$
\begin{aligned}
w_{n+1}^R &= T_{11}w_1^L + T_{12}\psi_1^L + T_{13}M_1^L + T_{14}Q_1^L, \\
\psi_{n+1}^R &= T_{21}w_1^L + T_{22}\psi_1^L + T_{23}M_1^L + T_{24}Q_1^L, \\
M_{n+1}^R &= T_{31}w_1^L + T_{32}\psi_1^L + T_{33}M_1^L + T_{34}Q_1^L, \\
Q_{n+1}^R &= T_{41}w_1^L + T_{42}\psi_1^L + T_{43}M_1^L + T_{44}Q_1^L.
\end{aligned}
\tag{6.174}
$$

As an illustration of the application of (6.174) let us consider a *cantilever beam*, fixed at the end $x = 0$, which corresponds to station 1, and free at the end $x = L$, which corresponds to station $n + 1$. The boundary conditions for this system are

$$
w_1^L = w_1^R = w_1 = 0, \qquad \psi_1^L = \psi_1^R = \psi_1 = 0,
$$
$$
M_{n+1}^R = 0, \qquad Q_{n+1}^R = 0,
\tag{6.175}
$$

As m_1 cannot move, it must be set equal to zero, so $[T_s]_1$ becomes the identity matrix. Equations (6.174) and (6.175) yield

$$
0 = T_{33}M_1^L + T_{34}Q_1^L, \qquad 0 = T_{43}M_1^L + T_{44}Q_1^L,
\tag{6.176}
$$

giving the frequency equation

$$
\Delta(\omega) = \begin{vmatrix} T_{33} & T_{34} \\ T_{43} & T_{44} \end{vmatrix} = 0.
\tag{6.177}
$$

The method of obtaining the natural frequencies is the same as the one discussed in Section 6–11. The method of obtaining the natural modes, however, requires further elaboration. The state vector on the left side of station k can be written

$$
\begin{Bmatrix} w \\ \psi \\ M \\ Q \end{Bmatrix}_k^L = [T]_{k-1}[T]_{k-2}\cdots[T]_2[T]_1 \begin{Bmatrix} w \\ \psi \\ M \\ Q \end{Bmatrix}_1^L = \prod_{i=k-1}^{1} [T]_i \begin{Bmatrix} w \\ \psi \\ M \\ Q \end{Bmatrix}_1^L,
\tag{6.178}
$$

and this relation must hold true for any of the natural frequencies and associated natural modes

$$
\begin{Bmatrix} w^{(r)} \\ \psi^{(r)} \\ M^{(r)} \\ Q^{(r)} \end{Bmatrix}_k^L = \prod_{i=k-1}^{1} [T^{(r)}]_i \begin{Bmatrix} w^{(r)} \\ \psi^{(r)} \\ M^{(r)} \\ Q^{(r)} \end{Bmatrix}_1^L,
\tag{6.179}
$$

where the superscript r denotes the mode number. Note that the transfer matrix $[T^{(r)}]_i$ is obtained by substituting ω_r for ω in $[T]_i$. For the cantilever beam discussed above, we have $w_1^{(r)} = \psi_1^{(r)} = 0$ for any mode. In addition, we can choose arbitrarily $(M^{(r)})_1^L = 1$, and from the first of equations (6.176) (the second one could have been used just as easily), we obtain $(Q^{(r)})_1^L = -T_{33}^{(r)}/T_{34}^{(r)}$, which determines the state vector corresponding to the left side of station 1 (also the right side, in view of the fact that m_1 must be set equal to zero). This allows us to write (6.179) in the form

$$\begin{Bmatrix} w^{(r)} \\ \psi^{(r)} \\ M^{(r)} \\ Q^{(r)} \end{Bmatrix}_k^L = \prod_{i=k-1}^{1} [T^{(r)}]_i \begin{Bmatrix} 0 \\ 0 \\ 1 \\ -T_{33}^{(r)}/T_{34}^{(r)} \end{Bmatrix}. \qquad (6.180)$$

Equation (6.180) enables us to calculate the state vector corresponding to the left side of each station. Proper allowance must be made when the state vector on the right side of a station is desired. This is left to the reader as an exercise.

Myklestad[†] suggested a solution of the problem by means of recurrence formulas. Thomson[‡] was the first to set up the problem in matrix form using transfer matrices without introducing the concept of station and field transfer matrices. The discussion presented here is closer to the treatment of Pestel and Leckie,[§] who applied the concept of transfer matrices to a large number of problems, including branched torsional systems and framed structures.

6-13 LUMPED-PARAMETER METHOD EMPLOYING INFLUENCE COEFFICIENTS

In Section 5–15 we derived the eigenvalue problem for a continuous system in the integral form

$$w(P) = \lambda \int_D G(P, Q)m(Q)w(Q) \, dD(Q), \qquad (6.181)$$

where $G(P, Q) = a(P, Q)$ is Green's function or the flexibility influence function, which is the displacement at point P caused by a unit load at point Q.

Quite often, and this is particularly true in the case of nonuniform systems,

† See N. O. Myklestad, *Fundamentals of Vibration Analysis*, McGraw-Hill, New York, 1956.

‡ See W. T. Thomson, *J. Appl. Mech.*, 337–339 (1950).

§ See E. C. Pestel and F. A. Leckie, *Matrix Methods in Elastomechanics*, McGraw-Hill, New York, 1963.

it is not a pleasant task to use (6.181), and an approximate method should be considered. We have already discussed some approximate methods in conjunction with the integral formulation. Specifically we have discussed the assumed-modes method, Galerkin's method, and the collocation method. All these methods treat a continuous system as an n-degree-of-freedom system by assuming the solution in the form of a linear combination of n known functions. The lumped-parameter method is very simple in principle, and it resembles slightly the methods of Holzer and Myklestad. The method consists of lumping the system into discrete masses at given points, but there the similarity ends. Whereas the Holzer and Myklestad methods assume the stiffness properties between stations as uniform and use a step-by-step approach, the present method essentially converts the integral (6.181) into a finite sum. To this end the domain is divided into finite portions $\Delta D(Q)$, and the corresponding lumped mass is denoted m_j and given by

$$m(Q)\,\Delta D(Q) \cong m_j. \tag{6.182}$$

In addition we denote

$$w(P) = w_i, \qquad w(Q) = w_j, \tag{6.183}$$

where i and j denote the position of the stations just as P and Q denote the position of points in a continuous system. With this notation (6.181) becomes simply

$$w_i = \lambda \sum_{j=1}^{n} a_{ij} m_j w_j, \tag{6.184}$$

which can be written in the matrix form

$$\lambda [a][m]\{w\} = \{w\} \tag{6.185}$$

and represents an eigenvalue problem for a discrete system encountered before (see Section 4–4). The method of obtaining the eigenvalues and eigenvectors remains as discussed previously.

Although the method is very simple to apply, the influence coefficients a_{ij} may be difficult to obtain if the system configuration or the boundary conditions are complicated. The methods of obtaining the influence coefficients are covered adequately in texts on solid mechanics.

6–14   LUMPED-PARAMETER METHOD. SEMIDEFINITE SYSTEMS

The lumped-parameter method employing influence coefficients may be used also in the case of semidefinite systems. It was pointed out in Section 4–9 that for semidefinite systems the flexibility influence coefficients are not defined. We can, however, modify the approach by considering the rigid-body modes of the unrestrained system and measuring the elastic deformations from a reference position defined by the rigid-body motion. Although the potential energy expression can be written in terms of elastic deformation alone by ignoring the rigid-body motion, the kinetic energy expression must be written in terms of absolute velocities. Denoting by w_i the elastic deformation measured relative to the rigid-body motion and by W_i the absolute displacement of mass m_i of a lumped system, we have

$$T = \tfrac{1}{2}\{\dot{W}\}^T [m]\{\dot{W}\}, \tag{6.186}$$

$$V = \tfrac{1}{2}\{w\}^T [k]\{w\}, \tag{6.187}$$

where $[m]$ is the diagonal mass matrix and $[k]$ is the symmetric matrix of the stiffness influence coefficients.

To illustrate the approach let us consider the problem of a free-free beam in bending as shown in Figure 6.7. The assumption is made that the

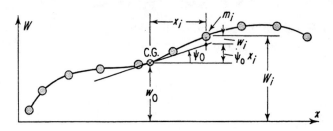

FIGURE 6.7

displacements are sufficiently small that the position of the center of mass does not change with respect to the beam and that the rotation ψ_0 does not introduce longitudinal components of motion. If w_0 denotes the transverse displacement of the center of mass C.G., ψ_0 denotes the angle that the tangent to the beam at the center of mass makes with the x axis, and w_i is the displacement of point i relative to that tangent, the absolute displacement of any point i can be written

$$W_i = w_0 + \psi_0 x_i + w_i. \tag{6.188}$$

In the above w_0 and ψ_0 can be regarded as rigid-body motion and w_i as the elastic displacement of point i obtained by regarding the beam as clamped at the center of mass. Equation (6.188) can be written in the matrix form

$$\{W\} = w_0\{1\} + \psi_0\{x\} + \{w\}, \tag{6.189}$$

where $\{1\}$ is a column matrix with all its elements unity and $\{x\}$ is a column matrix with its elements equal to the distances x_i from the center of mass of the beam to the masses m_i $(i = 1, 2, \ldots, n)$.

Let the first two modes of the beam be equal to the rigid-body modes

$$\{W^{(0)}\} = w_0\{1\}, \qquad \{W^{(1)}\} = \psi_0\{x\}, \tag{6.190}$$

so the remaining modes must be orthogonal to them. This is just another way of saying that for the free-free beam the linear and angular momenta are zero. The orthogonality relations are

$$\{W^{(0)}\}^T[m]\{W\} = 0, \qquad \{W^{(1)}\}^T[m]\{W\} = 0. \tag{6.191}$$

Introducing (6.189) and (6.190) in the above equations we obtain

$$w_0\{1\}^T[m]\{1\} + \psi_0\{1\}^T[m]\{x\} + \{1\}^T[m]\{w\} = 0,$$
$$w_0\{x\}^T[m]\{1\} + \psi_0\{x\}^T[m]\{x\} + \{x\}^T[m]\{w\} = 0. \tag{6.192}$$

But

$$\{1\}^T[m]\{1\} = \sum_{i=1}^{n} m_i = M,$$

$$\{1\}^T[m]\{x\} = \{x\}^T[m]\{1\} = 0, \tag{6.193}$$

$$\{x\}^T[m]\{x\} = \sum_{i=1}^{n} m_i x_i^2 = I_c,$$

where M is the total mass and I_c may be looked upon as the mass moment of inertia of the beam with respect to the center of mass of the beam. The second triple product in (6.193) is zero, because x_i is measured from the center of mass. Hence (6.192) and (6.193) yield

$$w_0 = -\frac{1}{M}\{1\}^T[m]\{w\}, \qquad \psi_0 = -\frac{1}{I_c}\{x\}^T[m]\{w\}, \tag{6.194}$$

so (6.189) can be written

$$\{W\} = -\frac{1}{M}\{1\}\{1\}^T[m]\{w\} - \frac{1}{I_c}\{x\}\{x\}^T[m]\{w\} + \{w\}. \tag{6.195}$$

Introducing a constraint matrix in the form

$$[c] = [I] - \frac{1}{M}\{1\}\{1\}^T[m] - \frac{1}{I_c}\{x\}\{x\}^T[m], \tag{6.196}$$

where $[I]$ is the unit matrix, (6.195) assumes the compact form

$$\{W\} = [c]\{w\}, \tag{6.197}$$

which allows us to express the absolute motion in terms of the motion relative to the rigid-body motion. It follows that the kinetic energy expression can be rewritten

$$T = \tfrac{1}{2}\{\dot{W}\}^T[m]\{\dot{W}\} = \tfrac{1}{2}\{\dot{w}\}^T[c]^T[m][c]\{\dot{w}\}$$

$$= \tfrac{1}{2}\{\dot{w}\}[m']\{\dot{w}\}, \tag{6.198}$$

where $$[m'] = [c]^T[m][c] \tag{6.199}$$

is a symmetric matrix.

Using Lagrange's equations in conjunction with (6.87) and (6.98) we obtain the eigenvalue problem

$$[m']\{w\} = \frac{1}{\omega^2}[k]\{w\}, \tag{6.200}$$

where $[m']$ and $[k]$ are positive definite matrices. The eigenvalue problem can also be written in the more general form

$$[m'][w] = [k][w][\omega^{-2}]. \tag{6.201}$$

Its solution calls for the calculation of the stiffness influence coefficients, which may be an unpleasant pursuit. As it turns out, this can be avoided.

The problem in the form (6.200) may be solved by premultiplying both sides of the equation by $[k]^{-1} = [a]$ and using the matrix iteration based on the sweeping procedure. The flexibility influence coefficients a_{ij} are obtained by considering the beam as clamped at the center of mass. Note that the coefficients relating the displacements on one side of the C.G. to the forces on the other side of the C.G. are zero.

It is also possible, and quite often recommended, to solve the problem in the form (6.201) by using the matrix iteration method based on the matrix diagonalization by successive rotations. This will necessitate the computation of $[a]^{1/2}$ (see Problem 6.15).

The eigenvalue problem (6.201) yields $n - 2$ eigenvalues and associated eigenvectors. The eigenvalues are related to the natural frequencies, whereas the eigenvectors, when introduced in (6.197), yield the natural modes of the system. Of course, in addition to these modes, there are two rigid-body modes, (6.190), with zero natural frequencies associated.

Example 6.7

Obtain the natural frequencies and the natural modes of vibration associated with the free-free bar shown in Figure 6.8. The mass and stiffness

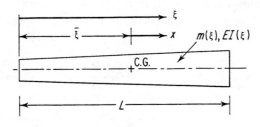

FIGURE 6.8

distributions are

$$m(\xi) = \frac{4}{5}\frac{M}{L}\left(1 + \frac{\xi}{2L}\right), \qquad EI(\xi) = \frac{4}{5}EI\left(1 + \frac{\xi}{2L}\right), \tag{a}$$

where M is the total mass and L the length of the bar. The center of mass of the system is located at a distance $\bar{\xi}$ from the left end, where $\bar{\xi}$ is given by

$$\bar{\xi} = \frac{1}{M}\int_0^L \xi m(\xi)\, d\xi = \frac{8}{15}L. \tag{b}$$

Next let us assume that the mass of the bar is lumped into n discrete masses such that the mass m_i $(i = 1, 2, \ldots, n)$ is equal to the mass in the segment $(i - 1)(L/n) \leq \xi \leq i(L/n)$ and is located at the corresponding center of mass. Hence the masses m_i have the values

$$m_i = \int_{(i-1)(L/n)}^{i(L/n)} m(\xi)\, d\xi = \frac{M}{5n^2}(4n + 2i - 1), \quad i = 1, 2, \ldots, n, \tag{c}$$

and their locations are at distances ξ_i from the left end, where the ξ_i are given by

$$\xi_i = \frac{1}{m_i}\int_{(i-1)(L/n)}^{i(L/n)} \xi m(\xi)\, d\xi = \frac{2L}{3n(4n + 2i - 1)}[3n(2i - 1) + 3i(i - 1) + 1],$$
$$i = 1, 2, \ldots, n. \tag{d}$$

When measured from the center of mass, instead of the left end, these locations are

$$x_i = \xi_i - \bar{\xi}, \qquad i = 1, 2, \ldots, n, \tag{e}$$

and, similarly, the stiffness has the form

$$EI(x) = \frac{2}{5}\frac{EI}{L}(2L + \bar{\xi} - x). \tag{f}$$

To obtain the influence coefficients a_{ij} we shall use the area-moment method. To this end we regard the bar as fixed at the center of mass, $\xi = \bar{\xi}$, and cantilevered in each side as shown in Figure 6.9(a). Figure 6.9(b) shows

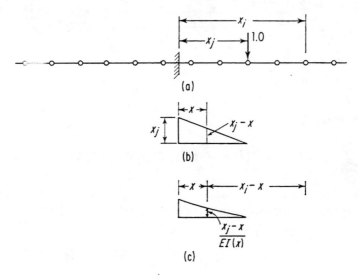

(a)

(b)

(c)

FIGURE 6.9

the bending moment diagram due to a unit load applied at $x = x_j$, and Figure 6.9(c) shows the bending moment divided by the stiffness $EI(x)$. The slope of the deflection curve is zero at the fixed end, $x = 0$, so the area-moment method gives the deflection at the point x_i due to a unit load at x_j in the form of the moment with respect to point x_i of the area of the bending moment divided by $EI(x)$ diagram. Hence, for $x_j > 0$ and $x_i > x_j$, we have

$$a_{ij} = \int_0^{x_j} \frac{(x_j - x)(x_i - x)}{EI(x)} dx$$

$$= \frac{5L}{2EI}\{[(x_i + x_j + 2L + \bar{\xi})(2L + \bar{\xi}) + x_i x_j]\ln\frac{x_j + 2L + \bar{\xi}}{2L + \bar{\xi}}$$

$$- (x_i + x_j + 2L + \bar{\xi})x_j - \tfrac{1}{2}x_j^2\}, \qquad i \geq j. \tag{g}$$

For $x_i > 0$ and $x_j < 0$ or $x_i < 0$ and $x_j > 0$, we obtain

$$a_{ij} = 0, \tag{h}$$

and an expression similar to (g) can be written for the case in which $x_i < 0$ and $x_j < 0$. Furthermore, the influence coefficients are symmetric, $a_{ij} = a_{ji}$.

The coefficients a_{ij} can be arranged in the form of a symmetric matrix which is the flexibility matrix $[a]$. The diagonal mass matrix, $[m]$, is obtained from (c), whereas (d) and (e) yield the column matrix $\{x\}$. This, in turn, allows us to compute the constraint matrix $[c]$ according to (6.196), and subsequently the modified mass matrix $[m']$ according to (6.199). The natural frequencies and natural modes are obtained by solving the eigenvalue problem (6.201).

A numerical solution of the eigenvalue problem (6.201) has been worked out for $n = 20$ by means of an electronic computer. The solution yields 18 eigenvalues and eigenvectors $w^{(r)}$ $(r = 2, 3, \ldots, 20)$. The eigenvalues are related to the natural frequencies ω_r, whereas the eigenvectors are substituted in (6.197), which enables us to obtain the natural modes $W^{(r)}$ $(r = 2, 3, \ldots, 20)$. It should be recalled that the remaining two modes are the rigid-body $W^{(0)}$ and $W^{(1)}$ [see (6.190)] with corresponding natural frequencies equal to zero. The numerical results have been used to plot the modes $W^{(2)}$, $W^{(3)}$, and $W^{(4)}$ in Figure 6.10, where the associated natural frequencies are also given.

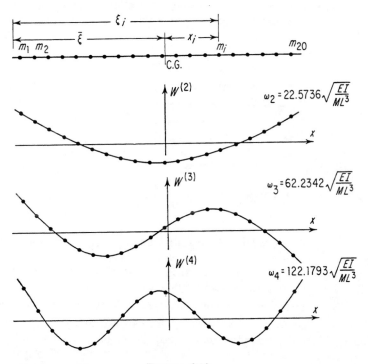

FIGURE 6.10

It should be noted that the natural frequencies and natural modes are reasonably close to the ones of a uniform free-free bar of total mass M and stiffness EI, as is to be expected. Also one must note that smaller displacements occur at the heavier, stiffer end of the bar, which agrees with the expectations.

Problems

6.1   A bar hinged at both ends has the mass per unit length

$$m(x) = \frac{m}{3}\left[1 + 12\frac{x}{L} - 12\left(\frac{x}{L}\right)^2\right]$$

and the stiffness distribution

$$EI(x) = \frac{EI_0}{3}\left[1 + 12\frac{x}{L} - 12\left(\frac{x}{L}\right)^2\right].$$

Use Rayleigh's energy method and estimate the fundamental frequency for the bending vibration of the bar.

6.2   Estimate the fundamental frequency for the bending vibration of the system shown in Figure 6.11, where the restoring force in the spring k_V is proportional to the linear deflection and the restoring moment in the spring k_M is proportional to the angular deflection of the end $x = L$. Use Rayleigh's energy method.

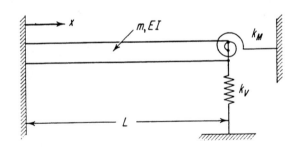

FIGURE 6.11

6.3   Consider the system of Problem 6.1 and use the Rayleigh-Ritz method to formulate the eigenvalue problem. Use as comparison functions the eigenfunctions of a uniform bar hinged at both ends. Use in sequence $n = 1$, $n = 2$, and $n = 3$ and compare the results obtained in each case.

6.4   Show that the eigenfunctions (6.33) are orthogonal in a generalized sense.

6.5   Show that for a system with symmetric mass and stiffness distributions, implied by the satisfaction of (6.41) through (6.44), it is possible, in using the Rayleigh-Ritz method, to reduce the eigenvalue problem to two independent eigenvalue problems for even and odd modes, respectively.

6.6 Consider the system of Figure 6.2 and formulate the eigenvalue problem by means of the Rayleigh-Ritz method. Let the bar be uniform and use as admissible functions the eigenfunctions of a uniform clamped-free bar. Let $n = 3$.

6.7 Formulate the eigenvalue problem of Problem 6.3 by means of the assumed-modes method (Section 6–5).

6.8 Formulate the eigenvalue problem of Problem 6.6 by means of the assumed-modes method (Section 6–5).

6.9 Formulate the eigenvalue problem of Problem 6.3 by means of Galerkin's method (Section 6–6).

6.10 Formulate the eigenvalue problem of Problem 6.3 by means of the colloca-tion method. Use the interior method and take three stations: $s_1 = L/4$, $s_2 = L/2$, and $s_3 = 3L/4$ (Section 6–7).

6.11 Extend Example 6.6 by selecting two additional admissible functions $\phi_3(x)$ and $\phi_4(x)$, representing the third and fourth modes. Select ϕ_3 and ϕ_4 such that they are even and odd functions of the variable $x - (L/2)$, respectively. Solve the eigenvalue problem and obtain the natural frequencies and the natural modes. Draw conclusions.

6.12 A nonuniform circular shaft is clamped at one end and has a torsional spring at the other end as shown in Figure 6.12. The mass polar moment of inertia and torsional stiffness distributions are

$$I(x) = \frac{6I_0}{5}\left[1 - \frac{1}{2}\left(\frac{x}{L}\right)^2\right], \qquad GJ(x) = \frac{6GJ_0}{5}\left[1 - \frac{1}{2}\left(\frac{x}{L}\right)^2\right].$$

Lump the shaft into three disks of polar mass moments of inertia I_i located at $x = x_i$ and with uniform torsional stiffnesses GJ_{i-1} in the intervals $x_{i-1} < x < x_i$. This is to be done for two cases:

(a) $I_1 = \displaystyle\int_0^{L/3} I(x)\,dx, \qquad x_1 = \frac{L}{6}, \qquad\qquad GJ_0 = GJ(0), \qquad 0 < x < \frac{L}{6},$

$\quad\ I_2 = \displaystyle\int_{L/3}^{2L/3} I(x)\,dx, \qquad x_2 = \frac{L}{2}, \qquad\qquad GJ_1 = GJ\!\left(\frac{L}{3}\right), \qquad \frac{L}{6} < x < \frac{L}{2},$

$\quad\ I_3 = \displaystyle\int_{2L/3}^{L} I(x)\,dx, \qquad x_3 = \frac{5L}{6}. \qquad\qquad GJ_2 = GJ\!\left(\frac{2L}{3}\right), \qquad \frac{L}{2} < x < \frac{5L}{6},$

$\qquad\qquad\qquad\qquad\qquad\qquad\qquad\qquad\qquad\qquad GJ_3 = GJ(L), \qquad \frac{5L}{6} < x < L.$

(b) $I_1 = \displaystyle\int_{L/6}^{L/2} I(x)\,dx, \qquad x_1 = \frac{L}{3}, \qquad\qquad GJ_0 = GJ\!\left(\frac{L}{6}\right), \qquad 0 < x < \frac{L}{3},$

$\quad\ I_2 = \displaystyle\int_{L/2}^{5L/6} I(x)\,dx, \qquad x_2 = \frac{2L}{3}, \qquad\qquad GJ_1 = GJ\!\left(\frac{L}{2}\right), \qquad \frac{L}{3} < x < \frac{2L}{3},$

$\quad\ I_3 = \displaystyle\int_{5L/6}^{L} I(x)\,dx, \qquad x_3 = L. \qquad\qquad GJ_2 = GJ\!\left(\frac{5L}{6}\right), \qquad \frac{2L}{3} < x < L.$

Derive the frequency equations for both cases if $k = GJ_0/4L$ (Section 6–11). Let $k = 0$ in the frequency equations and obtain the first natural frequencies and natural modes for both cases. Explain the difference in the calculated fundamental frequencies corresponding to the two cases.

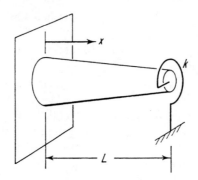

FIGURE 6.12

6.13 A nonuniform bar is clamped at $x = 0$ and free at $x = L$. The mass per unit length and stiffness distributions are

$$m(x) = m\left(1 - \frac{1}{4}\frac{x}{L}\right), \qquad EI(x) = EI\left[1 - \frac{1}{2}\frac{x}{L} + \frac{1}{16}\left(\frac{x}{L}\right)^2\right].$$

Divide the bar into lumps so as to yield a three-degree-of-freedom system and obtain the natural frequencies and natural modes corresponding to the bending vibration of the bar (Section 6–12).

6.14 Consider a uniform string fixed at both ends, divide the string into four lumps, and formulate the eigenvalue problem by means of the lumped-parameter method, employing influence coefficients. Obtain the first natural frequency and first natural mode.

6.15 Show how to solve the eigenvalue problem (6.201) by means of the matrix diagonalization by the successive rotations procedure. Use this procedure and obtain the natural frequencies and natural modes for the bending vibration of a uniform bar hinged at one end and free at the other end. Use three lumps.

Selected Readings

Bisplinghoff, R. L., H. Ashley, and R. L. Halfman, *Aeroelasticity*, Addison-Wesley, Reading, Mass., 1957.

Collatz, L., *Eigenwertprobleme*, Chelsea, New York, 1948.

Collatz, L., *The Numerical Treatment of Differential Equations*, Springer, Berlin, 1960.

Crandall, S. H., *Engineering Analysis*, McGraw-Hill, New York, 1956.

Hildebrand, F. B., *Methods of Applied Mathematics*, Prentice-Hall, Englewood Cliffs, N.J., 1960.

Hurty, W. C., and M. F. Rubinstein, *Dynamics of Structures*, Prentice-Hall, Englewood Cliffs, N.J., 1964.

Myklestad, N. O., "A New Method of Calculating Normal Modes of Uncoupled Bending Vibration of Airplane Wings and Other Types of Beams," *J. Aeron. Sci.*, **11**, 153–162 (1944).

Pestel, E. C., and F. A. Leckie, *Matrix Methods in Elastomechanics*, McGraw-Hill, New York, 1963.

Rayleigh, Lord, *The Theory of Sound*, Vol. 1, Dover, New York, 1945.

Temple, G., and W. G. Bickley, *Rayleigh's Principle and Its Applications to Engineering*, Dover, New York, 1956.

Thomson, W. T., "Matrix Solution for the Vibration of Nonuniform Beams," *J. Appl. Mech.*, **17**, 337–339 (1950).

Tong, K. N., *Theory of Mechanical Vibration*, Wiley, New York, 1960.

Weinstock, R., *Calculus of Variations*, McGraw-Hill, New York, 1952.

CHAPTER 7

UNDAMPED
SYSTEM
RESPONSE

7-1 GENERAL CONSIDERATIONS

In Chapter 1 we defined the response of a system as the behavior of a system when subjected to a certain excitation. By system behavior we mean, for the most part, the displacement of a point in the system at any time, although quantities such as the velocity and the acceleration can be used to describe the response. In the cases in which the excitation is deterministic, the response can be given in a deterministic form. In random vibration it is not practical, and in some cases not possible, to give the response in the form of a time-dependent function. In these cases quantities such as the mean square response may be more meaningful. In this chapter we shall be concerned exclusively with deterministic excitations. The response to random excitations will be discussed in Chapter 11.

The excitation can be divided into forcing functions, initial displacements and velocities, and moving supports. The vibration resulting from the action of forcing functions upon a system is known as *forced vibration*, and the one resulting from initial conditions is called *free vibration*. Moving supports result in forcing functions in the form of inertia forces and elastic forces and, as such, they lead to forced vibration problems.

In the case of discrete systems the response consists of n coordinates, where n is the degree of freedom of the system. These coordinates are obtained by solving the equations of motion of the system which consist of n, generally coupled, ordinary differential equations of second order. One can attempt to obtain a solution of the coupled equations, but this procedure becomes progressively cumbersome as the degree of freedom of the system increases. Nevertheless, it is feasible to solve the coupled equations when the excitation is periodic. When the excitation is not periodic a different approach is recommended. This approach consists of giving the equations of motion a simpler form by expressing the equations in a different set of coordinates. This is accomplished by means of a linear coordinate transformation which uncouples the system of equations; that is, it yields n independent equations. The linear transformation is obtained by assuming that the response is a superposition of the normal modes of the system multiplied by corresponding time-dependent generalized coordinates. This procedure is called *modal analysis* and was mentioned in Chapter 4 in connection with the expansion theorem. Modal analysis is particularly suitable for undamped systems, although in some special cases it can be used for damped systems. Independent equations of motion resemble, in structure, the equation of a single-degree-of-freedom system and can be solved without difficulty.

The problem of obtaining the response of a continuous system involves a larger degree of subtlety. For the most part the problems consist of a partial differential equation that the function describing the system response must satisfy throughout a given domain and, in addition, the function must also satisfy associated boundary conditions at every point on the boundaries of the domain. The satisfaction of the boundary conditions renders a solution unique. Such problems are called boundary-value problems, and there are various approaches to the solution of such problems. In some cases it might be possible to obtain a solution by means of an integral transformation such as the Laplace or Fourier transformations. In other cases one may be able to assume a solution in the form of an infinite series. The latter approach is modal analysis applied to continuous systems and leads to an infinite set of uncoupled ordinary differential equations. This approach is possible if the separation of variables method can be used to obtain an eigenvalue problem and, furthermore, one is able to solve the eigenvalue problem. When it is not possible to obtain an exact solution of the eigenvalue problem, one may be content with an approximate solution, in which case one can still use modal analysis. Approximate methods, in essence, look upon a continuous system as a finite-degree-of-freedom system and the formulation leads to a set of coupled ordinary differential equations. To obtain an approximate solution of a continuous system by modal analysis it is necessary to solve the eigenvalue problem of a finite-degree-of-freedom system which, for the most part, consists of symmetric matrices. For the collocation method, however, the matrices, in general, are not symmetric, and one must use the special type of expansion theorem that was presented in Section 4–12.

In the present chapter we discuss the response of undamped systems to deterministic excitations. The response of damped systems to deterministic excitations is presented in Chapter 9 and the response to random excitation is treated in Chapter 11. Part A of this chapter is concerned with the response of discrete systems. Excitations in the form of forcing functions and initial conditions are discussed. Methods of solution in which one solves a set of coupled equations of motion are shown. Subsequently, the treatment of the most general type of excitation by means of modal analysis is presented. Part B discusses the response of continuous systems exclusively by modal analysis. Part C discusses the response of continuous systems by means of approximate methods.

In Chapter 8 solutions by means of integral transform methods will be shown and the subject of standing and traveling waves discussed.

Part A—Response of Undamped Discrete Systems

7-2   GENERAL FORMULATION. LAPLACE TRANSFORM SOLUTION

The equation of motion of an undamped n-degree-of-freedom system can be written in the matrix form

$$[m]\{\ddot{q}(t)\} + [k]\{q(t)\} = \{Q(t)\}, \tag{7.1}$$

where $\{q(t)\}$ and $\{Q(t)\}$ are column matrices of generalized displacements and generalized forces, respectively.

The above equations may be solved by means of the Laplace transform method. To this end let

$$\bar{q}_i(s) = \mathscr{L}q_i(t) = \int_0^\infty e^{-st}q_i(t)\,dt, \tag{7.2}$$

$$\bar{Q}_i(s) = \mathscr{L}Q_i(t) = \int_0^\infty e^{-st}Q_i(t)\,dt \tag{7.3}$$

be the Laplace transforms of $q_i(t)$ and $Q_i(t)$, respectively. In addition, we have

$$\mathscr{L}\ddot{q}_i(t) = s^2\bar{q}_i(s) - sq_i(0) - \dot{q}_i(0), \tag{7.4}$$

where $q_i(0)$ and $\dot{q}_i(0)$ are the initial generalized displacement and initial generalized velocity associated with the ith coordinate. Transforming both sides of (7.1) and considering (7.2) through (7.4), we have

$$[m]\{s^2\bar{q}(s) - sq(0) - \dot{q}(0)\} + [k]\{\bar{q}(s)\} = \{\bar{Q}(s)\}. \tag{7.5}$$

Now define the generalized mechanical impedance matrix of the undamped system as

$$[Z(s)] = s^2[m] + [k], \tag{7.6}$$

and in addition denote

$$\{q(0)\} = \{q_0\}, \qquad \{\dot{q}(0)\} = \{\dot{q}_0\}, \tag{7.7}$$

so that (7.5) becomes

$$[Z(s)]\{\bar{q}\} = \{\bar{Q}\} + s[m]\{q_0\} + [m]\{\dot{q}_0\} = \{\bar{R}\}, \tag{7.8}$$

where

$$\{\bar{R}\} = \{\bar{Q}\} + s[m]\{q_0\} + [m]\{\dot{q}_0\} \tag{7.9}$$

is a more general transformed input in the sense that it includes not only the exciting forces but also the initial conditions. Note that the impedance relates the transformed outputs to transformed inputs. Substituting $\pm i\omega$ for s in $[Z(s)]$, one obtains the harmonic impedance. The solution of (7.8) is simply

$$\{\bar{q}\} = [Z(s)]^{-1}\{\bar{R}\}. \tag{7.10}$$

Recalling the definition (1.41) of the transfer function $G_{i,j}(s)$, we can relate the transformed output to the transformed input by

$$\{\bar{q}\} = [G(s)]\{\bar{R}\}, \tag{7.11}$$

where $[G(s)]$ is the matrix of the transfer functions. Comparing (7.10) with (7.11) we conclude that

$$[G(s)] = [Z(s)]^{-1}, \tag{7.12}$$

or the impedance matrix and transfer functions matrix are the reciprocal of each other.

Equation (7.10) can be written

$$\{\bar{q}\} = \frac{[A(s)]}{|Z(s)|}\{\bar{R}\}, \tag{7.13}$$

where $[A(s)]$ is the adjoint and $|Z(s)|$ the determinant of the impedance matrix $[Z(s)]$. To obtain the response $\{q(t)\}$ it is necessary to evaluate the inverse Laplace transform of $\{\bar{q}(s)\}$.

It will prove convenient to treat the initial conditions and the various types of forcing functions separately.

7-3 RESPONSE TO INITIAL DISPLACEMENTS AND VELOCITIES

When the external forces are absent, (7.13) reduces to the special form

$$\{\bar{q}\} = \frac{[A(s)]}{|Z(s)|} [m]\{s\{q_0\} + \{\dot{q}_0\}\}. \tag{7.14}$$

All elements $\bar{q}_i$ have poles at the roots of the equation

$$|Z(s)| = 0. \tag{7.15}$$

Substituting $s = \pm i\omega$ in Eq. (7.15) we obtain the characteristic or frequency equation, which is satisfied by the natural frequencies of the system. If all the natural frequencies are distinct, we obtain simple poles at the roots $s = \pm i\omega_r$, where ω_r are the natural frequencies of the system. If the characteristic equation has repeated roots, there will be poles of higher order. Usually, however, all roots are distinct, in which case (7.15) can be written

$$|Z(s)| = (s - i\omega_1)(s + i\omega_1)(s - i\omega_2)\cdots(s + i\omega_n)$$

$$= \prod_{r=1}^{n} (s^2 + \omega_r^2), \tag{7.16}$$

where $\prod_{r=1}^{n}$ is the standard product notation. The response is obtained by taking the inverse transformation of (7.14) by the method of partial fractions[†] If there are only simple poles, then

$$\{q(t)\} = \sum_{k=1}^{n} \{(s - i\omega_k)\{\bar{q}\}e^{st}|_{s=i\omega_k} + (s + i\omega_k)\{\bar{q}\}e^{st}|_{s=-i\omega_k}\}$$

$$= \sum_{k=1}^{n} \left\{ \frac{[A(s)][m]\{s\{q_0\} + \{\dot{q}_0\}\}e^{st}}{(s + i\omega_k) \prod_{\substack{r=1 \\ r\neq k}}^{n} (s^2 + \omega_r^2)} \right|_{s=i\omega_k}$$

$$+ \frac{[A(s)][m]\{s\{q_0\} + \{\dot{q}_0\}\}e^{st}}{(s - i\omega_k) \prod_{\substack{r=1 \\ r\neq k}}^{n} (s^2 + \omega_r^2)} \left.\vphantom{\frac{1}{1}}\right|_{s=-i\omega_k} \right\}. \tag{7.17}$$

† Those familiar with the complex-variable theory may wish to use the residue theorem for the evaluation of the inversion integral (see Section 8–6).

As

$$[A(i\omega_k)] = [A(-i\omega_k)],\tag{7.18}$$

(7.17) reduces to

$$\{q(t)\} = \sum_{k=1}^{n} \frac{[A(i\omega_k)][m]}{\prod\limits_{\substack{r=1 \\ r \neq k}}^{n} (\omega_r^2 - \omega_k^2)} \left\{ \{q_0\} \cos \omega_k t + \{\dot{q}_0\} \frac{\sin \omega_k t}{\omega_k} \right\}.\tag{7.19}$$

The solution (7.19) implies that if the system is subjected to some initial excitation it will oscillate indefinitely. The motion consists of a superposition of harmonic oscillations of frequencies equal to the natural frequencies. In a later section this point will be brought into sharper focus, when modal analysis is discussed.

Of course, the oscillation can take place indefinitely only if the system has no damping whatsoever, which was one of our basic assumptions. When one is interested in the response over a time interval that is short in comparison with the natural periods of the system, the assumption is justifiable. When the response over a longer interval of time is sought, one must remember that in any real system a slight amount of damping is present. If the analogy with the single-degree-of-freedom system is invoked, the poles for this case are given by $s = -\zeta_r \omega_r \pm i\omega_{rd}$, which gives a response in the form of exponentially decaying harmonics which eventually disappear (see Section 9–3). Equation (7.19) is, in essence, the solution of the homogeneous equations, $\{Q(t)\} = \{0\}$, and for the reasons just explained it is called the *starting transient response*. When the system is subjected to external forces persisting over a large time duration, the starting transient response is generally ignored.

7–4   RESPONSE TO HARMONIC EXCITATION

Let the system be acted upon by n harmonic forcing functions given by the *real part* of

$$Q_j(t) = Q_{0j} e^{i\Omega_j t}, \quad j = 1, 2, \ldots, n,\tag{7.20}$$

where Q_{0j} is the complex amplitude which contains the information concerning the amplitude and the phase angle and Ω_j is the driving circular frequency associated with the forcing function Q_j. Equation (7.20) can be written as a column matrix

$$\{Q(t)\} = \{Q_0 e^{i\Omega t}\},\tag{7.21}$$

which has the Laplace transformation

$$\{\bar{Q}(s)\} = \left\{\frac{Q_0}{s - i\Omega}\right\}. \tag{7.22}$$

Introducing (7.22) in (7.13) and setting the initial conditions equal to zero, we obtain

$$\{\bar{q}(s)\} = \frac{[A(s)]}{|Z(s)|}\left\{\frac{Q_0}{s - i\Omega}\right\}. \tag{7.23}$$

The adjoint of the impedance matrix may be written in terms of column matrices in the form

$$[A(s)] = [\{A_1(s)\}\{A_2(s)\}\cdots\{A_n(s)\}], \tag{7.24}$$

so (7.23) can be rewritten

$$\{\bar{q}(s)\} = \sum_{j=1}^{n} \frac{\{A_j(s)\}}{|Z(s)|}\frac{Q_{0j}}{s - i\Omega_j}. \tag{7.25}$$

Every term of matrix $\{\bar{q}\}$ has simple poles at $s = i\Omega_j$ $(j = 1, 2, \ldots, n)$. In addition, there are poles at $s = \pm i\omega_r$ which are the roots of the polynomial in s,

$$|Z(s)| = 0, \tag{7.26}$$

where ω_r $(r = 1, 2, \ldots, n)$ are the natural frequencies of the system. Equation (7.25) can be readily inverted by summing up the contributions of all poles. The contribution of the poles $s = \pm i\omega_r$ gives the starting transient response discussed previously, which is zero for zero initial conditions. The contribution of the poles $s = i\Omega_j$ is in the form of a series of harmonics of frequency Ω_j and is called the *steady-state response*. Concentrating on the steady-state response the inverse transformation of (7.25) is

$$\{q(t)\} = \sum_{j=1}^{n} \frac{\{A_j(i\Omega_j)\}}{|Z(i\Omega_j)|} Q_{0j}e^{i\Omega_j t} = \frac{[A(i\Omega)]}{|Z(i\Omega)|}\{Q_0 e^{i\Omega t}\}, \tag{7.27}$$

where only the *real part* of the above must be retained.

When one of the exciting frequencies Ω_j approaches one of the natural frequencies of the system, the determinant $|Z(i\Omega_j)|$ approaches zero and the elements of $\{q(t)\}$ become very large, primarily due to the jth term in the series. Hence the column matrix $\{A_j(i\Omega_j)\}/|Z(i\Omega_j)|$ may be looked upon as a *matrix of magnification factors*.

When any of the frequencies Ω_j coincides with one of the natural frequencies of the system, say $\Omega_j = \omega_k$, the response becomes infinite. This phenomenon is known as the *resonance condition*. In this case, however, we obtain a double pole at $s = i\Omega_j = i\omega_k$, and the previous analysis no longer holds. Accordingly, if we concentrate on the contribution of the pole $s = i\Omega_j = i\omega_k$ only, the response becomes

$$\{q(t)\} = \frac{d}{ds}\left[(s - i\omega_k)^2 \frac{\{A_j(s)\}}{|Z(s)|} \frac{Q_{0j}}{s - i\Omega_j} e^{st}\right]\Bigg|_{s=i\omega_k}$$

$$= Q_{0j}\frac{d}{ds}\left(\frac{\{A_j(s)\}e^{st}}{(s + i\Omega_j)\prod\limits_{\substack{r=1 \\ r \neq k}}^{n}(s^2 + \omega_r^2)}\right)\Bigg|_{s=i\Omega_j}$$

$$= \frac{Q_{0j}e^{i\Omega_j t}}{2i\Omega_j \sum\limits_{\substack{r=1 \\ r \neq k}}^{n}(\omega_r^2 - \Omega_j^2)}$$

$$\times \left\{tA_j(i\Omega_j) + A_j'(i\Omega_j) - \left(\frac{1}{2i\Omega_j} + \sum\limits_{\substack{r=1 \\ r \neq k}}^{n}\frac{2i\Omega_j}{\omega_r^2 - \Omega_j^2}\right)A_j(i\Omega_j)\right\}.$$

$$(7.28)$$

It is easy to see that the response increases continuously with time. It should be pointed out that the only conclusion we can draw is that the response is diverging, because before long the amplitudes of the displacements will exceed the linear range of the system, at which point the linear theory ceases to be valid.

7-5 RESPONSE TO PERIODIC EXCITATION. FOURIER SERIES

A function reproducing itself at equal intervals of time, $T = 2\pi/\Omega_0$, is said to be periodic and of period T. A harmonic function is periodic but a periodic function is not necessarily harmonic. The periodicity of a function is expressed by

$$Q(t) = Q(t + T). \tag{7.29}$$

A periodic function may be represented by a Fourier series of harmonic components in the form

$$Q(t) = \sum_{p=-\infty}^{\infty} A_p e^{ip\Omega_0 t}, \qquad \Omega_0 = \frac{2\pi}{T}, \tag{7.30}$$

where the complex coefficients A_p are given by

$$A_p = \frac{1}{T} \int_{-T/2}^{T/2} Q(\tau) e^{-ip\Omega_0\tau} \, d\tau, \tag{7.31}$$

where τ is a dummy variable of integration. In practice, one can limit the series (7.30) to a finite number of harmonics and still obtain a good representation of the function $Q(t)$.

Hence a periodic excitation $Q(t)$ can be represented by a train of harmonics, and the response to each of these harmonics is obtained by the analysis of Section 7–4. Thus we simply obtain the response to one of the harmonic components and sum the result over p. From (7.27) we write directly

$$\{q(t)\} = \sum_{p=-\infty}^{\infty} \frac{[A(ip\Omega)]}{|Z(ip\Omega)|} \{A_p e^{ip\Omega t}\}. \tag{7.32}$$

Note that each of the functions $Q_j(t)$ may have a different period.

7-6 RESPONSE OF DISCRETE SYSTEMS BY MODAL ANALYSIS

Although it is possible to pursue the previous type of analysis and obtain the general response to a nonperiodic excitation, it will prove more expedient to adopt the normal modes approach, which will be referred to as *modal analysis*. The procedure is analogous to the Fourier analysis and can be easily applied to obtain the responses to initial conditions, harmonic excitations, and periodic excitations which were obtained in the preceding sections by using a different approach. In fact, we shall derive a general expression covering all the cases just mentioned.

Consider again the equations of motion

$$[m]\{\ddot{q}\} + [k]\{q\} = \{Q\}, \tag{7.33}$$

where the excitation functions $Q_j(t)$ are arbitrary functions of time (periodic excitations are just special cases). To use the modal analysis it is necessary first to solve the eigenvalue problem

$$[m][u][\omega^2] = [k][u] \tag{7.34}$$

associated with the system described by (7.33). The solution of the eigenvalue problem (7.34) yields the modal matrix $[u]$ and the diagonal matrix of the eigenvalues $[\omega^2]$. Using the *expansion theorem* the response may be described as a superposition of the normal modes in the form

$$\{q(t)\} = [u]\{\eta(t)\}, \tag{7.35}$$

where $\{\eta(t)\}$ is a column matrix consisting of a set of time-dependent generalized coordinates. From (7.35) it follows that

$$\{\ddot{q}\} = [u]\{\ddot{\eta}\}, \tag{7.36}$$

so that introducing (7.35) and (7.36) in (7.33), we obtain

$$[m][u]\{\ddot{\eta}\} + [k][u]\{\eta\} = \{Q\}. \tag{7.37}$$

Premultiply both sides of (7.37) by $[u]^T$ and write

$$[u]^T[m][u]\{\ddot{\eta}\} + [u]^T[k][u]\{\eta\} = [u]^T\{Q\}. \tag{7.38}$$

But the normal modes are such that

$$[u]^T[m][u] = [I], \qquad [u]^T[k][u] = \lceil\omega^2\rfloor, \tag{7.39}$$

where $[I]$ is the identity matrix. In addition, introduce a column matrix of generalized forces $N_r(t)$ associated with the generalized coordinates $\eta_r(t)$ and related to the forces $Q_s(t)$ by

$$\{N\} = [u]^T\{Q\}. \tag{7.40}$$

In view of (7.39) and (7.40), (7.38) can be rewritten

$$\{\ddot{\eta}\} + \lceil\omega^2\rfloor\{\eta\} = \{N\}, \tag{7.41}$$

which represents a set of n *uncoupled* differential equations of the type

$$\ddot{\eta}_r(t) + \omega_r^2\eta_r(t) = N_r(t), \qquad r = 1, 2, \ldots, n, \tag{7.42}$$

which have precisely the form of the differential equation describing the motion of an undamped single-degree-of-freedom system. Hence *modal analysis consists of uncoupling the equations of motion by means of a linear coordinate transformation; the transformation matrix is just the modal matrix* $[u]$. Of course, the solution of the uncoupled equations of motion, (7.42), is considerably easier to obtain than the solution of the coupled equations, (7.33).

The solution of (7.42) may be obtained by means of the Laplace transform method. Transforming both sides of (7.42), we obtain

$$s^2\bar{\eta}_r(s) - s\eta_r(0) - \dot{\eta}_r(0) + \omega_r^2\bar{\eta}_r(s) = \bar{N}_r(s), \tag{7.43}$$

where $\bar{\eta}_r(s)$ and $\bar{N}_r(s)$ are the Laplace transforms of $\eta_r(t)$ and $N_r(t)$, respectively, and $\eta_r(0)$ and $\dot{\eta}_r(0)$ are the initial values associated with the generalized coordinate $\eta_r(t)$. The subsidiary equation is

$$\bar{\eta}_r(s) = \frac{\bar{N}_r(s)}{s^2 + \omega_r^2} + \frac{s}{s^2 + \omega_r^2} \eta_r(0) + \frac{1}{s^2 + \omega_r^2} \dot{\eta}_r(0), \qquad (7.44)$$

which can be readily inverted by using Borel's theorem (see Appendix B) so that the rth generalized coordinate becomes

$$\eta_r(t) = \frac{1}{\omega_r} \int_0^t N_r(\tau) \sin \omega_r(t - \tau) \, d\tau + \eta_r(0) \cos \omega_r t + \dot{\eta}_r(0) \frac{\sin \omega_r t}{\omega_r},$$

$$r = 1, 2, \ldots, n. \quad (7.45)$$

The integral in (7.45) is known as the *convolution integral* and was encountered previously (see Section 1–7). The initial generalized displacement $\eta_r(0)$ and initial generalized velocity $\dot{\eta}_r(0)$ are obtained from the expressions

$$\{\eta(0)\} = [u]^T[m]\{q(0)\}, \qquad \{\dot{\eta}(0)\} = [u]^T[m]\{\dot{q}(0)\}, \qquad (7.46)$$

where $\{q(0)\}$ and $\{\dot{q}(0)\}$ are column matrices of initial displacements and velocities, respectively. Introducing (7.45) together with the initial conditions (7.46) in (7.35), the response $\{q(t)\}$ is obtained.

Note that the above formulation holds true regardless of whether the excitations $Q_j(t)$ are harmonic, periodic, or nonperiodic.

Example 7.1

Consider the system shown in Figure 7.1 and obtain its response to the arbitrary forces $F_1(t)$, $F_2(t)$, and $F_3(t)$ by means of the modal analysis. Let the initial conditions be zero.

The differential equations of motion of the system are

$$m\ddot{x}_1(t) + 2kx_1(t) - kx_2(t) \qquad\qquad = F_1(t),$$

$$m\ddot{x}_2(t) - kx_1(t) + 2kx_2(t) - kx_3(t) = F_2(t), \qquad (a)$$

$$2m\ddot{x}_3(t) \qquad\qquad - kx_2(t) + 3kx_3(t) = F_3(t),$$

which can be written in the matrix form

$$[m]\{\ddot{x}\} + [k]\{x\} = \{F\}, \qquad (b)$$

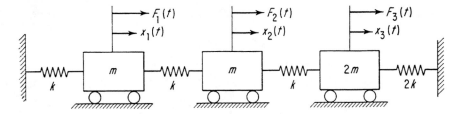

FIGURE 7.1

where the mass and stiffness matrices are

$$[m] = m \begin{bmatrix} 1 & 0 & 0 \\ 0 & 1 & 0 \\ 0 & 0 & 2 \end{bmatrix}, \qquad [k] = k \begin{bmatrix} 2 & -1 & 0 \\ -1 & 2 & -1 \\ 0 & -1 & 3 \end{bmatrix}. \tag{c}$$

To obtain the response of the system by means of modal analysis, we must first solve the eigenvalue problem associated with the system, which can be written

$$[m][u][\omega^2] = [k][u], \tag{d}$$

where $[m]$ and $[k]$ are given by (c). The eigenvalue problem (d) was in fact solved before (see Examples 4.1, 4.2, and 4.3). The results, consisting of the matrix of the natural frequencies $[\omega^2]$ and the modal matrix $[u]$, are†

$$[\omega^2] = \frac{k}{m} \begin{bmatrix} 0.6594 & 0 & 0 \\ 0 & 1.6789 & 0 \\ 0 & 0 & 3.1619 \end{bmatrix},$$

$$[u] = m^{-1/2} \begin{bmatrix} 0.4959 & 0.6074 & 0.6209 \\ 0.6646 & 0.1949 & -0.7215 \\ 0.3954 & -0.5446 & 0.2171 \end{bmatrix}. \tag{e}$$

From (7.40) we can write the generalized forces

$$N_1(t) = \{u^{(1)}\}^T\{F\} = m^{-1/2}[0.4959F_1(t) + 0.6646F_2(t) + 0.3954F_3(t)],$$

$$N_2(t) = \{u^{(2)}\}^T\{F\} = m^{-1/2}[0.6074F_1(t) + 0.1949F_2(t) - 0.5446F_3(t)], \tag{f}$$

$$N_3(t) = \{u^{(3)}\}^T\{F\} = m^{-1/2}[0.6209F_1(t) - 0.7215F_2(t) + 0.2171F_3(t)],$$

† The results are taken from Example 4.3.

so (7.42) become

$$\ddot{\eta}_1(t) + 0.6594\,\frac{k}{m}\,\eta_1(t) = N_1(t),$$

$$\ddot{\eta}_2(t) + 1.6789\,\frac{k}{m}\,\eta_2(t) = N_2(t), \tag{g}$$

$$\ddot{\eta}_3(t) + 3.1619\,\frac{k}{m}\,\eta_3(t) = N_3(t).$$

The solutions of (g) are obtained in the form of convolution integrals:

$$\eta_1(t) = \frac{1}{\omega_1}\int_0^t N_1(\tau)\sin\omega_1(t-\tau)\,d\tau$$

$$= \frac{k^{-1/2}}{0.8120}\int_0^t [0.4959F_1(\tau) + 0.6646F_2(\tau)$$

$$+ 0.3954F_3(\tau)]\sin 0.8120\sqrt{\frac{k}{m}}\,(t-\tau)\,d\tau, \tag{h}$$

and, similarly,

$$\eta_2(t) = \frac{k^{-1/2}}{1.2957}\int_0^t [0.6074F_1(\tau) + 0.1949F_2(\tau)$$

$$- 0.5446F_3(\tau)]\sin 1.2957\sqrt{\frac{k}{m}}\,(t-\tau)\,d\tau,$$

$$\tag{i}$$

$$\eta_3(t) = \frac{k^{-1/2}}{1.7782}\int_0^t [0.6209F_1(\tau) - 0.7215F_2(\tau)$$

$$+ 0.2171F_3(\tau)]\sin 1.7782\sqrt{\frac{k}{m}}\,(t-\tau)\,d\tau.$$

Finally, using (7.35), we obtain the response

$$x_1(t) = \sum_{j=1}^{3} u_1^{(j)}\eta_j(t) = m^{-1/2}[0.4959\eta_1(t) + 0.6074\eta_2(t) + 0.6209\eta_3(t)],$$

$$x_2(t) = \sum_{j=1}^{3} u_2^{(j)}\eta_j(t) = m^{-1/2}[0.6646\eta_1(t) + 0.1949\eta_2(t) - 0.7215\eta_3(t)], \tag{j}$$

$$x_3(t) = \sum_{j=1}^{3} u_3^{(j)}\eta_j(t) = m^{-1/2}[0.3954\eta_1(t) - 0.5446\eta_2(t) + 0.2171\eta_3(t)],$$

where the generalized coordinates $\eta_r(t)$ ($r = 1, 2, 3$) are given by (h) and (i).

Part B—Response of Undamped Continuous Systems

7-7 GENERAL FORMULATION. MODAL ANALYSIS

Consider a continuous system described by the partial differential equation

$$L[w(P, t)] + M(P) \frac{\partial^2 w(P, t)}{\partial t^2} = f(P, t) + F_j(t)\, \delta(P - P_j) \qquad (7.47)$$

over domain D. In the above L is a linear homogeneous self-adjoint differential operator consisting of derivatives through order $2p$ with respect to the spatial coordinates P but not with respect to time t. The operator L contains the information concerning the stiffness distribution and the mass distribution of the system is given by the function $M(P)$. The excitation consists of the distributed force $f(P, t)$ and the concentrated forces of amplitude $F_j(t)$ and acting at points $P = P_j$. The symbol $\delta(P - P_j)$ indicates a spatial Dirac's delta function defined by

$$\delta(P - P_j) = 0, \qquad P \neq P_j$$

$$\int_D \delta(P - P_j)\, dD(P) = 1, \qquad\qquad (7.48)$$

so in effect $F_j(t)\, \delta(P - P_j)$ may be looked upon as distributed forces which are distributed in the neighborhood of the points $P = P_j$. Note that $\delta(P - P_j)$ has units of $(\text{length})^{-n}$, where n is the number of spatial coordinates defining domain D. At every point of the boundary there are p boundary conditions of the type

$$B_i[w(P, t)] = 0, \qquad i = 1, 2, \ldots, p, \qquad\qquad (7.49)$$

where B_i are linear homogeneous differential operators containing derivatives normal to the boundary and along the boundary of order through $2p - 1$.

The normal modes analysis calls for the solution of the special eigenvalue problem consisting of the differential equation

$$L[w] = \lambda M w = \omega^2 M w \qquad\qquad (7.50)$$

to be satisfied over domain D, where w is subject to the boundary conditions

$$B_i[w] = 0, \qquad i = 1, 2, \ldots, p. \qquad\qquad (7.51)$$

The solution of the special eigenvalue problem consists of an infinite set of denumerable eigenfunctions $w_r(P)$ with corresponding natural frequencies ω_r.

The eigenfunctions are orthogonal, and if they are normalized such that

$$\int_D M(P)w_r(P)w_s(P)\,dD(P) = \delta_{rs},\qquad(7.52)$$

it follows that

$$\int_D w_r(P)L[w_s(P)]\,dD(P) = \omega_r^2\,\delta_{rs}.\qquad(7.53)$$

Using the *expansion theorem* we write the solution of (7.47) as a super-position of the normal modes $w_r(P)$ multiplying corresponding time-dependent generalized coordinates $\eta_r(t)$. Hence

$$w(P,\,t) = \sum_{r=1}^{\infty} w_r(P)\eta_r(t),\qquad(7.54)$$

and introducing (7.54) in (7.47), we obtain

$$L\left[\sum_{r=1}^{\infty} w_r(P)\eta_r(t)\right] + M(P)\frac{\partial^2}{\partial t^2}\sum_{r=1}^{\infty} w_r(P)\eta_r(t) = f(P,\,t) + F_j(t)\,\delta(P - P_j),$$

$$(7.55)$$

which reduces to

$$\sum_{r=1}^{\infty}\eta_r(t)L[w_r(P)] + \sum_{r=1}^{\infty}\ddot{\eta}_r(t)M(P)w_r(P) = f(P,t) + F_j(t)\,\delta(P - P_j).$$

$$(7.56)$$

Now multiply (7.56) by $w_s(P)$ and integrate over domain D to obtain

$$\sum_{r=1}^{\infty}\eta_r(t)\int_D w_s(P)L[w_r(P)]\,dD(P) + \sum_{r=1}^{\infty}\ddot{\eta}_r(t)\int_D w_s(P)M(P)w_r(P)\,dD(P)$$

$$= \int_D w_s(P)(f(P,\,t) + F_j(t)\,\delta(P - P_j))\,dD(P). \quad(7.57)$$

Let us say that there are a number l of concentrated forces acting throughout domain D, so recalling the definition of the delta function we can write

$$N_r(t) = \int_D w_r(P)(f(P,\,t) + F_j(t)\,\delta(P - P_j))\,dD(P)$$

$$= \int_D w_r(P)f(P,\,t)\,dD(P) + \sum_{j=1}^{l} w_r(P_j)F_j(t),\qquad(7.58)$$

where $N_r(t)$ denotes a generalized force associated with the generalized coordinate $\eta_r(t)$.

In view of (7.52), (7.53), and (7.58), (7.57) becomes

$$\ddot{\eta}_r(t) + \omega_r^2 \eta_r(t) = N_r(t), \qquad r = 1, 2, \ldots, \tag{7.59}$$

which represents an infinite set of *uncoupled* ordinary differential equations similar to the single-degree-of-freedom-system equation. Equations (7.59) have precisely the same form as (7.42), and it follows that their solutions are

$$\eta_r(t) = \frac{1}{\omega_r} \int_0^t N_r(\tau) \sin \omega_r(t - \tau)\, d\tau + \eta_r(0) \cos \omega_r t$$

$$+ \dot{\eta}_r(0) \frac{\sin \omega_r t}{\omega_r}, \qquad r = 1, 2, \ldots, \tag{7.60}$$

where $\eta_r(0)$ and $\dot{\eta}_r(0)$ are the initial generalized displacement and initial generalized velocity given by

$$\left.\begin{aligned}
\eta_r(0) &= \int_D M(P) w_r(P) w(P, 0)\, dD(P)\\[4pt]
\dot{\eta}_r(0) &= \int_D M(P) w_r(P) \dot{w}(P, 0)\, dD(P)
\end{aligned}\right\} \qquad r = 1, 2, \ldots. \tag{7.61}$$

The response $w(P, t)$ is obtained by introducing (7.60) in conjunction with (7.61) in (7.54).

7–8   RESPONSE OF AN UNRESTRAINED ROD IN LONGITUDINAL MOTION

Consider an unrestrained uniform rod lying at rest on a smooth surface when a longitudinal force $F(t)$ is applied at the end $x = 0$ (Figure 7.2). The free longitudinal vibration of an unrestrained uniform rod was discussed

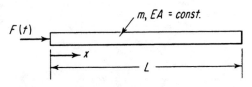

FIGURE 7.2

previously (see Section 5–7) and the following normal modes and natural frequencies were obtained:

$$U_0(x) = \sqrt{\frac{1}{mL}}, \qquad \omega_0 = 0,$$

$$U_r(x) = \sqrt{\frac{2}{mL}} \cos \frac{r\pi x}{L}, \qquad \omega_r = r\pi \sqrt{\frac{EA}{mL^2}}, \qquad r = 1, 2, \ldots. \tag{7.62}$$

Following the procedure outlined in Section 7–7, we write the longitudinal displacement of the bar in the form

$$u(x, t) = \sum_{r=0}^{\infty} U_r(x)\eta_r(t), \tag{7.63}$$

where the normal modes $U_r(x)$ are given by (7.62). The generalized coordinates $\eta_r(t)$ are computed from

$$\eta_r(t) = \frac{1}{\omega_r} \int_0^t N_r(\tau) \sin \omega_r(t - \tau)\, d\tau, \qquad r = 0, 1, 2, \ldots, \tag{7.64}$$

where the generalized force $N_r(t)$ is obtained by means of (7.58),

$$N_r(t) = U_r(0)F(t), \qquad r = 0, 1, 2, \ldots. \tag{7.65}$$

Hence, the general response is

$$u(x, t) = \sum_{r=0}^{\infty} \frac{U_r(x)U_r(0)}{\omega_r} \int_0^t F(\tau) \sin \omega_r(t - \tau)\, d\tau. \tag{7.66}$$

As an illustration let the applied force be in the form of a step function of magnitude F_0,

$$F(t) = F_0 \alpha(t) \tag{7.67}$$

and evaluate the integral

$$\int_0^t \alpha(\tau) \sin \omega_r(t - \tau)\, d\tau = -\int_0^t \sin \omega_r(\tau - t)\, d\tau$$

$$= \frac{\cos \omega_r(t - \tau)}{\omega_r}\Big|_0^t = \frac{1}{\omega_r}(1 - \cos \omega_r t), \tag{7.68}$$

so that

$$u(x, t) = F_0 \sum_{r=0}^{\infty} \frac{U_r(x)U_r(0)}{\omega_r^2} (1 - \cos \omega_r t). \tag{7.69}$$

As $\omega_0 = 0$, it would appear that the response is infinite. This is not the case actually, because

$$\lim_{\omega_0 \to 0} \frac{1 - \cos \omega_0 t}{\omega_0^2} = \frac{1}{2} t^2. \tag{7.70}$$

It follows that

$$u(x, t) = \frac{F_0}{mL} \frac{t^2}{2} + \frac{2F_0 L}{\pi^2 EA} \sum_{r=1}^{\infty} \frac{1}{r^2} \cos \frac{r\pi x}{L} (1 - \cos \omega_r t). \tag{7.71}$$

Furthermore,[†]

$$\sum_{r=1}^{\infty} \frac{1}{r^2} \cos \frac{r\pi x}{L} = \frac{\pi^2}{2L^2} \left[\frac{(L - x)^2}{2} - \frac{1}{6} L^2 \right], \tag{7.72}$$

so, finally,

$$u(x, t) = \frac{1}{2} \frac{F_0}{mL} t^2 + \frac{F_0}{EAL} \left[\frac{(L - x)^2}{2} - \frac{1}{6} L^2 \right]$$

$$- \frac{2F_0 L}{\pi^2 EA} \sum_{r=1}^{\infty} \frac{1}{r^2} \cos \frac{r\pi x}{L} \cos \omega_r t. \tag{7.73}$$

The first term in (7.73) is recognized as the rigid-body motion, and it is the only one to survive if the stiffness is increased indefinitely. The second term in (7.73) may be looked upon as the static deformation; the series represents the harmonic oscillation terms. The first two terms can be interpreted as an average position about which the harmonic oscillation takes place.

The same system can be looked upon as force-free with a time-dependent boundary condition at the end $x = 0$. This approach will be discussed in Section 7–14.

7–9   RESPONSE OF A SIMPLY SUPPORTED BEAM TO INITIAL DISPLACEMENTS

Let a uniform beam, of length L and mass m, which is simply supported at both ends, be deflected into the position[‡]

$$w(x, 0) = A \left(\frac{x}{L} - 2 \frac{x^3}{L^3} + \frac{x^4}{L^4} \right) \tag{7.74}$$

[†] See B. O. Peirce and R. M. Foster, *A Short Table of Integrals*, 4th ed., Ginn, Boston, 1957, Formulas 889 and 891.

[‡] Note that $w(x, 0)$ is symmetric with respect to $x = L/2$.

and then released. The initial velocity is zero and there are no external forces. The normal modes and natural frequencies of a uniform beam, simply supported at both ends, are

$$w_r(x) = \sqrt{\frac{2}{mL}} \sin \frac{r\pi x}{L}, \quad \omega_r = (r\pi)^2 \sqrt{\frac{EI}{mL^4}}, \quad r = 1, 2, \ldots \quad (7.75)$$

The response to the initial excitation, (7.74), is given by

$$w(x, t) = \sum_{r=1}^{\infty} w_r(x)\eta_r(t), \tag{7.76}$$

where the generalized coordinates $\eta_r(t)$ are obtained from (7.60) and the first of equations (7.61) in the form

$$\eta_r(t) = \eta_r(0) \cos \omega_r t = \left[\int_0^L m(x)w_r(x)w(x, 0)\, dx \right] \cos \omega_r t,$$
$$r = 1, 2, \ldots \quad (7.77)$$

Introducing (7.74) and (7.75) in the integral in (7.77), we write

$$\eta_r(0) = \int_0^L m(x)w_r(x)w(x, 0)\, dx = A\sqrt{\frac{2m}{L}} \int_0^L \sin \frac{r\pi x}{L} \left(\frac{x}{L} - 2\frac{x^3}{L^3} + \frac{x^4}{L^4}\right) dx,$$
$$r = 1, 2, \ldots \quad (7.78)$$

Evaluation of the integral yields

$$\eta_r(0) = A\sqrt{2mL} \frac{24}{r^5\pi^5} [1 - (-1)^r]. \tag{7.79}$$

When r is *even*,

$$\eta_r(0) = 0, \tag{7.80}$$

and when r is *odd*,

$$\eta_r(0) = \frac{48A}{r^5\pi^5} \sqrt{2mL}. \tag{7.81}$$

Combining the above results, the response to the initial displacement, (7.74), becomes

$$w(x, t) = \frac{96A}{\pi^5} \sum_{n=0}^{\infty} \frac{1}{(2n + 1)^5} \sin \frac{(2n + 1)\pi x}{L} \cos \omega_n t, \tag{7.82}$$

where
$$\omega_n = [(2n + 1)\pi]^2 \sqrt{\frac{EI}{mL^4}}, \quad n = 0, 1, 2, \ldots \tag{7.83}$$

Examining (7.82) we note that the terms in the series are symmetric with respect to the middle of the beam. This should not be surprising, because the displacement (7.74) is symmetric; in fact, it represents the static deflection caused by a uniformly distributed load. We must also note that the amplitude of the second harmonic is only 0.41 per cent of the amplitude of the first harmonic, so the motion resembles the first mode very closely. This is to be expected, because the initial displacement resembles the first mode.

7–10   RESPONSE OF A BEAM TO A TRAVELING FORCE

Let a concentrated force travel along a uniform, simply supported beam at a uniform velocity v in the positive x direction. The force can be described by

$$
\begin{aligned}
f(x, t) &= F(t)\,\delta(x - vt) && \text{for } 0 \le vt \le L, \\
&= 0 && \text{for } \quad vt > L,
\end{aligned}
\tag{7.84}
$$

where $F(t)$ describes the time dependence of the force.

The solution follows the same pattern as in Section 7–9. The response is assumed in the form

$$
w(x, t) = \sum_{r=1}^{\infty} w_r(x)\eta_r(t),
\tag{7.85}
$$

where the normal modes are as given by (7.75). The generalized coordinates $\eta_r(t)$ can be calculated by means of (7.60):

$$
\eta_r(t) = \frac{1}{\omega_r} \int_0^t N_r(\tau) \sin \omega_r(t - \tau)\, d\tau, \qquad r = 1, 2, \ldots,
\tag{7.86}
$$

where the generalized forces $N_r(t)$ are given by

$$
\begin{aligned}
N_r(t) &= \int_0^L w_r(x)F(t)\,\delta(x - vt)\, dx, \\
&= w_r(vt)F(t) && \text{for } 0 \le t \le L/v, \\
&= 0 && \text{for } t > L/v.
\end{aligned}
\tag{7.87}
$$

Equations (7.87) can be combined into

$$
N_r(t) = w_r(vt)F(t)\left[\alpha(t) - \alpha\left(t - \frac{L}{v}\right)\right],
\tag{7.88}
$$

where $\alpha(t)$ is a unit step function applied at $t = 0$ and $\alpha[t - (L/v)]$ is a unit step function applied at $t = L/v$.

The response for $0 \le t \le L/v$ becomes

$$w(x, t) = \sum_{r=1}^{\infty} \sqrt{\frac{2}{mL}} \sin \frac{r\pi x}{L} \frac{1}{\omega_r} \int_0^t F(\tau) \sqrt{\frac{2}{mL}} \sin \frac{r\pi v\tau}{L} \sin \omega_r(t - \tau) \, d\tau$$

$$= \frac{2}{\pi^2} \sqrt{\frac{L^2}{mEI}} \sum_{r=1}^{\infty} \frac{1}{r^2} \sin \frac{r\pi x}{L} \int_0^t F(\tau) \sin \frac{r\pi v\tau}{L} \sin r^2\pi^2 \sqrt{\frac{EI}{mL^4}} (t - \tau) \, d\tau$$

$$(7.89)$$

Equation (7.89) is quite general and is valid for any function $F(t)$.

Consider the case in which the function $F(t)$ is a step function of amplitude F_0,

$$F(t) = F_0\alpha(t) \tag{7.90}$$

and denote

$$\frac{r\pi v}{L} = \alpha, \qquad r^2\pi^2 \sqrt{\frac{EI}{mL^4}} = \beta. \tag{7.91}$$

Using integrations by parts it can be shown that

$$\int_0^t \alpha(\tau) \sin \alpha\tau \sin \beta(t - \tau) \, d\tau = \frac{1}{\alpha^2 - \beta^2} (\alpha \sin \beta t - \beta \sin \alpha t). \tag{7.92}$$

Introducing (7.90) through (7.92) in (7.89) we obtain the response

$$w(x, t) = \frac{2F_0 L^3}{\pi^2} \sum_{r=1}^{\infty} \frac{\sin (r\pi x/L)}{(L^2 v^2 m - r^2\pi^2 EI)}$$

$$\times \left(\frac{1}{r^3} \frac{v}{L\pi} \sqrt{\frac{EI}{mL^4}} \sin r^2\pi^2 \sqrt{\frac{EI}{mL^4}} t - \frac{1}{r^2} \sin \frac{r\pi v}{L} t \right). \tag{7.93}$$

Equation (7.93) is valid for $0 \le t \le L/v$ provided

$$v \ne r\pi \sqrt{\frac{EI}{mL^2}}, \qquad r = 1, 2, \ldots, \tag{7.94}$$

because otherwise a situation resembling resonance develops. This can be shown by letting $\alpha = \beta$ and evaluating the integral in the left side of (7.92). The contribution of the term for which $v = r\pi\sqrt{EI/mL^2}$ to the response is

$$\frac{F_0 L^3}{\pi^3 EI} \frac{1}{r^4} \left(\sin \frac{r\pi v}{L} t - \frac{r\pi v}{L} t \cos \frac{r\pi v}{L} t \right) \sin \frac{r\pi x}{L}, \tag{7.95}$$

and it is obvious that this term increases indefinitely.

7-11 RESPONSE OF A CIRCULAR MEMBRANE

Consider a circular membrane of uniform thickness, clamped at the boundary $r = a$. The membrane is subjected to a time-dependent uniform pressure over a circular area $0 \leq r \leq b$ as shown in Figure 7.3. The pressure can be written

$$p(r, \theta, t) = \begin{cases} p_0 f(t), & 0 \leq r \leq b, \\ 0, & b < r \leq a. \end{cases} \tag{7.96}$$

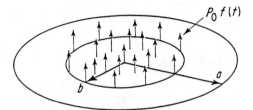

FIGURE 7.3

The general formulation of Section 7–7 can also be used in this case. The response of the membrane can be written in the form of a series

$$w(r, \theta, t) = \sum_{m=0}^{\infty} \sum_{n=1}^{\infty} W_{mn}(r, \theta) \eta_{mn}(t), \tag{7.97}$$

where $W_{mn}(r, \theta)$ are the normal modes of a uniform membrane clamped at $r = a$ which were obtained in Section 5–11. Recalling the degeneracy of the modes for which $m \neq 0$, we write the series (7.97) in the form

$$w(r, \theta, t) = \sum_{n=1}^{\infty} W_{0n}(r, \theta) \eta_{0n}(t) + \sum_{m=1}^{\infty} \sum_{n=1}^{\infty} W_{mnc}(r, \theta) \eta_{mnc}(t)$$

$$+ \sum_{m=1}^{\infty} \sum_{n=1}^{\infty} W_{mns}(r, \theta) \eta_{mns}(t), \tag{7.98}$$

where the generalized coordinates are given by

$$\eta_{0n}(t) = \frac{1}{\omega_{0n}} \int_0^t N_{0n}(\tau) \sin \omega_{0n}(t - \tau) \, d\tau,$$

$$\eta_{mnc}(t) = \frac{1}{\omega_{mn}} \int_0^t N_{mnc}(\tau) \sin \omega_{mn}(t - \tau) \, d\tau, \tag{7.99}$$

$$\eta_{mns}(t) = \frac{1}{\omega_{mn}} \int_0^t N_{mns}(\tau) \sin \omega_{mn}(t - \tau) \, d\tau.$$

The natural frequencies ω_{mn} have the values

$$\omega_{mn} = c\beta_{mn}, \tag{7.100}$$

where $\beta_{mn}a$ are the zeros of the Bessel functions $J_m(\beta a)$ and the wave propagation velocity c is given by

$$c = \sqrt{\frac{T}{\rho}}, \tag{7.101}$$

where T is the tension per unit length of membrane and ρ is the mass per unit area of membrane.

Using (5.232) and (5.233), the generalized forces take the specific form

$$N_{0n}(t) = \int_0^{2\pi} \int_0^a W_{0n}(r, \theta)p(r, \theta, t)r \, dr \, d\theta$$

$$= \frac{2\pi p_0 f(t)}{\sqrt{\pi}\rho a J_1[(\omega_{0n}/c)a]} \frac{bc}{\omega_{0n}} J_1\left(\frac{\omega_{0n}}{c}b\right) \tag{7.102}$$

$$N_{mnc}(t) = \int_0^{2\pi} \int_0^a W_{mnc}(r, \theta)p(r, \theta, t)r \, dr \, d\theta = 0, \tag{7.103}$$

and, similarly,

$$N_{mns}(t) = 0. \tag{7.104}$$

Hence

$$\eta_{0n}(t) = \frac{2\pi p_0 bc J_1[(\omega_{0n}/c)b]}{\sqrt{\pi}\rho a \omega_{0n}^2 J_1[(\omega_{0n}/c)a]} \int_0^t f(\tau) \sin \omega_{0n}(t - \tau) \, d\tau, \tag{7.105}$$

$$\eta_{mnc}(t) = \eta_{mns}(t) = 0. \tag{7.106}$$

Introducing (5.232), (7.105), and (7.106) in the series (7.98) we obtain the transverse displacement of the membrane,

$$w(r, \theta, t) = \frac{2p_0 bc}{\rho a^2} \sum_{n=1}^{\infty} \frac{J_1[(\omega_{0n}/c)b]J_0[(\omega_{0n}/c)r]}{\omega_{0n}^2 J_1^2[(\omega_{0n}/c)a]} \int_0^t f(\tau) \sin \omega_{0n}(t - \tau) \, d\tau. \tag{7.107}$$

It appears that only Bessel functions of zero order, $m = 0$, participate in the motion. This is to be expected, owing to the nature of the load. The load is radially symmetric, so there cannot be any trigonometric functions present. Furthermore, the Bessel functions of order higher than zero have a zero at $r = 0$ and they are antisymmetric with respect to the vertical through $r = 0$. In essence, the radially symmetric load prevents the modes with diametrical nodes from taking part in the motion.

7-12 RESPONSE OF A SIMPLY SUPPORTED RECTANGULAR PLATE

Consider a uniform rectangular plate extending over the domain $0 < x < a$, $0 < y < b$, and simply supported along the boundaries $x = 0,a$ and $y = 0,b$ (Figure 7.4). The plate is subjected to a concentrated force at the point

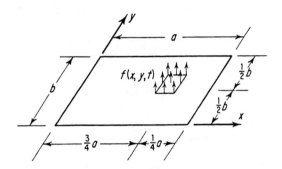

FIGURE 7.4

$x = \frac{3}{4}a$, $y = \frac{1}{2}b$. The force may be described mathematically as a distributed force given by

$$f(x, y, t) = F(t)\, \delta(x - \tfrac{3}{4}a, y - \tfrac{1}{2}b), \tag{7.108}$$

where $F(t)$ is the time-dependent amplitude of the concentrated force and $\delta(x - \tfrac{3}{4}a, y - \tfrac{1}{2}b)$ is a two-dimensional spatial Dirac's delta function defined by

$$\delta(x - \tfrac{3}{4}a, y - \tfrac{1}{2}b) = 0, \qquad x \neq \tfrac{3}{4}a \quad \text{and/or} \quad y \neq \tfrac{1}{2}b,$$

$$\int_0^a \int_0^b \delta(x - \tfrac{3}{4}a, y - \tfrac{1}{2}b)\, dx\, dy = 1. \tag{7.109}$$

The normal modes of the simply supported uniform plate are

$$W_{mn}(x, y) = \frac{2}{\sqrt{\rho ab}} \sin\frac{m\pi x}{a} \sin\frac{n\pi y}{b}, \qquad m, n = 1, 2, \ldots, \tag{7.110}$$

with corresponding natural frequencies

$$\omega_{mn} = \pi^2 \sqrt{\frac{D_E}{\rho}} \left[\left(\frac{m}{a}\right)^2 + \left(\frac{n}{b}\right)^2 \right], \qquad m, n = 1, 2, \ldots, \tag{7.111}$$

where D_E is the plate flexural rigidity and ρ is the mass per unit area of plate.

Using the expansion theorem, the transverse displacement of the plate is

$$w(x, y, t) = \sum_{m=1}^{\infty} \sum_{n=1}^{\infty} W_{mn}(x, y)\eta_{mn}(t), \qquad (7.112)$$

where the normal modes $W_{mn}(x, y)$ are given by (7.110) and $\eta_{mn}(t)$ are the corresponding generalized coordinates which, according to the general derivations of Section 7–7, are obtained from the expression

$$\eta_{mn}(t) = \frac{1}{\omega_{mn}} \int_0^t N_{mn}(\tau) \sin \omega_{mn}(t - \tau) \, d\tau. \qquad (7.113)$$

The generalized forces $N_{mn}(t)$ are given by

$$N_{mn}(t) = \int_0^a \int_0^b W_{mn}(x, y)f(x, y, t) \, dx \, dy$$

$$= \frac{2F(t)}{\sqrt{\rho ab}} \int_0^a \int_0^b \sin \frac{m\pi x}{a} \sin \frac{n\pi y}{b} \, \delta(x - \tfrac{3}{4}a, y - \tfrac{1}{2}b) \, dx \, dy$$

$$= \frac{2F(t)}{\sqrt{\rho ab}} \sin \frac{3m\pi}{4} \sin \frac{n\pi}{2}. \qquad (7.114)$$

It follows that

$$\eta_{mn}(t) = \frac{2}{\omega_{mn}\sqrt{\rho ab}} \sin \frac{3m\pi}{4} \sin \frac{n\pi}{2} \int_0^t F(\tau) \sin \omega_{mn}(t - \tau) \, d\tau, \qquad (7.115)$$

so the response reduces to

$$w(x, y, t) = \frac{4}{\rho ab} \sum_{m=1}^{\infty} \sum_{n=1}^{\infty} \frac{\sin (3m\pi/4) \sin (n\pi/2)}{\omega_{mn}} \sin \frac{m\pi x}{a} \sin \frac{n\pi y}{b}$$

$$\times \int_0^t F(\tau) \sin \omega_{mn}(t - \tau) \, d\tau, \qquad (7.116)$$

where the frequencies ω_{mn} are given by (7.111).

It can be easily checked that for any m which is an integer multiple of 4, the corresponding term in the series in (7.116) vanishes. This is consistent with the fact that a concentrated force applied at $x = \tfrac{3}{4}a$ cannot excite the modes $\sin (4\pi x/a)$, $\sin (8\pi x/a)$, etc., which have nodes at that point. The same argument explains why all the terms for which n is an even number vanish.

7-13 RESPONSE OF A SYSTEM WITH MOVING SUPPORTS

Let us consider the case in which the supports of a system undergo a known translational motion $w_0(t)$. The absolute displacement of any point can be written $w_0(t) + w(P, t)$, where $w_0(t)$ can be regarded as a rigid-body translation and $w(P, t)$ as an elastic deformation measured relative to the rigid-body motion. If there are no external forces applied, (7.47) can be written

$$L[w_0(t) + w(P, t)] + M(P) \frac{\partial^2}{\partial t^2} (w_0(t) + w(P, t)) = 0, \qquad (7.117)$$

where L is a differential operator defined previously. But L is such that

$$L[w_0(t)] = A_1(P)w_0(t), \qquad (7.118)$$

where $A_1(P)$ is a function of the spatial coordinates, so (7.117) can be written

$$L[w(P, t)] + M(P) \frac{\partial^2 w(P, t)}{\partial t^2} = -M(P) \frac{\partial^2 w_0(t)}{\partial t^2} - A_1(P)w_0(t)$$

$$= f(P, t), \qquad (7.119)$$

where
$$f(P, t) = -M(P) \frac{\partial^2 w_0(t)}{\partial t^2} - A_1(P)w_0(t) \qquad (7.120)$$

is a known function of time and space. The first term can be identified as a distributed inertia force and the second one as a distributed restoring force.†

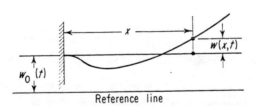

FIGURE 7.5

The solution of (7.119) can be obtained by following the procedure outlined in Section 7-7. As an illustration, consider the cantilever beam shown in Figure 7.5. The equation of motion (7.119) reduces to the form

$$\frac{\partial^2}{\partial x^2} \left[EI(x) \frac{\partial^2 w(x, t)}{\partial x^2} \right] + m(x) \frac{\partial^2 w(x, t)}{\partial t^2} = -m(x) \frac{\partial^2 w_0(t)}{\partial t^2}, \qquad (7.121)$$

---

† As an example consider a system which, in addition to its end supports, is elastically supported throughout the domain.

where $EI(x)$ is the bending stiffness and $m(x)$ is the distributed mass. The solution of (7.121) can be written

$$w(x, t) = \sum_{r=1}^{\infty} w_r(x)\eta_r(t), \tag{7.122}$$

where $w_r(x)$ $(r = 1, 2, \ldots)$ are the eigenfunctions corresponding to the cantilever beam and $\eta_r(t)$ are associated generalized coordinates given by

$$\eta_r(t) = \frac{1}{\omega_r} \int_0^t \left[\int_0^L - w_r(x)m(x) \, dx \right] \frac{d^2 w_0(\tau)}{d\tau^2} \sin \omega_r(t - \tau) \, d\tau$$

$$+ \eta_r(0) \cos \omega_r t + \dot{\eta}_r(0) \frac{\sin \omega_r t}{\omega_r}, \qquad r = 1, 2, \ldots, \tag{7.123}$$

where ω_r are the natural frequencies of the cantilever beam and

$$\left.\begin{aligned} \eta_r(0) &= \int_0^L m(x)w_r(x)w(x, 0) \, dx \\ \dot{\eta}_r(0) &= \int_0^L m(x)w_r(x)\dot{w}(x, 0) \, dx \end{aligned}\right\} \qquad r = 1, 2, \ldots \tag{7.124}$$

are the initial generalized displacement and initial generalized velocity associated with the elastic motion.

The absolute motion of any point is obtained by adding the rigid-body translation $w_0(t)$ to (7.122).

7-14   VIBRATION OF A SYSTEM WITH TIME-DEPENDENT BOUNDARY CONDITIONS

In Section 7–7 we presented the normal modes method for obtaining the response of a boundary-value problem consisting of a nonhomogeneous differential equation of motion with homogeneous boundary conditions. According to this method one first obtains the solution of the homogeneous boundary-value problem, which is done by separating the time and spatial dependence of the solution. This leads to an eigenvalue problem yielding the normal modes and the associated natural frequencies of the system. The solution of the nonhomogeneous differential equation is obtained by means of the expansion theorem, which assumes that the solution can be expressed as a superposition of normal modes.

In many cases the boundary conditions are not homogeneous but time-dependent. In general, in these cases the approach of Section 7–7 will not work, and a different method must be adopted. Solutions of boundary-value problems with time-dependent boundary conditions can be obtained by means of integral transform methods such as Laplace transform and Fourier transforms. Solutions by these methods will be discussed in Chapter 8. In this section we wish to modify the approach of Section 7–7 to enable us to use the normal modes method. This modified approach is based on the fact that a boundary-value problem consisting of a homogeneous differential equation with nonhomogeneous boundary conditions can be transformed into a problem consisting of a nonhomogeneous differential equation with homogeneous boundary conditions.† The latter problem can be solved by modal analysis (Section 7–7).

Let us consider a one-dimensional system described over the domain $0 < x < L$ by the differential equation of motion

$$L[w(x, t)] + M(x) \frac{\partial^2 w(x, t)}{\partial t^2} = F(x, t), \qquad (7.125)$$

where L is a linear homogeneous differential operator of order $2p$, and by the time-dependent boundary conditions

$$B_i[w(x, t)]\Big|_{x=0} = e_i(t), \qquad i = 1, 2, \ldots, p, \qquad (7.126)$$

$$B_j[w(x, t)]\Big|_{x=L} = f_j(t), \qquad j = 1, 2, \ldots, p, \qquad (7.127)$$

where B_i and B_j are linear homogeneous differential operators of order $2p - 1$ or lower. Let us assume, for simplicity, that the initial conditions are zero,

$$w(x, 0) = \frac{\partial w(x, t)}{\partial t}\bigg|_{t=0} = 0. \qquad (7.128)$$

It is not difficult to see that the differential equation of motion as well as the boundary conditions are nonhomogeneous. We shall attempt a solution of the problem by transforming it into a problem consisting of a non-homogeneous differential equation with homogeneous boundary conditions. To this end let us assume a solution of the boundary-value problem described by (7.125) through (7.127) in the form

$$w(x, t) = v(x, t) + \sum_{i=1}^{p} g_i(x)e_i(t) + \sum_{j=1}^{p} h_j(x)f_j(t), \qquad (7.129)$$

† See R. Courant and D. Hilbert, *Methods of Mathematical Physics*, Vol. 1, Interscience, New York, 1961, p. 277.

so that, in this manner, we transform the boundary-value problem for the variable $w(x, t)$ into a boundary-value problem for the variable $v(x, t)$. The functions $g_i(x)$ and $h_j(x)$ are chosen to render the boundary conditions for the variable $v(x, t)$ homogeneous.†

Introducing (7.129) in (7.126) and (7.127) we obtain the boundary conditions

$$B_r[w(x, t)]\Big|_{x=0} = B_r[v(x, t)]\Big|_{x=0} + \sum_{i=1}^{p} e_i(t)B_r[g_i(x)]\Big|_{x=0}$$

$$+ \sum_{j=1}^{p} f_j(t)B_r[h_j(x)]\Big|_{x=0} = e_r(t),$$

$$r = 1, 2, \ldots, p, \quad (7.130)$$

$$B_s[w(x, t)]\Big|_{x=L} = B_s[v(x, t)]\Big|_{x=L} + \sum_{i=1}^{p} e_i(t)B_s[g_i(x)]\Big|_{x=L}$$

$$+ \sum_{j=1}^{p} f_j(t)B_s[h_j(x)]\Big|_{x=L} = f_s(t),$$

$$s = 1, 2, \ldots, p. \quad (7.131)$$

The functions $g_i(x)$ and $h_j(x)$ must be chosen so that the boundary conditions for $v(x, t)$ are homogeneous; that is,

$$B_i[v(x, t)]\Big|_{x=0} = 0, \quad i = 1, 2, \ldots, p, \quad (7.132)$$

$$B_j[v(x, t)]\Big|_{x=L} = 0, \quad j = 1, 2, \ldots, p. \quad (7.133)$$

Examination of (7.130) and (7.131) shows that to satisfy these conditions we must have

$$\left.\begin{array}{l} B_r[g_i(x)]\Big|_{x=0} = \delta_{ir} \\[2mm] B_r[h_j(x)]\Big|_{x=0} = 0 \end{array}\right\} \quad i, j, r = 1, 2, \ldots, p, \quad (7.134)$$

$$\left.\begin{array}{l} B_s[g_i(x)]\Big|_{x=L} = 0 \\[2mm] B_s[h_j(x)]\Big|_{x=L} = \delta_{js} \end{array}\right\} \quad i, j, s = 1, 2, \ldots, p. \quad (7.135)$$

† The functions $g_i(x)$ and $h_j(x)$ are not unique and several choices may be acceptable. The corresponding results should be equivalent, however.

Introducing (7.129) in (7.125) we obtain the nonhomogeneous differential equation

$$L[v(x, t)] + M(x)\frac{\partial^2 v(x, t)}{\partial t^2} = F(x, t) - \sum_{i=1}^{p} (e_i(t)L[g_i(x)] + \ddot{e}_i(t)M(x)g_i(x))$$

$$- \sum_{j=1}^{p} (f_j(t)L[h_j(x)] + \ddot{f}_j(t)M(x)h_j(x)).$$

(7.136)

Using modal analysis, we first solve the eigenvalue problem consisting of the differential equation

$$L[v(x)] - \omega^2 M(x)v(x) = 0 \tag{7.137}$$

and the boundary conditions

$$B_i[v(x)]\Big|_{x=0} = 0, \qquad i = 1, 2, \ldots, p,$$

$$B_j[v(x)]\Big|_{x=L} = 0, \qquad j = 1, 2, \ldots, p.$$

(7.138)

The solution of the eigenvalue problem (7.137) and (7.138) yields an infinite set of natural modes $v_n(x)$ and associated natural frequencies ω_n. The modes are orthogonal and, in addition, we normalize them so that the orthonormal set of modes $v_n(x)$ satisfies the relation

$$\int_0^L M(x)v_n(x)v_m(x)\, dx = \delta_{mn}, \qquad m, n = 1, 2, \ldots. \tag{7.139}$$

Using the expansion theorem we assume a solution of (7.136) in the form

$$v(x, t) = \sum_{n=1}^{\infty} v_n(x)\eta_n(t), \tag{7.140}$$

where $\eta_n(t)$ $(n = 1, 2, \ldots)$ are time-dependent generalized coordinates. Introducing (7.140) in (7.136) we obtain

$$\sum_{n=1}^{\infty} (\eta_n(t)L[v_n(x)] + \ddot{\eta}_n(t)M(x)v_n(x))$$

$$= F(x, t) - \sum_{i=1}^{p} (e_i(t)L[g_i(x)] + \ddot{e}_i(t)M(x)g_i(x))$$

$$- \sum_{j=1}^{p} (f_j(t)L[h_j(x)] + \ddot{f}_j(t)M(x)h_j(x)), \tag{7.141}$$

and since $v_n(x)$ and ω_n satisfy (7.137), (7.141) reduces to

$$\sum_{n=1}^{\infty} (\ddot{\eta}_n(t) + \omega_n^2 \eta_n(t)) M(x) v_n(x)$$

$$= F(x, t) - \sum_{i=1}^{p} (e_i(t) L[g_i(x)] + \ddot{e}_i(t) M(x) g_i(x))$$

$$- \sum_{j=1}^{p} (f_j(t) L[h_j(x)] + \ddot{f}_j(t) M(x) h_j(x)). \quad (7.142)$$

Equation (7.142) contains all the generalized coordinates $\eta_n(t)$, so, in effect, it is a coupled equation. To uncouple it we multiply both sides of the equation by $v_m(x)$ and integrate with respect to x over the domain. If this is done and, in addition, we introduce the notation

$$\left. \begin{array}{l} G_{ni} = \displaystyle\int_0^L v_n(x) M(x) g_i(x)\, dx \\[2mm] G_{ni}^* = \displaystyle\int_0^L v_n(x) L[g_i(x)]\, dx \end{array} \right\} \quad i = 1, 2, \ldots, p; \quad n = 1, 2, \ldots, \quad (7.143)$$

$$\left. \begin{array}{l} H_{nj} = \displaystyle\int_0^L v_n(x) M(x) h_j(x)\, dx \\[2mm] H_{nj}^* = \displaystyle\int_0^L v_n(x) L[h_j(x)]\, dx \end{array} \right\} \quad j = 1, 2, \ldots, p; \quad n = 1, 2, \ldots, \quad (7.144)$$

$$F_n(t) = \int_0^L v_n(x) F(x, t)\, dx, \qquad n = 1, 2, \ldots, \tag{7.145}$$

in view of (7.139), we obtain an infinite set of uncoupled ordinary differential equations

$$\ddot{\eta}_n(t) + \omega_n^2 \eta_n(t) = N_n(t), \qquad n = 1, 2, \ldots, \tag{7.146}$$

where $N_n(t)$ is a generalized force associated with the nth generalized coordinate and has the form

$$N_n(t) = F_n(t) - \sum_{i=1}^{p} (G_{ni}^* e_i(t) + G_{ni} \ddot{e}_i(t)) - \sum_{i=1}^{p} (H_{nj}^* f_j(t) + H_{nj} \ddot{f}_j(t)),$$

$$n = 1, 2, \ldots. \tag{7.147}$$

The solution of (7.146) for zero initial conditions, is given by the convolution integral

$$\eta_n(t) = \frac{1}{\omega_n} \int_0^t N_n(\tau) \sin \omega_n(t - \tau)\, d\tau. \tag{7.148}$$

Equation (7.148), when introduced in (7.140), yields the solution $v(x, t)$ of the transformed problem and, subsequently, by using (7.129), the solution $w(x, t)$ of the original problem is obtained.

Example 7.2

As an illustration of the method described above, let us consider the longitudinal vibration problem of a uniform bar clamped at the end $x = 0$ and with a time-dependent tensile force, $P(t)$, at the end $x = L$. The differential equation of motion is

$$EA \frac{\partial^2 w(x, t)}{\partial x^2} = m \frac{\partial^2 w(x, t)}{\partial t^2}, \tag{a}$$

and the boundary conditions are

$$w(0, t) = 0, \qquad EA \left. \frac{\partial w(x, t)}{\partial x} \right|_{x=L} = P(t), \tag{b}$$

so we have a homogeneous differential equation with one homogeneous and one nonhomogeneous boundary condition.

Let us assume a solution of (a) in the form

$$w(x, t) = v(x, t) + h(x)P(t), \tag{c}$$

so the boundary conditions for $v(x, t)$ are

$$v(0, t) = -h(0)P(t),$$
$$EA \left. \frac{\partial v(x, t)}{\partial x} \right|_{x=L} = P(t) - P(t)EA \left. \frac{dh(x)}{dx} \right|_{x=L}. \tag{d}$$

To render the boundary conditions (d) homogeneous, we must have

$$h(0) = 0, \qquad EA \left. \frac{dh(x)}{dx} \right|_{x=L} = 1. \tag{e}$$

The second of boundary conditions (e) can be written

$$\frac{dh(x)}{dx} = \frac{1}{EA} u(x - (L - \epsilon)), \tag{f}$$

where $\alpha(x - (L - \epsilon))$ is a spatial step function† and ϵ is a small quantity. In view of the first of equations (e), (f) has the solution

$$h(x) = \frac{1}{EA}(x - (L - \epsilon))\alpha(x - (L - \epsilon)), \qquad (g)$$

and we note that $h(x)$ is zero over the domain $0 \le x \le L - \epsilon$.

The transformed problem consists of the nonhomogeneous differential equation

$$-EA\frac{\partial^2 v(x, t)}{\partial x^2} + m\frac{\partial^2 v(x, t)}{\partial t^2} = EA\frac{d^2 h(x)}{dx^2}P(t) - mh(x)\ddot{P}(t) \qquad (h)$$

and the homogeneous boundary conditions

$$v(0, t) = 0, \qquad EA\frac{\partial v(x, t)}{\partial x}\bigg|_{x=L} = 0. \qquad (i)$$

The corresponding eigenvalue problem consists of the differential equation

$$EA\frac{d^2 v(x)}{dx^2} + \omega^2 mv(x) = 0 \qquad (j)$$

and the boundary conditions

$$v(0) = 0, \qquad EA\frac{dv(x)}{dx}\bigg|_{x=L} = 0. \qquad (k)$$

Its solution was obtained previously (Section 5–7). The eigenfunctions are

$$v_n(x) = \sqrt{\frac{2}{mL}}\sin(2n - 1)\frac{\pi}{2}\frac{x}{L}, \qquad n = 1, 2, \ldots, \qquad (l)$$

and the corresponding eigenvalues are

$$\omega_n = (2n - 1)\frac{\pi}{2}\sqrt{\frac{EA}{mL^2}} = (2n - 1)\frac{\pi}{2}\frac{c}{L}, \qquad n = 1, 2, \ldots, \qquad (m)$$

where c is the wave velocity.

---

† The unit step function and Dirac's delta function are called *symbolic functions* by B. Friedman, a mathematician, who discusses their use in connection with boundary-value problems of the type considered here in *Principles and Techniques of Applied Mathematics*, Wiley, New York, 1956, p. 137.

Using (7.144) and dropping the second subscript, we write

$$
H_n = \int_0^L v_n(x)M(x)h(x)\, dx = m \int_0^L v_n(x)h(x)\, dx
$$

$$
= \frac{m}{EA} \sqrt{\frac{2}{mL}} \int_0^L \sin (2n - 1)\frac{\pi}{2}\frac{x}{L} \cdot (x - (L - \epsilon)) \mathcal{u}(x - (L - \epsilon))\, dx = 0,
$$
$$
n = 1, 2, \ldots \quad \text{(n)}
$$

$$
H_n^* = \int_0^L v_n(x)L[h(x)]\, dx = -EA \int_0^L v_n(x)\frac{d^2h(x)}{dx^2}\, dx
$$

$$
= -\sqrt{\frac{2}{mL}} \int_0^L \sin (2n - 1)\frac{\pi}{2}\frac{x}{L} \cdot \delta(x - (L - \epsilon))\, dx
$$

$$
= (-1)^n \sqrt{\frac{2}{mL}} \cos (2n - 1)\frac{\pi}{2}\frac{\epsilon}{L}, \qquad n = 1, 2, \ldots, \quad \text{(o)}
$$

where $\delta(x - (L - \epsilon))$ is a spatial Dirac's delta function.

Because (a) is homogeneous, $F_n(t) = 0$, so (7.147) gives the generalized forces

$$
N_n(t) = -H_n^* P(t) = (-1)^{n-1} \sqrt{\frac{2}{mL}} P(t) \cos (2n - 1)\frac{\pi}{2}\frac{\epsilon}{L},
$$
$$
n = 1, 2, \ldots, \quad \text{(p)}
$$

which, when introduced in (7.148), yield the generalized coordinates

$$
\eta_n(t) = \frac{(-1)^{n-1} \cos (2n - 1)(\pi\epsilon/2L)}{(2n - 1)(\pi c/2L)} \sqrt{\frac{2}{mL}} \int_0^t P(\tau)
$$

$$
\times \sin (2n - 1)\frac{\pi}{2}\frac{c}{L}(t - \tau)\, d\tau, \qquad n = 1, 2, \ldots. \quad \text{(q)}
$$

Finally, using (7.129) we obtain the solution

$$
w(x, t) = \sum_{n=1}^{\infty} v_n(x)\eta_n(t) + h(x)P(t)
$$

$$
= \frac{4c}{\pi EA} \sum_{n=1}^{\infty} \frac{(-1)^{n-1} \cos (2n - 1)(\pi\epsilon/2L)}{(2n - 1)}
$$

$$
\times \sin (2n - 1)\frac{\pi}{2}\frac{x}{L} \int_0^t P(\tau) \sin (2n - 1)\frac{\pi}{2}\frac{c}{L}(t - \tau)\, d\tau
$$

$$
+ \frac{P(t)}{EA}(x - (L - \epsilon))\mathcal{u}(x - (L - \epsilon)). \quad \text{(r)}
$$

Although the last term in (r) is zero for $0 \le x \le L - \epsilon$ and small for $L - \epsilon \le x \le L$, it must be retained, because its derivatives are neither zero nor small for $L - \epsilon < x < L$ and its presence ensures the satisfaction of the boundary condition at $x = L$.

Part C—Response of Undamped Continuous Systems. Approximate Methods

7-15 ASSUMED-MODES METHOD. SYSTEM RESPONSE

The assumed-modes method, discussed in conjunction with the free vibration problem, can be extended to the forced vibration case. The response of a continuous system is assumed in the form

$$w_n(P, t) = \sum_{i=1}^{n} \phi_i(P)q_i(t), \qquad (7.149)$$

where $\phi_i(P)$ are admissible functions, which are functions of the spatial coordinates P and satisfy the geometric boundary conditions of the system, and $q_i(t)$ are time-dependent generalized coordinates. In this manner a continuous system is approximated by an n-degree-of-freedom system. Consequently, we can write the kinetic energy of the system in the form

$$T(t) = \tfrac{1}{2} \sum_{i=1}^{n} \sum_{j=1}^{n} m_{ij} \dot{q}_i(t) \dot{q}_j(t), \qquad (7.150)$$

where the mass coefficients m_{ij} depend on the mass distribution $M(P)$ and the admissible functions $\phi_i(P)$. The potential energy can be written

$$V(t) = \tfrac{1}{2} \sum_{i=1}^{n} \sum_{j=1}^{n} k_{ij} q_i(t) q_j(t), \qquad (7.151)$$

where the stiffness coefficients k_{ij} depend on the stiffness properties of the system and the admissible functions $\phi_i(P)$ and its derivatives.

If any nonconservative forces are present, Lagrange's equations of motion have the form

$$\frac{d}{dt}\left(\frac{\partial T}{\partial \dot{q}_r}\right) - \frac{\partial T}{\partial q_r} + \frac{\partial V}{\partial q_r} = Q_r, \qquad r = 1, 2, \ldots, \qquad (7.152)$$

where $Q_r(t)$ are time-dependent nonconservative generalized forces. To obtain the nonconservative generalized forces we make use of the virtual work expression. To this end let $f(P, t) + F_j(t)\delta(P - P_j)$ be the distributed nonconservative forces, so that the virtual work expression is

$$
\begin{aligned}
\delta W(t) &= \int_D [f(P, t) + F_j(t)\delta(P - P_j)]\delta w_n(P, t)\, dD(P) \\
&= \sum_{r=1}^{n} \int_D [f(P, t) + F_j(t)\delta(P - P_j)]\phi_r(P)\delta q_r(t)\, dD(P) \\
&= \sum_{r=1}^{n} \left[\int_D f(P, t)\phi_r(P)\, dD(P) + \sum_{j=1}^{l} F_j(t)\phi_r(P_j) \right] \delta q_r(t) \\
&= \sum_{r=1}^{n} Q_r(t)\delta q_r(t),
\end{aligned}
\tag{7.153}
$$

from which we obtain the generalized nonconservative forces

$$
Q_r(t) = \int_D f(P, t)\phi_r(P)\, dD(P) + \sum_{j=1}^{l} F_j(t)\phi_r(P_j), \qquad r = 1, 2, \ldots, n,
\tag{7.154}
$$

where l is the number of concentrated forces.

Lagrange's equations of motion become

$$
\sum_{j=1}^{n} m_{rj}\ddot{q}_j(t) + \sum_{j=1}^{n} k_{rj}q_j(t) = Q_r(t), \qquad r = 1, 2, \ldots, n,
\tag{7.155}
$$

which can be written in the matrix form

$$
[m]\{\ddot{q}(t)\} + [k]\{q(t)\} = \{Q(t)\}.
\tag{7.156}
$$

Equation (7.156) has precisely the same structure as (7.1) describing a discrete system. The general method of solution of (7.156) was given in Part A.

Example 7.3   Torsional Vibration of a Nonuniform Circular Shaft

As an illustration of the assumed-modes approach let us consider the torsional vibration of a nonuniform shaft clamped at both ends (Figure 7.6), which has the mass polar moment of inertia per unit length

$$
I(x) = \frac{I_0}{3L}\left[1 + 12\frac{x}{L} - 12\left(\frac{x}{L}\right)^2 \right],
\tag{7.157}
$$

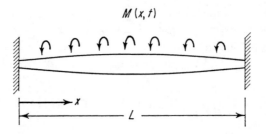

$$M(x, t)$$

FIGURE 7.6

where I_0 is the mass polar moment of inertia of the entire shaft; the torsional rigidity is

$$GJ(x) = \frac{GJ_0}{3}\left[1 + 12\frac{x}{L} - 12\left(\frac{x}{L}\right)^2\right]. \tag{7.158}$$

If $\theta(x, t)$ denotes the angular displacement of the shaft, the kinetic energy for the entire shaft has the form

$$T(t) = \tfrac{1}{2}\int_0^L I(x)\left[\frac{\partial\theta(x, t)}{\partial t}\right]^2 dx \tag{7.159}$$

and the potential energy for the entire shaft is

$$V(t) = \tfrac{1}{2}\int_0^L GJ(x)\left[\frac{\partial\theta(x, t)}{\partial x}\right]^2 dx. \tag{7.160}$$

We shall assume a solution in the form

$$\theta(x, t) = \sum_{r=1}^{n} \Theta_r(x)q_r(t), \tag{7.161}$$

where $\Theta_r(x)$ are admissible functions. Introducing (7.161) in (7.159) we obtain

$$T(t) = \tfrac{1}{2}\int_0^L I(x)\left[\sum_{i=1}^{n}\Theta_i(x)\dot{q}_i(t)\right]\left[\sum_{j=1}^{n}\Theta_j(x)\dot{q}_j(t)\right] dx$$

$$= \tfrac{1}{2}\sum_{i=1}^{n}\sum_{j=1}^{n} m_{ij}\dot{q}_i(t)\dot{q}_j(t), \tag{7.162}$$

where the inertia coefficients m_{ij} are given by

$$m_{ij} = \int_0^L I(x)\Theta_i(x)\Theta_j(x)\, dx. \tag{7.163}$$

In a similar way we obtain the potential energy

$$V(t) = \tfrac{1}{2} \sum_{i=1}^{n} \sum_{j=1}^{n} k_{ij} q_i(t) q_j(t), \tag{7.164}$$

where the stiffness coefficients k_{ij} have the form

$$k_{ij} = \int_0^L GJ(x) \frac{d\Theta_i(x)}{dx} \frac{d\Theta_j(x)}{dx} \, dx. \tag{7.165}$$

As admissible functions let us select the eigenfunctions of a uniform shaft clamped at both ends and of mass polar moment of inertia per unit length I_0/L. These eigenfunctions are

$$\Theta_r(x) = \sqrt{\frac{2}{I_0}} \sin r \frac{\pi x}{L}, \qquad r = 1, 2, \ldots, \tag{7.166}$$

so the inertia coefficients become

$$\begin{aligned} m_{ij} &= \frac{2}{3L} \int_0^L \left[1 + 12 \frac{x}{L} - 12\left(\frac{x}{L}\right)^2\right] \sin \frac{i\pi x}{L} \sin \frac{j\pi x}{L} \, dx \\ &= \tfrac{2}{3} \int_0^1 (1 + 12\zeta - 12\zeta^2) \sin i\pi\zeta \sin j\pi\zeta \, d\zeta. \end{aligned} \tag{7.167}$$

When $i = j$, (7.167) becomes

$$m_{ii} = \tfrac{2}{3} \int_0^1 (1 + 12\zeta - 12\zeta^2)(1 - \cos 2i\pi\zeta) \, d\zeta = 1 + \frac{2}{(i\pi)^2}, \tag{7.168}$$

and when $i \neq j$, we obtain

$$\begin{aligned} m_{ij} &= \tfrac{2}{3} \int_0^1 (1 + 12\zeta - 12\zeta^2)\tfrac{1}{2}[\cos (i - j)\pi\zeta - \cos (i + j)\pi\zeta] \, d\zeta \\ &= 4\left\{\frac{1}{(i + j)^2\pi^2} [(-1)^{i+j} + 1] - \frac{1}{(i - j)^2\pi^2} [(-1)^{i-j} + 1]\right\}. \end{aligned} \tag{7.169}$$

In a similar manner the stiffness coefficients become

$$\begin{aligned} k_{ij} &= \frac{GJ_0}{3} \frac{2}{I_0} \left(\frac{i\pi}{L}\right)\left(\frac{j\pi}{L}\right) \int_0^L \left[1 + 12 \frac{x}{L} - 12\left(\frac{x}{L}\right)^2\right] \cos \frac{i\pi x}{L} \cos \frac{j\pi x}{L} \, dx \\ &= \tfrac{2}{3}(i\pi)(j\pi) \frac{GJ_0}{LI_0} \int_0^1 (1 + 12\zeta - 12\zeta^2) \cos i\pi\zeta \cos j\pi\zeta \, d\zeta. \end{aligned} \tag{7.170}$$

For $i = j$ (7.170) reduces to

$$k_{ii} = \tfrac{1}{3}(i\pi)^2 \frac{GJ_0}{LI_0} \int_0^1 (1 + 12\zeta - 12\zeta^2)(1 + \cos 2i\pi\zeta)\, d\zeta = \frac{GJ_0}{LI_0}\left[(i\pi)^2 - 2\right],$$

$$(7.171)$$

and for $i \neq j$ we obtain

$$k_{ij} = \tfrac{2}{3}(i\pi)(j\pi)\frac{GJ_0}{LI_0}\int_0^1 (1 + 12\zeta - 12\zeta^2)\tfrac{1}{2}[\cos(i-j)\pi\zeta + \cos(i+j)\pi\zeta]\, d\zeta$$

$$= -4(i\pi)(j\pi)\frac{GJ_0}{LI_0}\left\{\frac{1}{(i+j)^2\pi^2}[(-1)^{i+j} + 1] + \frac{1}{(i-j)^2\pi^2}[(-1)^{i-j} + 1]\right\}.$$

$$(7.172)$$

Using (7.154) the generalized nonconservative forces become

$$Q_i(t) = \sqrt{\frac{2}{I_0}} \int_0^L M(x, t) \sin i\frac{\pi x}{L}\, dx, \qquad i = 1, 2, \ldots, n. \quad (7.173)$$

Let us use the first three modes, $n = 3$, so the inertia matrix becomes

$$[m] = \begin{bmatrix} 1 + \dfrac{2}{\pi^2} & 0 & -\dfrac{3}{2\pi^2} \\[2mm] 0 & 1 + \dfrac{1}{2\pi^2} & 0 \\[2mm] -\dfrac{3}{2\pi^2} & 0 & 1 + \dfrac{2}{9\pi^2} \end{bmatrix}, \qquad (7.174)$$

the stiffness matrix takes the form

$$[k] = \frac{GJ_0}{LI_0}\begin{bmatrix} \pi^2 - 2 & 0 & -\tfrac{15}{2} \\[2mm] 0 & 4\pi^2 - 2 & 0 \\[2mm] -\tfrac{15}{2} & 0 & 9\pi^2 - 2 \end{bmatrix}, \qquad (7.175)$$

and the problem reduces to

$$[m]\{\ddot{q}(t)\} + [k]\{q(t)\} = \{Q(t)\}, \qquad (7.176)$$

which represents a set of three coupled ordinary differential equations. The methods of solving (7.176) were discussed in Part A.

7-16 GALERKIN'S METHOD. SYSTEM RESPONSE

Although Galerkin's method adopts a different point of view than the assumed-modes method, it yields the same results.

Let us consider a continuous system described by the differential equation

$$L[w(P, t)] + M(P)\frac{\partial^2 w(P, t)}{\partial t^2} = f(P, t), \tag{7.177}$$

where the operator L was defined in Section 7–7 and $f(P, t)$ is a distributed force. For simplicity we shall ignore concentrated forces, although it is not difficult to include them. The function w, which must satisfy (7.177) throughout domain D, is subject to given boundary conditions that do not contain $\partial w/\partial t$ or $\partial^2 w/\partial t^2$. We shall be interested in the cases in which it is not possible to obtain an exact solution of (7.177) and we must be content with an approximate solution.

We assume that an approximate solution of (7.177), which also satisfies the associated boundary conditions, has the form

$$w_n(P, t) = \sum_{j=1}^{n} \Phi_j(P)q_j(t), \tag{7.178}$$

where $\Phi_j(P)$ $(j = 1, 2, \ldots, n)$ are comparison functions depending on the spatial coordinates and $q_j(t)$ are time-dependent generalized coordinates. Because $w_n(P, t)$ is only an approximate solution it will not satisfy (7.177). Nevertheless, we assume that the difference between the approximate and the exact solution is small and denote that difference by $\epsilon(P, t)$, such that

$$\epsilon(P, t) = L[w_n(P, t)] + M(P)\frac{\partial^2 w_n(P, t)}{\partial t^2} - f(P, t), \tag{7.179}$$

where $\epsilon(P, t)$ is referred to as the *error*. According to the Galerkin method we must insist that the *weighted* error integrated over the domain be zero. The weighting functions are the comparison function $\Phi_j(P)$, such that

$$\int_D \epsilon(P, t)\Phi_r(P) \, dD(P) = 0, \qquad r = 1, 2, \ldots, n, \tag{7.180}$$

where r is a dummy index.

Since L is a linear operator, (7.178) through (7.180) lead to

$$\sum_{j=1}^{n} \ddot{q}_j(t) \int_D M(P)\Phi_r(P)\Phi_j(P) \, dD(P) + \sum_{j=1}^{n} q_j(t) \int_D \Phi_r(P)L[\Phi_j(P)] \, dD(P)$$

$$= \int_D f(P, t) \, \Phi_r(P)dD(P), \qquad r = 1, 2, \ldots, n, \quad (7.181)$$

and introducing the notation

$$\int_D M(P)\Phi_r(P)\Phi_j(P) \, dD(P) = m_{rj},$$

$$\int_D \Phi_r(P)L[\Phi_j(P)] \, dD(P) = k_{rj}, \qquad (7.182)$$

$$\int_D f(P, t)\Phi_r(P) \, dD(P) = Q_r(t),$$

(7.181) reduces to

$$\sum_{j=1}^{n} m_{rj}\ddot{q}_j(t) + \sum_{j=1}^{n} k_{rj}q_j(t) = Q_r(t), \qquad r = 1, 2, \ldots, n, \quad (7.183)$$

where $Q_r(t)$ are recognized as being the generalized forces associated with the generalized displacements $q_r(t)$. It must be noted that for a self-adjoint system the mass and stiffness coefficients are symmetric, $m_{rj} = m_{jr}$ and $k_{rj} = k_{jr}$. Equation (7.183) can be written in the matrix form

$$[m]\{\ddot{q}(t)\} + [k]\{q(t)\} = \{Q(t)\}, \qquad (7.184)$$

where $[m]$ and $[k]$ are symmetric matrices, assuming the system is self-adjoint. Equation (7.184) entirely resembles (7.156) obtained by the assumed-modes method, and its solution follows the same pattern as the one described in Part A.

Example 7.4 Axial Vibration of a Nonuniform Rod

As an example let us consider the system of Example 6.2. The differential equation describing the system is

$$-\frac{\partial}{\partial x}\left[EA(x)\frac{\partial w(x, t)}{\partial x}\right] + m(x)\frac{\partial^2 w(x, t)}{\partial t^2} = f(x, t), \qquad \text{(a)}$$

where $\qquad EA(x) = 2EA\left(1 - \frac{x}{L}\right), \qquad m(x) = 2m\left(1 - \frac{x}{L}\right),$

$$f(x, t) = f_0\left(1 - 2\frac{x}{L} + \frac{x^2}{L^2}\right)\delta(t) \qquad \text{(b)}$$

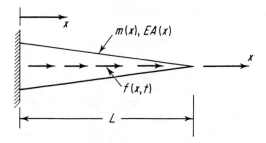

FIGURE 7.7

are the stiffness, mass, and force distributions, respectively (see Figure 7.7). It should be noted that f_0 is the force amplitude having units $FL^{-1}T$ and $\delta(t)$ is the Dirac delta function.

We shall limit the series (7.178) to three terms and choose as comparison functions the eigenfunctions of a uniform rod clamped at $x = 0$ and free at $x = L$,

$$\Phi_r(x) = \sin (2r - 1)\frac{\pi x}{2L}, \qquad r = 1, 2, 3. \tag{c}$$

Using the third of equations (7.182) we obtain the generalized forces

$$Q_r(t) = \int_0^L f(x, t)\Phi_r(x)\, dx = f_0\delta(t) \int_0^L \left(1 - \frac{x}{L} + \frac{x^2}{L^2}\right) \sin (2r - 1)\frac{\pi x}{2L}\, dx$$

$$= \frac{2Lf_0\delta(t)}{(2r - 1)\pi} \left[1 - \frac{8}{(2r - 1)^2\pi^2}\right], \qquad r = 1, 2, 3. \tag{d}$$

The eigenvalue problem corresponding to (7.184) was derived in Example 6.2 and has the form

$$\frac{EA}{2L}
\begin{bmatrix}
1 + \dfrac{\pi^2}{4} & 3 & \dfrac{5}{9} \\[2ex]
3 & 1 + \dfrac{9\pi^2}{4} & 15 \\[2ex]
\dfrac{5}{9} & 15 & 1 + \dfrac{25\pi^2}{4}
\end{bmatrix}
\begin{Bmatrix} u_1 \\[1ex] u_2 \\[1ex] u_3 \end{Bmatrix}$$

$$= \omega^2 \frac{mL}{2}
\begin{bmatrix}
1 - \dfrac{4}{\pi^2} & \dfrac{4}{\pi^2} & -\dfrac{4}{9\pi^2} \\[2ex]
\dfrac{4}{\pi^2} & 1 - \dfrac{4}{9\pi^2} & \dfrac{4}{\pi^2} \\[2ex]
-\dfrac{4}{9\pi^2} & \dfrac{4}{\pi^2} & 1 - \dfrac{4}{25\pi^2}
\end{bmatrix}
\begin{Bmatrix} u_2 \\[1ex] u_2 \\[1ex] u_3 \end{Bmatrix}. \tag{e}$$

Its solution yields the natural frequencies

$$\omega_1 = 2.4049 \sqrt{\frac{EA}{mL^2}}, \qquad \omega_2 = 5.5216 \sqrt{\frac{EA}{mL^2}}, \qquad \omega_3 = 8.6646 \sqrt{\frac{EA}{mL^2}}, \quad \text{(f)}$$

and the natural modes, which, upon normalization, can be arranged in the matrix form

$$[u] = \frac{1}{\sqrt{mL}} \begin{bmatrix} 1.8719 & 1.1414 & 0.8348 \\ -0.0579 & -1.7623 & -1.0109 \\ -0.0135 & 0.0760 & 1.6921 \end{bmatrix}. \tag{g}$$

Introducing the linear transformation

$$\{q(t)\} = [u]\{\eta(t)\} \tag{h}$$

in (7.184) and premultiplying the result by $[u]^T$, we obtain

$$\{\ddot{\eta}(t)\} + [\omega^2]\{\eta(t)\} = \{N(t)\}, \tag{i}$$

where $\{\eta(t)\}$ is a column matrix of generalized displacements and

$$\{N(t)\} = [u]^T\{Q(t)\} \tag{j}$$

is a column matrix of associated generalized forces. The values of the generalized forces are

$$N_1(t) = \{u^{(1)}\}^T\{Q(t)\} = \frac{2Lf_0\delta(t)}{\pi\sqrt{mL}} \left[1.8719 \left(1 - \frac{8}{\pi^2} \right) - 0.0579 \frac{1}{3} \left(1 - \frac{8}{9\pi^2} \right) \right.$$

$$\left. - 0.0135 \frac{1}{5} \left(1 - \frac{8}{25\pi^2} \right) \right] = 0.3338 \frac{2Lf_0\delta(t)}{\pi\sqrt{mL}}, \tag{k}$$

and, similarly,

$$N_2(t) = \{u^{(2)}\}^T\{Q(t)\} = -0.3036 \frac{2Lf_0\delta(t)}{\pi\sqrt{mL}},$$

$$N_3(t) = \{u^{(3)}\}^T\{Q(t)\} = 0.1789 \frac{2Lf_0\delta(t)}{\pi\sqrt{mL}}. \tag{l}$$

The generalized coordinates $\eta_r(t)$ are obtained by means of the convolution integral

$$\eta_1(t) = \frac{1}{\omega_1} \int_0^t N_1(\tau) \sin \omega_1(t - \tau) \, d\tau$$

$$= \frac{0.3338}{2.4049} \frac{2Lf_0}{\pi\sqrt{mL}} \sqrt{\frac{mL^2}{EA}} \int_0^t \delta(\tau) \sin 2.4049 \sqrt{\frac{EA}{mL^2}} (t - \tau) \, d\tau$$

$$= 0.0884 f_0 L \sqrt{\frac{L}{EA}} \sin 2.4049 \sqrt{\frac{EA}{mL^2}} t, \tag{m}$$

and, similarly,

$$\eta_2(t) = \frac{1}{\omega_2} \int_0^t N_2(\tau) \sin \omega_2(t - \tau) d\tau = -0.0350 f_0 L \sqrt{\frac{L}{EA}} \sin 5.5216 \sqrt{\frac{EA}{mL^2}} t,$$

$$\text{(n)}$$

$$\eta_3(t) = \frac{1}{\omega_3} \int_0^t N_3(\tau) \sin \omega_3(t - \tau) d\tau = 0.0131 f_0 L \sqrt{\frac{L}{EA}} \sin 8.6646 \sqrt{\frac{EA}{mL^2}} t.$$

Using (h) we obtain the generalized coordinates $q_j(t)$,

$$q_1(t) = \sum_{r=1}^{3} u_1^{(r)} \eta_r(t) = \frac{1}{\sqrt{mL}} [1.8719\eta_1(t) + 1.1414\eta_2(t) + 0.8348\eta_3(t)],$$

$$q_2(t) = \sum_{r=1}^{3} u_2^{(r)} \eta_r(t) = \frac{1}{\sqrt{mL}} [-0.0597\eta_1(t) - 1.7623\eta_2(t) - 1.0109\eta_3(t)],$$

$$\text{(o)}$$

$$q_3(t) = \sum_{r=1}^{3} u_3^{(r)} \eta_r(t) = \frac{1}{\sqrt{mL}} [-0.0135\eta_1(t) + 0.0760\eta_2(t) + 1.6921\eta_3(t)].$$

Finally, using (7.178), we obtain the approximate solution of the differential equation (a) in the form

$$w(x, t) = \sum_{j=1}^{3} \Phi_j(x)q_j(t) = q_1(t) \sin \frac{\pi x}{2L} + q_2(t) \sin \frac{3\pi x}{2L} + q_3(t) \sin \frac{5\pi x}{2L}$$

$$= \frac{1}{\sqrt{mL}} \left[\left(1.8719 \sin \frac{\pi x}{2L} - 0.0597 \sin \frac{3\pi x}{2L} - 0.0135 \sin \frac{5\pi x}{2L} \right) \eta_1(t) \right.$$

$$+ \left(1.1414 \sin \frac{\pi x}{2L} - 1.7623 \sin \frac{3\pi x}{2L} + 0.0760 \sin \frac{5\pi x}{2L} \right) \eta_2(t)$$

$$+ \left. \left(0.8348 \sin \frac{\pi x}{2L} - 1.0109 \sin \frac{3\pi x}{2L} + 1.6921 \sin \frac{5\pi x}{2L} \right) \eta_3(t) \right]$$

$$= \frac{Lf_0}{\sqrt{mEA}} \left[\left(0.1655 \sin \frac{\pi x}{2L} - 0.0053 \sin \frac{3\pi x}{2L} \right. \right.$$

$$\left. - 0.0012 \sin \frac{5\pi x}{2L} \right) \sin 2.4049 \sqrt{\frac{EA}{mL^2}} t$$

$$+ \left(-0.0399 \sin \frac{\pi x}{2L} + 0.0617 \sin \frac{3\pi x}{2L} \right.$$

$$\left. - 0.0027 \sin \frac{5\pi x}{2L} \right) \sin 5.5216 \sqrt{\frac{EA}{mL^2}} t$$

$$+ \left(0.0109 \sin \frac{\pi x}{2L} - 0.0132 \sin \frac{2\pi x}{2L} \right.$$

$$\left. \left. + 0.0222 \sin \frac{5\pi x}{2L} \right) \sin 8.6646 \sqrt{\frac{EA}{mL^2}} t \right], \quad \text{(p)}$$

where the subscript of $w(x, t)$ has been dropped to avoid confusion.

7–17 COLLOCATION METHOD. SYSTEM RESPONSE

The procedure of evaluating the response of a continuous system by the collocation method is distinctly different from the procedure used by either the assumed-modes method or the Galerkin method. It must be noted that all these methods assume the response in the form of a series leading to a set of ordinary differential equations with constant coefficients m_{rj} and k_{rj}. Whereas the evaluation of these coefficients by either the assumed-modes method or the Galerkin method requires integrations, in the case of the collocation method it does not. On the other hand, the first two methods yield symmetric coefficients, $m_{rj} = m_{jr}$ and $k_{rj} = k_{jr}$, whereas the collocation method, in general, does not. Hence the advantage of the collocation method lies in the easiness with which the coefficients m_{rj} and k_{rj} are obtained. This advantage may be lost, however, if the solution of the resulting differential equations proves excessively laborious. Of course, if the solution is obtained by automatic means, the collocation method retains much appeal.

We shall be interested in the solution of the partial differential equation

$$L[w(P, t)] + M(P)\frac{\partial^2 w(P, t)}{\partial t^2} = f(P, t) \qquad (7.185)$$

by means of the interior method (see Section 6–7). The problem under conconsideration is the same as the problem of Section 7–16 (see the beginning of that section for a fuller description of the problem).

To account for the boundary conditions we shall assume a solution of (7.185) in the form

$$w_n(P, t) = \sum_{j=1}^{n} \Phi_j(P)q_j(t), \qquad (7.186)$$

where $\Phi_j(P)$ $(j = 1, 2, \ldots, n)$ are a set of comparison functions. As in Galerkin's method, we shall denote the difference between the approximate and the exact solution by $\epsilon(P, t)$, such that

$$\epsilon(P, t) = L[w_n(P, t)] + M(P)\frac{\partial^2 w_n(P, t)}{\partial t^2} - f(P, t). \qquad (7.187)$$

At this point the similarity between the Galerkin and collocation methods ends, because in contrast with the Galerkin method, the collocation method assumes that the error vanishes at selected points $P = P_r$ $(r = 1, 2, \ldots, n)$, called stations. Introducing (7.186) in (7.187) and letting the error be zero at n such points, we obtain a set of n ordinary differential equations

$$\sum_{j=1}^{n} \ddot{q}_j(t)(M(P_r)\Phi_j(P_r)) + \sum_{j=1}^{n} q_j(t)L[\Phi_j(P_r)] = f(P_r, t), \qquad r = 1, 2, \ldots, n.$$
$$(7.188)$$

Next let us adopt the notation

$$M(P_r)\Phi_j(P_r) = m_{rj}, \qquad L[\Phi_j(P_r)] = k_{rj}, \qquad f(P_r, t) = Q_r(t), \quad (7.189)$$

and note that, in general, the coefficients m_{rj} and k_{rj} are not symmetric. With this notation (7.188) can be written

$$\sum_{j=1}^{n} m_{rj}\ddot{q}_j(t) + \sum_{j=1}^{n} k_{rj}q_j(t) = Q_r(t), \qquad r = 1, 2, \ldots, n, \quad (7.190)$$

which leads to the matrix form

$$[m]\{\ddot{q}(t)\} + [k]\{q(t)\} = \{Q(t)\}, \tag{7.191}$$

where, in general, the matrices $[m]$ and $[k]$ are not symmetric.

Equation (7.191) represents a set of n coupled ordinary differential equations. We shall be interested in a solution of these equations by means of modal analysis, to which end we must use the expansion theorem described in Section 4–12. This requires the solution of two adjoint eigenvalue problems

$$[D]\{u\} = \lambda\{u\}, \tag{7.192}$$

$$[D]^T\{v\} = \lambda\{v\}, \tag{7.193}$$

where

$$[D] = [k]^{-1}[m]. \tag{7.194}$$

The solution of the eigenvalue problems (7.192) and (7.193) was discussed in Section 4–12 and an example was given in Section 6–7. The solution consists of two adjoint sets of eigenvectors $\{u^{(i)}\}$ $(i = 1, 2, \ldots, n)$ and $\{v^{(j)}\}$ $(j = 1, 2, \ldots, n)$, which satisfy the biorthogonality relation

$$\{u^{(i)}\}^T\{v^{(j)}\} = 0 \qquad \text{for } i \neq j. \tag{7.195}$$

The eigenvectors for which $i = j$ are called *conjugates* and the corresponding eigenvalues are identical. For convenience the two sets of eigenvectors are normalized so that the product of two conjugate vectors is equal to 1. This can be written in the compact form

$$[u]^T[v] = [v]^T[u] = [I],\qquad(7.196)$$

where $[I]$ is the identity matrix.

Premultiply (7.191) by $[k]^{-1}$ and obtain

$$[D]\{\ddot{q}(t)\} + \{q(t)\} = [k]^{-1}\{Q(t)\}.\qquad(7.197)$$

We shall assume that the solution of (7.197) has the form

$$\{q(t)\} = [u]\{\eta(t)\},\qquad(7.198)$$

where $[u]$ is a square matrix of eigenvectors and $\{\eta(t)\}$ is a column matrix of generalized coordinates, so that

$$[D][u]\{\ddot{\eta}(t)\} + [u]\{\eta(t)\} = [k]^{-1}\{Q(t)\};\qquad(7.199)$$

premultiplying both sides of (7.199) by $[v]^T$ we obtain

$$[v]^T[D][u]\{\ddot{\eta}(t)\} + [v]^T[u]\{\eta(t)\} = [v]^T[k]^{-1}\{Q(t)\}.\qquad(7.200)$$

But from (7.192) and (7.196) we conclude that

$$[v]^T[D][u] = [\lambda],\qquad(7.201)$$

where the eigenvalues are related to the natural frequencies by $\lambda_i = \omega_i^{-2}$. Hence (7.200) and (7.201) lead to

$$\{\ddot{\eta}(t)\} + [\omega^2]\{\eta(t)\} = \{N(t)\},\qquad(7.202)$$

where
$$\{N(t)\} = [\omega^2][v]^T[k]^{-1}\{Q(t)\}\qquad(7.203)$$

is a column vector of generalized forces. Evidently (7.202) represents a set of uncoupled ordinary differential equations which can be easily solved for the coordinates $\eta_r(t)$ $(r = 1, 2, \ldots, n)$ (see Section 7–6).

The above result can be viewed from a slightly different angle. To this end we note that (7.198) can be written

$$q_j(t) = \sum_{i=1}^{n} u_j^{(i)} \eta_i(t), \qquad j = 1, 2, \ldots, n,\qquad(7.204)$$

so (7.186) leads to

$$w_n(P, t) = \sum_{j=1}^{n} \Phi_j(P)q_j(t) = \sum_{j=1}^{n} \Phi_j(P) \sum_{i=1}^{n} u_j^{(i)}\eta_i(t)$$

$$= \sum_{i=1}^{n} \left[\sum_{j=1}^{n} u_j^{(i)}\Phi_j(P) \right] \eta_i(t) = \sum_{i=1}^{n} w_u^{(i)}(P)\eta_i(t), \qquad (7.205)$$

where

$$w_u^{(i)}(P) = \sum_{j=1}^{n} u_j^{(i)}\Phi_j(P), \qquad j = 1, 2, \ldots, n \qquad (7.206)$$

are the eigenfunctions of the continuous system corresponding to the set of eigenvectors $\{u^{(i)}\}$ (see Example 6.5).

Example 7.5   Axial Vibration of a Nonuniform Rod

As an illustration consider the system of Example 6.5 (also Example 7.4). The system is described by the differential equation

$$-\frac{\partial}{\partial x}\left[2EA \left(1 - \frac{x}{L}\right) \frac{\partial w(x, t)}{\partial x} \right] + 2m \left(1 - \frac{x}{L}\right) \frac{\partial^2 w(x, t)}{\partial t^2}$$

$$= f(x, t) = f_0\left(1 - 2\frac{x}{L} + \frac{x^2}{L^2}\right)\delta(t). \quad (a)$$

As comparison functions we again use

$$\Phi_r(x) = \sin(2r - 1)\frac{\pi x}{2L}, \qquad r = 1, 2, 3 \qquad (b)$$

and select the stations at

$$P_r = r\frac{L}{4}, \qquad r = 1, 2, 3. \qquad (c)$$

From the third of equations (7.189) we obtain the generalized forces

$$Q_1(t) = f\left(\frac{L}{4}, t\right) = \frac{9}{16}f_0\delta(t),$$

$$Q_2(t) = f\left(\frac{L}{2}, t\right) = \frac{1}{4}f_0\delta(t), \qquad (d)$$

$$Q_3(t) = f\left(\frac{3L}{4}, t\right) = \frac{1}{16}f_0\delta(t).$$

The matrix $[D]$, associated with the system, was obtained in Example 6.5 in the form

$$[D] = \frac{mL^2}{5\pi EA} \begin{bmatrix} 2.7786 & 1.4487 & -0.3989 \\ -0.0602 & 0.4903 & 0.2033 \\ -0.0115 & -0.0170 & 0.2076 \end{bmatrix}. \tag{e}$$

Also in Example 6.5 the eigenvalue problems (7.192) and (7.193) were solved and the following corresponding eigenvectors obtained:

$$[u] = \begin{bmatrix} 1.0000 & 1.0000 & 0.5207 \\ -0.0271 & -1.5449 & -0.6384 \\ -0.0043 & 0.0464 & 1.0000 \end{bmatrix},$$

$$[v] = \begin{bmatrix} 1.0173 & -0.0190 & 0.0043 \\ 0.6555 & -0.6476 & 0.0288 \\ -0.1079 & -0.3973 & 1.0162 \end{bmatrix}. \tag{f}$$

Because not enough significant figures were taken, the two eigenvalue problems yielded slightly different natural frequencies. Their average values are

$$\omega_1 = 2.3939\sqrt{\frac{EA}{mL^2}}, \quad \omega_2 = 5.4799\sqrt{\frac{EA}{mL^2}}, \quad \omega_3 = 8.6007\sqrt{\frac{EA}{mL^2}}. \tag{g}$$

Also from Example 6.5 we obtain

$$[k]^{-1} = \frac{L^2}{10\pi EA} \begin{bmatrix} 1.5932 & 2.6132 & 6.0503 \\ 0.3233 & 0.4059 & -1.2836 \\ 0.1689 & -0.3176 & 0.2256 \end{bmatrix}, \tag{h}$$

which allows us to compute the triple matrix product

$$[\omega^2][v]^T[k]^{-1} = \frac{1}{m} \begin{bmatrix} 0.3310 & 0.5397 & 0.9648 \\ -0.2932 & -0.1781 & 0.5990 \\ 0.4422 & -0.7059 & 0.5140 \end{bmatrix}. \tag{i}$$

Introducing (i) in (7.203) we obtain the generalized forces

$$N_1(t) = \frac{1}{m}[0.3310Q_1(t) + 0.5397Q_2(t) + 0.9648Q_3(t)] = 0.3814\frac{f_0}{m}\delta(t),$$

$$N_2(t) = \frac{1}{m}[-0.2932Q_1(t) - 0.1781Q_2(t) + 0.5990Q_3(t)]$$

$$= -0.1720\frac{f_0}{m}\delta(t),$$ (j)

$$N_3(t) = \frac{1}{m}[0.4422Q_1(t) - 0.7059Q_2(t) + 0.5140Q_3(t)] = 0.1043\frac{f_0}{m}\delta(t).$$

Equation (7.202), in conjunction with the convolution integral, leads to the generalized coordinates

$$\eta_1(t) = \frac{1}{\omega_1}\int_0^t N_1(\tau)\sin\omega_1(t-\tau)\,d\tau$$

$$= \frac{0.3814}{2.3939}\frac{f_0}{m}\sqrt{\frac{mL^2}{EA}}\int_0^t \delta(\tau)\sin 2.3939\sqrt{\frac{EA}{mL^2}}(t-\tau)\,d\tau$$

$$= 0.1593\frac{f_0}{m}\sqrt{\frac{mL^2}{EA}}\sin 2.3939\sqrt{\frac{EA}{mL^2}}\,t,$$ (k)

and, similarly,

$$\eta_2(t) = \frac{1}{\omega_2}\int_0^t N_2(\tau)\sin\omega_2(t-\tau)\,d\tau = -0.0314\frac{f_0}{m}\sqrt{\frac{mL^2}{EA}}\sin 5.4799\sqrt{\frac{EA}{mL^2}}\,t,$$

(l)

$$\eta_3(t) = \frac{1}{\omega_3}\int_0^t N_3(\tau)\sin\omega_3(t-\tau)\,d\tau = 0.0121\frac{f_0}{m}\sqrt{\frac{mL^2}{EA}}\sin 8.6007\sqrt{\frac{EA}{mL^2}}\,t.$$

Using (7.198) we obtain the generalized coordinates $q_j(t)$ in the form

$$q_1(t) = 1.0000\eta_1(t) + 1.0000\eta_2(t) + 0.5207\eta_3(t),$$

$$q_2(t) = -0.0271\eta_1(t) - 1.5449\eta_2(t) - 0.6384\eta_3(t),$$ (m)

$$q_3(t) = -0.0043\eta_1(t) + 0.0464\eta_2(t) + 1.0000\eta_3(t).$$

Finally, introducing (m) in (7.186), we obtain the approximate solution of (a) as

$$
w(x, t) = q_1(t) \sin \frac{\pi x}{2L} + q_2(t) \sin \frac{3\pi x}{2L} + q_3(t) \sin \frac{5\pi x}{2L}
$$

$$
= \left(1.0000 \sin \frac{\pi x}{2L} - 0.0271 \sin \frac{3\pi x}{2L} - 0.0043 \sin \frac{5\pi x}{2L} \right) \eta_1(t)
$$

$$
+ \left(1.0000 \sin \frac{\pi x}{2L} - 1.5449 \sin \frac{3\pi x}{2L} + 0.0464 \sin \frac{5\pi x}{2L} \right) \eta_2(t)
$$

$$
+ \left(0.5207 \sin \frac{\pi x}{2L} - 0.6384 \sin \frac{3\pi x}{2L} + 1.0000 \sin \frac{5\pi x}{2L} \right) \eta_3(t)
$$

$$
= \frac{Lf_0}{\sqrt{mEA}} \left[\left(0.1593 \sin \frac{\pi x}{2L} - 0.0046 \sin \frac{3\pi x}{2L} \right.\right.
$$

$$
\left. - 0.0007 \sin \frac{5\pi x}{2L} \right) \sin 2.3939 \sqrt{\frac{EA}{mL^2}} \, t
$$

$$
+ \left(-0.0314 \sin \frac{\pi x}{2L} + 0.0485 \sin \frac{3\pi x}{2L} \right.
$$

$$
\left. - 0.0015 \sin \frac{5\pi x}{2L} \right) \sin 5.4799 \sqrt{\frac{EA}{mL^2}} \, t
$$

$$
+ \left(0.0063 \sin \frac{\pi x}{2L} - 0.0077 \sin \frac{3\pi x}{2L} \right.
$$

$$
\left.\left. + 0.0121 \sin \frac{5\pi x}{2L} \right) \sin 8.6007 \sqrt{\frac{EA}{mL^2}} \, t \right], \quad \text{(n)}
$$

where the subscript of $w(x, t)$ has been dropped. Comparing (p) of Example 7.4 with the above result, we note that the agreement is relatively good in the predominant term. The agreement can be improved by using more terms in the series (7.178) and (7.186).

Problems

7.1 Consider the system shown in Figure 7.8 and calculate the impedance matrix and the matrix of transfer functions. Assume that initially the masses are at rest but displaced by amounts δ_1, δ_2, and δ_3 from the equilibrium position and that the mass m_3 is acted upon by a force $P(t)$. Calculate the transformed response.

7.2 Calculate the response of the system of Example 4.2 if the disks were initially imparted the angular velocities α_1, α_2, and α_3, respectively (Section 7–3). Use the results obtained in Example 4.2.

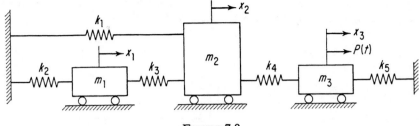

FIGURE 7.8

7.3 A harmonic force can be represented by either the real part of $Q_{0j}e^{i\Omega_j t}$, where Q_{0j} is a complex quantity, or $F_j \cos (\Omega_j t - \varphi_j)$, where F_j is the amplitude and φ_j is the phase angle, both real quantities. What is the relation between the two expressions? Let the system of Problem 7.2 be acted upon by the harmonic forces

$$Q_1 = 0, \qquad Q_2 = 0, \qquad Q_3 = Q_0 \sin \left(1.32 \sqrt{\frac{GJ}{LI_D}}\right) t$$

and calculate the subsequent motion. Take the initial conditions zero and draw conclusions as to the magnification factors.

7.4 What is the response of the system of Problem 7.3 if $Q_1 = Q_2 = 0$ and Q_3 is the periodic function shown in Figure 7.9? Take the initial conditions zero and use the analysis of Section 7–5. What can be said about the response obtained here as compared with the response obtained in Problem 7.3?

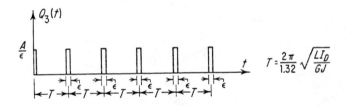

FIGURE 7.9

7.5 Repeat Problem 7.4 using modal analysis.

7.6 Obtain an expression for the longitudinal motion of a uniform bar clamped at the end $x = 0$ and free at the end $x = L$. The bar is subjected to a longitudinal distributed force $f(x, t) = F_0 \delta[x - (L - \epsilon)] u(t)$, where F_0 is a constant having units of force, $\delta[x - (L - \epsilon)]$ is a spatial Dirac's delta function, and $u(t)$ is a unit step function.

7.7 A concentrated moment of unit magnitude can be represented by two unit impulses acting in opposite directions, as shown in Figure 7.10. The function $u''(x - a)$ is called a *unit doublet* and has units length^{-2}. Hence a concentrated moment can be expressed as a distributed transverse force $p(x, t) = M(t)u''(x - a)$.

Use the unit doublet idea and calculate the response of a simply supported beam to an external moment applied at the end $x = L$.

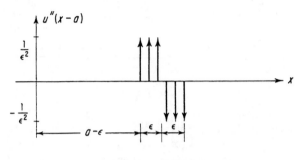

FIGURE 7.10

7.8 Obtain the response of a uniform rectangular membrane subjected to constant tension and clamped at the boundaries $x = 0,a$ and $y = 0,b$ to an initial displacement

$$f(x, y) = Axy(a - x)(b - y).$$

Discuss the mode participation in the response.

7.9 Obtain the response of a uniform circular plate clamped at the edge $r = a$ for the two cases:

(a) Two equal concentrated forces acting in the same direction are applied at the location shown in Figure 7.11(a).

(b) Two equal concentrated forces acting in opposite directions are applied at the location shown in Figure 7.11(b).

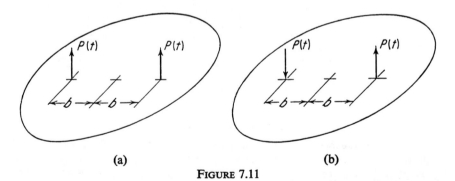

(a)   (b)

FIGURE 7.11

7.10 The end $x = 0$ of a uniform rod is imparted a harmonic longitudinal displacement $u(0, t) = u_0 \sin \Omega t$; the end $x = L$ is free. Determine the longitudinal motion, $u(x, t)$, of the rod.

7.11 A beam has both ends hinged. The end $x = L$ is subjected to a moment $M(t)$. Obtain the response by the method of Section 7–14.

7.12 Consider the system shown in Figure 7.12. Take advantage of the symmetric properties of the system and obtain the first four natural modes (two rigid and two elastic modes) for $k = 20EI/L^3$. Derive a general expression for the response to the force $P(t)$ as shown.

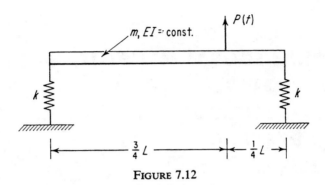

FIGURE 7.12

7.13 A nonuniform rectangular plate simply supported at the boundaries $x = 0,a$ and $y = 0,b$ has the mass and flexural rigidity distributions

$$m(x, y) = m\left[1 + 12\frac{x}{a} - 12\left(\frac{x}{a}\right)^2\right]\left[1 + 12\frac{y}{b} - 12\left(\frac{y}{b}\right)^2\right],$$

$$D_E(x, y) = D_E\left[1 + 12\frac{x}{a} - 12\left(\frac{x}{a}\right)^2\right]\left[1 + 12\frac{y}{b} - 12\left(\frac{y}{b}\right)^2\right].$$

Formulate the corresponding eigenvalue problem by means of an approximate method. Without solving the eigenvalue problem, write a general expression for the response to a distributed transverse load $p(x, t)$.

Selected Readings

Bisplinghoff, R. L., H. Ashley, and R. L. Halfman, *Aeroelasticity*, Addison-Wesley, Reading, Mass., 1957.

Courant, R., and D. Hilbert, *Methods of Mathematical Physics*, Vol. 1, Interscience, New York, 1961.

Friedman, B., *Principles and Techniques of Applied Mathematics*, Wiley, New York, 1956.

Hurty, W. C., and M. F. Rubinstein, *Dynamics of Structures*, Prentice-Hall, Englewood Cliffs, N.J., 1964.

Mindlin, R. D., and L. E. Goodman, "Beam Vibrations With Time-Dependent Boundary Conditions," *J. Appl. Mech.*, **17**, 377–380, 1950.

CHAPTER 8

TRANSFORM METHOD
SOLUTIONS OF
CONTINUOUS
SYSTEMS.
WAVE
SOLUTIONS

8-1 GENERAL CONSIDERATIONS

In Chapter 7 we treated the response of discrete as well as continuous systems. For discrete systems we obtained solutions directly from the coupled equations of motion by using the Laplace transform method. We also presented a particularly significant method for the solution of the equations of motion, namely the modal analysis. Modal analysis uses the modal matrix to represent a linear transformation uncoupling the equations of motion. Continuous systems were also treated in Chapter 7, and solutions were obtained exclusively by modal analysis.

When the normal modes can be obtained without much difficulty, modal analysis proves to be a very efficient method for the solution of boundary-value problems. It, essentially, yields a standing wave solution. In many cases in which there are time-dependent boundary conditions, if one insists on solving the boundary-value problem by modal analysis, it is necessary to use a transformation of the dependent variable such that a problem with nonhomogeneous boundary conditions is transformed into another problem with homogeneous boundary conditions. Quite often such cases are treated with much more ease by means of integral transform methods and in particular by means of the Laplace and Fourier transform methods. In vibration problems the independent variables are the spatial variables and time. If the integral transform is used to eliminate the time dependence, one must transform not only the differential equation but also the boundary conditions. In effect, one must solve a new boundary-value problem for the transformed dependent variable. The new boundary-value problem resembles a static problem, because the time variable has been eliminated. The solution of the transformed problem must be inverted to obtain the solution of the original problem. In the process the initial conditions are automatically taken into account. If one chooses to eliminate the spatial coordinate dependence from the problem, both the differential equation and the initial conditions must be transformed. The boundary conditions dictate the type of transform that can be used and are automatically provided for. The transformed problem resembles the initial-value problem associated with a single-degree-of-freedom system. Again one must perform an inversion of the transformed problem solution to obtain the solution of the original boundary-value problem.

In solving the wave equation, the Laplace transform method offers the choice between a standing wave and a traveling wave solution. There are various kinds of Fourier transforms available. The choice of a particular Fourier transform depends on the boundary conditions. If the domain is one-dimensional and finite, the choice is limited to finite sine and finite cosine Fourier transforms and, if any of these transforms can be used at all, the choice is further limited by the boundary conditions, as will be shown later. Except for the case of time-dependent boundary condition problems, in general, finite Fourier transforms do not present any particular advantage over modal analysis. In fact, the kernel of the transformation coincides with the normal modes of the system under consideration. For two-dimensional, finite domain problems the double Fourier transforms and Hankel transforms can be used at times with advantage. The decision as to which of these transforms should be used in a given problem must be made after considering the boundary conditions, which in two-dimensional problems must also specify the shape of the boundary curves. Both the Fourier and the Hankel transforms can be used to render the dependent variable a function of some parameters and time, instead of a function of the spatial variables and time, thereby necessitating solutions of ordinary differential equations instead of partial differential equations. In general, the solution of the resulting ordinary differential equation is relatively easy to obtain. The main difficulty in using transform methods, however, lies in the inversion process, which in most transforms involves the evaluation of an integral. The fact that in all cases treated here we are able to carry out the inversion and obtain a closed form solution should not lead one to believe that this is always possible; in many cases the inversion integral cannot be evaluated in closed form. If one is content to know the value of the function describing the system behavior for general values of the independent variable, one may be able to obtain an approximate solution. In this connection the method of steepest descent and numerical computation of integrals should be mentioned.†

The real advantage of integral transform methods lies in solving problems for which the domain is semi-infinite or infinite, because in such cases the normal modes lose their meaning. For problems involving infinite domains the Fourier transform is more general in scope, since the Hankel transform can be used only when the differential equation of the boundary-value problem can be reduced to a form involving Bessel equations. As a matter of interest the method of integration of Cauchy should be mentioned. This method makes use of Fourier integrals and is closely related to Fourier transform methods.‡

This chapter presents a variety of methods, other than modal analysis, for the solution of boundary-value problems. The wave equation is given particular attention, and the subject of traveling waves is discussed in detail. Solutions by Laplace and Fourier transform methods underscore the advantage of these methods in the treatment of problems involving time-dependent boundary conditions. The problem of traveling waves in a bar in transverse motion is discussed and concepts such as phase and group velocity are introduced. The dispersive character of such a medium is pointed out. The shortcomings of the elementary theory, which ignores the rotatory inertia and shear deformation effects, are revealed when this theory predicts that short waves travel with wave velocities that are not physically possible. Finally, problems associated with two-dimensional systems such as infinite membranes and plates are solved by means of Fourier and Hankel transforms.

8-2 THE WAVE EQUATION

The differential equation for the free transverse vibration of a string was shown previously [see (5.60)] to have the form

$$\frac{\partial}{\partial x}\left[T(x)\frac{\partial y(x,t)}{\partial x}\right] = \rho(x)\frac{\partial^2 y(x,t)}{\partial t^2}, \tag{8.1}$$

where $T(x)$ is the tension in the string and $\rho(x)$ is the mass per unit length of string. For constant tension and uniform mass distribution, (8.1) reduces to

$$\frac{\partial^2 y(x,t)}{\partial x^2} = \frac{1}{c^2}\frac{\partial^2 y(x,t)}{\partial t^2}, \qquad c = \sqrt{\frac{T}{\rho}}. \tag{8.2}$$

† See I. N. Sneddon, *Fourier Transforms*, McGraw-Hill, New York, 1951, p. 516.
‡ See A. G. Webster, *Partial Differential Equations of Mathematical Physics*, Dover, New York, 1955.

Equation (8.2) represents the one-dimensional wave equation in which c is the wave propagation velocity.

It is not difficult to verify that the general solution of (8.2) is

$$y(x, t) = F_1(x - ct) + F_2(x + ct), \qquad (8.3)$$

where F_1 and F_2 are arbitrary functions of their respective arguments. We see that $F_1(x - ct)$ represents a displacement wave of arbitrary shape *traveling* in the positive x direction at a constant velocity c *without altering the shape*. Similarly, $F_2(x + ct)$ is a displacement wave traveling in the negative x direction. Hence every motion of the string consists of a super-position of two waves traveling in opposite directions.

An important case of (8.3) is the sinusoidal wave traveling in the positive x direction,

$$y(x, t) = A \sin \frac{2\pi}{\lambda} (x - ct), \qquad (8.4)$$

where λ is the *wavelength* which is the distance between two successive crests. Its expression is

$$\lambda = c \frac{2\pi}{\omega} = c\tau, \qquad (8.5)$$

where ω is the *circular frequency* of the wave and τ is the *period* necessary for a complete wave to pass a given point. The number of waves in a unit distance is called the *wave number* and is given by

$$k = \frac{1}{\lambda}. \qquad (8.6)$$

With this notation (8.4) can be rewritten

$$y(x, t) = A \sin (2\pi kx - \omega t). \qquad (8.7)$$

Consider the special case in which two waves of equal amplitude and frequency travel in opposite directions. Their superposition yields

$$y(x, t) = A \sin (2\pi kx - \omega t) + A \sin (2\pi kx + \omega t)$$
$$= 2A \sin 2\pi kx \cos \omega t, \qquad (8.8)$$

and we conclude that the wave profile is not traveling. Hence in this instance, the two traveling waves combine into a *stationary* or *standing wave*. At the points where $\sin 2\pi kx = 0$ the two traveling waves cancel each other, giving a zero displacement. These points are called *nodes*. At the points where

$\sin 2\pi kx = \pm 1$ the two traveling waves add their effect yielding the greatest amplitude. These points, which are halfway between any two successive nodes, are called *loops* or *antinodes*.

Equation (8.8) admits any frequency ω. This is because no boundary conditions have been invoked. For a finite string, fixed at the ends $x = 0$ and $x = L$, the wave number k must assume the values

$$k = \frac{r}{2L}, \qquad r = 1, 2, \ldots, \tag{8.9}$$

which limits the harmonic waves to those which have nodes at $x = 0$ and $x = L$. The corresponding frequencies are

$$\omega_r = r\pi \sqrt{\frac{T}{\rho L^2}}, \qquad r = 1, 2, \ldots, \tag{8.10}$$

which are precisely the natural frequencies obtained previously by solving the eigenvalue problem.

This analysis is justified when the wavelengths are long compared with the thickness of the string. Otherwise, one must consider the bending stiffness also.

The subject of traveling waves will be pursued further in Section 8–3.

8–3   FREE VIBRATION OF AN INFINITE STRING. CHARACTERISTICS

Consider an infinite string, $-\infty < x < \infty$, free of any external forces and subjected to the initial conditions

$$y(x, 0) = f(x), \tag{8.11}$$

$$\left. \frac{\partial y(x, t)}{\partial t} \right|_{t=0} = \frac{\partial y(x, 0)}{\partial t} = g(x), \tag{8.12}$$

where $y(x, t)$ is the transverse displacement of the string.

When the tension T and the mass per unit length of string ρ are constant, the transverse displacement of the string must satisfy the wave equation

$$\frac{\partial^2 y(x, t)}{\partial x^2} = \frac{1}{c^2} \frac{\partial^2 y(x, t)}{\partial t^2}, \qquad c = \sqrt{\frac{T}{\rho}}, \tag{8.13}$$

where c is the wave propagation velocity.

We shall attempt a solution by means of a Fourier transform†; because the string is of infinite length, the complex Fourier transform is likely to suit our purposes. Thus, let the complex Fourier transform of $y(x, t)$ be

$$\bar{Y}(p, t) = \int_{-\infty}^{\infty} y(x, t)e^{ipx} \, dx. \tag{8.14}$$

The transform of the second derivative with respect to x is

$$\bar{Y}^{(2)}(p, t) = \int_{-\infty}^{\infty} \frac{\partial^2 y(x, t)}{\partial x^2} e^{ipx} \, dx = -p^2 \bar{Y}(p, t), \tag{8.15}$$

where on the basis of physical grounds it was assumed that $y(x, t)|_{x \to \pm \infty} = 0$ and $\partial y(x, t)/\partial x|_{x \to \pm \infty} = 0$. Furthermore, we can write

$$\int_{-\infty}^{\infty} \frac{\partial^2 y(x, t)}{\partial t^2} e^{ipx} \, dx = \frac{\partial^2}{\partial t^2} \int_{-\infty}^{\infty} y(x, t)e^{ipx} \, dx = \frac{d^2 \bar{Y}(p, t)}{dt^2}. \tag{8.16}$$

Transforming both sides of (8.13) we obtain the ordinary differential equation for $\bar{Y}(p, t)$,

$$\frac{d^2 \bar{Y}(p, t)}{dt^2} + (pc)^2 \bar{Y}(p, t) = 0, \tag{8.17}$$

where the transformed dependent variable, $\bar{Y}(p, t)$, is subject to the *transformed initial conditions*

$$\bar{Y}(p, 0) = \bar{F}(p) = \int_{-\infty}^{\infty} f(x)e^{ipx} \, dx, \tag{8.18}$$

$$\frac{d\bar{Y}(p, 0)}{dt} = \bar{G}(p) = \int_{-\infty}^{\infty} g(x)e^{ipx} \, dx. \tag{8.19}$$

The solution of (8.17), in conjunction with the transformed initial conditions (8.18) and (8.19), is

$$\bar{Y}(p, t) = \bar{F}(p) \cos pct + \frac{\bar{G}(p)}{pc} \sin pct. \tag{8.20}$$

The response $y(x, t)$ is obtained by taking the inverse transformation of

---

† The solution of this problem can be obtained directly by applying the initial conditions to (8.3) (see Problem 8.1). Here we choose the Fourier transform method to illustrate its usefulness. For details of the Fourier transforms see Appendix C.

(8.20). Hence, using the inversion formula for the complex Fourier transform, we write

$$y(x, t) = \frac{1}{2\pi} \int_{-\infty}^{\infty} \overline{Y}(p, t)e^{-ipx} \, dp$$

$$= \frac{1}{2\pi} \int_{-\infty}^{\infty} \left[\overline{F}(p) \cos pct + \frac{\overline{G}(p)}{pc} \sin pct \right] e^{-ipx} \, dp$$

$$= \frac{1}{4\pi} \int_{-\infty}^{\infty} \overline{F}(p)[e^{-ip(x-ct)} + e^{-ip(x+ct)}] \, dp$$

$$+ \frac{1}{4\pi c} \int_{-\infty}^{\infty} \frac{\overline{G}(p)}{ip} [e^{-ip(x-ct)} - e^{-ip(x+ct)}] \, dp. \qquad (8.21)$$

But if

$$g(u) = \frac{1}{2\pi} \int_{-\infty}^{\infty} \overline{G}(p)e^{-ipu} \, dp \qquad (8.22)$$

is the inverse complex Fourier transform of $\overline{G}(p)$, we can write

$$\int_{x-ct}^{x+ct} g(u) \, du = \frac{1}{2\pi} \int_{-\infty}^{\infty} \overline{G}(p)\left[\int_{x-ct}^{x+ct} e^{-ipu} \, du \right] dp$$

$$= \frac{1}{2\pi} \int_{-\infty}^{\infty} \frac{\overline{G}(p)}{ip} [e^{-ip(x-ct)} - e^{-ip(x+ct)}] \, dp. \qquad (8.23)$$

In addition,

$$\frac{1}{2\pi} \int_{-\infty}^{\infty} \overline{F}(p)e^{-ip(x \pm ct)} \, dp = f(x \pm ct), \qquad (8.24)$$

so (8.21) may be rewritten

$$y(x, t) = \tfrac{1}{2}[f(x - ct) + f(x + ct)] + \frac{1}{2c} \int_{x-ct}^{x+ct} g(u) \, du. \qquad (8.25)$$

Figure 8.1 shows the displacement at various instants for given initial displacement and velocity. Because of the initial displacement $f(x)$ shown on the left side of Figure 8.1 the response $y(x, t)$ is the superposition of two waves, $\tfrac{1}{2}f(x - ct)$ and $\tfrac{1}{2}f(x + ct)$, traveling with a velocity c in the positive and negative directions of the x axis. The response to the initial velocity $g(x)$ is slightly more involved. We note that for $t < a/c$ the largest displacement is Bt, in the interval $|x| \leq a - ct$. From the point $x = a - ct$ to $x = a + ct$, on the one hand, and from $x = -(a - ct)$ to $x = -(a + ct)$ on the other, the displacement decreases linearly to zero. For $t > a/c$ the largest displace-

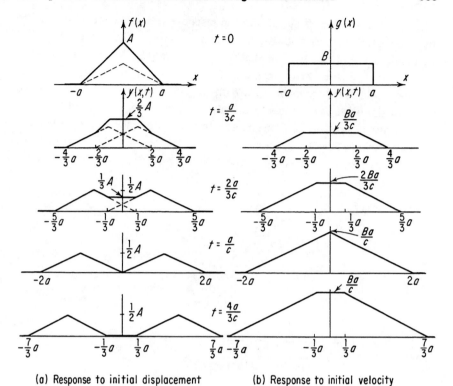

(a) Response to initial displacement   (b) Response to initial velocity

FIGURE 8.1

ment is Ba/c, in the interval $|x| < ct - a$. As $t \to \infty$, the entire string becomes displaced at a uniform distance Ba/c with respect to the original position.

It will prove of interest to show $y(x, t)$ in a three-dimensional plot. To this end let us concentrate on the response to the initial displacements only.

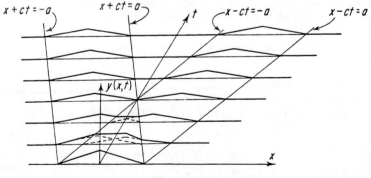

FIGURE 8.2

Figure 8.2 shows the xt plane, and the response $y(x, t)$ is plotted normal to that plane. It is possible to distinguish various regions in the xt plane bounded by the lines $x \pm ct = \pm a$. We shall call the waves $f(x - ct)$ and $f(x + ct)$ direct and opposite waves, respectively. We observe in Figure 8.2 three regions in which the response is zero, one region in which both the direct and the opposite wave contributes to the response, one region in which only the direct wave is present, and one region in which only the opposite wave is present. Perhaps we can develop a better understanding of the problem by considering the xt plane only. Each point in the xt plane represents the

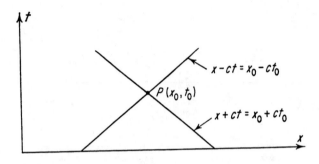

FIGURE 8.3

position of a point on the string at time t. The direct wave $f(x - ct)$, at point (x, t), propagating with velocity c, will reach a point x_0 at a time t_0 such that

$$(x - x_0) = c(t - t_0). \tag{8.26}$$

Similarly, for the opposite wave we have

$$-(x - x_0) = c(t - t_0). \tag{8.27}$$

Equations (8.26) and (8.27) can be written

$$x - ct = x_0 - ct_0, \qquad x + ct = x_0 + ct_0. \tag{8.28}$$

These equations represent two straight lines of slope c and $-c$ passing through the point $P(x_0, t_0)$. The straight lines as given by (8.28) are called the *characteristics*† of the partial differential equation (8.13). Along the

† For a more elaborate discussion of the method of characteristics, see F. B. Hilde-brand, *Advanced Calculus for Engineers*, Prentice-Hall, Englewood Cliffs, N.J., 1960, and A. G. Webster, *Partial Differential Equations of Mathematical Physics*, Dover, New York, 1955.

line $x - ct = x_0 - ct_0 = $ const, $f(x - ct)$ is constant, and it follows that along this characteristic all points have the same deflection. Hence we can say that the direct wave travels along this characteristic. A similar interpretation can be given for the opposite wave $f(x + ct)$ and the characteristic $x + ct = $ constant. Next let us consider how various points in the xt plane are affected by an initial displacement lying within the interval $x_1 \leq x \leq x_2$. The characteristics $x \pm ct = x_1$ and $x \pm ct = x_2$ divide the xt plane in six regions, as mentioned previously. Any point in region I will experience the effect of both the direct and opposite waves. A point in region II will feel the effect of the direct wave. For example, if we draw the characteristics

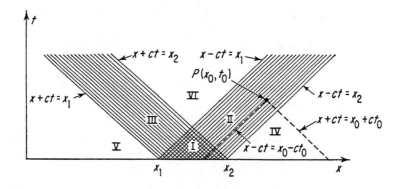

FIGURE 8.4

through point P (see Figure 8.4) their intersections with the x axis gives the points on the string for which the initial displacement produces a disturbance at P. It is easy to see that only the characteristic associated with the direct wave intersects the x axis in the interval (x_1, x_2). One can say that points on the string associated with region II experienced the opposite wave at an earlier time. A point in region III feels the effect of the opposite wave only, because the direct wave has passed. Points in region IV are at rest, because the direct wave has not arrived yet and the opposite wave has passed; the reverse is true for region V. Points in region VI are at rest, because both waves have passed. Points in region VI, however, can undergo displacements due to initial disturbance in the region $x_1 \leq x \leq x_2$ in the form of initial velocities, as can be concluded from Figure 8.1(b).

Boundaries complicate the above picture because waves are reflected at boundaries. The subject of reflection at boundaries will be discussed in Section 8–4.

8-4 FREE VIBRATION OF A SEMI-INFINITE STRING

Consider a semi-infinite string, $0 \leq x < \infty$, free of any external forces and subjected to the initial conditions

$$y(x, t)|_{t=0} = f(x), \tag{8.29}$$

$$\frac{\partial y(x, t)}{\partial t}\bigg|_{t=0} = g(x), \tag{8.30}$$

where $y(x, t)$ is the transverse displacement of the string.

When the tension T and the mass per unit length of string ρ are constant, the transverse displacement of the string must satisfy the wave equation

$$\frac{\partial^2 y(x, t)}{\partial x^2} = \frac{1}{c^2} \frac{\partial^2 y(x, t)}{\partial t^2}, \qquad c = \sqrt{\frac{T}{\rho}}, \tag{8.31}$$

where c is the wave propagation velocity.

Let the string be fixed at the origin so that $y(0, t) = 0$ at all times. On the other hand, we assume, on physical grounds, that $y(x, t)|_{x \to \infty} = 0$ and $\partial y(x, t)/\partial x|_{x \to \infty} = 0$. We shall attempt a solution by means of a Fourier transform, and because we do not know the value of $\partial y(x, t)/\partial x$ at $x = 0$, it appears that a Fourier sine transform, which does not require the knowledge of odd-order derivatives for an even-order differential equation (see Section C–4), is likely to suit our purpose. Thus let the Fourier sine transform of $y(x, t)$ be

$$\overline{Y}_s(p, t) = \int_0^\infty y(x, t) \sin px \, dx. \tag{8.32}$$

The transform of the second derivative with respect to x is

$$\overline{Y}_s^{(2)}(p, t) = \int_0^\infty \frac{\partial^2 y(x, t)}{\partial x^2} \sin px \, dx$$

$$= py(0, t) - p^2 \overline{Y}_s(p, t) = -p^2 \overline{Y}_s(p, t). \tag{8.33}$$

Furthermore, we can write

$$\int_0^\infty \frac{\partial^2 y(x, t)}{\partial t^2} \sin px \, dx = \frac{\partial^2}{\partial t^2} \int_0^\infty y(x, t) \sin px \, dx = \frac{d^2 \overline{Y}_s(p, t)}{dt^2}. \tag{8.34}$$

Transforming both sides of (8.31) we write

$$\frac{1}{c^2} \frac{d^2 \overline{Y}_s(p, t)}{dt^2} + p^2 \overline{Y}_s(p, t) = 0, \tag{8.35}$$

which has the solution

$$\overline{Y}_s(p, t) = \overline{F}_s(p) \cos pct + \frac{\overline{G}_s(p)}{pc} \sin pct, \tag{8.36}$$

where

$$\overline{F}_s(p) = \overline{Y}_s(p, 0) = \int_0^\infty f(x) \sin px \, dx \tag{8.37}$$

and

$$\overline{G}_s(p) = \frac{d\overline{Y}_s(p, t)}{dt}\bigg|_{t=0} = \int_0^\infty g(x) \sin px \, dx \tag{8.38}$$

are the Fourier sine transforms of the initial conditions (8.29) and (8.30), respectively.

The inverse Fourier sine transformation of (8.36) is

$$
\begin{aligned}
y(x, t) &= \frac{2}{\pi} \int_0^\infty \overline{Y}_s(p, t) \sin px \, dp \\
&= \frac{2}{\pi} \int_0^\infty \left[\overline{F}_s(p) \cos pct + \frac{\overline{G}_s(p)}{pc} \sin pct \right] \sin px \, dp \\
&= \frac{1}{2} \left\{ \frac{2}{\pi} \int_0^\infty \overline{F}_s(p)[\sin p(x + ct) + \sin p(x - ct)] \, dp \right. \\
&\left. \quad - \frac{1}{c}\frac{2}{\pi} \int_0^\infty \overline{G}_s(p)[\cos p(x + ct) - \cos p(x - ct)] \frac{dp}{p} \right\}. \tag{8.39}
\end{aligned}
$$

But if

$$g(u) = \frac{2}{\pi} \int_0^\infty \overline{G}_s(p) \sin up \, dp \tag{8.40}$$

is the inverse Fourier sine transform of $\overline{G}_s(p)$, we can write

$$
\begin{aligned}
\int_{x-ct}^{x+ct} g(u) \, du &= \frac{2}{\pi} \int_0^\infty \overline{G}_s(p) \left[\int_{x-ct}^{x+ct} \sin up \, du \right] dp \\
&= -\frac{2}{\pi} \int_0^\infty \overline{G}_s(p)[\cos p(x + ct) - \cos p(x - ct)] \frac{dp}{p}. \tag{8.41}
\end{aligned}
$$

In addition, for $x > 0$ the inverse Fourier sine transform of $\overline{F}_s(p)$ is

$$f(x) = \frac{2}{\pi} \int_0^\infty \overline{F}_s(p) \sin xp \, dp, \tag{8.42}$$

so for $x > ct$ we have

$$
\begin{aligned}
\frac{2}{\pi} \int_0^\infty \overline{F}_s(p) \sin p(x + ct) \, dp &= f(x + ct), \\
\frac{2}{\pi} \int_0^\infty \overline{F}_s(p) \sin p(x - ct) \, dp &= f(x - ct).
\end{aligned}
\tag{8.43}
$$

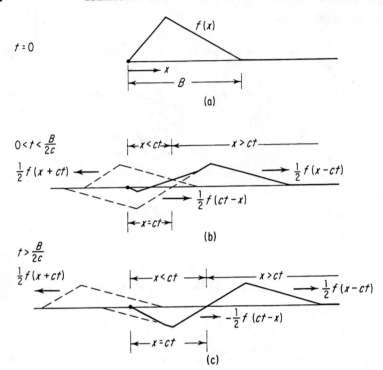

FIGURE 8.5

For $x < ct$ the first of equations (8.43) remains the same, but the second must be replaced by

$$\frac{2}{\pi} \int_0^\infty \bar{F}_s(p) \sin p(x - ct) \, dp = -f(ct - x). \tag{8.44}$$

Hence for $x > ct$ the response is

$$y(x, t) = \tfrac{1}{2}[f(x + ct) + f(x - ct)] + \frac{1}{2c} \int_{x-ct}^{x+ct} g(u) \, du, \tag{8.45}$$

and for $x < ct$ the response is

$$y(x, t) = \tfrac{1}{2}[f(ct + x) - f(ct - x)] + \frac{1}{2c} \int_{x-ct}^{x+ct} g(u) \, du. \tag{8.46}$$

We can interpret the function $f(x + ct)$ as a wave similar in shape to the function $f(x)$ and traveling in the negative x direction with velocity c, and $f(x - ct)$ may be interpreted as a wave traveling in the positive x direction. Furthermore, $-f(ct - x)$ may be interpreted as the antisymmetric image of $f(x - ct)$ and traveling in the positive x direction. If the initial velocity

$g(x)$ is zero, the response for $x > ct$ is simply the superposition of two waves, $\frac{1}{2}f(x + ct)$ and $\frac{1}{2}f(x - ct)$, the first traveling in the negative x direction and the second traveling in the positive x direction. For $x < ct$ the wave $\frac{1}{2}f(x - ct)$ does not enter the picture. The incident wave $\frac{1}{2}f(x + ct)$, after hitting the fixed end $x = 0$, returns in the reflected form $-\frac{1}{2}f(ct - x)$ traveling in the positive x direction. Of course, the displacement is the superposition of these two waves. The above interpretation is illustrated in Figure 8.5.

Another interpretation can be given by regarding the wave $-\frac{1}{2}f(ct - x)$ as the antisymmetric image of the wave $\frac{1}{2}f(x - ct)$ extending into the region $0 < x < ct$ and traveling with it in the same direction, while the tail $\frac{1}{2}f(x + ct)$ is to be superposed as before.

8–5 RESPONSE OF A FINITE STRING TO INITIAL EXCITATION

Consider a uniform string under constant tension T and fixed at the points $x = 0$ and $x = L$, so the boundary conditions are

$$y(0, t) = 0, \qquad y(L, t) = 0. \tag{8.47}$$

Let the initial displacement and initial velocity of the string be, respectively,

$$y(x, 0) = f(x), \tag{8.48}$$

$$\frac{\partial y(x, t)}{\partial t}\bigg|_{t=0} = g(x), \tag{8.49}$$

and consider the case in which there are no external forces applied on the string.

The transverse displacement $y(x, t)$ must satisfy the wave equation, and to solve for the response we choose a finite Fourier transform. A knowledge of even-order (zero order in this particular case) derivatives at the boundaries calls for a finite sine transform defined by (see Appendix C)

$$\bar{y}_s(n, t) = \int_0^L y(x, t) \sin \frac{n\pi x}{L}\, dx. \tag{8.50}$$

Taking into account the boundary conditions (8.47), we obtain the finite sine transform of $\partial^2 y(x, t)/\partial x^2$ in the form

$$\bar{y}_s^{(2)}(n, t) = \int_0^L \frac{\partial^2 y(x, t)}{\partial x^2} \sin \frac{n\pi x}{L}\, dx = -\left(\frac{n\pi}{L}\right)^2 \bar{y}_s(n, t). \tag{8.51}$$

Hence, transforming the wave equation (8.31), we obtain

$$\frac{d^2 \bar{y}_s(n, t)}{dt^2} + \left(\frac{n\pi c}{L}\right)^2 \bar{y}_s(n, t) = 0, \tag{8.52}$$

which is subject to the transformed initial conditions

$$\bar{y}_s(n, 0) = \bar{f}_s(n) = \int_0^L f(x) \sin \frac{n\pi x}{L} \, dx, \tag{8.53}$$

$$\frac{d\bar{y}_s(n, t)}{dt}\bigg|_{t=0} = \bar{g}_s(n) = \int_0^L g(x) \sin \frac{n\pi x}{L} \, dx. \tag{8.54}$$

It follows that the solution of (8.52), in conjunction with the transformed initial conditions (8.53) and (8.54) is

$$\bar{y}_s(n, t) = \bar{f}_s(n) \cos \frac{n\pi ct}{L} + \bar{g}_s(n) \frac{L}{n\pi c} \sin \frac{n\pi ct}{L}. \tag{8.55}$$

The response $y(x, t)$ is obtained by writing the inverse transformation of (8.55),

$$y(x, t) = \frac{2}{L} \sum_{n=1}^{\infty} \bar{y}_s(n, t) \sin \frac{n\pi x}{L}$$

$$= \frac{2}{L} \sum_{n=1}^{\infty} \sin \frac{n\pi x}{L} \cos \frac{n\pi ct}{L} \int_0^L f(u) \sin \frac{n\pi u}{L} \, du$$

$$+ \frac{2}{\pi c} \sum_{n=1}^{\infty} \frac{1}{n} \sin \frac{n\pi x}{L} \sin \frac{n\pi ct}{L} \int_0^L g(u) \sin \frac{n\pi u}{L} \, du, \tag{8.56}$$

where u is a dummy variable of integration. We must note that the solution, (8.56), is in terms of standing waves or normal modes.

Example 8.1

Consider the case in which the initial displacement is zero, $f(x) = 0$, and an impulsive force $\hat{I}$ is applied at the point $x = b$, resulting in an initial velocity

$$g(x) = A \, \delta(x - b). \tag{a}$$

The constant A may be obtained by evaluating the total impulse applied to the string, which is accomplished by integrating over the length of the string the linear momentum associated with each differential element of length.

If $\partial y(x, 0)/\partial t$ denotes the initial velocity at any point x, the impulse-momentum principle yields

$$\hat{I} = \int_{\text{mass}} \frac{\partial y(x, 0)}{\partial t}\, dm = A\rho \int_0^L \delta(x - b)\, dx = A\rho, \tag{b}$$

where ρ is the mass per unit length of string. Hence

$$g(x) = \frac{\hat{I}}{\rho} \delta(x - b). \tag{c}$$

Introducing (c) in (8.56) and recalling that $f(x) = 0$, we obtain

$$y(x, t) = \frac{2}{\pi c} \sum_{n=1}^{\infty} \frac{1}{n} \sin \frac{n\pi x}{L} \sin \frac{n\pi ct}{L} \int_0^L \frac{\hat{I}}{\rho} \delta(u - b) \sin \frac{n\pi u}{L}\, du$$

$$= \frac{2\hat{I}}{\pi c\rho} \sum_{n=1}^{\infty} \frac{1}{n} \sin \frac{n\pi b}{L} \sin \frac{n\pi x}{L} \sin \frac{n\pi ct}{L}. \tag{d}$$

Note that by letting $b = L/2$, all the terms for which n is even disappear. This result is to be expected, because the modes corresponding to even n have nodes at $x = L/2$ which cannot be excited.

One must observe the relative ease with which the response of a finite string was obtained by means of the finite sine transform. This is not surprising, because the kernel of the transformation is just the nth eigenfunction of the problem.

8–6 MOTION OF A BAR WITH A PRESCRIBED FORCE ON ONE END

The axial displacement $u(x, t)$ of a bar of uniform mass m and uniform stiffness EA (Figure 8.6) must satisfy the wave equation

$$\frac{\partial^2 u(x, t)}{\partial x^2} = \frac{1}{c^2} \frac{\partial^2 u(x, t)}{\partial t^2}, \tag{8.57}$$

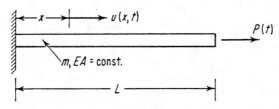

FIGURE 8.6

where c is the wave propagation velocity given by

$$c = \sqrt{\frac{EA}{m}} = \sqrt{\frac{E}{\rho}}, \tag{8.58}$$

where ρ denotes the mass density and E is the modulus of elasticity of the material. In deriving (8.57) an elementary theory has been used whose underlying assumptions were discussed in Chapter 5. According to the elementary theory a pulse propagates along the rod without changing its shape. A so-called "exact" theory, based on the linear theory of elasticity, is associated with the names of Pochhammer and Chree. This theory points out the dispersive nature of the elastic waves, which results in a change in the pulse shape as it propagates. The present discussion is confined to the elementary theory and the reader interested in the more refined theories is referred to a comprehensive survey paper by Miklowitz,† which discusses elastic wave propagation in rods, plates, circular cylindrical shells, half-spaces, and infinite medium. The paper contains a very extensive list of references.

Let us consider the bar of Figure 8.6 and assume that the bar is initially at rest,

$$u(x, 0) = 0, \qquad \frac{\partial u(x, t)}{\partial t}\bigg|_{t=0} = 0. \tag{8.59}$$

The boundary conditions are

$$u(0, t) = 0, \tag{8.60}$$

$$EA \frac{\partial u(x, t)}{\partial x}\bigg|_{x=L} = P(t). \tag{8.61}$$

We note that the differential equation (8.57), as well as the boundary condition (8.60), are homogeneous, whereas the boundary condition (8.61) is nonhomogeneous. The treatment of problems involving nonhomogeneous boundary conditions by means of modal analysis was discussed in Section 7–14. In fact, the present problem was solved there as an illustrative example. In that section it was indicated that boundary-value problems involving nonhomogeneous boundary conditions can be solved by means of integral transformations. In the present section we shall obtain the solutions $u(x, t)$ in three different ways: by means of a finite sine transform, by means of

† See J. Miklowitz, "Recent Developments in Elastic Wave Propagation," *Appl. Mech. Rev.*, **13**, 865–878 (1960).

Laplace transform with a solution in terms of standing waves, and by means of a Laplace transform in terms of traveling waves.

(a) Solution by a Finite Sine Transform

The particular nature of the boundary conditions requires the use of a small artifice. Let the finite sine transform of $u(x, t)$ be defined by

$$\bar{u}_s(p, t) = \int_0^L u(x, t) \sin px \, dx, \tag{8.62}$$

where p is a parameter to be determined. In addition,

$$\bar{u}_s^{(2)}(p, t) = \int_0^L \frac{\partial^2 u(x, t)}{\partial x^2} \sin px \, dx$$

$$= \left[\frac{\partial u(x, t)}{\partial x} \sin px - pu(x, t) \cos px \right]_0^L - p^2 \bar{u}_s(p, t). \tag{8.63}$$

Using the boundary conditions (8.60) and (8.61), (8.63) reduces to

$$\bar{u}_s^{(2)}(p, t) = \frac{P(t)}{EA} \sin pL - pu(L, t) \cos pL - p^2 \bar{u}_s(p, t). \tag{8.64}$$

The second term on the right side of (8.64) contains $u(L, t)$, which is not available. One can eliminate this term by letting $\cos pL = 0$, which leads to

$$p = (2n - 1) \frac{\pi}{2L}, \qquad n = 1, 2, \dots. \tag{8.65}$$

It follows that $\sin pL = \sin (2n - 1)\pi/2 = (-1)^{n-1}$. Transforming (8.57) we obtain

$$\frac{d^2 \bar{u}_s(n, t)}{dt^2} + \left[(2n - 1) \frac{\pi c}{2L} \right]^2 \bar{u}_s(n, t) = (-1)^{n-1} \frac{c^2}{EA} P(t). \tag{8.66}$$

Recalling the initial conditions (8.59), the solution of (8.66) can be written

$$\bar{u}_s(n, t) = \frac{(-1)^{n-1} 2Lc}{EA(2n - 1)\pi} \int_0^t P(\tau) \sin (2n - 1) \frac{\pi c}{2L} (t - \tau) \, d\tau. \tag{8.67}$$

The corresponding inversion formula is

$$u(x, t) = \frac{2}{L} \sum_{n=1}^{\infty} \bar{u}_s(n, t) \sin (2n - 1) \frac{\pi x}{2L}, \tag{8.68}$$

which yields

$$u\,(x,\,t) = \frac{4c}{\pi EA} \sum_{n=1}^{\infty} \frac{(-1)^{n-1}}{2n-1} \sin\,(2n-1)\frac{\pi x}{2L}$$

$$\times \int_0^t P(\tau)\sin\,(2n-1)\frac{\pi c}{2L}\,(t-\tau)\,d\tau. \quad (8.69)$$

It appears that this equation represents a normal modes solution.

As an illustration, let the prescribed force be in the form of a step function

$$P(t) = P_0 \alpha(t), \tag{8.70}$$

where $\alpha(t)$ is the unit step function. Introducing (8.70) in (8.69) we obtain

$$u(x,\,t) = \frac{8P_0L}{\pi^2 EA} \sum_{n=1}^{\infty} \frac{(-1)^{n-1}}{(2n-1)^2} \sin\,(2n-1)\frac{\pi x}{2L}\left[1-\cos\,(2n-1)\frac{\pi c}{2L}\,t\right].$$

$$\tag{8.71}$$

It is not difficult to show the following Fourier series representation:

$$\sum_{n=1}^{\infty} \frac{(-1)^{n-1}}{(2n-1)^2}\sin\,(2n-1)\frac{\pi x}{2L} = \frac{\pi^2}{8L}\,x, \qquad 0 < x < L,$$

so that (8.71) reduces to

$$u(x,\,t) = \frac{P_0}{EA}\,x - \frac{8P_0L}{\pi^2 EA}\sum_{n=1}^{\infty}\frac{(-1)^{n-1}}{(2n-1)^2}\sin\,(2n-1)\frac{\pi x}{2L}\cos\,(2n-1)\frac{\pi c}{2L}\,t.$$

$$\tag{8.72}$$

The first term on the right side of (8.72) is just the static solution, whereas the series represents a superposition of normal modes of amplitudes inversely proportional to $(2n-1)^2$. This shows that the amplitude of the second mode is about 11 per cent of the amplitude of the first mode and the amplitude of the third mode is 4 per cent of the first.

(b) Standing Wave Solution by Laplace Transformation

Using the Laplace transform method to solve the wave equation, it is possible to obtain a solution in terms of standing or traveling waves. The type of solution obtained depends on the manner in which the inverse transformation is carried out.

The Laplace transform of (8.57), in view of the initial conditions (8.59), is

$$\frac{d^2\bar{u}(x, s)}{dx^2} - \left(\frac{s}{c}\right)^2 \bar{u}(x, s) = 0, \tag{8.73}$$

where $\bar{u}(x, s)$ is the transformed displacement. In writing (8.73) it was assumed that

$$\mathscr{L} \frac{\partial^2 u(x, t)}{\partial x^2} = \int_0^\infty e^{-st} \frac{\partial^2 u(x, t)}{\partial x^2} \, dt = \frac{\partial^2}{\partial x^2} \int_0^\infty e^{-st} u(x, t) \, dt = \frac{d^2\bar{u}(x, s)}{dx^2},$$

which implies that the function $e^{-st}u(x, t)$ is such that interchange of the order of differentiation with respect to x and integration with respect to t is possible.† The function $\bar{u}(x, s)$ is subject to the transformed boundary conditions

$$\bar{u}(0, s) = 0, \tag{8.74}$$

$$EA \frac{d\bar{u}(x, s)}{dx} \bigg|_{x=L} = \bar{P}(s). \tag{8.75}$$

The general solution of (8.73) is

$$\bar{u}(x, s) = c_1 e^{(s/c)x} + c_2 e^{-(s/c)x}, \tag{8.76}$$

and if (8.74) and (8.75) are used, the solution reduces to

$$\bar{u}(x, s) = \frac{c}{EA} \frac{\sinh (s/c)x}{s \cosh (s/c)L} \bar{P}(s) = \frac{c}{EA} \bar{f}(x, s)\bar{P}(s), \tag{8.77}$$

where

$$\bar{f}(x, s) = \frac{\sinh (s/c)x}{s \cosh (s/c)L} = \frac{-i \sin i(s/c)x}{s \cos i(s/c)L}. \tag{8.78}$$

Equation (8.77) may be inverted by means of Borel's theory by regarding x as a parameter. It appears that the function $\bar{f}(x, s)$ has simple poles at the roots of $\cos i(s/c)L = 0$, or

$$i \frac{s}{c} L = \pm (2n - 1) \frac{\pi}{2}, \qquad n = 1, 2, \ldots. \tag{8.79}$$

Note that $s = 0$ is not a pole as one might be tempted to think. (The reader should verify this statement.)

---

† The conditions under which this interchange is possible are discussed in R. V. Churchill, *Operational Mathematics*, 2nd ed., McGraw-Hill, New York, 1958, Sec. 12.

But, in general, if

$$\bar{f}(s) = \frac{A(s)}{B(s)}, \tag{8.80}$$

where $B(s)$ has simple poles at $s = a_n$, it can be shown† that the inverse transformation of $\bar{f}(s)$ is

$$f(t) = \sum_n \lim_{s \to a_n} [(s - a_n)\bar{f}(s)]e^{st} = \sum_n \frac{A(a_n)}{B'(a_n)} e^{a_n t}, \tag{8.81}$$

provided $B'(a_n) = (d/ds)B(s)|_{s=a_n}$ does not vanish. Hence in our case

$$
\begin{aligned}
f(x, t) &= \sum_{n=1}^{\infty} \frac{-i \sin i(s/c)x}{\cos i(s/c)L - i(s/c)L \sin i(s/c)L} e^{st}\Big|_{i(s/c)L = (2n-1)\pi/2} \\
&+ \sum_{n=1}^{\infty} \frac{-i \sin i(s/c)x}{\cos i(s/c)L - i(s/c)L \sin i(s/c)L} e^{st}\Big|_{i(s/c)L = -(2n-1)\pi/2} \\
&= \frac{4}{\pi} \sum_{n=1}^{\infty} \frac{(-1)^{n-1}}{2n-1} \sin (2n-1)\frac{\pi x}{2L} \sin (2n-1)\frac{\pi c}{2L} t, \tag{8.82}
\end{aligned}
$$

so that by using Borel's theorem we obtain the response

$$
\begin{aligned}
u(x, t) &= \frac{c}{EA} \int_0^t P(\tau)f(x, t - \tau)\, d\tau \\
&= \frac{4c}{\pi EA} \sum_{n=1}^{\infty} \frac{(-1)^{n-1}}{2n-1} \sin (2n-1)\frac{\pi x}{2L} \int_0^t P(\tau) \sin (2n-1)\frac{\pi c}{2L}(t - \tau)\, d\tau, \tag{8.83}
\end{aligned}
$$

which is identical with (8.69), where the latter was obtained by means of a finite sine transform.

(c) Traveling Wave Solution by Laplace Transformation

Let us now write the function $\bar{f}(x, s)$ of (8.78) in a different form:

$$
\begin{aligned}
\bar{f}(x, s) &= \frac{\sinh (s/c)x}{s \cosh (s/c)L} = \frac{1}{s} \frac{e^{(s/c)x} - e^{-(s/c)x}}{e^{(s/c)L} + e^{-(s/c)L}} = \frac{e^{-(s/c)L}}{s} \frac{e^{(s/c)x} - e^{-(s/c)x}}{1 + e^{-2(s/c)L}} \\
&= \frac{e^{-(s/c)L}}{s} (e^{(s/c)x} - e^{-(s/c)x})(1 - e^{-2(s/c)L} + e^{-4(s/c)L} - e^{-6(s/c)L} + \cdots) \\
&= \frac{1}{s} [e^{-(s/c)(L - x)} - e^{-(s/c)(L + x)} - e^{-(s/c)(3L - x)} \\
&\qquad\qquad + e^{-(s/c)(3L + x)} + e^{-(s/c)(5L - x)} - \cdots]. \tag{8.84}
\end{aligned}
$$

---

† See W. T. Thomson, *Laplace Transformation*, 2nd ed., Prentice-Hall, Englewood Cliffs, N.J., 1960, p. 11; see also Sec. B–7 of this book.

According to the second shifting theorem (see Appendix B) the inverse transformation of (8.84) is

$$f(x, t) = u\left(t - \frac{L - x}{c}\right) - u\left(t - \frac{L + x}{c}\right) - u\left(t - \frac{3L - x}{c}\right) + \cdots, \quad (8.85)$$

where $u(t - (L - x)/c)$ denotes a unit step function which begins at $t = (L - x)/c$.

Using Borel's theorem, (8.77) and (8.85) yield the inverse transformation

$$u(x, t) = \frac{c}{EA} \int_0^t P(\tau) f(x, t - \tau)\, d\tau$$

$$= \frac{c}{EA} \int_0^t P(\tau) \left[u\left(t - \frac{L - x}{c} - \tau\right) - u\left(t - \frac{L + x}{c} - \tau\right) \right.$$

$$\left. - u\left(t - \frac{3L - x}{c} - \tau\right) + \cdots \right] d\tau. \quad (8.86)$$

But for $t < (L - x)/c$ we have

$$u\left(t - \frac{L - x}{c} - \tau\right) = 0 \qquad \text{for } \tau > 0$$

and for $t > (L - x)/c$,

$$u\left(t - \frac{L - x}{c} - \tau\right) = 1 \qquad \text{for } 0 < \tau < t - \frac{L - x}{c},$$

$$= 0 \qquad \text{for } \tau > t - \frac{L - x}{c}.$$

It follows that (8.86) can be written

$$u(x, t) = \frac{c}{EA} \left\{ u\left(t - \frac{L - x}{c}\right) \int_0^{t - (L - x)/c} P(\tau)\, d\tau \right.$$

$$- u\left(t - \frac{L + x}{c}\right) \int_0^{t - (L + x)/c} P(\tau)\, d\tau$$

$$- u\left(t - \frac{3L - x}{c}\right) \int_0^{t - (3L - x)/c} P(\tau)\, d\tau$$

$$\left. + u\left(t - \frac{3L + x}{c}\right) \int_0^{t - (3L + x)/c} P(\tau)\, d\tau + \cdots \right\}, \quad (8.87)$$

which is a traveling wave solution. The times $t = (L - x)/c$, $(L + x)/c$, $(3L - x)/c$, etc., are the arrival times.† Equation (8.87) indicates that the

---

† The time at which the traveling wave arrives at point x.

response is zero for $t < (L - x)/c$. At $t = (L - x)/c$, which is the first arrival time, a displacement wave, called an *incident wave*, whose amplitude is given by the first term on the right side of (8.87) traveling in the negative x direction, arrives at point x. The wave continues traveling past point x until it reaches the fixed end, $x = 0$, at which point it is reflected in the form of a negative displacement wave, called a *reflected wave*, whose amplitude is given by the second term on the right side of (8.87). During the time interval $(L - x)/c \le t \le (L + x)/c$ only the incident wave is sensed at point x, but in the time interval $(L + x)/c \le t \le (3L - x)/c$ both the incident and the reflected wave are sensed. The negative reflected wave upon arrival at the end $x = L$, which is geometrically a free end, is reflected without any sign change, as indicated by the third term in (8.87). One can generalize by stating that a displacement wave is reflected at a fixed boundary as a displacement wave of opposite sign and is reflected at a free boundary as a displacement wave of the same sign. In other words, a displacement wave is canceled at a fixed end and doubled at a free end. The shape of the wave does not change.

The force at any point x can be obtained by writing

$$P(x, t) = EA \frac{\partial u(x, t)}{\partial x}$$

$$= c\left\{ \frac{1}{c} \delta\left(t - \frac{L - x}{c}\right) \int_0^{t-(L-x)/c} P(\tau)\, d\tau \right.$$

$$+ u\left(t - \frac{L - x}{c}\right) \frac{\partial}{\partial x} \int_0^{t-(L-x)/c} P(\tau)\, d\tau$$

$$- \frac{1}{c} \delta\left(t - \frac{L + x}{c}\right) \int_0^{t-(L+x)/c} P(\tau)\, d\tau$$

$$\left. - u\left(t - \frac{L + x}{c}\right) \frac{\partial}{\partial x} \int_0^{t-(L+x)/c} P(\tau)\, d\tau - \cdots \right\}. \quad (8.88)$$

Using the Leibnitz rule for differentiation under the integral sign we can write

$$\frac{\partial}{\partial x} \int_0^{t-(L \pm x)/c} P(\tau)\, d\tau = \mp \frac{1}{c} P\left(t - \frac{L \pm x}{c}\right),$$

so (8.88) reduces to

$$P(x, t) = P\left(t - \frac{L - x}{c}\right) u\left(t - \frac{L - x}{c}\right) + P\left(t - \frac{L + x}{c}\right) u\left(t - \frac{L + x}{c}\right)$$

$$- P\left(t - \frac{3L - x}{c}\right) u\left(t - \frac{3L - x}{c}\right)$$

$$- P\left(t - \frac{3L + x}{c}\right) u\left(t - \frac{3L + x}{c}\right) + \cdots, \quad (8.89)$$

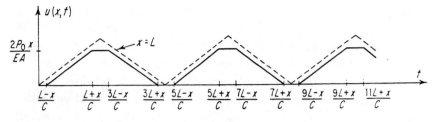

FIGURE 8.7

which expresses the force at any point as a superposition of traveling waves. Equation (8.89) indicates that the shape of the wave does not change. As opposed to the displacement wave, a force wave doubles at a fixed boundary and cancels out at a free boundary.

To gain further insight into the problem let us consider the case in which the force is the step function given by (8.70). Introducing (8.70) in the general solution for the displacement as given by (8.87) we immediately obtain

$$u(x, t) = \frac{P_0 c}{EA} \left[\left(t - \frac{L - x}{c} \right) u \left(t - \frac{L - x}{c} \right) - \left(t - \frac{L + x}{c} \right) u \left(t - \frac{L + x}{c} \right) \right.$$

$$- \left(t - \frac{3L - x}{c} \right) u \left(t - \frac{3L - x}{c} \right)$$

$$\left. + \left(t - \frac{3L + x}{c} \right) u \left(t - \frac{3L + x}{c} \right) + \cdots \right]. \quad (8.90)$$

Equation (8.90) is plotted in Figure 8.7 as a function of time.

Similarly, introducing (8.70) in (8.89) we obtain simply

$$P(x, t) = P_0 \left[u \left(t - \frac{L - x}{c} \right) + u \left(t - \frac{L + x}{c} \right) - u \left(t - \frac{3L - x}{c} \right) \right.$$

$$\left. - u \left(t - \frac{3L + x}{c} \right) + u \left(t - \frac{5L - x}{c} \right) + \cdots \right], \quad (8.91)$$

which is plotted in Figure 8.8.

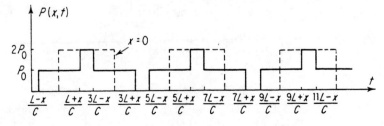

FIGURE 8.8

Equation (8.90) can be explained in terms of incident and reflected displacement waves by the scheme shown in Figure 8.9. The first wave on the right side represents the incident wave I; the first wave on the left side represents the reflected wave R_1 at the end $x = 0$. The second wave on the right side represents the reflection R_2 at the end $x = L$ of the reflected wave R_1. Similarly, R_3 is the reflection of R_2 at the end $x = 0$. The incident wave I and the reflected waves R_i ($i = 1, 2, 3, \ldots$) start traveling simultaneously with velocity c at the starting points and in the directions shown. Their

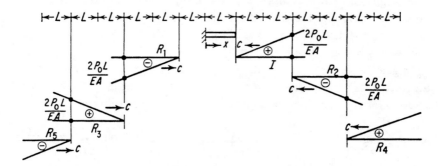

FIGURE 8.9

effects at a point x add linearly as they arrive. Note that at the fixed end the combination of incident and reflected waves results in zero displacement, whereas at the free end the displacement takes the maximum value $2P_0L/EA$ at times $t = 2L/c,\ 6L/c,\ 10L/c, \ldots$, and the minimum value zero at times $t = 0,\ 4L/c,\ 8L/c, \ldots$.

A similar scheme can be devised for force waves (see Problem 8.7).

8-7   WAVE MOTION OF A BAR IN BENDING VIBRATION

If the shear deformation and rotatory inertia effects are neglected, the differential equation for the free bending vibration of a bar takes the form

$$\frac{\partial^2}{\partial x^2}\left[EI(x)\frac{\partial^2 y(x, t)}{\partial x^2}\right] + m(x)\frac{\partial^2 y(x, t)}{\partial t^2} = 0, \tag{8.92}$$

and if the bar is uniform, (8.92) reduces to

$$\frac{\partial^4 y(x, t)}{\partial x^4} + \frac{1}{a^2}\frac{\partial^2 y(x, t)}{\partial t^2} = 0, \qquad a^2 = \frac{EI}{m}. \tag{8.93}$$

Equation (8.93) differs from the wave equation in two respects: It has a fourth derivative with respect to x instead of a second derivative; and the constant a does not possess dimensions of velocity and, therefore, does not represent a velocity. It is easy to see that the general solution of the wave equation, $y(x, t) = F_1(x - ct) + F_2(x + ct)$, *is not* a solution of (8.93), so we cannot conclude that, in general, the motion consists of waves traveling with constant velocity and without alteration of shape. The concept of wave velocity must be re-examined.

Let us assume that the solution of (8.93) is a simple harmonic wave traveling with velocity v in the positive x direction, so that its form is

$$y(x, t) = A \cos \frac{2\pi}{\lambda} (x - vt), \tag{8.94}$$

where λ is the wavelength. Substituting (8.94) in (8.93) we conclude that the velocity of propagation of the sinusoidal wave is

$$v = \frac{2\pi a}{\lambda} = \frac{2\pi}{\lambda} \sqrt{\frac{EI}{m}}, \tag{8.95}$$

and, therefore, the velocity of propagation of a sinusoidal flexural wave is not constant (as in the case of the transverse vibration of a string) but varies inversely as the wavelength. The velocity of propagation v of a simple harmonic wave is called the *wave velocity* or *phase velocity*. A nonharmonic flexural pulse may be regarded as consisting of a superposition of harmonic waves of different wavelengths. Each of these waves has a different phase velocity, so it follows that a flexural wave of arbitrary shape cannot propagate along a bar without *dispersion*, which results in a change in the shape of the pulse. A medium exhibiting a wave velocity $v(\lambda)$ depending on the wavelength is called a *dispersive medium*. The only flexural wave profile propagating in a uniform bar without altering its shape is the simple harmonic wave. A pulse consisting of a group of harmonic waves is called a *wave packet* and the velocity with which such a group of waves is propagated is called *group velocity* and denoted U. The group velocity is the velocity with which the energy is propagated. To illustrate the concept let us use the argument presented by Stephens and Bate.† To this end we consider a wave packet consisting of two simple harmonic waves of equal amplitudes, of wavelengths λ and $\lambda + \Delta\lambda$ and of wave velocities v and $v + \Delta v$, respectively. Let us denote by τ the time interval necessary for crest e to catch up with crest E, as shown in Figure 8.10. The new common positions are denoted e' and E', respectively.

† R. W. P. Stephens and A. E. Bate, *Wave Motion and Sound*, Arnold, London, 1950, p. 73.

Wave I travels with velocity v, wave II with velocity $v + \Delta v$, and the packet with group velocity U. From Figure 8.10 we obtain the two equations

$$\Delta \lambda + (v + \Delta v)\tau = v\tau, \qquad \lambda + v\tau = U\tau. \tag{8.96}$$

Combining (8.96) and letting the wavelength difference become very small, we obtain the group velocity expression

$$U = v - \lambda \frac{dv}{d\lambda}. \tag{8.97}$$

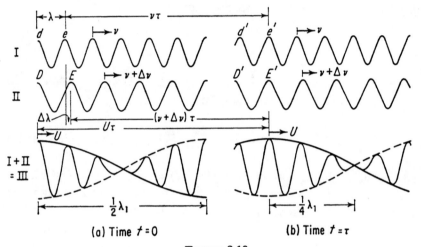

(a) Time $t = 0$ (b) Time $t = \tau$

FIGURE 8.10

The same result can be obtained by adding algebraically two simple harmonic waves of equal amplitude but with slightly different wavelength and phase velocity,

$$y_1(x, t) = A \cos \frac{2\pi}{\lambda} (x - vt) = A \cos (2\pi kx - \omega t),$$

$$y_2(x, t) = A \cos \frac{2\pi}{\lambda + \Delta\lambda} [x - (v + \Delta v)t]$$

$$= A \cos [2\pi(k + \Delta k)x - (\omega + \Delta\omega)t], \tag{8.98}$$

so that

$$y(x, t) = y_1(x, t) + y_2(x, t) = 2A \cos \tfrac{1}{2}(2\pi x \, \Delta k - t \, \Delta\omega) \cos (2\pi kx - \omega t)$$

$$= 2A \cos \frac{2\pi}{\lambda_1} (x - Ut) \cos \frac{2\pi}{\lambda} (x - vt), \tag{8.99}$$

where

$$\lambda_1 = \frac{2\lambda^2}{\Delta\lambda}, \qquad U = \frac{\Delta\omega}{2\pi \, \Delta k} = v - \lambda \frac{\Delta v}{\Delta\lambda}. \tag{8.100}$$

In deriving (8.99) and (8.100) it was assumed that the increments $\Delta\lambda$, Δk, and $\Delta\omega$ are small enough to be treated as differentials. The simple harmonic wave described by $\cos(2\pi/\lambda)(x - vt)$ is called the *carrier wave* and the factor $2A\cos(2\pi/\lambda_1)(x - Ut)$, representing the envelope in Figure 8.10 (III), can be regarded as a slowly varying amplitude moving forward with group velocity U. The wave as given by (8.99) is called an *amplitude-modulated carrier*. The motion can be interpreted as wavelets moving within the envelope with velocity v. When $U > v$, the wavelets are building up in front of the group and vanishing in the rear end of the group.

Again considering the propagation of flexural waves and introducing (8.95) in (8.97) we obtain

$$U = v - \lambda\frac{dv}{d\lambda} = v - \lambda\left(-\frac{2\pi a}{\lambda^2}\right) = 2v = \frac{4\pi}{\lambda}\sqrt{\frac{EI}{m}} \qquad (8.101)$$

and conclude that for flexural waves the group velocity is twice the wave velocity.

As the wavelength λ approaches zero, both the wave and group velocities approach infinity. The group velocity is the velocity with which energy propagates, so we must conclude that for a pulse consisting primarily of infinitely short waves, the energy is propagated with infinite velocity, which does not seem physically possible. The reason for this puzzling result lies in the fact that for very short waves one cannot neglect rotatory inertia. When the wavelength is of the same order of magnitude as the depth of the bar, the rotatory inertia effect is of the same order of magnitude as the translational inertia effect.

When the shear deformation effect is ignored but the rotatory inertia effect included in the formulation, the differential equation of motion becomes

$$\frac{\partial^2}{\partial x^2}\left[EI(x)\frac{\partial^2 y(x, t)}{\partial x^2}\right] + m(x)\frac{\partial^2 y(x, t)}{\partial t^2} - k^2(x)m(x)\frac{\partial^4 y(x, t)}{\partial x^2\,\partial t^2} = 0, \quad (8.102)$$

where $k(x)$ is the radius of gyration of the cross-sectional area. When the bar is uniform, (8.102) reduces to

$$\frac{\partial^4 y(x, t)}{\partial x^4} + \frac{1}{a^2}\frac{\partial^2 y(x, t)}{\partial t^2} - \frac{k^2}{a^2}\frac{\partial^4 y(x, t)}{\partial x^2\,\partial t^2} = 0, \qquad a^2 = \frac{EI}{m}. \quad (8.103)$$

Assuming a solution of (8.103) in the form of (8.94) we obtain the wave velocity

$$v = c\left(1 + \frac{\lambda^2}{4\pi^2 k^2}\right)^{-1/2}, \qquad c = \sqrt{\frac{E}{\rho}}, \qquad (8.104)$$

where c is the wave propagation velocity in the longitudinal vibration of bars. Similarly, the group velocity becomes

$$U = v - \lambda \frac{dv}{d\lambda} = c\left(1 + \frac{\lambda^2}{4\pi^2 k^2}\right)^{-1/2}\left(1 + \frac{1}{1 + 4\pi^2 k^2/\lambda^2}\right). \quad (8.105)$$

It is easy to see that as $\lambda \to 0$ both the wave and group velocities tend to the constant value c.

It appears that the wave propagation in a dispersive medium is a much more complex problem than the wave propagation in a nondispersive medium. We shall not pursue the topic further in this text, but the reader interested in this subject is referred to the survey paper by Miklowitz† for literature discussing this approach.

8-8 FREE BENDING VIBRATION OF AN INFINITE BAR

The differential equation for the free bending vibration of a uniform bar is

$$\frac{\partial^4 y(x, t)}{\partial x^4} + \frac{1}{a^2}\frac{\partial^2 y(x, t)}{\partial t^2} = 0, \qquad a^2 = \frac{EI}{m}, \quad (8.106)$$

and as pointed out previously, the constant a does not represent a velocity.

Let us consider an infinite bar, $-\infty < x < \infty$, with the initial conditions given by

$$y(x, 0) = f(x), \quad (8.107)$$

$$\left.\frac{\partial y(x, t)}{\partial t}\right|_{t=0} = a\frac{d^2 g(x)}{dx^2}, \quad (8.108)$$

where both $f(x)$ and $g(x)$ are given functions.

The beam is infinite, so there are no normal modes associated with the system and, therefore, modal analysis cannot be used to obtain the response. We wish to consider an integral transform solution, and to this end let us assume that the displacement and its derivatives up to third order are zero as $x \to \pm\infty$, which enables us to use a complex Fourier transform. Define the complex Fourier transform of $y(x, t)$ as

$$\bar{Y}(p, t) = \int_{-\infty}^{\infty} y(x, t)e^{ipx}\, dx, \quad (8.109)$$

† J. Miklowitz, *Appl. Mech. Rev.*, **13**, 865–878 (1960).

so that the transform of the fourth derivative with respect to x is

$$\overline{Y}^{(4)}(p, t) = \int_{-\infty}^{\infty} \frac{\partial^4 y(x, t)}{\partial x^4} e^{ipx}\, dx = p^4\, \overline{Y}(p, t). \qquad (8.110)$$

Transforming (8.106) we obtain

$$\frac{d^2\, \overline{Y}(p, t)}{dt^2} + a^2 p^4\, \overline{Y}(p, t) = 0, \qquad (8.111)$$

which is an ordinary differential equation subject to the transformed initial conditions

$$\overline{Y}(p, 0) = \overline{F}(p) = \int_{-\infty}^{\infty} y(x, 0)e^{ipx}\, dx = \int_{-\infty}^{\infty} f(x)e^{ipx}\, dx, \quad (8.112)$$

$$\frac{d\overline{Y}(p, t)}{dt}\bigg|_{t=0} = \int_{-\infty}^{\infty} \frac{\partial y(x, t)}{\partial t}\bigg|_{t=0} e^{ipx}\, dx = a \int_{-\infty}^{\infty} \frac{d^2 g(x)}{dx^2} e^{ipx}\, dx$$

$$= -ap^2 \int_{-\infty}^{\infty} g(x)e^{ipx}\, dx = -ap^2\overline{G}(p). \qquad (8.113)$$

Upon considering the transformed boundary conditions (8.112) and (8.113), the solution of (8.111) is

$$\overline{Y}(p, t) = \overline{F}(p) \cos ap^2 t - \overline{G}(p) \sin ap^2 t \qquad (8.114)$$

and the displacement $y(x, t)$ is obtain by inverting (8.114):

$$y(x, t) = \frac{1}{2\pi} \int_{-\infty}^{\infty} \overline{Y}(p, t)e^{-ipx}\, dp$$

$$= \frac{1}{2\pi} \int_{-\infty}^{\infty} \overline{F}(p) \cos (ap^2 t)e^{-ipx}\, dp - \frac{1}{2\pi} \int_{-\infty}^{\infty} \overline{G}(p) \sin (ap^2 t)e^{-ipx}\, dp.$$

$$(8.115)$$

Next consider the integral

$$\int_{-\infty}^{\infty} e^{-iap^2 t}e^{-ipx}\, dp = \int_{-\infty}^{\infty} \exp\left(-\left[\sqrt{iatp} + \frac{\sqrt{ix}}{\sqrt{4at}}\right]^2\right)e^{ix^2/4at}\, dp$$

$$= \frac{e^{-x^2/4iat}}{\sqrt{iat}} \int_{-\infty}^{\infty} \exp\left(-\left[\sqrt{iatp} + \frac{\sqrt{ix}}{\sqrt{4at}}\right]^2\right) d(\sqrt{iatp})$$

$$= \frac{e^{-x^2/4iat}}{\sqrt{iat}} \sqrt{\pi}, \qquad (8.116)$$

where the latter integral is recognized† as the *Gamma function* of argument $\frac{1}{2}$. But

$$\frac{1}{\sqrt{iat}} = \frac{1}{\sqrt{at}e^{i\pi/4}} = \frac{1}{\sqrt{at}}e^{-i(\pi/4)} = \frac{1-i}{\sqrt{2at}},$$

$$e^{-x^2/4iat} = e^{ix^2/4at} = \cos\frac{x^2}{4at} + i\sin\frac{x^2}{4at}.$$

Hence (8.116) reduces to

$$\int_{-\infty}^{\infty} e^{-iap^2t}e^{-ipx}\,dp = \int_{-\infty}^{\infty} \cos\,(ap^2t)e^{-ipx}\,dp - i\int_{-\infty}^{\infty} \sin\,(ap^2t)e^{-ipx}\,dp$$

$$= \sqrt{\frac{\pi}{2at}}\left[\left(\cos\frac{x^2}{4at} + \sin\frac{x^2}{4at}\right) - i\left(\cos\frac{x^2}{4at} - \sin\frac{x^2}{4at}\right)\right],$$

$$(8.117)$$

from which we conclude that

$$\int_{-\infty}^{\infty} \cos\,(ap^2t)e^{-ipx}\,dp = \sqrt{\frac{\pi}{2at}}\left(\cos\frac{x^2}{4at} + \sin\frac{x^2}{4at}\right), \qquad (8.118)$$

$$\int_{-\infty}^{\infty} \sin\,(ap^2t)e^{-ipx}\,dp = \sqrt{\frac{\pi}{2at}}\left(\cos\frac{x^2}{4at} - \sin\frac{x^2}{4at}\right). \qquad (8.119)$$

The integrals in (8.115) can be evaluated by means of the convolution theorem. To this end we use (8.118) and (8.119) and write

$$y(x, t) = \frac{1}{2\pi}\sqrt{\frac{\pi}{2at}}\left[\int_{-\infty}^{\infty} f(x - \eta)\left(\cos\frac{\eta^2}{4at} + \sin\frac{\eta^2}{4at}\right)d\eta\right.$$

$$\left. - \int_{-\infty}^{\infty} g(x - \eta)\left(\cos\frac{\eta^2}{4at} - \sin\frac{\eta^2}{4at}\right)d\eta\right]$$

$$= \frac{1}{\sqrt{2\pi}}\left[\int_{-\infty}^{\infty} f(x - 2a^{1/2}t^{1/2}u)(\cos u^2 + \sin u^2)\,du\right.$$

$$\left. - \int_{-\infty}^{\infty} g(x - 2a^{1/2}t^{1/2}u)(\cos u^2 - \sin u^2)\,du\right], \quad (8.120)$$

where the substitution $u^2 = \eta^2/4at$ has been made. This result was first obtained by Boussinesq.‡

As an illustration let us consider the initial conditions

$$f(x) = y_0e^{-x^2/4x_0^2}, \qquad g(x) = 0, \qquad (8.121)$$

---

† See F. B. Hildebrand, *Advanced Calculus for Engineers*, Prentice-Hall, Englewood Cliffs, N.J., 1960, p. 83.

‡ J. Boussinesq, *Application des potentiels*, Gauthier-Villars, Paris, 1885, p. 463.

for which it is possible to evaluate the integrals in (8.120). Introducing (8.121) in (8.120) the response becomes

$$y(x, t) = \frac{y_0}{\sqrt{2\pi}} \int_{-\infty}^{\infty} e^{-(x - 2a^{1/2}t^{1/2}u)^2/4x_0^2} (\cos u^2 + \sin u^2) \, du. \qquad (8.122)$$

Next consider the integral

$$I(x, t) = \int_{-\infty}^{\infty} e^{-(x - 2a^{1/2}t^{1/2}u)^2/4x_0^2 e^{iu^2}} \, du$$

$$= \int_{-\infty}^{\infty} e^{-[(x - 2a^{1/2}t^{1/2}u)^2 - 4iu^2 x_0^2]/4x_0^2} \, du$$

$$= e^{ix^2/(4at - 4ix_0^2)} \int_{-\infty}^{\infty} e^{-[u - 2xa^{1/2}t^{1/2}/(4at - 4ix_0^2)]^2 (at - ix_0^2)/x_0^2} \, du$$

$$= e^{ix^2/(4at - 4ix_0^2)} \frac{\sqrt{\pi} x_0}{\sqrt{at - ix_0^2}}$$

$$= e^{ix^2(at + ix_0^2)/4(a^2t^2 + x_0^4)} \frac{\sqrt{\pi} x_0}{\sqrt{at - ix_0^2}}.$$

But

$$(at - ix_0^2)^{1/2} = (-i)^{1/2}(x_0^2 + iat)^{1/2} = (e^{-\pi/2})^{1/2}(a^2t^2 + x_0^4)^{1/4} e^{i(1/2)\tan^{-1}(at/x_0^2)}$$

$$= \frac{1 - i}{\sqrt{2}} (a^2t^2 + x_0^4)^{1/4} e^{i(1/2)\tan^{-1}(at/x_0^2)},$$

so

$$I(x, t) = \int_{-\infty}^{\infty} e^{-(x - 2a^{1/2}t^{1/2}u)^2/4x_0^2 e^{iu^2}} \, du$$

$$= \sqrt{\frac{\pi}{2}} \frac{x_0(1 + i)}{(a^2t^2 + x_0^4)^{1/4}} e^{i[x^2(at + ix_0^2)/4(a^2t^2 + x_0^4) - (1/2)\tan^{-1}(at/x_0^2)]}$$

$$= \sqrt{\frac{\pi}{2}} \frac{e^{-x^2 x_0^2/4(a^2t^2 + x_0^4)}}{(1 + a^2t^2/x_0^4)^{1/4}} (1 + i) e^{i[x^2 at/4(a^2t^2 + x_0^4) - (1/2)\tan^{-1}(at/x_0^2)]}.$$

The integral in (8.122) is obtained by adding the real part and the imaginary part of the last expression,

$$y(x, t) = \frac{y_0}{\sqrt{2\pi}} [\text{Re } I(x, t) + \text{Im } I(x, t)]$$

$$= \frac{y_0 e^{-x^2 x_0^2/4(a^2t^2 + x_0^4)}}{(1 + a^2t^2/x_0^4)^{1/4}} \cos \left[\frac{x^2 at}{4(a^2t^2 + x_0^4)} - \frac{1}{2} \tan^{-1} \left(\frac{at}{x_0^2} \right) \right]. \qquad (8.123)$$

The response, (8.123), can be plotted as a function of x for various values of t (see Figure 8.11).

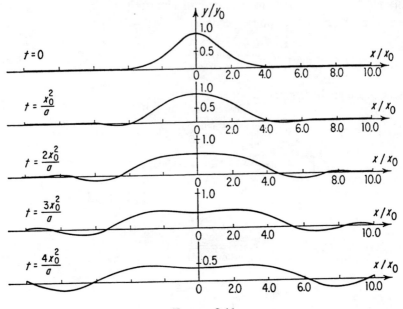

FIGURE 8.11

The response of a bar to an initial displacement $f(x)$ is quite different from that of a string. In the case of a string the response consists of the superposition of two waves, $\frac{1}{2}f(x - ct)$ and $\frac{1}{2}f(x + ct)$, traveling in opposite directions without alteration of shape. In the case of a bar one can regard the response as two packets of waves traveling in opposite directions. Because the harmonic components travel at speeds inversely proportional to their wavelengths, dispersion occurs, as mentioned in Section 8.7. One must note the ripples forming in the front of the wave packets. These are due to the short wavelength components, which travel faster than the longer wavelength components.

8-9   SEMI-INFINITE BEAM WITH PRESCRIBED END MOTION

Consider a semi-infinite beam, $0 \leq x < \infty$, with the end $x = 0$ hinged and undergoing a prescribed motion such that the boundary conditions at that end are

$$y(0, t) = f(t), \tag{8.124}$$

$$EI(x) \left. \frac{\partial^2 y(x, t)}{\partial x^2} \right|_{x=0} = 0. \tag{8.125}$$

Let the beam be originally at rest, so that the initial conditions are

$$y(x, 0) = \frac{\partial y(x, t)}{\partial t}\bigg|_{t=0} = 0. \tag{8.126}$$

If the bar is uniform and there are no external forces present, the displacement $y(x, t)$ must satisfy the differential equation (8.106).

Let us attempt a solution of the problem by means of a Fourier transformation. The bar is semi-infinite and the even order derivatives at the end $x = 0$ are known, so a Fourier sine transform appears in order. In view of (8.124) and (8.125) the Fourier sine transform of the fourth derivative of $y(x, t)$ with respect to x is

$$\overline{Y}_s^{(4)}(p, t) = \int_0^\infty \frac{\partial^4 y(x, t)}{\partial x^4} \sin px \, dx$$

$$= p \frac{\partial^2 y(x, t)}{\partial x^2}\bigg|_{x=0} - p^3 y(0, t) + p^4 \overline{Y}_s(p, t)$$

$$= -p^3 f(t) + p^4 \overline{Y}_s(p, t), \tag{8.127}$$

where $\overline{Y}_s(p, t)$ is the Fourier sine transform of $y(x, t)$. Hence, transforming (8.106), we obtain

$$\frac{d^2 \overline{Y}_s(p, t)}{dt^2} + a^2 p^4 \overline{Y}_s(p, t) = a^2 p^3 f(t), \tag{8.128}$$

which is an ordinary differential equation. The initial conditions are zero, so the solution of (8.128) is readily obtained in the form of the convolution integral

$$\overline{Y}_s(p, t) = ap \int_0^t f(\tau) \sin ap^2(t - \tau) \, d\tau. \tag{8.129}$$

The response is obtained by writing the inverse Fourier sine transform of (8.129),

$$y(x, t) = \frac{2}{\pi} \int_0^\infty \overline{Y}_s(p, t) \sin px \, dp$$

$$= \frac{2a}{\pi} \int_0^t f(\tau) \left[\int_0^\infty p \sin ap^2(t - \tau) \sin px \, dp \right] d\tau. \tag{8.130}$$

The real part of (8.119) (the imaginary part is zero) can be differentiated with respect to x within the integral sign, because the variable of integration is p. If this is done and if we note that the resulting integrand is an even function of p, we obtain

$$\int_0^\infty p \sin ap^2 t \sin px \, dp = \frac{\pi^{1/2} x}{2(2at)^{3/2}} \left(\sin \frac{x^2}{4at} + \cos \frac{x^2}{4at} \right), \quad (8.131)$$

so (8.130) reduces to

$$y(x, t) = \frac{x}{2} \frac{1}{(2\pi a)^{1/2}} \int_0^t f(\tau) \frac{1}{(t - \tau)^{3/2}} \left[\sin \frac{x^2}{4a(t - \tau)} + \cos \frac{x^2}{4a(t - \tau)} \right] d\tau.$$
$$(8.132)$$

Finally, letting $x^2/2a(t - \tau) = u^2$, (8.132) becomes

$$y(x, t) = \frac{1}{\sqrt{\pi}} \int_{x/(2at)^{1/2}}^\infty f\left(t - \frac{x^2}{2au^2} \right) (\sin \tfrac{1}{2} u^2 + \cos \tfrac{1}{2} u^2) \, du. \quad (8.133)$$

As was the general solution of Section 8–8, (8.120), solution (8.133) was also first obtained by Boussinesq.†

8–10 BENDING VIBRATION OF A BAR HINGED AT BOTH ENDS AND WITH A MOMENT APPLIED AT ONE END

Consider the case of a uniform bar hinged at both ends such that the supports undergo no transverse motion. In addition, let a time-dependent bending moment be applied at the end $x = L$ so that the boundary conditions are

$$y(x, t) = 0 \quad \text{at } x = 0, L, \quad (8.134)$$

$$EI(x) \frac{\partial^2 y(x, t)}{\partial x^2} \bigg|_{x=0} = 0, \quad (8.135)$$

$$EI(x) \frac{\partial^2 y(x, t)}{\partial x^2} \bigg|_{x=L} = M(t). \quad (8.136)$$

The bar is uniform, so the differential equation of motion remains in the form given by (8.106).

We wish to solve the problem by means of an integral transform method rather than by the method of Section 7–14. Because of the nature of the

† J. Boussinesq, *Application des potentiels*, Gauthier-Villars, Paris, 1885, p. 445.

boundary conditions, we shall attempt a solution by means of a finite sine transform defined by

$$\bar{y}_s(n, t) = \int_0^L y(x, t) \sin \frac{n\pi x}{L} \, dx. \tag{8.137}$$

The transform of the fourth derivative is

$$\bar{y}_s^{(4)}(n, t) = \int_0^L \frac{\partial^4 y(x, t)}{\partial x^4} \sin \frac{n\pi x}{L} \, dx$$

$$= \frac{n\pi}{L} \left[(-1)^{n+1} \frac{\partial^2 y(x, t)}{\partial x^2} \bigg|_{x=L} + \frac{\partial^2 y(x, t)}{\partial x^2} \bigg|_{x=0} \right]$$

$$- \left(\frac{n\pi}{L} \right)^3 [(-1)^{n+1} y(L, t) + y(0, t)] + \left(\frac{n\pi}{L} \right)^4 \bar{y}_s(n, t)$$

$$= (-1)^{n+1} \frac{n\pi}{EIL} M(t) + \left(\frac{n\pi}{L} \right)^4 \bar{y}_s(n, t). \tag{8.138}$$

A transformation of (8.106) leads to

$$\frac{d^2 \bar{y}_s(n, t)}{dt^2} + \left(\frac{n^2 \pi^2 a}{L^2} \right)^2 \bar{y}_s(n, t) = (-1)^n \frac{n\pi a^2}{EIL} M(t). \tag{8.139}$$

Assuming the bar is initially at rest, (8.139) has the solution

$$\bar{y}_s(n, t) = \frac{(-1)^n L a}{n\pi EI} \int_0^t M(\tau) \sin \frac{n^2 \pi^2 a}{L^2} (t - \tau) \, d\tau. \tag{8.140}$$

The response is obtained by taking the inverse transformation of (8.140),

$$y(x, t) = \frac{2}{L} \sum_{n=1}^{\infty} \bar{y}_s(n, t) \sin \frac{n\pi x}{L}$$

$$= \frac{2a}{\pi EI} \sum_{n=1}^{\infty} \frac{(-1)^n}{n} \sin \frac{n\pi x}{L} \int_0^t M(\tau) \sin \frac{n^2 \pi^2 a}{L} (t - \tau) \, d\tau. \tag{8.141}$$

8-11 FREE VIBRATION OF AN INFINITE MEMBRANE. FOURIER TRANSFORM SOLUTION

The differential equation for the free vibration of a thin uniform membrane, subjected to constant uniform tension T, is the two-dimensional wave equation which, in rectangular coordinates, has the form

$$\frac{\partial^2 w(x, y, t)}{\partial x^2} + \frac{\partial^2 w(x, y, t)}{\partial y^2} = \frac{1}{c^2} \frac{\partial^2 w(x, y, t)}{\partial t^2}, \qquad c^2 = \frac{T}{m}, \tag{8.142}$$

where w is the transverse displacement of the membrane, m the constant mass per unit area of membrane, and c the wave velocity.

Let us consider the case in which the membrane is initially imparted a displacement and a velocity of the form

$$w(x, y, 0) = f(x, y), \tag{8.143}$$

$$\left. \frac{\partial w(x, y, t)}{\partial t} \right|_{t=0} = g(x, y). \tag{8.144}$$

For a finite membrane it is theoretically possible to obtain the response in terms of the normal modes. In practice, however, it may be difficult to obtain the normal modes, because of the nature of the boundary conditions. If the membrane is infinite, the concept of normal modes loses its meaning and a different approach is advisable. We shall obtain the response of an infinite membrane by means of a double complex Fourier transform defined by

$$\bar{w}(\xi, \eta, t) = \int_{-\infty}^{\infty} \int_{-\infty}^{\infty} w(x, y, t) e^{i(\xi x + \eta y)} \, dx \, dy. \tag{8.145}$$

Assuming w, $\partial w/\partial x$, and $\partial w/\partial y$ tend to zero as x and y tend to infinity and assuming that the order of differentiation and integration is not important, we can write

$$\int_{-\infty}^{\infty} \int_{-\infty}^{\infty} \frac{\partial^2 w(x, y, t)}{\partial x^2} e^{i(\xi x + \eta y)} \, dx \, dy = \int_{-\infty}^{\infty} e^{i\xi x} \frac{\partial^2}{\partial x^2} \left[\int_{-\infty}^{\infty} w(x, y, t) e^{i\eta y} \, dy \right] dx$$

$$= \int_{-\infty}^{\infty} \frac{\partial^2 \bar{w}(x, \eta, t)}{\partial x^2} e^{i\xi x} \, dx$$

$$= -\xi^2 \bar{w}(\xi, \eta, t).$$

In a similar way we obtain

$$\int_{-\infty}^{\infty} \int_{-\infty}^{\infty} \frac{\partial^2 w(x, y, t)}{\partial y^2} e^{i(\xi x + \eta y)} \, dx \, dy = -\eta^2 \bar{w}(\xi, \eta, t),$$

$$\int_{-\infty}^{\infty} \int_{-\infty}^{\infty} \frac{\partial^2 w(x, y, t)}{\partial t^2} e^{i(\xi x + \eta y)} \, dx \, dy = \frac{d^2 \bar{w}(\xi, \eta, t)}{dt^2}.$$

Hence, transforming (8.142), we obtain the ordinary differential equation

$$\frac{d^2 \bar{w}(\xi, \eta, t)}{dt^2} + c^2(\xi^2 + \eta^2) \bar{w}(\xi, \eta, t) = 0, \tag{8.146}$$

which is subject to the transformed initial conditions

$$\bar{F}(\xi, \eta) = \bar{w}(\xi, \eta, 0) = \int_{-\infty}^{\infty} \int_{-\infty}^{\infty} f(x, y)e^{i(\xi x + \eta y)} \, dx \, dy, \tag{8.147}$$

$$\bar{G}(\xi, \eta) = \frac{d\bar{w}(\xi, \eta, t)}{dt}\bigg|_{t=0} = \int_{-\infty}^{\infty} \int_{-\infty}^{\infty} g(x, y)e^{i(\xi x + \eta y)} \, dx \, dy. \tag{8.148}$$

In view of (8.147) and (8.148) the solution of (8.146) is

$$\bar{w}(\xi, \eta, t) = \bar{F}(\xi, \eta) \cos [c(\xi^2 + \eta^2)^{1/2}t] + \frac{\bar{G}(\xi, \eta)}{c(\xi^2 + \eta^2)^{1/2}} \sin [c(\xi^2 + \eta^2)^{1/2}t], \tag{8.149}$$

and the response $w(x, y, t)$ is obtained by writing the inverse transformation defined by

$$w(x, y, t) = \frac{1}{(2\pi)^2} \int_{-\infty}^{\infty} \int_{-\infty}^{\infty} \bar{w}(\xi, \eta, t)e^{-i(\xi x + \eta y)} \, d\xi \, d\eta$$

$$= \frac{1}{(2\pi)^2} \int_{-\infty}^{\infty} \int_{-\infty}^{\infty} \bar{F}(\xi, \eta) \cos [c(\xi^2 + \eta^2)^{1/2}t]e^{-i(\xi x + \eta y)} \, d\xi \, d\eta$$

$$+ \frac{1}{(2\pi)^2} \int_{-\infty}^{\infty} \int_{-\infty}^{\infty} \frac{\bar{G}(\xi, \eta)}{c(\xi^2 + \eta^2)^{1/2}} \sin [c(\xi^2 + \eta^2)^{1/2}t]$$

$$\times e^{-i(\xi x + \eta y)} \, d\xi \, d\eta. \tag{8.150}$$

The general solution (8.150) can be simplified by using the convolution theorem (see Section C–7) and writing

$$w(x, y, t) = \int_{-\infty}^{\infty} \int_{-\infty}^{\infty} f(x', y')\varphi(x - x', y - y', t) \, dx' \, dy'$$

$$+ \int_{-\infty}^{\infty} \int_{-\infty}^{\infty} g(x', y')\psi(x - x', y - y', t) \, dx' \, dy', \tag{8.151}$$

where

$$\varphi(x, y, t) = \frac{1}{(2\pi)^2} \int_{-\infty}^{\infty} \int_{-\infty}^{\infty} \cos [c(\xi^2 + \eta^2)^{1/2}t]e^{-i(\xi x + \eta y)} \, d\xi \, d\eta, \tag{8.152}$$

$$\psi(x, y, t) = \frac{1}{(2\pi)^2} \int_{-\infty}^{\infty} \int_{-\infty}^{\infty} \frac{\sin [c(\xi^2 + \eta^2)^{1/2}t]}{c(\xi^2 + \eta^2)^{1/2}} e^{-i(\xi x + \eta y)} \, d\xi \, d\eta \tag{8.153}$$

are the Fourier transforms of the functions

$$\cos [c(\xi^2 + \eta^2)^{1/2}t] \quad \text{and} \quad \frac{\sin [c(\xi^2 + \eta^2)^{1/2}t]}{c(\xi^2 + \eta^2)^{1/2}},$$

respectively. These functions do not belong to the class of functions considered by the Fourier transform theorem, however, and we shall try to overcome this difficulty by multiplying these functions by a convergence factor $e^{-\epsilon(\xi^2 + \eta^2)^{1/2}}$, where ϵ is an arbitrary positive quantity. We shall thus consider the transforms of functions such as

$$e^{-\epsilon(\xi^2 + \eta^2)^{1/2}} \cos [c(\xi^2 + \eta^2)^{1/2}t] \quad \text{and} \quad \frac{e^{-\epsilon(\xi^2 + \eta^2)^{1/2}} \sin [c(\xi^2 + \eta^2)^{1/2}t]}{c(\xi^2 + \eta^2)^{1/2}}$$

and investigate their behavior as ϵ tends to zero. Before proceeding, however, let us make the following coordinate transformations:

$$\xi = \rho \cos \beta, \quad \eta = \rho \sin \beta, \quad d\xi \, d\eta = \rho \, d\rho \, d\beta$$

$$x = r \cos \theta, \quad y = r \sin \theta$$

which allow us to write

$$\xi x + \eta y = \rho r(\cos \beta \cos \theta + \sin \beta \sin \theta) = \rho r \cos (\beta - \theta).$$

Moreover, letting

$$\beta - \theta = \gamma, \quad d\beta = d\gamma,$$

we can write

$$\varphi(r, t) = \lim_{\epsilon \to 0} \frac{1}{(2\pi)^2} \int_0^{2\pi} \left[\int_0^\infty e^{-\epsilon\rho} \cos (c\rho t) e^{-i\rho r \cos \gamma} \rho \, d\rho \right] d\gamma, \quad (8.154)$$

$$\psi(r, t) = \lim_{\epsilon \to 0} \frac{1}{(2\pi)^2 c} \int_0^{2\pi} \left[\int_0^\infty e^{-\epsilon\rho} \sin (c\rho t) e^{-i\rho r \cos \gamma} \, d\rho \right] d\gamma. \quad (8.155)$$

Next consider†

$$\int_0^\infty e^{-\epsilon\rho} e^{\pm ic\rho t} e^{-i\rho r \cos \gamma} \rho \, d\rho = \int_0^\infty e^{[-\epsilon + i(\pm ct - r \cos \gamma)]\rho} \rho \, d\rho$$

$$= \frac{e^{[-\epsilon + i(\pm ct - r \cos \gamma)]\rho}}{[-\epsilon + i(\pm ct - r \cos \gamma)]^2}$$

$$\times \{[-\epsilon + i(\pm ct - r \cos \gamma)]\rho - 1\} \Big|_0^\infty$$

$$= \frac{1}{[-\epsilon + i(\pm ct - r \cos \gamma)]^2},$$

† See B. O. Peirce and R. M. Foster, *A Short Table of Integrals*, 4th ed., Ginn, Boston, 1957, Formula 413.

so we can write

$$\varphi(r, t) = -\frac{1}{2(2\pi)^2} \int_0^{2\pi} \left[\frac{1}{(ct - r \cos \gamma)^2} + \frac{1}{(ct + r \cos \gamma)^2} \right] d\gamma. \quad (8.156)$$

Similarly, consider

$$\int_0^\infty e^{-\epsilon\rho} e^{\pm ic\rho t} e^{-ir\rho \cos \gamma} \, d\rho = \frac{e^{[-\epsilon + i(\pm ct - r \cos \gamma)]\rho}}{-\epsilon + i(\pm ct - r \cos \gamma)} \Big|_0^\infty$$

$$= \frac{1}{\epsilon - i(\pm ct - r \cos \gamma)}$$

and obtain the expression

$$\psi(r, t) = \frac{1}{2i(2\pi)^2 c} \int_0^{2\pi} \left[\frac{1}{-i(ct - r \cos \gamma)} - \frac{1}{i(ct + r \cos \gamma)} \right] d\gamma$$

$$= \frac{1}{2c(2\pi)^2} \int_0^{2\pi} \left[\frac{1}{ct - r \cos \gamma} + \frac{1}{ct + r \cos \gamma} \right] d\gamma. \quad (8.157)$$

Performing the integration† with respect to γ in (8.156) and (8.157), we finally obtain

$$\varphi(x, y, t) = -\frac{1}{2(2\pi)^2} \int_0^{2\pi} \left[\frac{1}{(ct - r \cos \gamma)^2} + \frac{1}{(ct + r \cos \gamma)^2} \right] d\gamma$$

$$= -\frac{1}{2\pi} \frac{ct}{(c^2 t^2 - x^2 - y^2)^{3/2}} = \frac{1}{2\pi c} \frac{\partial}{\partial t} \frac{1}{(c^2 t^2 - x^2 - y^2)^{1/2}},$$

$$c^2 t^2 > x^2 + y^2, \quad (8.158)$$

$$\psi(x, y, t) = \frac{1}{2c(2\pi)^2} \int_0^{2\pi} \left(\frac{1}{ct - r \cos \gamma} + \frac{1}{ct + r \cos \gamma} \right) d\gamma$$

$$= \frac{1}{2\pi c} \frac{1}{(c^2 t^2 - x^2 - y^2)^{1/2}}, \quad c^2 t^2 > x^2 + y^2. \quad (8.159)$$

Introducing (8.158) and (8.159) in the convolution integral (8.151), we obtain

$$w(x, y, t) = \frac{1}{2\pi c} \left[\frac{\partial}{\partial t} \int_{-\infty}^\infty \int_{-\infty}^\infty \frac{f(x', y') \, dx' \, dy'}{[c^2 t^2 - (x - x')^2 - (y - y')^2]^{1/2}} \right.$$

$$\left. + \int_{-\infty}^\infty \int_{-\infty}^\infty \frac{g(x', y') \, dx' \, dy'}{[c^2 t^2 - (x - x')^2 - (y - y')^2]^{1/2}} \right],$$

$$c^2 t^2 > (x - x')^2 + (y - y')^2, \quad (8.160)$$

† Peirce and Foster, Formulas 317 and 309.

which is essentially the form obtained by Heins.† Making the substitution

$$x - x' = R \cos \vartheta,$$
$$y - y' = R \sin \vartheta,$$
$$dx'\, dy' = R\, dR\, d\vartheta,$$

we can rewrite (8.160)

$$w(x, y, t) = \frac{1}{2\pi c} \left[\frac{\partial}{\partial t} \int_0^{ct} \int_0^{2\pi} \frac{f(x', y')R\, dR\, d\vartheta}{(c^2 t^2 - R^2)^{1/2}} + \int_0^{ct} \int_0^{2\pi} \frac{g(x', y')R\, dR\, d\vartheta}{(c^2 t^2 - R^2)^{1/2}} \right].$$
(8.161)

The above expression is known as the *Poisson* and *Parseval formula*. The same result was derived by Webster‡ by different means.

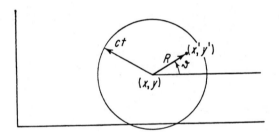

FIGURE 8.12

Equation (8.161) indicates that the disturbance of a membrane at the point (x, y) and time t depends on the initial values f and g not only at all points on the circumference of a circle of radius $R = ct$, but also on the values of f and g within that circle (see Figure 8.12). Hence the effect persists within that circle and while the crest of a traveling wave is losing amplitude it leaves a "tail" behind. This behavior, therefore, is different from the one encountered in the one-dimensional wave propagation, in which a pulse propagates with no change in shape and leaves no tail behind. The text by Morse contains some interesting illustrations comparing the propagation of pulses in a string and a membrane.§ A similar comparison can be found in Figure 8.13.

† A. E. Heins,"Applications of the Fourier Transform Theorem," *J. Math. Phys.*, **14**, 137 (1935).

‡ A. G. Webster, *Partial Differential Equations of Mathematical Physics*, Dover, New York, 1955, p. 187.

§ P. M. Morse, *Vibration and Sound*, 2nd ed., McGraw-Hill, New York, 1948, p. 184.

8–12 HANKEL TRANSFORM. RELATION BETWEEN HANKEL AND FOURIER TRANSFORMS

The two-dimensional complex Fourier transform is defined by

$$\bar{w}(\xi, \eta, t) = \int_{-\infty}^{\infty} \int_{-\infty}^{\infty} w(x, y, t)e^{i(\xi x + \eta y)} \, dx \, dy \tag{8.162}$$

and its inverse transformation has the form

$$w(x, y, t) = \frac{1}{(2\pi)^2} \int_{-\infty}^{\infty} \int_{-\infty}^{\infty} \bar{w}(\xi, \eta, t)e^{-i(\xi x + \eta y)} \, d\xi \, d\eta. \tag{8.163}$$

The transform pair, (8.162) and (8.163), can be used to construct another transform pair suitable for use in the case of a problem formulated in terms of polar coordinates. To this end let us make the following coordinate transformations:

$$x = r \cos \theta, \qquad y = r \sin \theta,$$
$$\xi = \rho \cos \beta, \qquad \eta = \rho \sin \beta,$$

so (8.162) can be written

$$\bar{w}(\rho, \beta, t) = \int_{0}^{\infty} \int_{0}^{2\pi} w(r, \theta, t)e^{ir\rho \cos(\theta - \beta)}r \, dr \, d\theta. \tag{8.164}$$

Let us consider the case in which the function $w(r, \theta, t)$ has the special form

$$w(r, \theta, t) = w(r, t)e^{-in\theta}, \tag{8.165}$$

where n is an integer, so that (8.164) becomes

$$\bar{w}(\rho, \beta, t) = \int_{0}^{\infty} rw(r, t) \left[\int_{0}^{2\pi} e^{-in\theta}e^{ir\rho \cos(\theta - \beta)} \, d\theta \right] dr. \tag{8.166}$$

Next let $\theta - \beta = \alpha - \pi/2$ and consider the integral

$$\int_{0}^{2\pi} e^{-in\theta}e^{ir\rho \cos(\theta - \beta)} \, d\theta = \int_{(\pi/2)-\beta}^{(5\pi/2)-\beta} e^{-in(\alpha + \beta - \pi/2)}e^{ir\rho \sin \alpha} \, d\alpha$$

$$= e^{in[(\pi/2)-\beta]} \int_{-\pi}^{\pi} e^{-in\alpha}e^{ir\rho \sin \alpha} \, d\alpha$$

$$= 2\pi e^{in[(\pi/2)-\beta]}J_n(r\rho), \tag{8.167}$$

where advantage has been taken of the fact that exponential functions are periodic and it was recognized that the latter integral in α is the integral representation† of the Bessel function $J_n(r\rho)$. Introducing (8.167) in (8.166) we have

$$\bar{w}(\rho, \beta, t) = 2\pi e^{in[(\pi/2) - \beta]} \int_0^\infty rw(r, t)J_n(r\rho)\, dr$$

$$= 2\pi e^{in[(\pi/2) - \beta]}\bar{w}(\rho, t), \tag{8.168}$$

where
$$\bar{w}(\rho, t) = \int_0^\infty rw(r, t)J_n(r\rho)\, dr. \tag{8.169}$$

Equation (8.163) can be written

$$w(r, \theta, t) = \frac{1}{(2\pi)^2} \int_0^\infty \int_0^{2\pi} \bar{w}(\rho, \beta, t)e^{-ir\rho\cos(\theta - \beta)}\rho\, d\rho\, d\beta$$

$$= \frac{1}{2\pi} \int_0^\infty \rho\bar{w}(\rho, t)\left[\int_0^{2\pi} e^{in[(\pi/2) - \beta]}e^{-ir\rho\cos(\theta - \beta)}\, d\beta\right] d\rho, \tag{8.170}$$

and making the substitution $\theta - \beta = -\gamma - \pi/2$, (8.170) reduces to

$$w(r, \theta, t) = \frac{1}{2\pi} \int_0^\infty \rho\bar{w}(\rho, t)\left[\int_{-(\pi/2) - \theta}^{(3\pi/2) - \theta} e^{-in(\theta + \gamma)}e^{ir\rho\sin\gamma}\, d\gamma\right] d\rho$$

$$= \frac{1}{2\pi} \int_0^\infty \rho\bar{w}(\rho, t)\left[e^{-in\theta} \int_{-\pi}^{\pi} e^{-in\gamma}e^{ir\rho\sin\gamma}\, d\gamma\right] d\rho$$

$$= e^{-in\theta} \int_0^\infty \rho\bar{w}(\rho, t)J_n(r\rho)\, d\rho. \tag{8.171}$$

Equation (8.169) is the definition of the Hankel transform of the function $w(r, t)$. From (8.165) and (8.171) we conclude that the inverse Hankel transformation is given by

$$w(r, t) = \int_0^\infty \rho\bar{w}(\rho, t)J_n(r\rho)\, d\rho. \tag{8.172}$$

8–13 RESPONSE OF AN INFINITE MEMBRANE. HANKEL TRANSFORM SOLUTION

The differential equation of motion of a thin uniform membrane, subjected to constant uniform tension T, can be written in terms of polar coordinates in the form

$$\frac{\partial^2 w(r, \theta, t)}{\partial r^2} + \frac{1}{r}\frac{\partial w(r, \theta, t)}{\partial r} + \frac{1}{r^2}\frac{\partial^2 w(r, \theta, t)}{\partial \theta^2} + \frac{p(r, \theta, t)}{T} = \frac{1}{c^2}\frac{\partial^2 w(r, \theta, t)}{\partial t^2}, \tag{8.173}$$

† See R. Courant and D. Hilbert, *Methods of Mathematical Physics*, Vol. I, Interscience, New York, 1961, p. 474.

where $p(r, \theta, t)$ is the external pressure. Let us consider the special case in which the external pressure is given by

$$p(r, \theta, t) = p(r, t)e^{-in\theta}, \tag{8.174}$$

where n is an integer and assume that the response $w(r, \theta, t)$ has the form

$$w(r, \theta, t) = w(r, t)e^{-in\theta}. \tag{8.175}$$

A substitution of (8.174) and (8.175) reduces (8.173) to

$$\frac{\partial^2 w(r, t)}{\partial r^2} + \frac{1}{r}\frac{\partial w(r, t)}{\partial r} - \frac{n^2}{r^2}w(r, t) + \frac{p(r, t)}{T} = \frac{1}{c^2}\frac{\partial^2 w(r, t)}{\partial t^2}. \tag{8.176}$$

Let us produce the solution of (8.176) by means of the Hankel transform. To this end it is necessary to calculate transforms of derivatives of $w(r, t)$. The transform of $\partial^2 w/\partial r^2$ can be obtained through an integration by parts,

$$\int_0^\infty r\frac{\partial^2 w(r, t)}{\partial r^2}J_n(\rho r)\,dr = r\frac{\partial w(r, t)}{\partial r}J_n(\rho r)\Big|_0^\infty - \int_0^\infty \frac{\partial w(r, t)}{\partial r}\frac{\partial}{\partial r}[rJ_n(\rho r)]\,dr$$

$$= r\frac{\partial w(r, t)}{\partial r}J_n(\rho r)\Big|_0^\infty$$

$$- \int_0^\infty \frac{\partial w(r, t)}{\partial r}\left[J_n(\rho r) + r\frac{\partial J_n(\rho r)}{\partial r}\right]dr.$$

If $w(r, t)$ is such that the product $r[\partial w(r, t)/\partial r]$ tends to zero as r tends to zero and infinity, we can rewrite the above expression

$$\int_0^\infty r\left[\frac{\partial^2 w(r, t)}{\partial r^2} + \frac{1}{r}\frac{\partial w(r, t)}{\partial r}\right]J_n(\rho r)\,dr = -\int_0^\infty r\frac{\partial w(r, t)}{\partial r}\frac{\partial J_n(\rho r)}{\partial r}\,dr.$$

One more integration by parts gives

$$\int_0^\infty r\frac{\partial w(r, t)}{\partial r}\frac{\partial J_n(\rho r)}{\partial r}\,dr = rw(r, t)\frac{\partial J_n(\rho r)}{\partial r}\Big|_0^\infty - \int_0^\infty w(r, t)\frac{\partial}{\partial r}\left[r\frac{\partial J_n(\rho r)}{\partial r}\right]dr$$

$$= -\int_0^\infty w(r, t)\left[\frac{\partial J_n(\rho r)}{\partial r} + r\frac{\partial^2 J_n(\rho r)}{\partial r^2}\right]dr.$$

But the Bessel function $J_n(z)$ satisfies the equation

$$J_n''(z) + \frac{1}{z}J_n'(z) + \left(1 - \frac{n^2}{z^2}\right)J_n(z) = 0,$$

where primes indicate differentiations with respect to the variable z, so we can write

$$\frac{\partial J_n(\rho r)}{\partial r} + r\frac{\partial^2 J_n(\rho r)}{\partial r^2} = -\rho^2 r\left(1 - \frac{n^2}{\rho^2 r^2}\right)J_n(\rho r).$$

It follows from the above that

$$\int_0^\infty r\left[\frac{\partial^2 w(r,t)}{\partial r^2} + \frac{1}{r}\frac{\partial w(r,t)}{\partial r} - \frac{n^2}{r^2}w(r,t)\right]J_n(\rho r)\,dr$$

$$= -\rho^2 \int_0^\infty rw(r,t)J_n(\rho r)\,dr = -\rho^2\bar{w}(\rho,t),$$

where $\bar{w}(\rho, t)$ is the Hankel transform of $w(r, t)$. Furthermore, we have

$$\int_0^\infty r\,\frac{\partial^2 w(r,t)}{\partial t^2}\,J_n(\rho r)\,dr = \frac{d^2\bar{w}(\rho,t)}{dt^2}.$$

Letting

$$\bar{p}(\rho, t) = \int_0^\infty rp(r,t)J_n(\rho r)\,dr \tag{8.177}$$

be the Hankel transform of $p(r, t)$, we can transform both sides of (8.176) and, after rearranging the terms, obtain the ordinary differential equation

$$\frac{d^2\bar{w}(\rho,t)}{dt^2} + c^2\rho^2\bar{w}(\rho,t) = \frac{c^2}{T}\bar{p}(\rho,t), \tag{8.178}$$

where ρ plays the role of a parameter.

Consider the case in which the initial conditions are of the special form

$$w(r, \theta, 0) = w(r, 0)e^{-in\theta} = f(r)e^{-in\theta},$$

$$\left.\frac{\partial w(r, \theta, t)}{\partial t}\right|_{t=0} = \left.\frac{\partial w(r, t)}{\partial t}\right|_{t=0}e^{-in\theta} = g(r)e^{-in\theta}, \tag{8.179}$$

and write the corresponding transforms

$$\bar{w}(\rho, 0) = \bar{f}(\rho) = \int_0^\infty rf(r)J_n(\rho r)\,dr,$$

$$\left.\frac{d\bar{w}(\rho, t)}{dt}\right|_{t=0} = \bar{g}(\rho) = \int_0^\infty rg(r)J_n(\rho r)\,dr. \tag{8.180}$$

In view of the transformed initial conditions (8.180), the solution of (8.178) can be written

$$\bar{w}(\rho, t) = \frac{c}{\rho T}\int_0^t \bar{p}(\rho, \tau)\sin c\rho(t - \tau)\,d\tau + \bar{f}(\rho)\cos c\rho t + \frac{\bar{g}(\rho)}{c\rho}\sin c\rho t, \tag{8.181}$$

which has the inverse transformation

$$w(r, t) = \int_0^\infty \rho\bar{w}(\rho, t)J_n(\rho r)\, d\rho = \frac{c}{T} \int_0^\infty J_n(\rho r)\left[\int_0^t \bar{p}(\rho, \tau) \sin c\rho(t - \tau)\, d\tau\right] d\rho$$

$$+ \int_0^\infty \rho\bar{f}(\rho) \cos (c\rho t)J_n(\rho r)\, d\rho + \frac{1}{c} \int_0^\infty \bar{g}(\rho) \sin (c\rho t)J_n(\rho r)\, d\rho.$$

$$(8.182)$$

The general solution, $w(r, \theta, t)$, is obtained by introducing (8.182) in (8.175).

As an illustration let us consider the response of an infinite membrane to an initial displacement depending only on the radial distance. Specifically,

$$p(r, \theta, t) = 0, \qquad \frac{\partial w(r, \theta, t)}{\partial t}\bigg|_{t=0} = 0, \qquad (8.183)$$

$$w(r, \theta, 0) = f(r) = \frac{A}{[1 + (r/a)^2]^{1/2}}, \qquad n = 0, \qquad (8.184)$$

so the response reduces to

$$w(r, \theta, t) = w(r, t) = \int_0^\infty \rho\bar{f}(\rho) \cos (c\rho t)J_0(\rho r)\, d\rho, \qquad (8.185)$$

where $\bar{f}(\rho)$ is the Hankel transform of $f(r)$ which, according to the first of equations (8.180), has the expression

$$\bar{f}(\rho) = \int_0^\infty rf(r)J_0(\rho r)\, dr = Aa \int_0^\infty \frac{r}{(a^2 + r^2)^{1/2}} J_0(\rho r)\, dr. \quad (8.186)$$

It can be shown,† however, that

$$\int_0^\infty e^{-a\rho}J_0(\rho r)\, d\rho = \frac{1}{(a^2 + r^2)^{1/2}}, \qquad (8.187)$$

and, if (8.187) is looked upon as a Hankel inversion formula, we conclude that

$$\bar{f}(\rho) = Aa \frac{e^{-a\rho}}{\rho}. \qquad (8.188)$$

---

† See N. W. McLachlan, *Bessel Functions for Engineers*, 2nd ed., Oxford Univ. Press, New York, 1955, p. 195, Formula 77.

Introducing (8.188) in (8.185), we obtain

$$w(r, t) = Aa \int_0^\infty e^{-a\rho} \cos (c\rho t) J_0(\rho r)\, d\rho$$

$$= \frac{Aa}{2} \int_0^\infty [e^{-\rho(a-ict)} + e^{-\rho(a+ict)}] J_0(\rho r)\, d\rho$$

$$= \frac{Aa}{2} \left\{ \frac{1}{[(a - ict)^2 + r^2]^{1/2}} + \frac{1}{[(a + ict)^2 + r^2]^{1/2}} \right\}$$

$$= \frac{A}{2} \frac{\left(1 + \dfrac{r^2 - c^2 t^2}{a^2} + \dfrac{2ict}{a}\right)^{1/2} + \left(1 + \dfrac{r^2 - c^2 t^2}{a^2} - \dfrac{2ict}{a}\right)^{1/2}}{\left[\left(1 + \dfrac{r^2 - c^2 t^2}{a^2}\right)^2 + \left(\dfrac{2ct}{a}\right)^2\right]^{1/2}}. \qquad (8.189)$$

Equation (8.189) can be given a more tractable form by making the substitution

$$\left[\left(1 + \frac{r^2 - c^2 t^2}{a^2}\right)^2 + \left(\frac{2ct}{a}\right)^2\right]^{1/2} = R,$$

$$1 + \frac{r^2 - c^2 t^2}{a^2} = R \cos 2\varphi, \qquad \frac{2ct}{a} = R \sin 2\varphi,$$

which reduces (8.189) to

$$w(r, t) = \frac{A}{2R} [R^{1/2}(\cos 2\varphi + i \sin 2\varphi)^{1/2} + R^{1/2}(\cos 2\varphi - i \sin 2\varphi)^{1/2}]$$

$$= \frac{A}{2R^{1/2}} (e^{i\varphi} + e^{-i\varphi}) = \frac{A}{R^{1/2}} \cos \varphi = \frac{A}{2^{1/2}} \left(\frac{1 + \cos 2\varphi}{R}\right)^{1/2}$$

$$= \frac{A}{2^{1/2}} \left\{ \frac{1}{\left[\left(1 + \dfrac{r^2 - c^2 t^2}{a^2}\right)^2 + \left(\dfrac{2ct}{a}\right)^2\right]^{1/2}} \right.$$

$$\left. + \frac{1 + \dfrac{r^2 - c^2 t^2}{a^2}}{\left(1 + \dfrac{r^2 - c^2 t^2}{a^2}\right)^2 + \left(\dfrac{2ct}{a}\right)^2} \right\}^{1/2}. \qquad (8.190)$$

The response $w(r, t)$ is plotted as a heavy line in Figure 8.13 for three values of time, $t = 0, 2a/c, 4a/c$. (Of course, the response for $t = 0$ is just the initial excitation.) For comparison purposes the response of a uniform string to the same initial excitation is plotted as a thin line.

We note that while the maximum displacement of the string remains $A/2$ as t tends to infinity, the displacement of the membrane decreases steadily as t increases and finally reduces to zero.

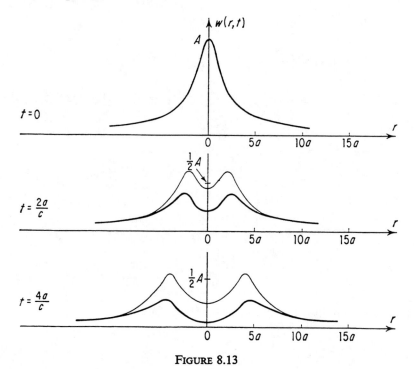

FIGURE 8.13

8–14   RESPONSE OF AN INFINITE PLATE. HANKEL TRANSFORM SOLUTION

The differential equation of motion of a thin uniform plate has the form

$$D_E \nabla^2 \nabla^2 w + m \frac{\partial^2 w}{\partial t^2} = p, \tag{8.191}$$

where D_E is the plate flexural rigidity, m the mass per unit area of plate, and p the transverse force per unit area of plate. Let us consider the case of an infinite plate and assume that the transverse force is given, in polar coordinates, by the product of a function of the radial distance and time, on the one hand, and a function of the angle θ, on the other. Specifically, let

$$p(r, \theta, t) = p(r, t)e^{-in\theta}, \tag{8.192}$$

where n is an integer, and, consequently, assume that the displacement $w(r, \theta, t)$ has a similar special form

$$w(r, \theta, t) = w(r, t)e^{-in\theta}. \tag{8.193}$$

Now consider

$$\nabla^2 w(r, \theta, t) = \left(\frac{\partial^2}{\partial r^2} + \frac{1}{r}\frac{\partial}{\partial r} + \frac{1}{r^2}\frac{\partial^2}{\partial \theta^2}\right)[w(r, t)e^{-in\theta}]$$

$$= e^{-in\theta}\left(\frac{\partial^2}{\partial r^2} + \frac{1}{r}\frac{\partial}{\partial r} - \frac{n^2}{r^2}\right)w(r, t) = W(r, t)e^{-in\theta}, \quad (8.194)$$

where

$$W(r, t) = \left(\frac{\partial^2}{\partial r^2} + \frac{1}{r}\frac{\partial}{\partial r} - \frac{n^2}{r^2}\right)w(r, t). \qquad (8.195)$$

It follows that

$$\nabla^2\nabla^2 w(r, \theta, t) = \left(\frac{\partial^2}{\partial r^2} + \frac{1}{r}\frac{\partial}{\partial r} + \frac{1}{r^2}\frac{\partial^2}{\partial \theta^2}\right)[W(r, t)e^{-in\theta}]$$

$$= e^{-in\theta}\left(\frac{\partial^2}{\partial r^2} + \frac{1}{r}\frac{\partial}{\partial r} - \frac{n^2}{r^2}\right)W(r, t). \qquad (8.196)$$

Defining a modified Laplace operator in the form

$$\Delta^2 = \frac{\partial^2}{\partial r^2} + \frac{1}{r}\frac{\partial}{\partial r} - \frac{n^2}{r^2}, \qquad (8.197)$$

and considering (8.192) and (8.193), (8.191) can be written

$$b^2\Delta^2\Delta^2 w(r, t) + \frac{\partial^2 w(r, t)}{\partial t^2} = \frac{p(r, t)}{m}, \qquad b^2 = \frac{D_E}{m}, \qquad (8.198)$$

and we note that the independent variable θ has been eliminated, which, of course, is possible because of the special type of load to which the plate is subjected. Furthermore, we shall assume that the initial displacement and velocity have the special forms

$$w(r, \theta, 0) = w(r, 0)e^{-in\theta} = f(r)e^{-in\theta}$$

$$\left.\frac{\partial w(r, \theta, t)}{\partial t}\right|_{t=0} = \left.\frac{\partial w(r, t)}{\partial t}\right|_{t=0}e^{-in\theta} = g(r)e^{-in\theta}, \qquad (8.199)$$

so, in effect, (8.198) is subject to the initial conditions $f(r)$ and $g(r)$.

We shall solve the problem by means of a Hankel transform, and to this end we recall from Section 8–13 that

$$\int_0^\infty r\left[\left(\frac{\partial^2}{\partial r^2} + \frac{1}{r}\frac{\partial}{\partial r} - \frac{n^2}{r^2}\right)z(r, t)\right]J_n(\rho r)\, dr = \int_0^\infty r\Delta^2 z(r, t)J_n(\rho r)\, dr = -\rho^2\bar{z}(\rho, t),$$

where $\bar{z}(\rho, t)$ is the Hankel transform of $z(r, t)$. This, in turn, enables us to write

$$\int_0^\infty r \Delta^2\Delta^2 w(r, t)J_n(\rho r)\, dr = \int_0^\infty r \Delta^2 W(r, t)J_n(\rho r)\, dr = -\rho^2\overline{W}(\rho, t)$$

$$= -\rho^2\int_0^\infty r \Delta^2 w(r, t)J_n(\rho r)\, dr = \rho^4\overline{w}(\rho, t).$$

Now transform both sides of (8.198) and obtain the ordinary differential equation

$$\frac{d^2\overline{w}(\rho, t)}{dt^2} + b^2\rho^4\overline{w}(\rho, t) = \frac{1}{m}\bar{p}(\rho, t), \qquad (8.200)$$

where $\bar{p}(\rho, t)$ is the Hankel transform of $p(r, t)$. Equation (8.200) is subject to the initial conditions $\bar{f}(\rho)$ and $\bar{g}(\rho)$, which are the Hankel transforms of $f(r)$ and $g(r)$, respectively. The solution of (8.200) is

$$\overline{w}(\rho, t) = \frac{1}{mb\rho^2}\int_0^t \bar{p}(\rho, \tau)\sin b\rho^2(t - \tau)\, d\tau + \bar{f}(\rho)\cos b\rho^2 t + \frac{\bar{g}(\rho)}{b\rho^2}\sin b\rho^2 t.$$

$$(8.201)$$

The response can be written in the general form

$$w(r, \theta, t) = w(r, t)e^{-in\theta} = e^{-in\theta}\int_0^\infty \rho\overline{w}(\rho, t)J_n(\rho r)\, d\rho, \qquad (8.202)$$

where the integral in (8.202) represents the inverse Hankel transform of the function $\overline{w}(\rho, t)$.

As an illustration let us obtain the response of an infinite plate to a radially symmetric, time-dependent, distributed force

$$p(r, \theta, t) = p(r, t) = P(t)p(r), \qquad n = 0, \qquad (8.203)$$

where $P(t)$ has units of force and $p(r)$ has units of $(\text{length})^{-2}$. The initial conditions are zero. The Hankel transform of the force is

$$\bar{p}(\rho, t) = \int_0^\infty rp(r, t)J_0(\rho r)\, dr$$

$$= P(t)\int_0^\infty \xi p(\xi)J_0(\rho\xi)\, d\xi = P(t)\bar{p}(\rho), \qquad (8.204)$$

where ξ is a dummy variable of integration replacing r and $\bar{p}(\rho)$ is the Hankel transform of $p(r)$. Introducing (8.204) in (8.201) and recalling that the initial conditions are zero, we obtain

$$\bar{w}(\rho, t) = \frac{1}{mb\rho^2} \int_0^t \bar{p}(\rho, \tau) \sin b\rho^2(t - \tau) \, d\tau = \frac{\bar{p}(\rho)}{mb\rho^2} \int_0^t P(\tau) \sin b\rho^2(t - \tau) \, d\tau.$$
(8.205)

The response is obtained by writing the inverse transformation

$$
\begin{aligned}
w(r, t) &= \int_0^\infty \rho \bar{w}(\rho, t) J_0(\rho r) \, d\rho \\
&= \frac{1}{mb} \int_0^\infty \frac{1}{\rho} \left[\int_0^\infty \xi p(\xi) J_0(\rho \xi) \, d\xi \int_0^t P(\tau) \sin b\rho^2(t - \tau) \, d\tau \right] J_0(\rho r) \, d\rho \\
&= \frac{1}{mb} \int_0^t P(\tau) \left\{ \int_0^\infty \xi p(\xi) \left[\int_0^\infty J_0(\rho r) J_0(\rho \xi) \sin b\rho^2(t - \tau) \, \frac{d\rho}{\rho} \right] d\xi \right\} d\tau.
\end{aligned}
$$
(8.206)

The general response (8.207) can be put into a simpler form. To this end write

$$\dot{w}(r, t) = \frac{1}{m} \int_0^t P(\tau) \left\{ \int_0^\infty \xi p(\xi) \left[\int_0^\infty \rho J_0(\rho r) J_0(\rho \xi) \cos b\rho^2(t - \tau) \, d\rho \right] d\xi \right\} d\tau,$$
(8.207)

where the Leibnitz rule for differentiation of an integral has been used. Next consider the integral

$$\int_0^\infty \rho J_n(\xi \rho) J_n(r\rho) e^{-k^2 \rho^2} \, d\rho = \frac{1}{2k^2} e^{-(\xi^2 + r^2)/4k^2} J_n\left(\frac{i\xi r}{2k^2}\right),$$

which is called *Webster's second exponential integral.*† Letting $n = 0$ and $k^2 = ib(t - \tau)$ in the above expression we obtain

$$\int_0^\infty \rho J_0(\xi \rho) J_0(r\rho) e^{-ib\rho^2(t-\tau)} \, d\rho = -\frac{i}{2b(t - \tau)} e^{i(\xi^2 + r^2)/4b(t-\tau)} J_0\left(\frac{\xi r}{2b(t - \tau)}\right),$$

and equating the real parts in both sides of the above equation we conclude that

$$\int_0^\infty \rho J_0(\xi \rho) J_0(r\rho) \cos b\rho^2(t - \tau) \, d\rho = \frac{1}{2b(t - \tau)} \sin \frac{\xi^2 + r^2}{4b(t - \tau)} J_0\left(\frac{\xi r}{2b(t - \tau)}\right).$$
(8.208)

† See G. N. Watson, *Theory of Bessel Functions*, 2nd ed., Macmillan, New York, 1948, p. 395.

Substitution of (8.208) in (8.207) yields

$$\dot{w}(r, t) = \frac{1}{2mb} \int_0^t \frac{P(\tau)}{t - \tau} \left[\int_0^\infty \xi p(\xi) J_0\!\left(\frac{\xi r}{2b(t - \tau)}\right) \sin \frac{\xi^2 + r^2}{4b(t - \tau)} \, d\xi \right] d\tau,$$

(8.209)

and the response can be written

$$w(r, t) = \int_0^t \dot{w}(r, \sigma) \, d\sigma.$$

(8.210)

As an example let us consider the case in which the load consists of a concentrated force applied at $r = 0$. Such a concentrated force can be represented as a distributed force of the form

$$p(r) = \delta(r),$$

(8.211)

where $\delta(r)$ is a two-dimensional Dirac's delta function defined by

$$\delta(r) = 0, \qquad r \neq 0,$$

$$\int_D \delta(r) \, dD(r) = 2\pi \int_0^\infty \delta(r) r \, dr = 1.$$

(8.212)

Note that $\delta(r)$ has units (length)$^{-2}$. Upon introducing (8.211) in (8.209), and in view of (8.212), we obtain

$$\dot{w}(r, t) = \frac{1}{2mb} \int_0^t \frac{P(\tau)}{t - \tau} \left[\int_0^\infty \xi \delta(\xi) J_0\!\left(\frac{\xi r}{2b(t - \tau)}\right) \sin \frac{\xi^2 + r^2}{4b(t - \tau)} \, d\xi \right] d\tau$$

$$= \frac{1}{4\pi mb} \int_0^t \frac{P(\tau)}{t - \tau} \sin \frac{r^2}{4b(t - \tau)} \, d\tau,$$

(8.213)

because $J_0(0) = 1$. Making the substitution

$$\frac{r^2}{4b(t - \tau)} = \alpha, \qquad \frac{d\tau}{t - \tau} = \frac{d\alpha}{\alpha},$$

we can rewrite (8.213)

$$\dot{w}(r, t) = \frac{1}{4\pi mb} \int_{r^2/4bt}^\infty P\!\left(t - \frac{r^2}{4b\alpha}\right) \frac{\sin \alpha}{\alpha} \, d\alpha,$$

(8.214)

and letting

$$R(t) = \int_0^t P(\sigma) \, d\sigma,$$

(8.215)

(8.214) leads to the response

$$w(r, t) = \frac{1}{4\pi mb} \int_{r^2/4bt}^{\infty} R\left(t - \frac{r^2}{4b\alpha}\right) \frac{\sin \alpha}{\alpha} \, d\alpha, \qquad (8.216)$$

which is essentially the result obtained by Boussinesq[†] by other means. Note that the functions P and R are defined only for a positive argument, so, in effect, the lower limits in (8.214) and (8.216) can be taken zero.

Let us now evaluate the response of the plate to a time-dependent variation in the form of a step function

$$P(t) = P_0 \mathscr{u}(t), \qquad (8.217)$$

so that

$$R(t) = P_0 t \mathscr{u}(t). \qquad (8.218)$$

Introducing (8.218) in (8.216) we obtain

$$w(r, t) = \frac{P_0}{4\pi mb} \int_{r^2/4bt}^{\infty} \left(t - \frac{r^2}{4b\alpha}\right) \frac{\sin \alpha}{\alpha} \, d\alpha$$

$$= \frac{P_0}{4\pi mb} \left[t \int_{r^2/4bt}^{\infty} \frac{\sin \alpha}{\alpha} \, d\alpha - \frac{r^2}{4b} \int_{r^2/4bt}^{\infty} \frac{\sin \alpha}{\alpha^2} \, d\alpha \right]$$

$$= \frac{P_0 t}{4\pi mb} \left[\frac{\pi}{2} - \text{Si}\left(\frac{r^2}{4bt}\right) + \frac{r^2}{4bt} \text{Ci}\left(\frac{r^2}{4bt}\right) - \sin \frac{r^2}{4bt} \right], \quad (8.219)$$

where the special functions

$$\text{Si}(x) = \int_0^x \frac{\sin \alpha}{\alpha} \, d\alpha = \sum_{n=0}^{\infty} \frac{(-1)^n x^{2n+1}}{(2n+1)(2n+1)!},$$

$$\text{Ci}(x) = \int_{\infty}^x \frac{\cos \alpha}{\alpha} \, d\alpha = \log x + 0.5772157 + \sum_{n=1}^{\infty} \frac{(-1)^n x^{2n}}{2n(2n)!},$$

are called sine integral and cosine integral, respectively.[‡] The sine and cosine integral functions are given in table form in various handbooks of mathematical functions.[§]

The reponse for various values of time is plotted in Figure 8.14. For convenience a unit length a has been introduced.

---

† J. Boussinesq, *Application des potentiels*, Gauthier-Villars, Paris, 1885, p. 470.

‡ See B. O. Peirce and R. M. Foster, *A Short Table of Integrals*, 4th ed., Ginn, Boston, 1957, Formulas 799, 800, 802, and 803.

§ For example, see M. Abramowitz and I. A. Stegun, eds., *Handbook of Mathematical Functions*, National Bureau of Standards, Washington, D.C., 1965, p. 238.

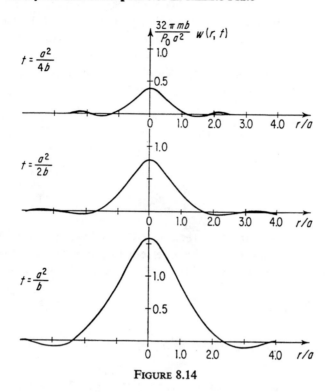

FIGURE 8.14

8-15   NONSYMMETRICAL RESPONSE OF AN INFINITE PLATE. FOURIER TRANSFORM SOLUTION

Section 8–14 discussed the response of an infinite plate for the special case in which the excitation could be separated into a product of a function of the radial distance and time, on the one hand, and a function of the angle θ, on the other. The present section approaches the same problem from a more general point of view.

The differential equation of motion of a thin uniform infinite plate can be written, in terms of rectangular coordinates, in the form

$$D_E \nabla^2 \nabla^2 w(x, y, t) + m \frac{\partial^2 w(x, y, t)}{\partial t^2} = p(x, y, t) \qquad (8.220)$$

The plate is subject to the initial conditions

$$w(x, y, 0) = f(x, y), \qquad \frac{\partial w(x, y, t)}{\partial t}\bigg|_{t=0} = g(x, y) \qquad (8.221)$$

A general solution of (8.220) can be made by means of a double complex Fourier transform. Assuming that all partial derivatives of the function $w(x, y, t)$ with respect to either x or y or both x and y and of order 3 and lower vanish as x or y approach infinity, it is not difficult to show that the transform of the biharmonic expression, $\nabla^2\nabla^2 w$, is

$$\int_{-\infty}^{\infty} \int_{-\infty}^{\infty} \nabla^2\nabla^2 w(x, y, t) e^{i(\xi x + \eta y)} \, dx \, dy = (\xi^2 + \eta^2)^2 \overline{w}(\xi, \eta, t), \quad (8.222)$$

where

$$\overline{w}(\xi, \eta, t) = \int_{-\infty}^{\infty} \int_{-\infty}^{\infty} w(x, y, t) e^{i(\xi x + \eta y)} \, dx \, dy \quad (8.223)$$

is the transform of the transverse displacement $w(x, y, t)$. Denoting by $\bar{p}(\xi, \eta, t)$ the transform of the transverse load $p(x, y, t)$ we can transform both sides of (8.220) and obtain the ordinary differential equation

$$\frac{d^2\overline{w}(\xi, \eta, t)}{dt^2} + b^2(\xi^2 + \eta^2)^2 \overline{w}(\xi, \eta, t) = \frac{1}{m} \bar{p}(\xi, \eta, t), \qquad b^2 = \frac{D_E}{m}, \quad (8.224)$$

which is subject to the transformed initial conditions

$$\overline{w}(\xi, \eta, 0) = \bar{f}(\xi, \eta), \qquad \frac{d\overline{w}(\xi, \eta, t)}{dt}\bigg|_{t=0} = \bar{g}(\xi, \eta), \quad (8.225)$$

where $\bar{f}(\xi, \eta)$ and $\bar{g}(\xi, \eta)$ are the transforms of $f(x, y)$ and $g(x, y)$, respectively. Considering (8.225), the general solution of (8.224) is

$$\overline{w}(\xi, \eta, t) = \frac{1}{mb(\xi^2 + \eta^2)} \int_0^t \bar{p}(\xi, \eta, \tau) \sin b(\xi^2 + \eta^2)(t - \tau) \, d\tau$$

$$+ \bar{f}(\xi, \eta) \cos b(\xi^2 + \eta^2)t + \frac{\bar{g}(\xi, \eta)}{b(\xi^2 + \eta^2)} \sin b(\xi^2 + \eta^2)t, \quad (8.226)$$

and the general response of the plate is obtained by writing the inverse transformation

$$w(x, y, t) = \frac{1}{(2\pi)^2} \int_{-\infty}^{\infty} \int_{-\infty}^{\infty} \overline{w}(\xi, \eta, t) e^{-i(\xi x + \eta y)} \, d\xi \, d\eta. \quad (8.227)$$

As an illustration let us consider the response of a plate, initially at rest, to two concentrated forces applied at two points $x = \pm a$ on the x axis (Figure 8.15). The forces can be represented as distributed forces in the form

$$p(x, y, t) = P(t)[\delta(x - a)\delta(y) + \delta(x + a)\delta(y)], \quad (8.228)$$

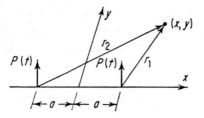

FIGURE 8.15

where the various δ's represent appropriate spatial Dirac delta functions. The transformed load becomes

$$\bar{p}(\xi, \eta, t) = \int_{-\infty}^{\infty} \int_{-\infty}^{\infty} p(x, y, t)e^{i(\xi x + \eta y)} \, dx \, dy$$

$$= P(t) \int_{-\infty}^{\infty} \int_{-\infty}^{\infty} [\delta(x - a)\delta(y) + \delta(x + a)\delta(y)]e^{i(\xi x + \eta y)} \, dx \, dy$$

$$= P(t)(e^{i\xi a} + e^{-i\xi a}). \qquad (8.229)$$

Substitution of (8.229) in (8.226) yields the transformed response

$$\bar{w}(\xi, \eta, t) = \frac{e^{i\xi a} + e^{-i\xi a}}{mb(\xi^2 + \eta^2)} \int_0^t P(\tau) \sin b(\xi^2 + \eta^2)(t - \tau) \, d\tau, \quad (8.230)$$

and the response $w(x, y, t)$ is obtained by writing the inverse transformation

$$w(x, y, t) = \frac{1}{(2\pi)^2} \int_{-\infty}^{\infty} \int_{-\infty}^{\infty} \bar{w}(\xi, \eta, t)e^{-i(\xi x + \eta y)} \, d\xi \, d\eta$$

$$= \frac{1}{(2\pi)^2 mb} \int_0^t P(\tau)\left[\int_{-\infty}^{\infty} \int_{-\infty}^{\infty} \frac{\sin b(\xi^2 + \eta^2)(t - \tau)}{\xi^2 + \eta^2}\right.$$

$$\times \left. \{e^{-i[\xi(x - a) + \eta y]} + e^{-i[\xi(x + a) + \eta y]}\} \, d\xi \, d\eta\right] d\tau. \quad (8.231)$$

To give (8.231) a simpler form we use the Leibnitz rule for the differentiation of an integral and write

$$\dot{w}(x, y, t) = \frac{1}{(2\pi)^2 m} \int_0^t P(\tau)\left[\int_{-\infty}^{\infty} \int_{-\infty}^{\infty} \cos b(\xi^2 + \eta^2)(t - \tau)\right.$$

$$\times \left. \{e^{-i[\xi(x - a) + \eta y]} + e^{-i[\xi(x + a) + \eta y]}\} \, d\xi \, d\eta\right] d\tau$$

$$= \frac{1}{(2\pi)^2 2m} \int_0^t P(\tau)\left[\int_{-\infty}^{\infty} \int_{-\infty}^{\infty} \{e^{ib(\xi^2 + \eta^2)(t - \tau)} + e^{-ib(\xi^2 + \eta^2)(t - \tau)}\}\right.$$

$$\times \left. \{e^{-i[\xi(x - a) + \eta y]} + e^{-i[\xi(x + a) + \eta y]}\} \, d\xi \, d\eta\right] d\tau. \quad (8.232)$$

To evaluate the double integral in (8.232) we use the technique employed in Section 8–11 and multiply the integrand by a convergence factor $e^{-\epsilon(\xi^2+\eta^2)}$ and investigate the behavior of the integral as ϵ tends to zero. To this end let us define the integral

$$I(\epsilon) = \int_{-\infty}^{\infty} \int_{-\infty}^{\infty} e^{-\epsilon(\xi^2+\eta^2)}\{e^{ib(\xi^2+\eta^2)(t-\tau)} + e^{-ib(\xi^2+\eta^2)(t-\tau)}\}$$

$$\times \{e^{-i[\xi(x-a)+\eta y]} + e^{-i[\xi(x+a)+\eta y]}\}\, d\xi\, d\eta. \quad (8.233)$$

First let us consider†

$$I_1(\epsilon) = \int_{-\infty}^{\infty} \int_{-\infty}^{\infty} e^{-\epsilon(\xi^2+\eta^2)} e^{ib(\xi^2+\eta^2)t} e^{-i(\xi x+\eta y)}\, d\xi\, d\eta$$

$$= \exp\left[-\frac{x^2+y^2}{4(\epsilon-ibt)}\right] \int_{-\infty}^{\infty} \exp\left\{-(\epsilon-ibt)\left[\xi+\frac{ix}{2(\epsilon-ibt)}\right]^2\right\}\, d\xi$$

$$\times \int_{-\infty}^{\infty} \exp\left\{-(\epsilon-ibt)\left[\eta+\frac{iy}{2(\epsilon-ib)}\right]^2\right\}\, d\eta$$

$$= \frac{\pi(\epsilon+ibt)}{\epsilon^2+b^2t^2} \exp\left[-\frac{x^2+y^2}{4(\epsilon-ibt)}\right],$$

from which it follows that

$$I(\epsilon) = \int_{-\infty}^{\infty} \int_{-\infty}^{\infty} e^{-\epsilon(\xi^2+\eta^2)}\{e^{ib(\xi^2+\eta^2)(t-\tau)} + e^{-ib(\xi^2+\eta^2)(t-\tau)}\}$$

$$\times \{e^{-i[\xi(x-a)+\eta y]} + e^{-i[\xi(x+a)+\eta y]}\}\, d\xi\, d\eta$$

$$= \frac{2\pi}{[\epsilon^2+b^2(t-\tau)^2]}$$

$$\times \left(\left[\exp\left\{-\frac{\epsilon[(x-a)^2+y^2]}{4[\epsilon^2+b^2(t-\tau)^2]}\right\}\right]\left\{\epsilon\cos\frac{b(t-\tau)[(x-a)^2+y^2]}{4[\epsilon^2+b^2(t-\tau)^2]}\right.\right.$$

$$\left.+ b(t-\tau)\sin\frac{b(t-\tau)[(x-a)^2+y^2]}{4[\epsilon^2+b^2(t-\tau)^2]}\right\}$$

$$+ \left[\exp\left\{-\frac{\epsilon[(x+a)^2+y^2]}{4[\epsilon^2+b^2(t-\tau)^2]}\right\}\right]\left\{\epsilon\cos\frac{b(t-\tau)[(x+a)^2+y^2]}{4[\epsilon^2+b^2(t-\tau)^2]}\right.$$

$$\left.\left.+ b(t-\tau)\sin\frac{b(t-\tau)[(x+a)^2+y^2]}{4[\epsilon^2+b^2(t-\tau)^2]}\right\}\right), \quad (8.234)$$

which, upon letting $\epsilon \to 0$, reduces to

$$I(0) = \frac{2\pi}{b(t-\tau)}\left[\sin\frac{r_1^2}{4b(t-\tau)} + \sin\frac{r_2^2}{4b(t-\tau)}\right], \quad (8.235)$$

† See E. C. Titchmarsh, *The Theory of Functions*, Oxford Univ. Press, New York, 1932, p. 148.

where $r_1 = [(x - a)^2 + y^2]^{1/2}$ and $r_2 = [(x + a)^2 + y^2]^{1/2}$ (8.236)

are the distances from the points $x = a$, $y = 0$ and $x = -a$, $y = 0$ to an arbitrary point (x, y). A combination of (8.232) and (8.235) yields

$$\dot{w}(x, y, t) = \frac{1}{4\pi bm} \int_0^t \frac{P(\tau)}{t - \tau} \left[\sin \frac{r_1^2}{4b(t - \tau)} + \sin \frac{r_2^2}{4b(t - \tau)} \right] d\tau, \quad (8.237)$$

which can be written

$$\dot{w}(x, y, t) = \frac{1}{4\pi bm} \left[\int_{r_1^2/4bt}^{\infty} P\left(t - \frac{r_1^2}{4b\alpha} \right) \frac{\sin \alpha}{\alpha} d\alpha + \int_{r_2^2/4bt}^{\infty} P\left(t - \frac{r_2^2}{4b\alpha} \right) \frac{\sin \alpha}{\alpha} d\alpha \right]$$
(8.238)

To obtain (8.237) we used substitutions analogous to those used in Section 8–14. Also, as in Section 8–14, we let

$$R(t) = \int_0^t P(\sigma) \, d\sigma \quad (8.239)$$

and obtain finally the response

$$w(x, y, t) = \frac{1}{4\pi bm} \left[\int_{r_1^2/4bt}^{\infty} R\left(t - \frac{r_1^2}{4bt} \right) \frac{\sin \alpha}{\alpha} d\alpha + \int_{r_2^2/4bt}^{\infty} R\left(t - \frac{r_2^2}{4b\alpha} \right) \frac{\sin \alpha}{\alpha} d\alpha \right].$$
(8.240)

Equation (8.240) can be obtained directly from (8.216) by an appropriate superposition of responses.

Problems

8.1 Let the solution of the wave equation (8.2) be in the form (8.3) and obtain the response of an infinite string to the initial conditions

$$y(x, 0) = f(x), \qquad \left. \frac{\partial y(x, t)}{\partial t} \right|_{t=0} = g(x).$$

(*Hint:* Adjust the solution to fit the initial conditions.)

8.2 Obtain the response of an infinite string to the initial velocity

$$g(x) = \begin{cases} A\left(1 - \frac{|x|}{|a|}\right), & |x| < a, \\ 0, & |x| > a. \end{cases}$$

Plot the response for $t = 0$, $\frac{1}{2}a/c$, a/c, and $\frac{3}{2}a/c$. What is the response for $t = \infty$? Draw conclusions.

8.3 Consider a semi-infinite string, $0 \leq x \leq \infty$, and obtain the response to a prescribed motion $y(0, t) = f(t)$ at the end $x = 0$ by means of the Laplace transform method. The initial conditions are zero.

8.4 Obtain the solution of Problem 8.3 by means of a Fourier transform instead of the Laplace transform.

8.5 A uniform rod of length L, mass per unit length ρ, and stiffness EA lies unrestrained on a frictionless plane. At $t = 0$, with the initial conditions zero, an axial force in the form of a step function is applied at the end $x = 0$. Obtain the subsequent motion $u(x, t)$, in terms of traveling waves, by means of the Laplace transform method. Plot $u(L/2, t)$ vs. t and compare the result with the motion of a rigid rod of the same mass. Draw conclusions.

8.6 The system of Problem 8.5 is acted upon by an impulsive force at the end $x = 0$. Obtain the response by means of a Fourier transform.

8.7 Consider the system of Section 8–6 and devise a scheme to represent force waves similar to the scheme for the displacement waves shown in Figure 8.9.

8.8 Two harmonic flexural waves of equal amplitudes A and wavelengths $\lambda_1 = 5\lambda$ and $\lambda_2 = 5.5\lambda$, respectively, are propagating in a uniform bar. If the rotatory inertia is neglected, derive expressions for the phase velocity and group velocity of the wave packet obtained by a superposition of the two harmonic waves. Consider a rectangular cross section with the ratio of depth to width $R = 2$, include the rotatory inertia effect, and plot the phase velocity v and the group velocity U as functions of the wavelength λ.

8.9 A uniform bar of length L, hinged at both ends, is imparted a motion $y(0, t) = f(t)$ at the end $x = 0$. Derive an expression for the bending deflection of the bar at any point.

8.10 Obtain the response of a thin uniform rectangular membrane, $0 \leq x \leq a$ and $0 \leq y \leq b$, to the initial conditions

$$w(x, y, 0) = f(x, y), \qquad \left.\frac{\partial w(x, y, t)}{\partial t}\right|_{t=0} = g(x, y).$$

8.11 Obtain the response of a thin uniform circular membrane, $r = a$, to the symmetric load

$$p(r, \theta, t) = P_0 \delta(r - b) \alpha(t), \qquad b < a,$$

where $\delta(r - b)$ is a spatial Dirac's delta function defined by

$$\delta(r - b) = 0, \qquad r \neq b,$$

$$\int_D \delta(r - b)\, dD = 1,$$

and $\alpha(t)$ is the unit step function. The initial conditions are zero.

8.12 Obtain the free vibration solution of a thin uniform infinite plate to the initial excitation

$$w(r, \theta, 0) = Ae^{-r^2/a^2}, \qquad \left.\frac{\partial w(r, \theta, t)}{\partial t}\right|_{t=0} = 0.$$

Plot the response for $t = 0$, $a^2/4b$, and $a^2/2b$.

Selected Readings

Brillouin, L., *Wave Propagation and Group Velocity*, Academic Press, New York, 1960.

Churchill, R. V., *Operational Mathematics*, 2nd ed., McGraw-Hill, New York, 1958.

Heins, A. E., "Applications of the Fourier Transform Theorem," *J. Math. Phys.*, **14**, 137–142, 1935.

Hildebrand, F. B., *Advanced Calculus for Engineers*, Prentice-Hall, Englewood Cliffs, N.J., 1960.

Kolsky, H., *Stress Waves in Solids*, Oxford Univ. Press, New York, 1953.

Morse, P. M., *Vibration and Sound*, 2nd ed., McGraw-Hill, New York, 1948.

Sneddon, I. N., *Fourier Transforms*, McGraw-Hill, New York, 1951.

Sneddon, I. N., "The Fourier Transform Solution of an Elastic Wave Equation," *Proc. Cambridge Phil. Soc.*, **41**, 239–243 (1945).

Sneddon, I. N., "The Symmetrical Vibrations of a Thin Elastic Plate," *Proc. Cambridge Phil. Soc.*, **41**, 27–43 (1945).

Sokolnikoff, I. S., and R. M. Redheffer, *Mathematics of Physics and Modern Engineering*, McGraw-Hill, New York, 1958.

Stephens, R. W. P., and A. E. Bate, *Wave Motion and Sound*, E. Arnold, London, 1950.

Thomson, W. T., *Laplace Transformation*, 2nd ed., Prentice-Hall, Englewood Cliffs, N.J., 1960.

Titchmarsh, E. C., *Introduction to the Theory of Fourier Integrals*, Oxford Univ. Press, New York, 1948.

Tranter, C. J., *Integral Transforms in Mathematical Physics*, Methuen, London, 1959.

Webster, A. G., *Partial Differential Equations of Mathematical Physics*, Dover, New York, 1955.

CHAPTER 9

DAMPED
SYSTEMS

9-1 GENERAL DISCUSSION

In many ways the assumption that systems possess no damping is a mathematical convenience rather than a reflection of physical evidence. In fact, if a system is set in motion and allowed to vibrate freely, the vibration will eventually die out; the rate of decay depends on the amount of damping. Nevertheless, the concept of undamped systems not only serves a useful purpose in analysis but can also be justified in certain circumstances. For example, if the damping is small and one is interested in the free vibration of a system over a short interval of time, there may not be sufficient time for the effect of damping to become noticeable. Similarly, for small damping, one may not be able to notice the effect of damping in the case of a system with harmonic excitation, provided the driving frequency is not in the neighborhood of any of the natural frequencies of a system. On the other hand there are cases in which the damping effect cannot be ignored. In a limited number of these cases, the analysis of the corresponding undamped systems can be used to obtain the response of damped systems.

There are many mathematical models representing damping. The most important type of damping in vibration study is *linear viscous damping*. According to this model the damping takes the form of a force proportional in magnitude to the velocity and acting in the direction opposite to the direction of the velocity. *Coulomb damping* also gives rise to a force opposing the motion, but, in contrast with viscous damping, it has a constant magnitude. This damping is also referred to as *dry friction*. Another widely used model is *structural damping*. It is associated with internal energy dissipation due to the hysteresis effect in cyclic stress, for which reason it is also called *hysteretic damping*.

The coupled equations of motion describing an undamped multi-degree-of-freedom system can be uncoupled by means of modal analysis, which uses a linear transformation to express the equations of motion in terms of a different set of coordinates, the principal coordinates. The linear transformation is represented in matrix form by the *classical modal matrix*, obtained from the eigenvalue problem associated with the undamped system. In some special cases the classical modal matrix can also be used successfully to uncouple the equations of motion of a viscously damped linear system. Unfortunately, this is not always possible. The general case of viscous damping can be treated by transforming a set of *n* ordinary differential equations of second order into a set of *2n* ordinary differential equations of first order. The eigenvalues and eigenvectors associated with the latter set of equations are, for the cases in which we are primarily interested, complex quantities.

The concept of structural damping can be used in conjunction with an analogy with viscous damping if sufficient care is exercised. In using the analogy with viscous damping, one should always remember the basic assumptions that the energy dissipated in one cycle of harmonic motion is proportional to the square of the amplitude and independent of the frequency of motion. Some materials may not conform with these assumptions.

Damped continuous systems can sometimes be treated by assuming a solution, in the form of a superposition of the classical normal modes, which leads to a set of ordinary differential equations.

In this chapter we first show under what conditions classical modal analysis can be used for the treatment of damped discrete systems. This approach is used to evaluate the response of viscously damped as well as structurally damped systems. The general case of damping, in which classical modal analysis fails to uncouple the equations of motion, is discussed and a solution of the coupled equations by means of the Laplace transform is presented. A modal analysis procedure designed to handle the general case of damping is discussed in detail. Solutions of the eigenvalue problem associated with a damped system by means of the characteristic determinant and matrix iteration method are shown. The expansion theorem is used to derive the forced vibration solution for the general case of damping. Finally, viscously damped and structurally damped continuous systems are discussed in terms of modal analysis.

9-2 EXISTENCE OF NORMAL MODES IN VISCOUSLY DAMPED DISCRETE SYSTEMS

The kinetic energy of an n-degree-of-freedom system has the form

$$T = \tfrac{1}{2} \sum_{i=1}^{n} \sum_{j=1}^{n} m_{ij}\dot{q}_i\dot{q}_j, \tag{9.1}$$

where the coefficients m_{ij} are symmetric inertia coefficients and $\dot{q}_i$ are generalized velocities. The kinetic energy, (9.1), is a positive definite quadratic expression. Similarly, the potential energy can be written

$$V = \tfrac{1}{2} \sum_{i=1}^{n} \sum_{j=1}^{n} k_{ij}q_i q_j, \tag{9.2}$$

where the coefficients k_{ij} are symmetric stiffness coefficients. The potential energy is a positive definite expression when the system is positive definite.

Next let us assume that among the various kinds of forces acting upon a system it is possible to recognize a special type of friction force arising from motion in a viscous medium. These are nonconservative, retarding forces and are assumed to be proportional to the velocities as discussed in Chapter 1. In deriving the equations of motion by means of Lagrange's equations it will prove convenient to introduce a function

$$F = \tfrac{1}{2} \sum_{i=1}^{n} \sum_{j=1}^{n} c_{ij}\dot{q}_i\dot{q}_j, \tag{9.3}$$

which was called the *dissipation function* by Lord Rayleigh.[†] It also has a positive definite quadratic form similar to the kinetic and potential energy expressions.

With this definition in mind, Lagrange's equations assume the form

$$\frac{d}{dt}\left(\frac{\partial T}{\partial \dot{q}_r}\right) - \frac{\partial T}{\partial q_r} + \frac{\partial F}{\partial \dot{q}_r} + \frac{\partial V}{\partial q_r} = Q_r, \qquad r = 1, 2, \ldots, n, \tag{9.4}$$

where Q_r are generalized forces acting upon the system; these forces are not provided for by the functions V and F. Introducing (9.1) through (9.3) in (9.4) we obtain a set of n coupled ordinary differential equations describing the motion of a viscously damped linear system. These equations can be written in the matrix form

$$[m]\{\ddot{q}\} + [c]\{\dot{q}\} + [k]\{q\} = \{Q\}, \tag{9.5}$$

† Lord Rayleigh, *The Theory of Sound*, Vol. 1, Dover, New York, 1945, p. 103.

where the symmetric matrices $[m]$, $[c]$, and $[k]$ are positive definite for a positive definite system.

Next we shall seek the general conditions under which the coupled system of equations of motion, (9.5), can be uncoupled by means of a linear transformation similar to the one used for undamped systems. First let us consider the linear transformation

$$\{q\} = [u]\{\eta\}, \tag{9.6}$$

where $[u]$ is the matrix of orthonormal modes associated with the eigenvalue problem

$$[m][u] = [k][u][\lambda]. \tag{9.7}$$

Although the linear transformation (9.6) eliminates the cross products in the kinetic and potential energy expressions, it does not eliminate the cross products in the dissipation function F, except in some special cases. Introducing (9.6) in (9.5) and premultiplying the result by $[u]^T$, we obtain

$$\{\ddot{\eta}\} + [C]\{\dot{\eta}\} + [\omega^2]\{\eta\} = \{N\}, \tag{9.8}$$

where
$$[C] = [u]^T[c][u] \tag{9.9}$$

is a symmetric but, in general, not diagonal matrix, and

$$\{N\} = [u]^T\{Q\} \tag{9.10}$$

is a column matrix of generalized forces associated with the generalized coordinates $\eta_r(t)$ $(r = 1, 2, \ldots, n)$.

In the special case in which the damping matrix $[c]$ is a linear combination of the $[m]$ and $[k]$ matrices, the matrix $[C]$ is diagonal, and the equations of motion (9.8) are an uncoupled set of equations. This fact was pointed out by Lord Rayleigh,† who stated that uncoupling is achieved when F is a linear function of T and V.

Uncoupling can also be achieved under slightly more general conditions. To show this, let us consider the free vibration problem

$$[m]\{\ddot{q}\} + [c]\{\dot{q}\} + [k]\{q\} = \{0\}. \tag{9.11}$$

If we substitute the linear transformation

$$\{q\} = [m]^{-1/2}\{x\} \tag{9.12}$$

† Lord Rayleigh, *The Theory of Sound*, Vol. 1, Dover, New York, 1945, p. 130.

in (9.11) and premultiply the result by $[m]^{-1/2}$, we obtain

$$[I]\{\ddot{x}\} + [A]\{\dot{x}\} + [B]\{x\} = \{0\}, \tag{9.13}$$

where $[I]$ is the identity matrix and

$$[A] = [m]^{-1/2}[c][m]^{-1/2}, \qquad [B] = [m]^{-1/2}[k][m]^{-1/2} \tag{9.14}$$

are symmetric and positive definite matrices.

Next we seek another linear transformation,

$$\{x\} = [\Phi]\{y\}, \tag{9.15}$$

where the square matrix $[\Phi]$ is such that

$$\lceil a \rfloor = [\Phi]^T[A][\Phi], \qquad \lceil b \rfloor = [\Phi]^T[B][\Phi], \tag{9.16}$$

where the matrices $\lceil a \rfloor$ and $\lceil b \rfloor$ are diagonal matrices. Since $[A]$ and $[B]$ are symmetric and positive definite matrices, such a transformation is always possible, as we have shown in Section 4–8. In addition, however, let us require that

$$[\Phi]^T[\Phi] = [I], \tag{9.17}$$

because only if $[\Phi]$ is such that (9.17) is satisfied will the linear transformation (9.15) uncouple (9.13). From (9.17) we observe that the transformation (9.15) is an orthonormal transformation. Furthermore, the condition that $[\Phi]$ satisfies (9.17) in addition to (9.16) places restrictions on the form of matrices $[A]$ and $[B]$ for which this is possible.

If such a linear transformation exists, it is precisely the same orthogonal transformation which uncouples the undamped system obtained by setting $[A]$ equal to zero.

Let us consider the case in which

$$\lceil a \rfloor = \lceil b \rfloor^p, \tag{9.18}$$

and, as (9.17) implies that $[\Phi]^T = [\Phi]^{-1}$, we can write

$$[A] = [\Phi]\lceil a \rfloor[\Phi]^{-1}. \tag{9.19}$$

Substituting (9.18) in (9.19) we conclude that

$$[A] = [B]^p. \tag{9.20}$$

According to the Cayley-Hamilton theorem,† every matrix satisfies its own

---

† See F. B. Hildebrand, *Methods of Applied Mathematics*, Prentice-Hall, Englewood Cliffs, N.J., 1960, p. 64.

characteristic equation, from which it follows that any power of a matrix can be expressed in terms of the $(n - j)$th $(j = 1, 2, \ldots, n)$ powers of that matrix, so p need only assume the values $0, 1, \ldots, n - 1$.

Hence we conclude that if matrix $[A]$ can be expressed as a polynomial in matrix $[B]$, the modal matrix obtained from the solution of the eigenvalue problem associated with the undamped system can be used as a transformation matrix uncoupling the equations of motion of the damped system.

In the case in which

$$\lceil a \rfloor = \lceil b \rfloor^{p/s} \tag{9.21}$$

we have

$$[A] = [\Phi]\lceil b \rfloor^{p/s}[\Phi]^{-1} \tag{9.22}$$

or

$$[A] = [B]_1^{p/s}, \tag{9.23}$$

where the subscript in (9.23) denotes one particular root of $[B]$. Because $\lceil a \rfloor$ as given by (9.21) is a diagonal matrix, the modal matrix associated with the undamped system can be used to uncouple (9.13) in this case, too—when $\lceil a \rfloor$ has the form indicated in (9.21). Furthermore, this is true for any linear combination of such terms, so the system does possess classical normal modes if $[A]$ can be written

$$[A] = \sum_{s=1}^{\infty} \sum_{p=0}^{n-1} \alpha_{sp}[B]_1^{p/s}, \tag{9.24}$$

which is a sufficient, but not necessary, condition. This development is due to Caughey.†

The linear transformation (9.15) leads to a set of n uncoupled equations

$$\{\ddot{y}\} + \lceil C \rfloor\{\dot{y}\} + \lceil \omega^2 \rfloor\{y\} = \{0\}, \tag{9.25}$$

where

$$\lceil C \rfloor = [u]^T[c][u] \tag{9.26}$$

is a diagonal matrix.

Letting $s = 1$ and $p = 0,1$ in (9.24) we obtain

$$[A] = \alpha_{10}[B]^0 + \alpha_{11}[B] = \alpha_{10}[I] + \alpha_{11}[B], \tag{9.27}$$

which can be premultiplied and postmultiplied by $[m]^{1/2}$ to give

$$[c] = \alpha_{10}[m] + \alpha_{11}[k], \tag{9.28}$$

which is the expression of Rayleigh's statement mentioned previously.

† T. K. Caughey, *J. Appl. Mech.*, **27**, 269–271 (1960).

In general, however, it is not possible to uncouple the equations of motion of an *n*-degree-of-freedom system by the method described here. When the damping coefficients are small, a meaningful approximation can be obtained by using the modal matrix as the transformation matrix and ignoring the terms $\{u^{(r)}\}^T[c]\{u^{(s)}\}$ for which $r \neq s$. This in effect implies that the uncoupled equations (9.25) can be used when damping is small without causing serious errors. Physically this means that damping is sufficiently small that coupling is a second order effect.

When the matrix $[c]$ cannot be reduced to a diagonal form by means of the modal matrix and damping is not small, classical modal analysis must be abandoned and a different procedure adopted. Example 9.1 is a case in which classical modal analysis does lead to uncoupled equations of motion.

A procedure treating cases in which the modal matrix cannot be used to uncouple the system of equations will be described later.

Example 9.1

Consider the system shown in Figure 9.1. The kinetic energy of the system is

$$T = \tfrac{1}{2}(m_1\dot{q}_1^2 + m_2\dot{q}_2^2 + m_3\dot{q}_3^2), \tag{a}$$

the potential energy has the form

$$V = \tfrac{1}{2}[k_1q_1^2 + k_2(q_2 - q_1)^2 + k_3(q_3 - q_2)^2 + k_4q_3^2], \tag{b}$$

and Rayleigh's dissipation function is

$$F = \tfrac{1}{2}[c_1\dot{q}_1^2 + c_2(\dot{q}_2 - \dot{q}_1)^2 + c_3(\dot{q}_3 - \dot{q}_2)^2 + c_4\dot{q}_3^2 + c_5(\dot{q}_3 - \dot{q}_1)^2 + c_6\dot{q}_2^2]. \tag{c}$$

Correspondingly, the differential equations of motion for free vibration can be written in the matrix form

$$[m]\{\ddot{q}\} + [c]\{\dot{q}\} + [k]\{q\} = \{0\}. \tag{d}$$

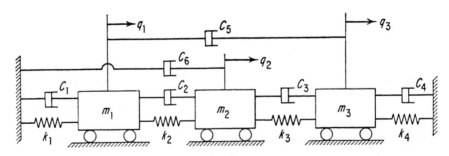

FIGURE 9.1

Letting $m_1 = m_2 = m$ and $m_3 = 2m$, we obtain the inertia matrix

$$[m] = \begin{bmatrix} m_1 & 0 & 0 \\ 0 & m_2 & 0 \\ 0 & 0 & m_3 \end{bmatrix} = m \begin{bmatrix} 1 & 0 & 0 \\ 0 & 1 & 0 \\ 0 & 0 & 2 \end{bmatrix}. \tag{e}$$

If $k_1 = k_2 = k_3 = k$ and $k_4 = 2k$, the stiffness matrix takes the form

$$[k] = \begin{bmatrix} k_1 + k_2 & -k_2 & 0 \\ -k_2 & k_2 + k_3 & -k_3 \\ 0 & -k_3 & k_3 + k_4 \end{bmatrix} = k \begin{bmatrix} 2 & -1 & 0 \\ -1 & 2 & -1 \\ 0 & -1 & 3 \end{bmatrix}. \tag{f}$$

For $c_1 = 2.9316c$, $c_2 = 0.3747c$, $c_3 = 0.4079c$, $c_4 = 5.9128c$, $c_5 = 0.0581c$, and $c_6 = 2.5511c$ the damping matrix becomes

$$[c] = \begin{bmatrix} c_1 + c_2 + c_5 & -c_2 & -c_5 \\ -c_2 & c_2 + c_3 + c_6 & -c_3 \\ -c_5 & -c_3 & c_3 + c_4 + c_5 \end{bmatrix}$$

$$= c \begin{bmatrix} 3.3644 & -0.3747 & -0.0581 \\ -0.3747 & 3.3337 & -0.4079 \\ -0.0581 & -0.4079 & 6.3788 \end{bmatrix}, \tag{g}$$

so it is easy to see that the system of equations of motion (d) is coupled.

The eigenvalue problem associated with the undamped system can be written

$$[m][u] = [k][u][\lambda], \tag{h}$$

where the matrices $[m]$ and $[k]$ are given by (e) and (f), and we obtain the normalized modal matrix,

$$[u] = m^{-1/2} \begin{bmatrix} 0.4959 & 0.6074 & 0.6209 \\ 0.6646 & 0.1949 & -0.7215 \\ 0.3954 & -0.5446 & 0.2171 \end{bmatrix} \tag{i}$$

and the associated matrix of natural frequencies,

$$[\omega^2] = \frac{k}{m} \begin{bmatrix} 0.6594 & 0 & 0 \\ 0 & 1.6789 & 0 \\ 0 & 0 & 3.1619 \end{bmatrix} \tag{j}$$

(see Example 4.3).

Introducing the linear transformation

$$\{q\} = [u]\{y\} \tag{k}$$

in (d) and premultiplying the result by $[u]^T$, we obtain

$$[I]\{\ddot{y}\} + [C]\{\dot{y}\} + [\omega^2]\{y\} = \{0\}, \tag{l}$$

and we note that the transformation (k) reduces the damping matrix to a diagonal form

$$[C] = [u]^T[c][u]$$

$$= \frac{c}{m} \begin{bmatrix} 0.4959 & 0.6646 & 0.3954 \\ 0.6074 & 0.1949 & -0.5446 \\ 0.6209 & -0.7215 & 0.2171 \end{bmatrix} \begin{bmatrix} 3.3644 & -0.3747 & -0.0581 \\ -0.3747 & 3.3337 & -0.4079 \\ -0.0581 & -0.4079 & 6.3788 \end{bmatrix}$$

$$\times \begin{bmatrix} 0.4959 & 0.6074 & 0.6209 \\ 0.6646 & 0.1949 & -0.7215 \\ 0.3954 & -0.5446 & 0.2171 \end{bmatrix} = \frac{c}{m} \begin{bmatrix} 2.8130 & 0 & 0 \\ 0 & 3.2957 & 0 \\ 0 & 0 & 3.7810 \end{bmatrix}. \tag{m}$$

The uncoupled equations of motion are

$$\ddot{y}_1 + 2.8130 \frac{c}{m} \dot{y}_1 + 0.6594 \frac{k}{m} y_1 = 0,$$

$$\ddot{y}_2 + 3.2957 \frac{c}{m} \dot{y}_2 + 1.6789 \frac{k}{m} y_2 = 0, \tag{n}$$

$$\ddot{y}_3 + 3.7810 \frac{c}{m} \dot{y}_3 + 3.1619 \frac{k}{m} y_3 = 0.$$

The uncoupling of the equations of motion can be shown, for this particular case, to be caused by the fact that the damping matrix (g) is of the type discussed in this section. To this end use the first of equations (9.14) and write

$$[A] = [m]^{-1/2}[c][m]^{-1/2}$$

$$= \frac{c}{m} \begin{bmatrix} 1 & 0 & 0 \\ 0 & 1 & 0 \\ 0 & 0 & 1/\sqrt{2} \end{bmatrix} \begin{bmatrix} 3.3644 & -0.3747 & -0.0581 \\ -0.3747 & 3.3337 & -0.4079 \\ -0.0581 & -0.4079 & 6.3788 \end{bmatrix} \begin{bmatrix} 1 & 0 & 0 \\ 0 & 1 & 0 \\ 0 & 0 & 1/\sqrt{2} \end{bmatrix}$$

$$= \frac{c}{m} \begin{bmatrix} 3.3644 & -0.3747 & -0.0411 \\ -0.3747 & 3.3337 & -0.2844 \\ -0.0411 & -0.2884 & 3.1894 \end{bmatrix}. \tag{o}$$

From the second of equations (9.14) we obtain

$$[B] = [m]^{-1/2}[k][m]^{-1/2} = \frac{k}{m} \begin{bmatrix} 2.0000 & -1.0000 & 0 \\ -1.0000 & 2.0000 & -0.7071 \\ 0 & -0.7071 & 1.5000 \end{bmatrix}, \quad (\text{p})$$

and we note that

$$[B]^{1/2} = \frac{k^{1/2}}{m^{1/2}} \begin{bmatrix} 1.3644 & -0.3747 & -0.0411 \\ -0.3747 & 1.3337 & -0.2884 \\ -0.0411 & -0.2884 & 1.1894 \end{bmatrix}, \quad (\text{q})$$

implying that

$$[A] = \frac{2c}{m} [B]^0 + \frac{c}{m^{1/2}k^{1/2}} [B]^{1/2}, \quad (\text{r})$$

which is of the form (9.24).

Hence the modal matrix associated with the undamped system can be used to uncouple the equations of motion (d), as we have already concluded.

9–3 FORCED VIBRATION OF VISCOUSLY DAMPED SYSTEMS. MODAL ANALYSIS

Let us consider again the differential equations of motion of a viscously damped system,

$$[m]\{\ddot{q}\} + [c]\{\dot{q}\} + [k]\{q\} = \{Q\}, \quad (9.29)$$

and assume that the normal modes associated with the undamped system can be used as a transformation matrix uncoupling these equations. Let this linear transformation be given by

$$\{q\} = [u]\{\eta\}, \quad (9.30)$$

where $[u]$ is the matrix of the orthonormal modes for which we have the relations

$$[u]^T[m][u] = [I], \qquad [u]^T[k][u] = \lceil\omega^2\rfloor, \quad (9.31)$$

where $[I]$ is the identity matrix and $\lceil\omega^2\rfloor$ is a diagonal matrix of the squares of the natural frequencies of the undamped system. In addition, let us introduce the notations

$$[u]^T[c][u] = \lceil2\zeta\omega\rfloor, \quad (9.32)$$

where ζ_r $(r = 1, 2, \ldots, n)$ are damping factors, and

$$[u]^T\{Q\} = \{N\}, \tag{9.33}$$

where N_r $(r = 1, 2, \ldots, n)$ are generalized forces.

Introducing (9.30) in (9.29), premultiplying the result by $[u]^T$, and considering (9.31) through (9.33), we obtain

$$\ddot{\eta}_r(t) + 2\zeta_r\omega_r\dot{\eta}_r(t) + \omega_r^2\eta_r(t) = N_r(t), \qquad r = 1, 2, \ldots, n, \tag{9.34}$$

which represents a set of n uncoupled equations having the same form as the equation describing the motion of a damped single-degree-of-freedom system [see (1.25)].

The solution of (9.34) can be readily obtained by means of the Laplace transformation. Transforming both sides of (9.34) we obtain

$$[s^2\bar{\eta}_r(s) - s\eta_r(0) - \dot{\eta}_r(0)] + 2\zeta_r\omega_r[s\bar{\eta}_r(s) - \eta_r(0)] + \omega_r^2\bar{\eta}_r(s) = \bar{N}_r(s),$$

where $\eta_r(0)$ and $\dot{\eta}_r(0)$ are the rth generalized displacement and velocity, respectively. The subsidiary equation becomes

$$\bar{\eta}_r(s) = \frac{\bar{N}_r(s)}{s^2 + 2\zeta_r\omega_r s + \omega_r^2} + \frac{s + 2\zeta_r\omega_r}{s^2 + 2\zeta_r\omega_r s + \omega_r^2}\eta_r(0)$$

$$+ \frac{1}{s^2 + 2\zeta_r\omega_r s + \omega_r^2}\dot{\eta}_r(0), \tag{9.35}$$

and the generalized displacement $\eta_r(t)$ is obtained by taking the inverse transformation of $\bar{\eta}_r(s)$. First denote

$$\bar{f}(s) = \frac{1}{s^2 + 2\zeta_r\omega_r s + \omega_r^2} = \frac{1}{(s + \zeta_r\omega_r)^2 + \omega_{rd}^2},$$

where $\omega_{rd} = \omega_r(1 - \zeta_r^2)^{1/2}$ can be regarded as a natural frequency associated with the damped system† The inverse transformation of $\bar{f}(s)$ is simply

$$f(t) = \mathscr{L}^{-1}\bar{f}(s) = \frac{e^{-\zeta_r\omega_r t}}{\omega_{rd}} \sin \omega_{rd}t,$$

and it can also be shown that

$$\mathscr{L}^{-1}\frac{s + 2\zeta_r\omega_r}{s^2 + 2\zeta_r\omega_r s + \omega_r^2} = e^{-\zeta_r\omega_r t}\cos \omega_{rd}t + \frac{\zeta_r}{(1 - \zeta_r^2)^{1/2}} e^{-\zeta_r\omega_r t} \sin \omega_{rd}t.$$

† We assume that the system is underdamped, $\zeta_r < 1$.

The first term on the right side of (9.35) can be inverted by means of Borel's theorem, so the inverse transformation can be written

$$\eta_r(t) = \frac{1}{\omega_{rd}} \int_0^t N_r(\tau)e^{-\zeta_r\omega_r(t-\tau)} \sin \omega_{rd}(t - \tau) \, d\tau$$

$$+ e^{-\zeta_r\omega_r t}\left[\cos \omega_{rd}t + \frac{\zeta_r}{(1 - \zeta_r^2)^{1/2}} \sin \omega_{rd}t\right]\eta_r(0)$$

$$+ \left[\frac{1}{\omega_{rd}} e^{-\zeta_r\omega_r t} \sin \omega_{rd}t\right]\dot{\eta}_r(0). \tag{9.36}$$

The last two terms on the right side of (9.36) are called *starting transients* and vanish as t increases indefinitely. Let us now define the *steady state* of a dynamic system as the state in which the dependent variables describing the system behavior are either invariable with time or are periodic functions of time. Any system not in steady state is said to be in *transient state*. According to this definition we obtain transient response not only from the initial conditions but also from the forcing functions.

As an illustration let the forcing functions be given by

$$Q_i(t) = \begin{cases} Q_0 \alpha(t) & \text{for } i = 1, \\ 0 & \text{for } i \neq 1, \end{cases} \tag{9.37}$$

where $\alpha(t)$ is the unit step function. It follows that

$$N_r(t) = Q_0 u_1^{(r)}\alpha(t), \tag{9.38}$$

where $u_1^{(r)}$ is the top element in the rth modal vector. Introducing (9.38) in (9.36) and ignoring the initial conditions, we obtain

$$\eta_r(t) = \frac{Q_0 u_1^{(r)}}{\omega_{rd}} \int_0^t \alpha(\tau)e^{-\zeta_r\omega_r(t-\tau)} \sin \omega_{rd}(t - \tau) \, d\tau$$

$$= \frac{Q_0 u_1^{(r)}}{\omega_r^2}\left[1 - \frac{1}{(1 - \zeta_r^2)^{1/2}} e^{-\zeta_r\omega_r t} \cos (\omega_{rd}t - \phi_r)\right], \tag{9.39}$$

where

$$\phi_r = \tan^{-1}\frac{\zeta_r}{(1 - \zeta_r^2)^{1/2}}. \tag{9.40}$$

We note that the steady-state response in this case is

$$\eta_{rss}(t) = \frac{Q_0 u_1^{(r)}}{\omega_r^2}, \tag{9.41}$$

which is simply the response obtained after the transient component

$$\eta_{rtr}(t) = -\frac{Q_0 u_i^{(r)}}{\omega_r \omega_{rd}} e^{-\zeta_r \omega_r t} \cos(\omega_{rd} t - \phi_r) \tag{9.42}$$

dies out.

9-4   THE CONCEPT OF STRUCTURAL DAMPING

The equation of motion of a viscously damped single-degree-of-freedom system subjected to a harmonic force can be written

$$\ddot{x}(t) + 2\zeta\omega_n\dot{x}(t) + \omega_n^2 x(t) = \omega_n^2 f(t) = \omega_n^2 A e^{i\omega t}, \tag{9.43}$$

where it is understood that the harmonic force is given by the real part of $F(t) = kf(t)$. Consequently, the steady-state response of the system to that harmonic excitation is given by the real part of

$$x(t) = \frac{Ae^{i\omega t}}{1 - (\omega/\omega_n)^2 + i2\zeta\omega/\omega_n} = H(\omega)Ae^{i\omega t}, \tag{9.44}$$

where $H(\omega)$ is the complex frequency response (see Section 1–8). We can write, however,

$$1 - \left(\frac{\omega}{\omega_n}\right)^2 + i2\zeta\frac{\omega}{\omega_n} = \left\{\left[1 - \left(\frac{\omega}{\omega_n}\right)^2\right]^2 + \left(2\zeta\frac{\omega}{\omega_n}\right)^2\right\}^{1/2} e^{i\varphi},$$

$$\varphi = \tan^{-1}\frac{2\zeta\omega/\omega_n}{1 - (\omega/\omega_n)^2}, \tag{9.45}$$

where φ is a phase angle, so (9.44) can be rewritten

$$x(t) = A|H(\omega)|e^{i(\omega t - \varphi)}, \tag{9.46}$$

from which we conclude that the complex vector describing the response $x(t)$ lags behind the complex vector describing the excitation $f(t)$ by the phase angle φ. The velocity is obtained by differentiating (9.46) with respect to time,

$$\dot{x}(t) = i\omega A|H(\omega)|e^{i(\omega t - \varphi)} = i\omega x(t), \tag{9.47}$$

and since we can write

$$i = \cos\frac{\pi}{2} + i\sin\frac{\pi}{2} = e^{i\pi/2},$$

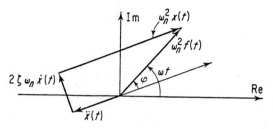

FIGURE 9.2

we conclude that the velocity vector is 90° ahead of the displacement vector and ω times as large. The second derivative of (9.46) gives

$$\ddot{x}(t) = (i\omega)^2 A |H(\omega)| e^{i(\omega t - \varphi)} = -\omega^2 x(t), \qquad (9.48)$$

so the acceleration vector is 180° ahead of the displacement vector and ω^2 as large. Equation (9.43) can be represented in a complex plane as shown in Figure 9.2. We note that all the complex vectors in Figure 9.2 are harmonic functions of time and their circular frequencies are all equal to the driving frequency ω. It follows that for steady-state harmonic motion the complex vectors rotate in the complex plane with the circular frequency ω while maintaining the same relative position. Considering only the real part of the excitation and the real part of the response is equivalent to taking the real axis projections in Figure 9.2. One could have represented the excitation and the response just as easily as the projections on the imaginary axis, or any other axis for that matter. This would have resulted in the addition of a phase angle to all the complex vectors without affecting their relative positions or their magnitudes.

The energy dissipated by a viscously damped single-degree-of-freedom system in one cycle of motion can be written

$$\Delta E_{\text{cyc}} = \int_{\text{cycle}} F \, dx = \int_0^{2\pi/\omega} F \dot{x} \, dt, \qquad (9.49)$$

where only the real part of the impressed force F and the velocity $\dot{x}$ must be considered. Regarding A as a real number,† we obtain

$$\Delta E_{\text{cyc}} = m\omega_n^2 \int_0^{2\pi/\omega} \text{Re} \, [f(t)] \, \text{Re} \, [\dot{x}(t)] \, dt$$

$$= -m\omega_n^2 A^2 |H(\omega)| \omega \int_0^{2\pi/\omega} \cos \omega t \sin (\omega t - \varphi) \, dt$$

$$= m\omega_n^2 A^2 |H(\omega)| \pi \sin \varphi = c\pi\omega X^2, \qquad (9.50)$$

† In view of the discussion of the complex vectors diagram of Figure 9.2 we realize that this does not affect the result.

where X is the maximum displacement amplitude and c is the damping coefficient.

With this brief introduction we are now in a position to discuss the concept of structural damping. Experience shows that energy is dissipated in all systems, including those generally considered conservative. In fact, conservative systems are more of a mathematical convenience than a physical reality. Although the simple mass-spring system is treated as a conservative system, its motion does not last forever but dies out eventually. This is due to the internal friction in the spring. Unlike viscous damping, this type of damping does not depend on the time rate of strain. Experiments performed by Kimball and Lovell[†] indicate that for a large variety of materials, such as

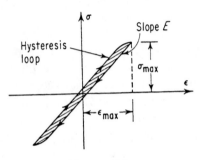

FIGURE 9.3

metals, glass, rubber, and maple wood, subjected to cyclic stress such that the strains remain below the elastic limit, the internal friction is entirely independent of the rate of strain. In fact, it was found that over a considerable frequency range the internal friction depended on the amplitude of oscillation. The energy loss per cycle of stress was found to be proportional to the amplitude squared,

$$\Delta E_{\text{cyc}} = \alpha X^2, \tag{9.51}$$

where α is a constant independent of the frequency of the harmonic oscillation. The same behavior is exhibited by any elastic piece of material. This type of damping is called *structural damping*, and it is generally attributed to the hysteresis[‡] of the elastic material. A piece of material in cyclic stress exhibits a stress-strain relation characterized by a hysteresis loop, as shown in Figure 9.3, even if the strain amplitude is well below the elastic limit of the material.

[†] A. L. Kimball and D. E. Lovell,"Internal Friction in Solids," *Phys. Rev.*, **30** (2nd Ser.), 948–959 (1927).

[‡] Structural damping is also attributed to friction between two components of a system.

The energy dissipated in one cycle is proportional to the area within the hysteresis loop. For that reason structural damping is often called *hysteretic damping*. In spite of the fact that this type of damping is independent of the frequency of the cycling stress, it is a phenomenon associated with cyclic stress. Hence, for systems subjected to harmonic excitation, one may treat a system with structural damping as a viscously damped system with an equivalent viscous damping coefficient inversely proportional to the driving frequency.

$$c_{eq} = \frac{\alpha}{\pi\omega}, \tag{9.52}$$

where the value of c_{eq} was obtained by simply comparing (9.50) and (9.51). It follows that, provided the excitation is harmonic, we can write the equation of motion of a single-degree-of-freedom system with structural damping in the form

$$m\ddot{x}(t) + \frac{\alpha}{\pi\omega}\,\dot{x}(t) + kx(t) = kAe^{i\omega t}, \tag{9.53}$$

and substitution of (9.47) in (9.53) yields

$$m\ddot{x}(t) + k(1 + i\gamma)x(t) = kAe^{i\omega t}, \tag{9.54}$$

where

$$\gamma = \frac{\alpha}{\pi k} \tag{9.55}$$

is called the *structural damping factor*. The quantity $k(1 + i\gamma)$ is called *complex stiffness*, and occasionally it is also called *complex damping*. In brief, structural damping is proportional to the amplitude of the displacement and opposite in direction to the velocity in harmonic oscillation.

The steady-state solution of (9.54) is

$$x(t) = \frac{Ae^{i\omega t}}{1 - (\omega/\omega_n)^2 + i\gamma}, \tag{9.56}$$

and we note that the maximum reponse for the structurally damped case is obtained for $\omega = \omega_n$, in contrast with the viscously damped case [see (1.75)].

9–5   STRUCTURALLY DAMPED DISCRETE SYSTEMS

The concept of structural damping can be applied to a multi-degree-of-freedom system provided the excitation forces are harmonic and of the same driving frequency Ω.

The equations of motion of a viscously damped system subjected to harmonic forces of equal frequency are

$$[m]\{\ddot{q}\} + [c]\{\dot{q}\} + [k]\{q\} = \{Q_0\}e^{i\Omega t}. \tag{9.57}$$

Consistent with (9.52) for a single-degree-of-freedom system, we can introduce equivalent viscous damping coefficients, representing structural damping, inversely proportional to the driving frequency

$$c_{ij} = \frac{1}{\pi\Omega}\,\alpha_{ij}. \tag{9.58}$$

This allows us to write the equations of motion of a structurally damped system subjected to harmonic forces in the form

$$[m]\{\ddot{q}\} + \frac{1}{\pi\Omega}[\alpha]\{\dot{q}\} + [k]\{q\} = \{Q_0\}e^{i\Omega t}, \tag{9.59}$$

where $(1/\pi\Omega)[\alpha]$ is a symmetric matrix often called a *hysteretic damping matrix*. It follows that the structurally damped system described by (9.59) can be treated in the same manner as the viscously damped system described by (9.57). We recall that in certain cases it is possible to uncouple the system of equations (9.57) with the help of the normal modes of the undamped system so that the method of solution of Sections 9–2 and 9–3 can be used to solve (9.59).

Next let us denote

$$\alpha_{ij} = \pi\gamma_{ij}k_{ij}, \tag{9.60}$$

where the coefficients γ_{ij} are structural damping factors. Furthermore, for steady-state harmonic oscillation, we have

$$\{\dot{q}\} = i\Omega\{q\}, \tag{9.61}$$

so equations (9.59) reduce to

$$[m]\{\ddot{q}\} + i[\gamma k]\{q\} + [k]\{q\} = \{Q_0\}e^{i\Omega t}. \tag{9.62}$$

It is customary to assume that the hysteretic damping matrix is proportional to the stiffness matrix, implying that all the coefficients γ_{ij} have the same value, $\gamma_{ij} = \gamma$, so that (9.62) can be written

$$[m]\{\ddot{q}\} + (1 + i\gamma)[k]\{q\} = \{Q_0\}e^{i\Omega t}. \tag{9.63}$$

Equations (9.63) can be solved with relative ease by using modal analysis. To this end we introduce the linear transformation

$$\{q\} = [u]\{\eta\}, \tag{9.64}$$

where $[u]$ is the matrix of the orthonormal modes associated with the un-damped system. Substitution of (9.64) in (9.63) and premultiplication of the result by $[u]^T$ yields a set of n uncoupled equations of the type

$$\ddot{\eta}_r(t) + (1 + i\gamma)\omega_r^2\eta_r(t) = N_r e^{i\Omega t}, \qquad r = 1, 2, \ldots, \tag{9.65}$$

where ω_r is the rth natural frequency associated with the undamped system and

$$N_r = \{u^{(r)}\}^T\{Q_0\} \tag{9.66}$$

is the amplitude of the generalized force associated with the rth generalized coordinate. Equations (9.65) have the appearance of a single-degree-of-freedom system equation. Their steady-state solutions resemble (9.56).

9–6 GENERAL CASE OF VISCOUSLY DAMPED DISCRETE SYSTEMS. LAPLACE TRANSFORM SOLUTION

The equations of motion of a viscously damped linear discrete system can be written

$$[m]\{\ddot{x}(t)\} + [c]\{\dot{x}(t)\} + [k]\{x(t)\} = \{F(t)\}, \tag{9.67}$$

where $[m]$, $[c]$, and $[k]$ are symmetric matrices defined previously. In Section 9–2 we explored the possibility of using the classical modal matrix $[u]$, associated with the undamped system, to represent a linear transformation uncoupling the equations of motion (9.67). Specifically we discussed the special form the damping matrix $[c]$ must have for the triple product $[u]^T[c][u]$ to be a diagonal matrix. In the general case, however, it is not possible to use the classical modal matrix $[u]$ to uncouple (9.67), and a different treatment must be adopted.

Equations (9.67) represent a system of coupled linear ordinary differential equations with constant coefficients. At least theoretically, a solution by means of the Laplace transform is possible. Letting $\{\bar{x}(s)\}$ and $\{\bar{F}(s)\}$ be the Laplace transforms of the column matrices $\{x(t)\}$ and $\{F(t)\}$, respectively, we can transform both sides of (9.67) and write

$$[m]\{s^2\bar{x}(s) - sx(0) - \dot{x}(0)\} + [c]\{s\bar{x}(s) - x(0)\} + [k]\{\bar{x}(s)\} = \{\bar{F}(s)\}, \tag{9.68}$$

where $x_i(0)$ and $\dot{x}_i(0)$ ($i = 1, 2, \ldots, n$) are initial displacements and velocities, respectively. Equation (9.68) leads to the transformed solution

$$\{\bar{x}(s)\} = [Z(s)]^{-1}\{\bar{R}(s)\}, \tag{9.69}$$

where
$$[Z(s)] = s^2[m] + s[c] + [k] \tag{9.70}$$

is the mechanical impedance of the viscously damped system and

$$\{\bar{R}(s)\} = \{\bar{F}(s)\} + [s[m] + [c]]\{x(0)\} + [m]\{\dot{x}(0)\} \tag{9.71}$$

is a more general transformed input which includes the effect of both the excitation forces and initial conditions. As for the undamped case, we can define a matrix of transfer functions given by

$$[G(s)] = [Z(s)]^{-1} = \frac{[A(s)]}{|Z(s)|}, \tag{9.72}$$

where $[A(s)]$ is the adjoint and $|Z(s)|$ is the determinant of the impedance matrix $[Z(s)]$. Introducing (9.72) in (9.69) we can write the transformed response in the form

$$\{\bar{x}(s)\} = [G(s)]\{\bar{R}(s)\}, \tag{9.73}$$

and the response is obtained by writing the inverse Laplace transformation,

$$\{x(t)\} = \mathscr{L}^{-1}\{\bar{x}(s)\} = \mathscr{L}^{-1}[G(s)]\{\bar{R}(s)\}. \tag{9.74}$$

To perform the inverse transformation one must know the poles of all the transformed outputs $\bar{x}_i(s)$. These poles can result from two sources: the transformed inputs $\bar{F}_i(s)$ and the transfer functions $G_{ij}(s)$, where the latter depend on the system characteristics. The poles of $G_{ij}(s)$ are obtained from the equation

$$|Z(s)| = 0, \tag{9.75}$$

which is recognized as the characteristic equation. In contrast to the undamped case, in the damped case the roots can be real, pure imaginary, or complex. If they are pure imaginary or complex, the roots appear in pairs of complex conjugates; the pairs of complex conjugates possess negative real parts.

Of course, the main difficulty lies in obtaining the roots of the characteristic equation. For undamped systems the order of the characteristic equation coincides with the degree of freedom of the system, because the parameter s does not appear in odd powers. This is not the case for damped systems, as one can conclude from (9.70). There are various schemes for obtaining the roots of an algebraic equation of order $2n$. Such schemes can be found in many texts on numerical analysis. Graeffe's root-squaring method can be used for equations known to have roots consisting of complex conjugate pairs. In this case, however, because the absolute values of a pair of complex conjugate roots are equal, $|\lambda_1| = |\lambda_2|$, the root-squaring method separates one pair of complex conjugates from another. This in effect separates from the original equation a quadratic factor which enables us to obtain the pair of complex conjugates with the largest magnitude first. For a description of the procedure the reader is referred to the texts by Scanlan and Rosenbaum[†] and Hildebrand.[‡]

Having obtained the poles of all the functions $\bar{x}_i(s)$, there is no difficulty, in principle, in carrying out the inversion and obtaining the responses $x_i(t)$.

Example 9.2

For the system shown in Figure 9.4 it is required to obtain the responses $x_1(t)$ and $x_2(t)$ by means of the Laplace transformation. The equations of

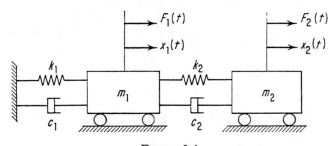

FIGURE 9.4

motion of the system are in the form given by (9.67), where the inertia, damping, and stiffness matrices have the form

$$[m] = \begin{bmatrix} m_1 & 0 \\ 0 & m_2 \end{bmatrix}, \qquad [c] = \begin{bmatrix} c_1 + c_2 & -c_2 \\ -c_2 & c_2 \end{bmatrix}, \qquad [k] = \begin{bmatrix} k_1 + k_2 & -k_2 \\ -k_2 & k_2 \end{bmatrix}$$

$$\text{(a)}$$

and the displacement and force vectors are

$$\{x(t)\} = \begin{Bmatrix} x_1(t) \\ x_2(t) \end{Bmatrix}, \qquad \{F(t)\} = \begin{Bmatrix} F_1(t) \\ F_2(t) \end{Bmatrix}. \qquad \text{(b)}$$

The impedance matrix can be written

$$[Z(s)] = \begin{bmatrix} m_1 s^2 + (c_1 + c_2)s + (k_1 + k_2) & -(c_2 s + k_2) \\ -(c_2 s + k_2) & m_2 s^2 + c_2 s + k_2 \end{bmatrix}, \qquad \text{(c)}$$

† R. H. Scanlan and R. Rosenbaum, *Introduction to the Study of Aircraft Vibration and Flutter*, Macmillan, New York, 1962, p. 50.

‡ F. B. Hildebrand, *Introduction to Numerical Analysis*, McGraw-Hill, New York, 1956, p. 462.

and the determinant of the impedance matrix is

$$\Delta(s) = |Z(s)|$$

$$= [s^2 m_1 + s(c_1 + c_2) + (k_1 + k_2)](s^2 m_2 + sc_2 + k_2) - (sc_2 + k_2)^2$$

$$= m_1 m_2 s^4 + [m_1 c_2 + m_2(c_1 + c_2)]s^3 + [m_1 k_2 + m_2(k_1 + k_2) + c_1 c_2]s^2$$

$$+ (c_1 k_2 + c_2 k_1)s + k_1 k_2, \tag{d}$$

which is a polynomial of fourth order in s. The matrix of the transfer functions becomes

$$[G(s)] = \frac{1}{\Delta(s)} \begin{bmatrix} m_2 s^2 + c_2 s + k_2 & c_2 s + k_2 \\ c_2 s + k_2 & m_1 s^2 + (c_1 + c_2)s + (k_1 + k_2) \end{bmatrix}. \tag{e}$$

Letting the initial conditions be zero, we have the transformed input

$$\{\bar{R}(s)\} = \begin{cases} \bar{F}_1(s) \\ \bar{F}_2(s) \end{cases} \tag{f}$$

so the transformed responses are obtained by writing

$$\bar{x}_1(s) = G_{11}(s)\bar{F}_1(s) + G_{12}(s)\bar{F}_2(s)$$

$$= \frac{m_2 s^2 + c_2 s + k_2}{\Delta(s)} \bar{F}_1(s) + \frac{c_2 s + k_2}{\Delta(s)} \bar{F}_2(s),$$

$$\tag{g}$$

$$\bar{x}_2(s) = G_{21}(s)\bar{F}_1(s) + G_{22}(s)\bar{F}_2(s)$$

$$= \frac{c_2 s + k_2}{\Delta(s)} \bar{F}_1(s) + \frac{m_1 s^2 + (c_1 + c_2)s + (k_1 + k_2)}{\Delta(s)} \bar{F}_2(s).$$

Letting $m_1 = m$, $m_2 = 2m$, $c_1 = c_2 = c$, $k_1 = k$, and $k_2 = 4k$, as well as $c = 0.2m\omega$ and $k = m\omega^2$, equations (g) reduce to

$$\bar{x}_1(s) = \frac{2m(s^2 + 0.1s\omega + 2\omega^2)}{\Delta(s)} \bar{F}_1(s) + \frac{2m(0.1s\omega + 2\omega^2)}{\Delta(s)} \bar{F}_2(s),$$

$$\tag{h}$$

$$\bar{x}_2(s) = \frac{2m(0.1s\omega + 2\omega^2)}{\Delta(s)} \bar{F}_1(s) + \frac{m(s^2 + 0.4s\omega + 5\omega^2)}{\Delta(s)} \bar{F}_2(s),$$

where the characteristic determinant is

$$\Delta(s) = |Z(s)| = 2m^2(s^4 + 0.50s^3\omega + 7.02s^2\omega^2 + 0.50s\omega^3 + 2\omega^4). \tag{i}$$

The inverse transformation of (h) may be carried out by means of Borel's theorem,

$$x_1(t) = \int_0^t F_1(\tau)G_{11}(t - \tau)\, d\tau + \int_0^t F_2(\tau)G_{12}(t - \tau)\, d\tau,$$

$$x_2(t) = \int_0^t F_1(\tau)G_{21}(t - \tau)\, d\tau + \int_0^t F_2(\tau)G_{22}(t - \tau)\, d\tau,$$
(j)

where $\qquad\qquad G_{ij}(t) = \mathscr{L}^{-1}G_{ij}(s), \qquad i, j = 1, 2$ (k)

are the inverse transformations of the transfer functions $G_{ij}(s)$. As an illustration let us explore the procedure for obtaining $G_{11}(t)$. To this end write $G_{11}(s)$ in the form

$$G_{11}(s) = \frac{2m(s^2 + 0.1s\omega + 2\omega^2)}{\Delta(s)} = \frac{s^2 + 0.1s\omega + 2\omega^2}{m(s - s_1)(s - s_2)(s - s_3)(s - s_4)}, \quad (l)$$

where s_i ($i = 1, 2, 3, 4$) are the roots of the characteristic equation, $\Delta(s) = 0$, in other words, the poles of $G_{11}(s)$. Assuming all the s_i to be distinct, the inverse transformation becomes

$$G_{11}(t) = \sum_{i=1}^{4} \lim_{s \to s_i} [(s - s_i)G_{11}(s)e^{st}]$$

$$= \frac{1}{m}\frac{s_1^2 + 0.1s_1\omega + 2\omega^2}{(s_1 - s_2)(s_1 - s_3)(s_1 - s_4)} e^{s_1 t}$$

$$+ \frac{1}{m}\frac{s_2^2 + 0.1s_2\omega + 2\omega^2}{(s_2 - s_1)(s_2 - s_3)(s_2 - s_4)} e^{s_2 t}$$

$$+ \frac{1}{m}\frac{s_3^2 + 0.1s_3\omega + 2\omega^2}{(s_3 - s_1)(s_3 - s_2)(s_3 - s_4)} e^{s_3 t}$$

$$+ \frac{1}{m}\frac{s_4^2 + 0.1s_4\omega + 2\omega^2}{(s_4 - s_1)(s_4 - s_2)(s_4 - s_3)} e^{s_4 t}$$

$$= A_1 e^{s_1 t} + A_2 e^{s_2 t} + A_3 e^{s_3 t} + A_4 e^{s_4 t}, \quad (m)$$

where the quantities A_i are complex numbers. It turns out that the roots of the determinantal equation are complex conjugates,[†]

$$\left.\begin{matrix} s_1 \\ s_2 \end{matrix}\right\} = (-0.2226 \pm 2.5783i)\omega, \qquad \left.\begin{matrix} s_3 \\ s_4 \end{matrix}\right\} = (-0.0274 \pm 0.5458i)\omega,$$

† The roots were obtained by means of a computer.

so we must conclude that A_1 and A_2, on the one hand, and A_3 and A_4, on the other, are complex conjugates. This allows us to write

$$G_{11}(t) = 2e^{-0.2226\omega t}(\text{Re } A_1 \cos 2.5783\omega t - \text{Im } A_1 \sin 2.5783\omega t)$$

$$+ 2e^{-0.0274\omega t}(\text{Re } A_3 \cos 0.5458\omega t - \text{Im } A_3 \sin 0.5458\omega t), \quad \text{(n)}$$

which is typical of an underdamped system. The remaining functions $G_{ij}(t)$ are obtained in a similar way. They are all of the same type as $G_{11}(t)$ and only the constants A_i are different. The responses $x_1(t)$ and $x_2(t)$ are obtained by introducing the functions $G_{ij}(t)$ in (j) and performing the indicated integrations. Of course, before performing the integrations one must know the excitation functions $F_1(t)$ and $F_2(t)$.

9-7 GENERAL CASE OF DAMPING. INTRODUCTION TO MODAL ANALYSIS

Consider again the equations of motion of a viscously damped system

$$[m]\{\ddot{q}(t)\} + [c]\{\dot{q}(t)\} + [k]\{q(t)\} = \{Q(t)\}, \tag{9.76}$$

where $[m]$, $[c]$, and $[k]$ are symmetric inertia, damping, and stiffness matrices, respectively.

Next consider the free vibration case, $\{Q(t)\} = \{0\}$, and assume a solution of the homogeneous set of equations in the form

$$\{q(t)\} = \{\psi\}e^{\lambda t}, \tag{9.77}$$

so that we obtain a set of n homogeneous algebraic equations representing the eigenvalue problem

$$[\lambda^2[m] + \lambda[c] + [k]]\{\psi\} = [f(\lambda)]\{\psi\} = \{0\}, \tag{9.78}$$

where $[f(\lambda)]$ is a square matrix depending on λ (sometimes referred to as a lambda matrix). Equations (9.78) have a nontrivial solution only if the determinant of the coefficients is zero,

$$\Delta(\lambda) = |f(\lambda)| = 0, \tag{9.79}$$

which is the characteristic equation or determinantal equation—an algebraic equation of order $2n$ in λ. As mentioned in preceding sections, the roots can be real, purely imaginary, or complex. If the roots are real they must be negative, which corresponds to an overdamped system for which an aperiodic decaying motion is obtained. If the roots are complex they must appear in pairs of complex conjugates with negative real part. The corresponding modal columns $\{\psi\}$ must also be complex conjugates. A pair of complex conjugate modes multiplied by the corresponding time-dependent exponential functions can be combined to obtain a damped oscillatory motion. This is the case in which the system is underdamped. For undamped systems one obtains purely imaginary roots which appear in pairs of complex conjugates.

Assuming that all roots are distinct, a constituent solution of the homogeneous equation is

$$\{q(t)\} = \{\psi^{(r)}\}e^{\lambda_r t}, \qquad r = 1, 2, \ldots, 2n, \tag{9.80}$$

where $\{\psi^{(r)}\}$ is a column of constants called the rth modal column. As in the undamped case, $\{\psi^{(r)}\}$ can be obtained by taking any of the nonvanishing columns of the matrix $[F(\lambda_r)]$, which is the adjoint of the matrix $[f(\lambda_r)]$. The most general solution of the homogeneous equation can be written as a linear superposition of the constituent solutions, (9.80), multiplied by arbitrary constants c_r, which may be real or complex numbers. Hence, in matrix form, we have

$$\{q(t)\} = [\psi]\{ce^{\lambda t}\}, \tag{9.81}$$

where $[\psi]$ is an $n \times 2n$ rectangular modal matrix. Unfortunately the modal matrix $[\psi]$ cannot be used as a transformation matrix of the form

$$\{q(t)\} = [\psi]\{\eta(t)\} \tag{9.82}$$

to obtain a solution of the nonhomogeneous problem. The reason is that there are $2n$ modes $\{\psi^{(r)}\}$ and consequently $2n$ coordinates $\eta_r(t)$, but there are only n coordinates $q_i(t)$. One can overcome this difficulty by introducing a set of auxiliary variables and convert a Lagrangian set of n second order ordinary differential equations into an equivalent set of $2n$ first order ordinary differential equations known as *Hamilton's canonical equations*. The auxiliary set of variables are the generalized momenta $\{p\} = [m]\{\dot{q}\}$. An equivalent approach is employed by Frazer, Duncan, and Collar,† in which the generalized velocities $\{\dot{q}\}$ are used as auxiliary variables. This leads to a set of $2n$ first order ordinary differential equations

$$[M]\{\dot{y}(t)\} + [K]\{y(t)\} = \{Y(t)\}, \tag{9.83}$$

where $\qquad \{y(t)\} = \begin{Bmatrix} \{\dot{q}(t)\} \\ \{q(t)\} \end{Bmatrix} \qquad$ and $\qquad \{Y(t)\} = \begin{Bmatrix} \{0\} \\ \{Q(t)\} \end{Bmatrix} \qquad$ (9.84)

† R. A. Frazer, W. J. Duncan, and A. R. Collar, *Elementary Matrices*, Cambridge Univ. Press, New York, 1957, p. 289.

are column matrices consisting of $2n$ elements representing generalized coordinates and generalized forces, respectively, and

$$[M] = \begin{bmatrix} [0] & [m] \\ [m] & [c] \end{bmatrix} \quad \text{and} \quad [K] = \begin{bmatrix} -[m] & [0] \\ [0] & [k] \end{bmatrix} \quad (9.85)$$

are real symmetric matrices of order $2n$, because $[m]$, $[c]$, and $[k]$ are real symmetric matrices.

This formulation has the advantage that the modes obtained from the solution of the homogeneous equations, obtained by letting $\{Y(t)\} = \{0\}$ in (9.83), are orthogonal, and hence can be used in conjunction with the expansion theorem to obtain the solution of the nonhomogeneous problem.

9–8   SOLUTION OF THE EIGENVALUE PROBLEM. CHARACTERISTIC DETERMINANT

The solution of (9.83) by modal analysis follows closely the pattern established for undamped systems. First we must solve the homogeneous set of equations from which we obtain the eigenvalues and eigenvectors. The solution of the nonhomogeneous set of differential equations is assumed as a superposition of the eigenvectors multiplied by time-dependent generalized coordinates. The procedure leads to an uncoupled set of differential equations for the generalized coordinates which can be solved with relative ease, thus completing the solution.

In this section we shall present the solution of the eigenvalue problem based on the solution of the characteristic equation. Later we shall show a method based on the sweeping procedure.

From (9.83) we obtain the homogeneous set of differential equations which can be written in the matrix form

$$[M]\{\dot{y}(t)\} + [K]\{y(t)\} = \{0\}. \quad (9.86)$$

The solution of (9.86) is obtained by letting

$$\{y(t)\} = e^{\alpha t}\{\Phi\}, \quad (9.87)$$

where $\{\Phi\}$ represents a vector consisting of $2n$ constant elements. Equation (9.87), when introduced in (9.86), leads to the eigenvalue problem

$$\alpha[M]\{\Phi\} + [K]\{\Phi\} = \{0\}, \quad (9.88)$$

which can be written in the familiar form

$$[D]\{\Phi\} = \frac{1}{\alpha}\{\Phi\}, \tag{9.89}$$

where $\quad [D] = -[K]^{-1}[M] = \begin{bmatrix} [m]^{-1} & [0] \\ [0] & -[k]^{-1} \end{bmatrix}\begin{bmatrix} [0] & [m] \\ [m] & [c] \end{bmatrix}$

$$= \begin{bmatrix} [0] & [I] \\ -[k]^{-1}[m] & -[k]^{-1}[c] \end{bmatrix} \tag{9.90}$$

plays the role of a dynamical matrix. Note that the multiplication of partitioned matrices is handled as if the submatrices were ordinary elements. Equation (9.89) can be rewritten

$$[f(\alpha)]\{\Phi\} = \{0\}, \tag{9.91}$$

where $\qquad\qquad [f(\alpha)] = [D] - \frac{1}{\alpha}[I]. \tag{9.92}$

Note that the identity matrix in (9.92) is of order $2n$, whereas the one in (9.90) is only of order n.

The eigenvalue problem (9.91) has a nontrivial solution only if the characteristic determinant, i.e. the determinant of the matrix $[f(\alpha)]$, vanishes:

$$\Delta(\alpha) = |f(\alpha)| = 0, \tag{9.93}$$

which is the characteristic equation. Equation (9.93) is an algebraic equation of order $2n$ in α and its solution yields a set of $2n$ eigenvalues α_r ($r = 1, 2, \ldots, 2n$). The corresponding eigenvectors $\{\Phi^{(r)}\}$ are determined by substituting the eigenvalues α_r in (9.92) and writing the matrix $[f(\alpha_r)]$; the eigenvector $\{\Phi^{(r)}\}$ is obtained by taking any nonvanishing column of the adjoint matrix $[F(\alpha_r)]$ of the matrix $[f(\alpha_r)]$. The proof of this statement is the same as for undamped systems.

The solution of the eigenvalue problem yields $2n$ eigenvalues α_r and eigenvectors

$$\{\Phi^{(r)}\} = \begin{Bmatrix} \alpha_r\{\phi^{(r)}\} \\ \{\phi^{(r)}\} \end{Bmatrix}, \qquad r = 1, 2, \ldots, 2n, \tag{9.94}$$

where n is the degree of freedom of the system.

At this point a discussion of the eigenvectors $\{\Phi^{(r)}\}$ seems in order. Looking at the adjoint matrix $[F(\alpha_r)]$ one would find it difficult to conclude that all of its columns represent the same vector—the eigenvector $\{\Phi^{(r)}\}$. One must recall, however, that for free vibration the modes are not determined in an absolute sense. In the case of undamped systems the amplitude is arbitrary, and the mode is determined within a multiplicative constant. For damped systems not only the amplitude but also the phase angle is arbitrary. The columns are said to represent the same vector if the ratios of the magnitudes of corresponding complex elements are the same for every column and in addition the difference in phase angles of the complex elements are the same for every column. This, in effect, means that if one factors out of every column a complex quantity the resulting columns are identical.

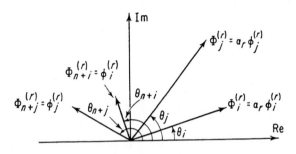

FIGURE 9.5

This is best illustrated by Figure 9.5. For every column of the matrix $[F(\alpha_r)]$ we must have the relations

$$\frac{|\Phi_i^{(r)}|}{|\Phi_j^{(r)}|} = \frac{|\Phi_{n+i}^{(r)}|}{|\Phi_{n+j}^{(r)}|} = \text{const}, \qquad \theta_j - \theta_i = \theta_{n+j} - \theta_{n+i} = \text{const}, \quad (9.95)$$

where $|\Phi_i^{(r)}|$ is the magnitude and θ_i is the phase angle of the ith complex element of the eigenvector $\{\Phi^{(r)}\}$.

A constituent solution of the homogeneous equation is

$$\{y(t)\} = e^{\alpha_r t}\{\Phi^{(r)}\}, \qquad r = 1, 2, \ldots, 2n, \qquad (9.96)$$

where the complex root α_r can be written

$$\alpha_r = -\zeta_r + i\omega_r. \qquad (9.97)$$

If we now plot the vector $\{y(t)\}$ in a complex diagram we obtain a diagram similar to the one of Figure 9.5, where all the complex elements of the modal vector $\{\Phi^{(r)}\}$ are multiplied by $e^{\alpha_r t}$. The effect of this multiplication is to reduce the magnitudes exponentially and to rotate the entire diagram in the complex plane with an angular velocity ω_r while the relative position of the complex elements is preserved. The motion in this mode, although synchronous, it not as easy to recognize as in the case of undamped systems. Every complex element possesses a different phase angle, so the corresponding coordinate will reach its maximum excursion at a different time than the remaining coordinates. However, the sequence in which the coordinates reach their maximum remains the same for each cycle; furthermore, after one complete cycle the coordinates are in the same position as at the beginning of the cycle. Therefore, the nodes continuously change their position during one cycle, but during the next cycle the pattern repeats itself. Of course, the maximum excursions decay exponentially from cycle to cycle. This motion is distinctly different than the synchronous motion of the modes associated with undamped systems. For undamped systems the elements are no longer complex and thus the phase angle is either zero or 180° (if the element is of opposite sign). All the elements reach their maximum excursions simultaneously and these maximum excursions remain the same cycle after cycle; the nodes are stationary.

As mentioned previously, the modes $\{\Phi^{(r)}\}$ are orthogonal. The orthogonality relation will be shown in Section 9–9. The modes can be normalized, which in this case means removing the arbitrariness both from the magnitudes and phase angles of the complex elements of the eigenvector $\{\Phi^{(r)}\}$.

Example 9.3

Set up the eigenvalue problem for the system of Figure 9.4 and obtain its solution by means of the characteristic determinant.

The $[M]$ and $[K]$ matrices for the system of Figure 9.4 are

$$[M] = \begin{bmatrix} [0] & [m] \\ [m] & [c] \end{bmatrix} = \left[\begin{array}{cc:cc} 0 & 0 & m_1 & 0 \\ 0 & 0 & 0 & m_2 \\ \hdashline m_1 & 0 & c_1 + c_2 & -c_2 \\ 0 & m_2 & -c_2 & c_2 \end{array} \right], \tag{a}$$

$$[K] = \begin{bmatrix} -[m] & [0] \\ [0] & [k] \end{bmatrix} = \left[\begin{array}{cc:cc} -m_1 & 0 & 0 & 0 \\ 0 & -m_2 & 0 & 0 \\ \hdashline 0 & 0 & k_1 + k_2 & -k_2 \\ 0 & 0 & -k_2 & k_2 \end{array} \right]. \tag{b}$$

The dynamical matrix $[D]$ can be obtained by premultiplying $[M]$ by $-[K]^{-1}$. Instead of using (a) and (b), however, we shall calculate $[D]$ by means of (9.90). To this end we calculate

$$[k]^{-1} = \cfrac{\begin{bmatrix} k_2 & k_2 \\ k_2 & k_1 + k_2 \end{bmatrix}}{\begin{vmatrix} k_1 + k_2 & -k_2 \\ -k_2 & k_2 \end{vmatrix}} = \begin{bmatrix} k_1^{-1} & k_1^{-1} \\ k_1^{-1} & k_1^{-1} + k_2^{-1} \end{bmatrix} \tag{c}$$

which enables us to write

$$[D] = -[K]^{-1}[M] = \begin{bmatrix} [0] & [I] \\ -[k]^{-1}[m] & -[k]^{-1}[c] \end{bmatrix}$$

$$= \left[\begin{array}{cc:cc} 0 & 0 & 1 & 0 \\ 0 & 0 & 0 & 1 \\ \hdashline -m_1 k_1^{-1} & -m_2 k_1^{-1} & -c_1 k_1^{-1} & 0 \\ -m_1 k_1^{-1} & -(m_2 k_1^{-1} + m_2 k_2^{-1}) & -(c_1 k_1^{-1} - c_2 k_2^{-1}) & -c_2 k_2^{-1} \end{array}\right].$$

$$\tag{d}$$

Next we write

$$[f(\alpha)] = [D] - \frac{1}{\alpha}[I]$$

$$= \begin{bmatrix} -\alpha^{-1} & 0 & 1 & 0 \\ 0 & -\alpha^{-1} & 0 & 1 \\ -m_1 k_1^{-1} & -m_2 k_1^{-1} & -c_1 k_1^{-1} - \alpha^{-1} & 0 \\ -m_1 k_1^{-1} & -(m_2 k_1^{-1} + m_2 k_2^{-1}) & -(c_1 k_1^{-1} - c_2 k_2^{-1}) & -c_2 k_2^{-1} - \alpha^{-1} \end{bmatrix},$$

$$\tag{e}$$

which has the determinant

$$\Delta(\alpha) = |f(\alpha)| = \frac{1}{k_1 k_2}\{m_1 m_2 \alpha^4 + [m_1 c_1 + m_2(c_1 + c_2)]\alpha^3$$

$$+ [m_1 k_2 + m_2(k_1 + k_2) + c_1 c_2]\alpha^2$$

$$+ (c_1 k_2 + c_2 k_1)\alpha + k_1 k_2\}. \tag{f}$$

Note that we obtained essentially the same determinant as in Example 9.2. This is to be expected.

The solution of the characteristic determinant, obtained by setting $\Delta(\alpha) = 0$, yields the roots α_r ($r = 1, 2, 3, 4$). The matrix $[f(\alpha_r)]$ is obtained by replacing

α in (e) by α_r. Let us choose the values $m_1 = m$, $m_2 = 2m$, $c_1 = c_2 = 0.2m\omega$, $k_1 = m\omega^2$, and $k_2 = 4m\omega^2$, so that the matrix $[f(\alpha_r)]$ takes the form

$$[f(\alpha_r)] = -\frac{1}{\omega^2\alpha_r}\begin{bmatrix} \omega^2 & 0 & -\omega^2\alpha_r & 0 \\ 0 & \omega^2 & 0 & -\omega^2\alpha_r \\ \alpha_r & 2\alpha_r & 0.2\omega\alpha_r + \omega^2 & 0 \\ \alpha_r & 2.5\alpha_r & 0.15\omega\alpha_r & 0.05\omega\alpha_r + \omega^2 \end{bmatrix}. \tag{g}$$

Letting

$$\alpha_r = \omega\beta_r, \tag{h}$$

where β_r is a nondimensional quantity, (h) reduces to

$$[f(\beta_r)] = -\frac{1}{\omega^2\beta_r}\begin{bmatrix} \omega & 0 & -\omega^2\beta_r & 0 \\ 0 & \omega & 0 & -\omega^2\beta_r \\ \beta_r & 2\beta_r & 0.2\omega\beta_r + \omega & 0 \\ \beta_r & 2.5\beta_r & 0.15\omega\beta_r & 0.05\omega\beta_r + \omega \end{bmatrix}. \tag{i}$$

The adjoint matrix is

$$[F(\beta_r)] = -\frac{1}{\omega^3\beta_r^3}$$

$$\times \begin{bmatrix} (1+0.25\beta_r+2.51\beta_r^2+0.20\beta_r^3)\omega & -(2\beta_r^2+0.10\beta_r^3)\omega \\ -(\beta_r^2+0.05\beta_r^3)\omega & (1+0.25\beta_r+1.01\beta_r^2+0.05\beta_r^3)\omega \\ -(\beta_r+0.05\beta_r^2+0.50\beta_r^3) & -(2\beta_r+0.10\beta_r^2) \\ -(\beta_r+0.05\beta_r^2) & -(2.50\beta_r+0.20\beta_r^2+0.50\beta_r^3) \end{bmatrix}$$

$$\begin{bmatrix} (\beta_r+0.05\beta_r^2+2.50\beta_r^3)\omega & -2\beta_r^3\omega \\ -(0.15\beta_r^2+\beta_r^3)\omega & (\beta_r+0.20\beta_r^2+\beta_r^3)\omega \\ 1+0.05\beta_r+2.50\beta_r^2 & -2\beta_r^2 \\ -(0.15\beta_r+\beta_r^2) & 1+0.20\beta_r+\beta_r^2 \end{bmatrix}. \tag{j}$$

The values β_r are obtained by solving the characteristic equation

$$\beta^4 + 0.50\beta^3 + 7.02\beta^2 + 0.50\beta + 2 = 0, \tag{k}$$

which has the roots

$$\left.\begin{matrix}\beta_1\\\beta_2\end{matrix}\right\} = -0.2226 \pm 2.5783i \quad \text{and} \quad \left.\begin{matrix}\beta_3\\\beta_4\end{matrix}\right\} = -0.0274 \pm 0.5458i. \tag{l}$$

As an illustration we shall calculate the first eigenvector. Introducing β_1 into (j) we obtain

$$[F(\alpha_1)] = -\frac{1}{\alpha_1^3} \begin{bmatrix} \alpha_1(-1.6615 + 5.8569i) & \alpha_1(1.1050 - 5.0418i) \\ \alpha_1(0.5525 - 2.5209i) & \alpha_1(-0.3379 + 2.1617i) \\ -1.6615 + 5.8569i & 1.1050 - 5.0418i \\ 0.5525 - 2.5209i & -0.3379 + 2.1617i \end{bmatrix}$$

$$\begin{matrix} \alpha_1(-15.5059 - 2.7406i) & \alpha_1(13.1958 + 2.2956i) \\ \alpha_1(6.6313 + 0.7611i) & \alpha_1(-5.6424 - 0.6321i) \\ -15.5059 - 2.7406i & 13.1958 + 2.2956i \\ 6.6313 + 0.7611i & -5.6424 - 0.6321i \end{matrix} \Bigg]. \quad \text{(m)}$$

Equation (m) gives no indication that all the columns of the matrix $[F(\alpha_1)]$ represent the same eigenvector, but they do. To show this let us represent the elements of the matrix $[F(\alpha_1)]$ in terms of complex vectors given by their magnitudes and phase angles:

$$[F(\alpha_1)] = -\frac{1}{\alpha_1^3} \begin{bmatrix} 15.7772\omega \ \underline{/200°\ 46'} & 13.3719\,\omega\ \underline{/\ 17°\ 18'} \\ 6.6858\,\omega\ \underline{/\ 17°\ 18'} & 5.6682\,\omega\ \underline{/193°\ 49'} \\ 6.0880\ \underline{/105°\ 50'} & 5.1615\ \underline{/282°\ 22'} \\ 2.5807\ \underline{/282°\ 22'} & 2.1879\ \underline{/\ 98°\ 53'} \end{bmatrix}$$

$$\begin{matrix} 40.7937\,\omega\ \underline{/284°\ 57'} & 34.6998\,\omega\ \underline{/104°\ 48'} \\ 17.2924\,\omega\ \underline{/101°\ 28'} & 14.7092\,\omega\ \underline{/281°\ 19'} \\ 15.7462\ \underline{/190°\ 01'} & 13.3940\ \underline{/\ 9°\ 52'} \\ 6.6748\ \underline{/\ 6°\ 32'} & 5.6777\ \underline{/186°\ 23'} \end{matrix} \Bigg]. \quad \text{(n)}$$

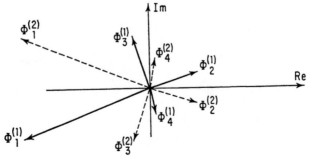

FIGURE 9.6

Now, it is easy to recognize that the four columns of matrix $[F(\alpha_1)]$ represent the same eigenvector. To obtain the first column we must multiply the magnitudes of all the elements of the second, third, and fourth columns by 1.1799, 0.3868, and 0.4547, respectively, and advance the corresponding phase angles by $183° 28'$, $-84° 11'$, and $95° 58'$. The first column of $[F(\alpha_1)]$ is plotted in Figure 9.6. Note that the complex quantity $-\alpha_1^{-3}$ multiplying the matrix $[F(\alpha_1)]$ was ignored.

The eigenvectors $\{\Phi^{(2)}\}$, $\{\Phi^{(3)}\}$, and $\{\Phi^{(4)}\}$ are obtained by introducing the eigenvalues α_2, α_3, and α_4, respectively, in the adjoint matrix $[F(\alpha)]$. Note that it is not really necessary to calculate $\{\Phi^{(2)}\}$, because it is just the complex conjugate of $\{\Phi^{(1)}\}$.† Similarly, $\{\Phi^{(3)}\}$ and $\{\Phi^{(4)}\}$ are complex conjugates, so in effect one need calculate only $\{\Phi^{(3)}\}$.

9–9 ORTHOGONALITY OF MODES

The eigenvalue α_r and the corresponding eigenvector $\{\Phi^{(r)}\}$ are such that they satisfy (9.88),

$$\alpha_r[M]\{\Phi^{(r)}\} + [K]\{\Phi^{(r)}\} = \{0\}, \qquad (9.98)$$

where $[M]$ and $[K]$ are symmetric matrices. Consider another solution of the eigenvalue problem, consisting of the eigenvalue α_s and eigenvector $\{\Phi^{(s)}\}$. The equation corresponding to this solution, in transposed form, can be written

$$\alpha_s\{\Phi^{(s)}\}^T[M] + \{\Phi^{(s)}\}^T[K] = \{0\}^T, \qquad (9.99)$$

where advantage has been taken of the symmetry of $[M]$ and $[K]$. Premultiply (9.98) by $\{\Phi^{(s)}\}^T$, postmultiply (9.99) by $\{\Phi^{(r)}\}$, and write

$$\alpha_r\{\Phi^{(s)}\}^T[M]\{\Phi^{(r)}\} + \{\Phi^{(s)}\}^T[K]\{\Phi^{(r)}\} = 0, \qquad (9.100)$$

$$\alpha_s\{\Phi^{(s)}\}^T[M]\{\Phi^{(r)}\} + \{\Phi^{(s)}\}^T[K]\{\Phi^{(r)}\} = 0. \qquad (9.101)$$

Subtraction of (9.101) from (9.100) yields

$$(\alpha_r - \alpha_s)\{\Phi^{(s)}\}^T[M]\{\Phi^{(r)}\} = 0. \qquad (9.102)$$

If the eigenvalues are distinct, $\alpha_r \neq \alpha_s$, we obtain the orthogonality relation

$$\{\Phi^{(s)}\}^T[M]\{\Phi^{(r)}\} = 0, \qquad \alpha_r \neq \alpha_s, \qquad (9.103)$$

† The elements of the eigenvector $\{\Phi^{(2)}\}$ are shown as dashed lines in Figure 9.6.

and either (9.100) or (9.101) indicates that the orthogonality relation can also be written

$$\{\Phi^{(s)}\}^T[K]\{\Phi^{(r)}\} = 0, \qquad \alpha_r \neq \alpha_s. \tag{9.104}$$

The matrix iteration method, using the sweeping technique to obtain the solution of the eigenvalue problem, is based on the orthogonality relation. For semidefinite systems, when rigid-body modes corresponding to zero eigenvalues are present, one can construct constraint matrices to remove the rigid-body motion from the absolute motion.

9–10 SOLUTION OF THE EIGENVALUE PROBLEM. MATRIX ITERATION METHOD

When the degree of freedom of the system is relatively large, solution of the eigenvalue problem by the characteristic determinant becomes impracticable. In such a case a matrix iteration procedure is more advisable. In Section 9–8 it was shown that the eigenvalue problem associated with a viscously damped n-degree-of-freedom system can be written

$$[D]\{\Phi\} = \lambda\{\Phi\}, \qquad \lambda = \frac{1}{\alpha}, \tag{9.105}$$

where the $2n \times 2n$ dynamical matrix $[D]$ is given by (9.90). Equation (9.105) can be used as the basis of a matrix iteration solution which will yield the eigenvalues λ in decreasing order of magnitude. For underdamped systems, however, we obtain pairs of complex conjugates as eigenvalues and eigenvectors and there may be some difficulty in interpreting the results of the iteration. In fact, we must develop new criteria enabling us to establish when convergence has been achieved. This will necessitate a slight digression into matrix algebra.

Let us consider the matrix

$$[f(\lambda)] = [D] - \lambda[I], \tag{9.106}$$

which has the reciprocal

$$[f(\lambda)]^{-1} = \frac{[F(\lambda)]}{\Delta(\lambda)} = \frac{[F(\lambda)]}{(\lambda - \lambda_1)(\lambda - \lambda_2)\cdots(\lambda - \lambda_{2n})}, \tag{9.107}$$

where λ_r $(r = 1, 2, \ldots, 2n)$ are the eigenvalues of the matrix $[D]$, $[F(\lambda)]$ the adjoint, and $\Delta(\lambda)$ the determinant of matrix $[f(\lambda)]$. Equation (9.107) can be written in terms of the partial-fraction expansion

$$[f(\lambda)]^{-1} = \frac{[A_1]}{\lambda - \lambda_1} + \frac{[A_2]}{\lambda - \lambda_2} + \cdots + \frac{[A_{2n}]}{\lambda - \lambda_{2n}} = \sum_{r=1}^{2n} \frac{[A_r]}{\lambda - \lambda_r}, \quad (9.108)$$

where the matrices $[A_r]$ $(r = 1, 2, \ldots, 2n)$ can be obtained from

$$[A_r] = (\lambda - \lambda_r)[f(\lambda)]^{-1}\Big|_{\lambda = \lambda_r} = \frac{[F(\lambda_r)]}{\Delta^{(1)}(\lambda_r)} = [B(\lambda_r)], \quad (9.109)$$

in which

$$\Delta^{(1)}(\lambda_r) = \frac{d\Delta(\lambda)}{d\lambda}\Big|_{\lambda = \lambda_r}. \quad (9.110)$$

Sylvester's theorem† states that if the eigenvalues λ_r of a matrix $[D]$ are all distinct, then any polynomial $P([D])$ of the matrix $[D]$ can be written

$$P([D]) = \sum_{r=1}^{2n} P(\lambda_r)[B(\lambda_r)], \quad (9.111)$$

where $P(\lambda_r)$ is a scalar polynomial in λ_r of the same form as $P([D])$.

Consider two complex conjugate eigenvalues, $\lambda_1 = \mu_1 + i\omega_1$ and $\lambda_2 = \mu_1 - i\omega_1$, possessing the largest magnitude of all the eigenvalues associated with the matrix $[D]$. Then, if $P_0([D])$ is a polynomial independent of a given integer s, the polynomial $P([D])$ can be written

$$P([D]) = [D]^s P_0([D]) \cong \lambda_1^s P_0(\lambda_1)[B(\lambda_1)] + \lambda_2^s P_0(\lambda_2)[B(\lambda_2)], \quad (9.112)$$

where the integer s is sufficiently large that the remainder of the terms in the series can be neglected. If $P_0([D])$ is of the special form

$$P_0([D]) = ([D] - \lambda_1[I])([D] - \lambda_2[I]), \quad (9.113)$$

it follows that $P_0(\lambda_1) = P_0(\lambda_2) = 0$, so (9.112) reduces to

$$[D]^s([D] - \lambda_1[I])([D] - \lambda_2[I]) \cong [0], \quad (9.114)$$

which leads to the algebraic equation

$$(\mu_1^2 + \omega_1^2)E_s - 2\mu_1 E_{s+1} + E_{s+2} = 0, \quad (9.115)$$

† See R. A. Frazer, W. J. Duncan, and A. R. Collar, *Elementary Matrices*, Cambridge Univ. Press, New York, 1957, p. 78.

where E_s, E_{s+1}, and E_{s+2} are homologous† elements of the matrices $[D]^s$, $[D]^{s+1}$, and $[D]^{s+2}$, respectively. Replacing s by $s + 1$ in (9.115) we obtain

$$(\mu_1^2 + \omega_1^2)E_{s+1} - 2\mu_1 E_{s+2} + E_{s+3} = 0. \tag{9.116}$$

Equations (9.115) and (9.116), when solved for $\mu_1^2 + \omega_1^2$ and $2\mu_1$, yield

$$\mu_1^2 + \omega_1^2 = \frac{\begin{vmatrix} E_{s+1} & E_{s+2} \\ E_{s+2} & E_{s+3} \end{vmatrix}}{\begin{vmatrix} E_s & E_{s+1} \\ E_{s+1} & E_{s+2} \end{vmatrix}} = \frac{E_{s+1}E_{s+3} - E_{s+2}^2}{E_s E_{s+2} - E_{s+1}^2} = \frac{F_{s+1}}{F_s},$$

$$\hfill (9.117)$$

$$2\mu_1 = \frac{\begin{vmatrix} E_s & E_{s+2} \\ E_{s+1} & E_{s+3} \end{vmatrix}}{\begin{vmatrix} E_s & E_{s+1} \\ E_{s+1} & E_{s+2} \end{vmatrix}} = \frac{E_s E_{s+3} - E_{s+1}E_{s+2}}{E_s E_{s+2} - E_{s+1}^2} = \frac{G_s}{F_s}.$$

Equations (9.117) yield the magnitude and the real part of the complex conjugate eigenvalues λ_1 and λ_2, thus determining them completely.

As the integer s assumes larger and larger values, the quantities $\mu_1^2 + \omega_1^2$ and $2\mu_1$ tend to constant values, at which time convergence is said to be achieved. The iteration uses trial vectors consisting of real numbers. To begin the iteration we begin with a trial vector $\{\Phi\}_0$, with one element unity and the remaining elements zero, and write

$$\{\Phi\}_1 = [D]\{\Phi\}_0,$$

$$\{\Phi\}_2 = [D]\{\Phi\}_1 = [D]^2\{\Phi\}_0,$$

$$\{\Phi\}_3 = [D]\{\Phi\}_2 = [D]^3\{\Phi\}_0, \tag{9.118}$$

$$\vdots$$

$$\{\Phi\}_s = [D]\{\Phi\}_{s-1} = [D]^s\{\Phi\}_0,$$

$$\vdots$$

The elements E_s, E_{s+1}, ... are the elements in the columns $\{\Phi\}_s$, $\{\Phi\}_{s+1}$, ... corresponding to the position of the nonzero element in the first trial vector $\{\Phi\}_0$. The iteration uses only real numbers; however, the eigenvectors are known to be complex. Next we must find a way to obtain the imaginary part. In Section 9.8 we pointed out that the eigenvectors of damped systems are determined within a multiplicative complex number. For convenience we assume that the multiplicative complex number is such that the column $\{\Phi\}_p$ resulting from the pth iteration, after convergence was achieved, is

---

† Having the same relative position.

equal to half the sum of the complex conjugate eigenvectors $\{\Phi^{(1)}\}$ and $\{\Phi^{(2)}\}$,

$$\{\Phi\}_p = \tfrac{1}{2}\{\Phi^{(1)}\} + \tfrac{1}{2}\{\Phi^{(2)}\}$$
$$= \tfrac{1}{2}\{\xi^{(1)} + i\eta^{(1)}\} + \tfrac{1}{2}\{\xi^{(1)} - i\eta^{(1)}\} = \{\xi^{(1)}\}, \qquad (9.119)$$

where $\{\xi^{(1)}\}$ is the real and $\{\eta^{(1)}\}$ the imaginary part of the eigenvector $\{\Phi^{(1)}\}$. The $(p + 1)$th iteration yields

$$\{\Phi\}_{p+1} = [D]\{\Phi\}_p = \lambda_1\{\Phi^{(1)}\}, \qquad (9.120)$$

and equating the real parts in (9.120) we conclude that

$$\{\Phi\}_{p+1} = \mu_1\{\xi^{(1)}\} - \omega_1\{\eta^{(1)}\} = \mu_1\{\Phi\}_p - \omega_1\{\eta^{(1)}\}, \qquad (9.121)$$

from which we obtain $\{\eta^{(1)}\}$. Hence to obtain the complete eigenvector $\{\Phi^{(1)}\}$ we must use two vectors obtained from two consecutive iterations after convergence has been achieved. Obviously this iteration sequence yields the first pair of complex conjugate modes.

To obtain the second pair of complex conjugate modes we must use the orthogonality relation derived in Section 9–9. To this end we choose again a trial vector consisting only of real numbers and write the orthogonality relations

$$\{\Phi^{(1)}\}^T[M]\{\Phi\} = \{\Phi^{(1)}\}^T[M]\{\xi\} = 0,$$
$$\{\Phi^{(2)}\}^T[M]\{\Phi\} = \{\Phi^{(2)}\}^T[M]\{\xi\} = 0, \qquad (9.122)$$

which leads to the relations

$$\{\xi^{(1)}\}^T[M]\{\xi\} = 0, \qquad \{\eta^{(1)}\}^T[M]\{\xi\} = 0. \qquad (9.123)$$

Introducing (9.119) and (9.121) in the orthogonality relations (9.123), we obtain

$$\{\Phi\}_p^T[M]\{\xi\} = 0, \qquad \{\Phi\}_{p+1}^T[M]\{\xi\} = 0, \qquad (9.124)$$

which indicates that the orthogonality relations can be written in terms of the vectors $\{\Phi\}_p$ and $\{\Phi\}_{p+1}$ obtained from two successive iterations after convergence has been established.

Equations (9.124) represent two homogeneous equations in terms of $2n$ unknowns ξ_i $(i = 1, 2, \ldots, 2n)$. These equations can be used to construct a constraint matrix, called a *sweeping matrix*, enabling us to construct a modified dynamical matrix to be used for iteration to the second pair of complex conjugate eigenvalues, which are second largest in magnitude. The procedure is similar to the one used for undamped systems (see Section 4–5).

Example 9.4

Solve the eigenvalue problem of Example 9.3 by the matrix iteration method.

In Example 9.3 we obtained the dynamical matrix

$$[D] = \begin{bmatrix} 0 & 0 & 1 & 0 \\ 0 & 0 & 0 & 1 \\ -m_1 k_1^{-1} & -m_2 k_1^{-1} & -c_2 k_2^{-1} & 0 \\ -m_1 k_1^{-1} & -(m_2 k_1^{-1} + m_2 k_2^{-1}) & -(c_1 k_1^{-1} - c_2 k_2^{-1}) & -c_2 k_2^{-1} \end{bmatrix}. \quad \text{(a)}$$

Choosing the values $m_1 = m$, $m_2 = 2m$, $c_1 = c_2 = 0.2m$, $k_1 = m$, and $k_2 = 4m$, the dynamical matrix reduces to

$$[D] = \begin{bmatrix} 0 & 0 & 1.00 & 0 \\ 0 & 0 & 0 & 1.00 \\ -1.00 & -2.00 & -0.20 & 0 \\ -1.00 & -2.50 & -0.15 & -0.05 \end{bmatrix}. \quad \text{(b)}$$

We begin the iteration with the trial vector

$$\{\Phi\}_0 = \begin{Bmatrix} 0 \\ 0 \\ 0 \\ 1 \end{Bmatrix}, \quad \text{(c)}$$

TABLE 9.1

s	E_s	$F_s = E_s E_{s+2}$ $- E_{s+1}^2$	$G_s = E_s E_{s+3}$ $- E_{s+1} E_{s+2}$	$\mu^2 + \omega^2 = \dfrac{F_{s+1}}{F_s}$	$2\mu = \dfrac{G_s}{F_s}$
0	1.0000	−2.5000	—	2.5060	—
1	−0.0500	−6.2650	—	3.2937	—
2	−2.4975	−20.6353	—	3.3004	—
3	0.5499	−68.1043	—	3.3455	—
4	8.1413	−227.8457	—	3.3465	—
5	−3.3161	−762.4790	—	3.3483	—
6	−26.6357	−2,553.0209	—	3.3484	—
7	15.9882	−8,548.5277	—	3.3485	—
8	86.2526	−28,624.4762	5,253.2676	3.3485	−0.1835
9	−69.3647	−95,848.3384	—	—	—
10	−276.0846	—	—	—	—
11	282.9340	—	—	—	—

so the homologous elements E_s will be the last elements in the vectors $\{\Phi\}_s$. The first iteration is

$$
\begin{bmatrix}
0 & 0 & 1.00 & 0 \\
0 & 0 & 0 & 1.00 \\
-1.00 & -2.00 & -0.20 & 0 \\
-1.00 & -2.50 & -0.15 & -0.05
\end{bmatrix}
\begin{Bmatrix}
0 \\ 0 \\ 0 \\ 1
\end{Bmatrix}
=
\begin{Bmatrix}
0 \\ 1.00 \\ 0 \\ -0.05
\end{Bmatrix},
\qquad \text{(d)}
$$

so $E_1 = -0.05$. The subsequent matrix multiplications are not shown; we shall tabulate the results instead. We note from Table 9.1 that it is not necessary to calculate the ratio G_s/F_s, except when the ratio F_{s+1}/F_s has reached a constant value. From the table we obtain

$$
\mu_1 = \frac{-0.1835}{2} = -0.09175,
$$

$$
\omega_1 = \pm\sqrt{3.3485 - \mu_1^2} = \pm\sqrt{3.3485 - 0.09175^2} = \pm 1.8276,
$$

so the first pair of eigenvalues is

$$
\begin{matrix}
\lambda_1 \\ \lambda_2
\end{matrix}\Bigg\} = -0.09175 \pm 1.8276i. \qquad \text{(e)}
$$

This set corresponds to the eigenvalues β_3 and β_4 in Example 9.3. Note that according to our notation $\lambda = \beta^{-1}$.

The vectors corresponding to $s = 10$ and $s = 11$ are

$$
\{\Phi\}_{10} =
\begin{Bmatrix}
-60.7093 \\
-69.3647 \\
-233.3910 \\
-276.0846
\end{Bmatrix}
\quad \text{and} \quad
\{\Phi\}_{11} =
\begin{Bmatrix}
-233.3910 \\
-276.0846 \\
246.1169 \\
282.9340
\end{Bmatrix},
\qquad \text{(f)}
$$

so using (9.119) and (9.121) we obtain the modes

$$
\{\Phi^{(1)}\} = \{\xi^{(1)}\} + i\{\eta^{(1)}\}
\quad \text{and} \quad
\{\Phi^{(2)}\} = \{\xi^{(1)}\} - i\{\eta^{(1)}\}, \qquad \text{(g)}
$$

where
$$
\{\xi^{(1)}\} =
\begin{Bmatrix}
-60.7093 \\
-69.3647 \\
-233.3910 \\
-276.0846
\end{Bmatrix}
\quad \text{and} \quad
\{\eta^{(1)}\} =
\begin{Bmatrix}
130.7522 \\
154.5473 \\
-122.9516 \\
-140.9536
\end{Bmatrix}.
\qquad \text{(h)}
$$

To obtain the second pair of modes we must write the orthogonality relations. First we need the matrix $[M]$, which is obtained from Example 9.3. For the values used in this example the $[M]$ matrix becomes

$$[M] = \begin{bmatrix} 0 & 0 & m_1 & 0 \\ 0 & 0 & 0 & m_2 \\ m_1 & 0 & c_1 + c_2 & -c_2 \\ 0 & m_2 & -c_2 & c_2 \end{bmatrix} = m \begin{bmatrix} 0 & 0 & 1.0 & 0 \\ 0 & 0 & 0 & 2.0 \\ 1.0 & 0 & 0.4 & -0.2 \\ 0 & 2.0 & -0.2 & 0.2 \end{bmatrix}. \quad (i)$$

Using (9.124) we obtain

$$\{\Phi\}_{10}^T[M]\{\xi\} = (-233.3910\xi_1 - 552.1692\xi_2 - 98.8488\xi_3 - 147.2681\xi_4)m = 0 \quad (j)$$

$$\{\Phi\}_{11}^T[M]\{\xi\} = (246.1169\xi_1 + 565.8680\xi_2 - 191.5310\xi_3 - 544.8058\xi_4)m = 0.$$

We choose arbitrarily to solve for ξ_1 and ξ_3 in terms of ξ_2 and ξ_4 and subsequently write the sweeping matrix

$$[S^{(1)}] = \begin{bmatrix} 0 & -2.342358 & 0 & 0.371535 \\ 0 & 1 & 0 & 0 \\ 0 & -0.055479 & 0 & -2.367059 \\ 0 & 0 & 0 & 1 \end{bmatrix}, \quad (k)$$

which enables us to construct a new dynamical matrix,

$$[D^{(2)}] = [D][S^{(1)}] = \begin{bmatrix} 0 & -0.055479 & 0 & -2.367059 \\ 0 & 0 & 0 & 1 \\ 0 & 0.353444 & 0 & 0.101877 \\ 0 & -0.149328 & 0 & -0.066476 \end{bmatrix}. \quad (l)$$

TABLE 9.2

s	E_s	$F_s = E_s E_{s+2}$ $- E_{s+1}^2$	$G_s = E_s E_{s+3}$ $- E_{s+1}E_{s+2}$	$\mu^2 + \omega^2 = \dfrac{F_{s+1}}{F_s}$	$2\mu = \dfrac{G_s}{F_s}$
0	1.0000	-1.4932×10^{-1}	0.0993×10^{-1}	0.1493	—
1	-0.0665	-0.2230×10^{-1}	0.1484×10^{-2}	0.1493	-0.0665
2	-0.1449	-0.0333×10^{-1}	—	—	—
3	0.1957×10^{-1}	—	—	—	—
4	0.2033×10^{-1}	—	—	—	—

To obtain the second and last pair of modes there is in fact no iteration involved, although we shall follow the iteration procedure. We again choose a trial vector $\{\Phi\}_0$ with the last element unity and the remaining elements zero. The results are shown in Table 9.2. The second pair of eigenvalues is

$$\left.\begin{matrix} \lambda_3 \\ \lambda_4 \end{matrix}\right\} = -0.03325 \pm 0.38510i. \tag{m}$$

The vectors corresponding to $s = 3$ and $s = 4$ are

$$\{\Phi\}_3 = \begin{Bmatrix} 0.34668 \\ -0.14490 \\ -0.03827 \\ 0.01957 \end{Bmatrix} \quad \text{and} \quad \{\Phi\}_4 = \begin{Bmatrix} -0.03827 \\ 0.01957 \\ -0.04922 \\ 0.02033 \end{Bmatrix}, \tag{n}$$

from which we obtain the second pair of modes

$$\{\Phi^{(3)}\} = \{\xi^{(3)}\} + i\{\eta^{(3)}\} \quad \text{and} \quad \{\Phi^{(4)}\} = \{\xi^{(3)}\} - i\{\eta^{(3)}\}, \tag{o}$$

where

$$\{\xi^{(3)}\} = \begin{Bmatrix} 0.34668 \\ -0.14490 \\ -0.03827 \\ 0.01957 \end{Bmatrix} \quad \text{and} \quad \{\eta^{(3)}\} = \begin{Bmatrix} -0.01999 \\ 0.00743 \\ 0.01608 \\ -0.00697 \end{Bmatrix}. \tag{p}$$

9-11 FORCED VIBRATION OF VISCOUSLY DAMPED DISCRETE SYSTEMS. GENERAL CASE

Let us return to the equations of motion for the forced vibration of a viscously damped discrete system. In Section 9–7 we derived a set of $2n$ ordinary differential equations of first order to replace a set of n equations of second order. With the notation used there we write the set of $2n$ equations in the matrix form

$$[M]\{\dot{y}(t)\} + [K]\{y(t)\} = \{Y(t)\}, \tag{9.125}$$

where $\{y(t)\}$ and $\{Y(t)\}$ are $2n \times 1$ matrices defined by (9.84) and $[M]$ and $[K]$ are $2n \times 2n$ symmetric matrices defined by (9.85).

We shall seek to uncouple (9.125) by assuming a solution in the form

$$\{y(t)\} = \sum_{r=1}^{2n} \{\Phi^{(r)}\}z_r(t) = [\Phi]\{z(t)\}, \tag{9.126}$$

where $[\Phi]$ is a $2n \times 2n$ matrix having as its columns the eigenvectors $\{\Phi^{(r)}\}$ obtained from the solution of the homogeneous problem. The elements of the column matrix $\{z(t)\}$ consist of $2n$ generalized coordinates to be determined. Introducing (9.126) in (9.125) and premultiplying the result by $[\Phi]^T$, we obtain

$$[\Phi]^T[M][\Phi]\{\ddot{z}\} + [\Phi]^T[K][\Phi]\{z\} = [\Phi]^T\{Y\}. \tag{9.127}$$

In view of the orthogonality relations (9.103) and (9.104), we can define diagonal matrices $[M^*]$ and $[K^*]$ given by

$$[M^*] = [\Phi]^T[M][\Phi] \quad \text{and} \quad [K^*] = [\Phi]^T[K][\Phi]. \tag{9.128}$$

In addition, let us denote a matrix of generalized forces by

$$\{Z(t)\} = [\Phi]^T\{Y(t)\}, \tag{9.129}$$

so (9.127) can be written

$$[M^*]\{\ddot{z}\} + [K^*]\{z\} = \{Z\}, \tag{9.130}$$

which represents an uncoupled set of equations of the type

$$M_r^*\dot{z}_r + K_r^*z_r = Z_r, \qquad r = 1, 2, \ldots, 2n. \tag{9.131}$$

Letting $s = r$ in (9.100) we conclude that

$$K_r^* = -\alpha_r M_r^*, \tag{9.132}$$

where α_r is the rth eigenvalue of the system so that (9.131) can be rewritten

$$\dot{z}_r - \alpha_r z_r = \frac{1}{M_r^*} Z_r, \qquad r = 1, 2, \ldots, 2n. \tag{9.133}$$

Assuming that the initial conditions are zero, the solution of (9.133) can be written in terms of the convolution integral

$$z_r(t) = \frac{1}{M_r^*} \int_0^t Z_r(\tau)e^{\alpha_r(t-\tau)} \, d\tau. \tag{9.134}$$

Recalling that

$$\{y(t)\} = \left\{ \begin{matrix} \{\dot{q}(t)\} \\ \{q(t)\} \end{matrix} \right\} = [\Phi]\{z(t)\} = \left[\begin{matrix} [\alpha\phi] \\ [\phi] \end{matrix} \right] \{z(t)\}, \tag{9.135}$$

where the $n \times 2n$ rectangular matrix $[\phi]$ is the lower half of the modal matrix $[\Phi]$, it follows that the displacement vector $\{q(t)\}$ can be written

$$\{q(t)\} = [\phi]\{z(t)\} = \sum_{r=1}^{2n} \{\phi^{(r)}\}z_r(t) = \sum_{r=1}^{2n} \{\phi^{(r)}\} \frac{1}{M_r^*} \int_0^t Z_r(\tau)e^{\alpha_r(t-\tau)} \, d\tau. \tag{9.136}$$

Let us assume that the modes of the system appear in pairs of complex conjugates and let the $(r + 1)$th mode be the complex conjugate of the rth mode. Introducing the complex vector notation we can write

$$M_r^* = |M_r^*|e^{i\theta_m^r}, \qquad M_{r+1}^* = |M_r^*|e^{-i\theta_m^r},$$

$$\{\phi_j^{(r)}\} = \{|\phi_j^{(r)}|e^{i\theta_j^r}\}, \qquad \{\phi_j^{(r+1)}\} = \{|\phi_j^{(r)}|e^{-i\theta_j^r}\},$$

$$Z_r(t) = |Z_r(t)|e^{i\theta_z^r}, \qquad Z_{r+1}(t) = |Z_r(t)|e^{-i\theta_z^r},$$

$$\alpha_r = \frac{1}{\mu_r + i\omega_r} = |\alpha_r|^2(\mu_r - i\omega_r), \qquad \alpha_{r+1} = |\alpha_r|^2(\mu_r + i\omega_r),$$

(9.137)

so grouping pairs of complex conjugates, we obtain

$$\{q_j(t)\} = \sum_{r=1,3,\dots}^{2n-1} \frac{1}{|M_r^*|} \int_0^t |Z_r(\tau)| \exp\left[|\alpha_r|^2\mu_r(t - \tau)\right]$$

$$\times \{|\phi_j^{(r)}|(\exp i[-|\alpha_r|^2\omega_r(t - \tau) - \theta_m^r + \theta_z^r + \theta_j^r]$$

$$+ \exp i[|\alpha_r|^2\omega_r(t - \tau) + \theta_m^r - \theta_z^r - \theta_j^r])\} \, d\tau$$

$$= \sum_{r=1,3,\dots}^{2n-1} \frac{2}{|M_r^*|} \int_0^t |Z_r(\tau)| \exp\left[|\alpha_r|^2\mu_r(t - \tau)\right]$$

$$\times \{|\phi_j^{(r)}| \cos\left[|\alpha_r|^2\omega_r(t - \tau) + \theta_m^r - \theta_z^r - \theta_j^r\right]\} \, d\tau.$$

(9.138)

Note that for mechanical systems $\mu_r < 0$. The above result was first reported by Foss.[†]

9–12   DAMPED CONTINUOUS SYSTEMS. VISCOUS DAMPING

Let us consider a continuous system subjected to forces resisting the motion, so that at any point P the forces depend on the velocity and are opposite in the direction to the velocity

$$F_c(P, t) = -C\left[\frac{\partial w(P, t)}{\partial t}\right] = -\frac{\partial}{\partial t} C[w(P, t)],$$

(9.139)

where $F_c(P, t)$ has units of distributed force and C is a linear homogeneous differential operator consisting of derivatives through order $2p$ with respect to the spatial coordinates P but not with respect to time t. In fact, C is an operator similar to the operator L defined by (5.37).

[†] K. A. Foss, *J. Appl. Mech.*, **25**, 361–364 (1958).

The following discussion will parallel the discussion of Section 7–7, where the response of undamped continuous systems was presented. One can refer to that section for definitions and symbols if they are not defined here. The partial differential equation (7.47), describing the motion of an undamped system, can be modified to include the damping force. Adding the effect of $F_c(P, t)$ to that equation we obtain

$$L[w(P, t)] + \frac{\partial}{\partial t} C[w(P, t)] + M(P) \frac{\partial^2 w(P, t)}{\partial t^2} = F(P, t), \quad (9.140)$$

where
$$F(P, t) = f(P, t) + F_j(t)\delta(P - P_j) \quad (9.141)$$

is an external distributed force also taking into account concentrated forces but excluding the damping forces provided for by F_c.

For simplicity we assume that the boundary conditions are all homogeneous, so that at every point on the boundaries of domain D we must satisfy p boundary conditions

$$B_i[w(P, t)] = 0, \quad i = 1, 2, \ldots, p, \quad (9.142)$$

where B_i are operators defined in Section 7–7.

We shall seek a solution of (9.140) by means of modal analysis. To this end we consider the eigenvalue problem, associated with the homogeneous equation of the undamped system, defined by the differential equation

$$L[w] = \lambda M w = \omega^2 M w \quad (9.143)$$

and the boundary conditions

$$B_i[w] = 0, \quad i = 1, 2, \ldots, p. \quad (9.144)$$

Let us assume that we are able to solve the eigenvalue problem and obtain an infinite set of denumerable natural frequencies ω_r and characteristic functions $w_r(P)$. Let us further assume that the eigenfunctions are orthogonal and we shall normalize them to obtain an orthonormal set satisfying the relations

$$\int_D M(P) w_r(P) w_s(P) \, dD(P) = \delta_{rs},$$

$$\int_D w_r(P) L[w_s(P)] \, dD(P) = \omega_r^2 \delta_{rs}, \quad (9.145)$$

where δ_{rs} is the Kronecker delta.

Using the expansion theorem we assume the solution of (9.140) in the form of an infinite series of the normal modes multiplied by time-dependent generalized coordinates

$$w(P, t) = \sum_{r=1}^{\infty} w_r(P)\eta_r(t).$$
(9.146)

Introducing (9.146) in (9.140) and recalling that the operators L and C are linear, we obtain

$$\sum_{r=1}^{\infty} \eta_r(t)L[w_r(P)] + \sum_{r=1}^{\infty} \dot{\eta}_r(t)C[w_r(P)] + \sum_{r=1}^{\infty} \ddot{\eta}_r(t)M(P)w_r(P) = F(P, t).$$
(9.147)

Next we multiply (9.147) by $w_s(P)$ and integrate over domain D. In view of (9.145), the result is

$$\ddot{\eta}_r(t) + \sum_{s=1}^{\infty} c_{rs}\dot{\eta}_s(t) + \omega_r^2\eta_r(t) = N_r(t), \qquad r = 1, 2, \ldots,$$
(9.148)

where
$$c_{rs} = \int_D w_r(P)C[w_s(P)] \, dD(P), \qquad r, s = 1, 2, \ldots$$
(9.149)

are viscous damping coefficients and

$$N_r(t) = \int_D w_r(P)F(P, t) \, dD(P), \qquad r = 1, 2, \ldots$$
(9.150)

are generalized forces.

Equation (9.148) represents an infinite set of coupled ordinary differential equations, so that, in general, damping produces coupling of the normal coordinates. In some special cases there is no coupling introduced by damping. That is when the operator C is a linear combination of the operator L and the mass function M,

$$C = a_1 L + a_2 M,$$
(9.151)

where a_1 and a_2 are constant coefficients, in which case damping does not couple the normal coordinates, because

$$c_{rs} = c_r\delta_{rs} = 2\zeta_r\omega_r\delta_{rs}, \qquad r, s = 1, 2, \ldots.$$
(9.152)

The notation in (9.152) was chosen to render (9.149) similar in structure to the equation of a single-degree-of-freedom system. Hence, (9.148) becomes

$$\ddot{\eta}_r(t) + 2\zeta_r\omega_r\dot{\eta}_r(t) + \omega_r^2\eta_r(t) = N_r(t), \qquad r = 1, 2, \ldots.$$
(9.153)

The solution of the above equation was shown in Section 9–3.

The above analysis certainly holds true when the system is uniform and the operator C is just a constant. This case is not unimportant.

When the operator C does produce coupling we must explore approximate methods. First we shall consider the case in which the operator C is self-adjoint,

$$\int_D uC[v]\, dD = \int_D vC[u]\, dD, \qquad (9.154)$$

where u and v are comparison functions (see the definition in Chapter 5), and seek an approximate solution of (9.140) by assuming a solution in the form of a series similar to (9.146) but limiting it to n terms,

$$w(P, t) = \sum_{r=1}^{n} w_r(P)\eta_r(t). \qquad (9.155)$$

This leads to a set of n coupled ordinary differential equations having the matrix form

$$[I]\{\ddot{\eta}(t)\} + [c]\{\dot{\eta}(t)\} + [\omega^2]\{\eta(t)\} = \{N(t)\}, \qquad (9.156)$$

which can be looked upon as the equations of a discrete system and treated by the methods of Sections 9–7 through 9–11 for the general case of damping. Note that the matrix $[c]$ is symmetric if the operator C is self-adjoint.

Finally, in the case in which damping is light, it is possible to obtain an approximate solution by considering the coupling due to damping as a secondary effect and simply ignore the off-diagonal elements in the damping matrix $[c]$.

9–13 DAMPED CONTINUOUS SYSTEMS. STRUCTURAL DAMPING

The same expression, (9.139), used to represent damping forces can be used to represent forces arising from structural damping. Again, as for lumped systems, we shall confine our analysis to the steady-state harmonic case in which every point P in the system executes harmonic motion of frequency Ω. Furthermore the materials must be such that the energy dissipated by a unit volume of material is proportional to the stress amplitude squared and independent of the frequency.

Let us consider again the partial differential equation describing the forced vibration of a damped continuous system and investigate the case in which the external force is harmonic:

$$L[w(P, t)] + \frac{\partial}{\partial t} C[w(P, t)] + M(P)\frac{\partial^2 w(P, t)}{\partial t^2} = A(P)e^{i\Omega t}, \qquad (9.157)$$

where $A(P)$ is a complex function depending on the position P. Since the motion is harmonic, we have

$$\frac{\partial w(P, t)}{\partial t} = i\Omega w(P, t), \tag{9.158}$$

so (9.157) becomes

$$L[w(P, t)] + i\Omega C[w(P, t)] + M(P)\frac{\partial^2 w(P, t)}{\partial t^2} = A(P)e^{i\Omega t}. \tag{9.159}$$

Analogous to the assumption that structural damping is proportional to stiffness in the case of lumped systems, it is customary to assume in the case of continuous systems that the operator C is proportional to the operator L:

$$C = \frac{\gamma}{\Omega}L, \tag{9.160}$$

where γ is a structural damping factor. Introducing (9.160) in (9.159), we obtain

$$(1 + i\gamma)L[w(P, t)] + M(P)\frac{\partial^2 w(P, t)}{\partial t^2} = A(P)e^{i\Omega t}. \tag{9.161}$$

Equation (9.160) implies that $a_2 = 0$ in (9.151), and it is obvious that the normal modes of the undamped system can be used to obtain an infinite set of uncoupled ordinary differential equations. Indeed, letting the solution of (9.161) have the form

$$w(P, t) = \sum_{r=1}^{\infty} w_r(P)\eta_r(t), \tag{9.162}$$

where $w_r(P)$ $(r = 1, 2, \ldots)$ are the orthonormal functions associated with the undamped system and $\eta_r(t)$ are the corresponding generalized coordinates, the standard modal analysis leads to the infinite set of equations

$$\ddot{\eta}_r(t) + (1 + i\gamma)\omega_r^2\eta_r(t) = N_r e^{i\Omega t}, \tag{9.163}$$

where ω_r $(r = 1, 2, \ldots)$ are the natural frequencies of the undamped system and

$$N_r = \int_D w_r(P)A(P)\, dD(P), \qquad r = 1, 2, \ldots \tag{9.164}$$

are generalized forces with complex amplitudes.

Equation (9.163) has precisely the same form as the equation of a structurally damped single-degree-of-freedom system, which was discussed in Section 9–4.

Problems

9.1   Consider the system of Example 9.1, assign the values $m_1 = m_2 = \dot{m}$, $m_3 = 2m$, $c_1 = 3c$, $c_2 = c_3 = c$, $c_4 = 6c$, $c_5 = 0$, $c_6 = 2c$, $k_1 = k_2 = k_3 = k$, and $k_4 = 2k$, and show that the classical modal matrix can be used to represent a linear transformation uncoupling the system differential equations of motion. Give the reason this is possible. Obtain the set of uncoupled differential equations.

9.2   Obtain the response of the system of Problem 9.1 to the forces $F_1 = 0$, $F_2 = 0$, and $F_3 = F_0 \alpha(t)$, where $\alpha(t)$ is the unit step function. Assume an underdamped system.

9.3   Check whether it is possible to use the concept of structural damping to write the equation of motion of a single-degree-of-freedom system in the form of (9.54) if the excitation consists of two harmonic forces of different frequencies.

9.4   Assume that the system of Example 4.1 is structurally damped and obtain the response of the system to a torque $M_T = A e^{i\omega t}$ applied to the center disk. The structural damping factor is $\gamma = 0.1$ and the driving frequency is

$$\omega = 0.8 \sqrt{\frac{GJ}{LI_D}}.$$

9.5   The differential equations of motion of a discrete system can be written in the matrix form

$$[m]\{\ddot{x}(t)\} + [c]\{\dot{x}(t)\} + [k]\{x(t)\} = \{F(t)\}.$$

Consider a three-degree-of-freedom system and derive the corresponding transfer functions $G_{ij}(s)$ $(i, j = 1, 2, 3)$. Show the general equations for the response (Section 9–6).

9.6   Consider the system of Example 9.2 and obtain the response $x_1(t)$ if the applied forces are

$$F_1(t) = 0, \qquad F_2(t) = F_0 \alpha(t),$$

where $\alpha(t)$ is the unit step function. The initial displacements are $x_1(0)$ and $x_2(0)$ and the initial velocities are $v_1(0)$ and $v_2(0)$.

9.7   Obtain the roots of the equation

$$\beta^4 + 0.50\beta^3 + 7.02\beta^2 + 0.50\beta + 2 = 0.$$

Describe the method chosen and show every computational step.

9.8   Complete the solution of the eigenvalue problem of Example 9.3 by obtaining the eigenvectors $\{\Phi^{(3)}\}$ and $\{\Phi^{(4)}\}$. Verify that the modes are orthogonal.

9.9   Consider the system shown in Figure 9.7 and derive the differential equations of motion. Let $m_1 = 2m$, $m_2 = m_3 = m$, $c_1 = c_2 = 0.10m$, $k_1 = 2m$, and $k_2 = m$, and obtain the solution of eigenvalue problem by the matrix iteration method.

9.10   The system of Example 9.3 is acted upon by the forces

$$\begin{aligned} &\text{(a)} \quad F_1 = F_0 \alpha(t), \qquad F_2 = 0, \\ &\text{(b)} \quad F_1 = 0, \qquad\qquad F_2 = I_0 \delta(t), \end{aligned}$$

where $\alpha(t)$ and $\delta(t)$ are the unit step function and unit impulse, respectively. Obtain the response for cases (a) and (b).

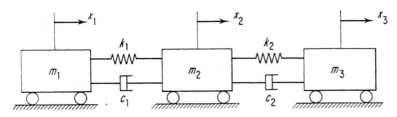

FIGURE 9.7

9.11 Derive the general equations for the response of a damped system to initial displacements and velocities (Section 9–11).

9.12 Obtain the response of a uniform rod in axial vibration to the initial conditions

$$u(x, 0) = f(x), \qquad \left.\frac{\partial u(x, t)}{\partial t}\right|_{t=0} = g(x).$$

Assume that the rod is subjected to damping forces proportional to the velocity. The proportionality constant is c. Let the rod be fixed at the end $x = 0$ and free at the end $x = L$.

Selected Readings

Caughey, T. K., Classical Normal Modes in Damped Linear Dynamic Systems, *J. Appl. Mech.*, **27**, 269–271 (1960).

Foss, K. A., Co-Ordinates Which Uncouple the Equations of Motion of Damped Linear Dynamic Systems, *J. Appl. Mech.*, **25**, 361–364 (1958).

Frazer, R. A., W. J. Duncan, and A. R. Collar, *Elementary Matrices*, Cambridge Univ. Press, New York, 1957.

Hurty, W. C., and M. F. Rubinstein, *Dynamics of Structures*, Prentice-Hall, Englewood Cliffs, N.J., 1964.

Kimball, A. L., and D. E. Lovell, Internal Friction in Solids, *Phys. Rev.*, **30** (2nd ser.), 948–959 (1927).

Rayleigh, Lord, *The Theory of Sound*, Vol. 1, Dover, New York, 1945.

Tong, K. N., *Theory of Mechanical Vibration*, Wiley, New York, 1960.

CHAPTER 10

VIBRATION
UNDER
COMBINED
EFFECTS

10–1 INTRODUCTION

In previous discussions we confined ourselves almost exclusively to the treatment of continuous systems for which the vibration was due to a single source. As an example, in studying the axial vibration of a rod we assumed that the excitation and the resulting motion was in the longitudinal direction only. For such systems the linear differential operator L consists of only one term. In this chapter we shall consider additional factors that may affect the system behavior by studying selected problems of current interest.

The subject of elastically supported systems was mentioned previously. An elastic foundation can be regarded as a distributed spring which, in general, affects the natural modes and natural frequencies of a system in transverse vibration. In the special case of a uniform bar in transverse vibration which, in addition, is supported by a uniform spring distributed over the entire length of the bar, the natural modes are not affected by the distributed spring, but the natural frequencies increase.

In many practical applications the effect of axial forces upon the transverse vibration of bars is of particular interest. The boundary-value problem associated with such a system is formulated by means of the extended Hamilton principle. In the case of a rotating bar the axial forces are tensile or compressive, depending on the way the bar is supported. A tensile axial force tends to render the bar "stiffer" and, in general, such a force distorts the shape of the natural modes and increases the natural frequencies. The eigen-value problem associated with a rotating bar is set up by means of Galerkin's method. A relatively elementary model of a flight vehicle consists of a bar with a compressive axial force at one end and free at the other end. The eigenvalue problem associated with such a system is formulated by means of the assumed-modes method for the case in which the compressive force is constant. A constant compressive force at one end, in addition to distorting the natural modes, tends to decrease the natural frequencies as compared to the natural frequencies of a free-free bar in bending vibration. For certain critical values of the compressive force the natural frequencies reduce to zero, at which point stability problems arise. The problem is related to the buckling of a bar problem. The buckling problem associated with the system is solved in closed form and the first several critical values of the compressive force obtained.

For various bars of nonsymmetric cross section the locus of the shear centers does not coincide with the locus of mass centers. For such bars the vibration consists of a combination of flexure and torsion and, furthermore, these effects cannot be separated. Because this problem arises, for the most part, in the case of nonuniform bars, the problem is formulated by regarding the continuous system as a lumped system.

Finally, the natural modes of a bar in transverse vibration, where, in addition to the axial force effect, the effects of rotatory inertia and shear deformation are included, are of much interest. In this case, too, we choose to formulate the problem by treating the continuous system as a lumped system.

10–2   TRANSVERSE VIBRATION OF A BAR ON AN ELASTIC FOUNDATION

The subject of elastically supported systems was mentioned briefly in Section 7–7. In a one-dimensional elastically supported system in which the restoring force is proportional to the displacement, the restoring force can be expressed

$$F_{\text{rest}}(x, t) = -k(x)w(x, t), \tag{10.1}$$

where $k(x)$ plays the role of a distributed spring and $w(x, t)$ is the absolute displacement at any point x. In the case of a bar in transverse vibration, the differential equation of motion can be written

$$\frac{\partial^2}{\partial x^2}\left[EI(x)\frac{\partial^2 w(x, t)}{\partial x^2}\right] + k(x)w(x, t) + m(x)\frac{\partial^2 w(x, t)}{\partial t^2} = p(x, t), \tag{10.2}$$

where $p(x, t)$ is the transverse load per unit length. Hence, in this case, the linear differential operator L has the form

$$L = \frac{\partial^2}{\partial x^2}\left[EI(x)\frac{\partial^2}{\partial x^2}\right] + k(x). \tag{10.3}$$

Note that the effect of the restoring force, which is proportional to the displacement, is provided for in the definition of L, (5.37), by the function $A_1(P)$.

Let us now consider the free vibration problem of a uniform bar hinged at both ends and supported by a uniformly distributed elastic support so that the equation of motion reduces to

$$\frac{\partial^4 w(x, t)}{\partial x^4} + \frac{k}{EI}w(x, t) + \frac{m}{EI}\frac{\partial^2 w(x, t)}{\partial t^2} = 0. \tag{10.4}$$

The corresponding boundary conditions are

$$w(x, t) = 0 \quad \text{and} \quad EI\frac{\partial^2 w(x, t)}{\partial x^2} = 0 \qquad \text{at } x = 0, L. \tag{10.5}$$

Equations (10.4) and (10.5) lead to an eigenvalue problem defined by the differential equation

$$\frac{d^4 w(x)}{dx^4} - \beta^4 w(x) = 0, \qquad \beta^4 = \frac{\omega^2 m}{EI} - \frac{k}{EI} \tag{10.6}$$

and the boundary conditions

$$w(x) = 0 \quad \text{and} \quad \frac{d^2 w(x)}{dx^2} = 0 \qquad \text{at } x = 0, L. \tag{10.7}$$

The general solution of (10.6) is

$$w(x) = c_1 \sin \beta x + c_2 \cos \beta x + c_3 \sinh \beta x + c_4 \cosh \beta x, \tag{10.8}$$

and using the boundary conditions (10.7) we obtain the frequency equation

$$\sin \beta L = 0, \tag{10.9}$$

which yields the denumerable infinite set of eigenvalues

$$\beta_r = \frac{r\pi}{L}, \qquad r = 1, 2, \ldots. \tag{10.10}$$

Corresponding to these eigenvalues we have an infinite set of eigenfunctions which upon normalization assume the form

$$w_r(x) = \sqrt{\frac{2}{mL}} \sin \beta_r x = \sqrt{\frac{2}{mL}} \sin \frac{r\pi x}{L}, \qquad r = 1, 2, \dots . \quad (10.11)$$

We note that the eigenvalues and the eigenfunctions are the same as for a bar hinged at both ends and with no elastic support. On the other hand, the natural frequencies are

$$\omega_r = \beta_r^2 \sqrt{\frac{EI}{m} \left(1 + \frac{k}{EI} \frac{1}{\beta_r^4}\right)} = \frac{r^2\pi^2}{L^2} \sqrt{\frac{EI}{m} \left(1 + \frac{kL^4}{r^4\pi^4 EI}\right)}, \qquad r = 1, 2, \dots,$$

$$(10.12)$$

which are different from the natural frequencies of the same bar without the elastic support. Note that the effect of the elastic support upon the natural frequencies diminishes as the mode number increases.

Returning to the forced vibration case, we write the differential equation in the form

$$\frac{EI}{m} \frac{\partial^4 w(x, t)}{\partial x^4} + \frac{k}{m} w(x, t) + \frac{\partial^2 w(x, t)}{\partial t^2} = \frac{1}{m} p(x, t). \quad (10.13)$$

To solve (10.13) we follow the standard procedure used in modal analysis. Calling upon the expansion theorem we assume a solution in the form of a superposition of the normal modes

$$w(x, t) = \sum_{r=1}^{\infty} w_r(x)\eta_r(t) \quad (10.14)$$

and recall that

$$\frac{d^4 w_r(x)}{dx^4} = \left(\frac{\omega_r^2 m}{EI} - \frac{k}{EI}\right) w_r(x). \quad (10.15)$$

The procedure leads to an infinite set of uncoupled ordinary differential equations

$$\ddot{\eta}_r(t) + \omega_r^2 \eta_r(t) = N_r(t), \qquad r = 1, 2, \dots, \quad (10.16)$$

where ω_r is given by (10.12) and the generalized force has the form

$$N_r(t) = \int_0^L w_r(x) p(x, t) \, dx. \quad (10.17)$$

The general solution of (10.16) was given previously [see (7.45)]. In conclusion, the only effect of the elastic support on the system, in this particular case, is to increase the natural frequencies of vibration.

10–3   EFFECT OF AXIAL FORCES ON THE BENDING VIBRATION OF A BAR. GENERAL EQUATIONS

In Chapter 5 we studied the transverse and axial vibration of a bar as two separate problems; in the case of transverse vibration no axial forces were considered and in the case of axial vibration there were no transverse forces present. In some cases it is not possible to ignore the effect of the axial forces upon the bending vibration of bars. We shall consider a case, however, in which it is possible, because of the axial stiffness, to ignore the axial elastic motion. Hence we wish to investigate the effect of an axial force $P(x, t)$ upon the bending vibration of a bar (Figure 10.1). For the purpose of derivation we assume

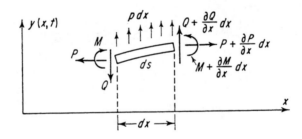

FIGURE 10.1

that $P(x, t)$ is a tensile force. It is further assumed that rotatory inertia and shear deformation effects can be neglected.

We shall derive the differential equation of motion and the associated boundary conditions by means of the extended Hamilton principle

$$\int_{t_1}^{t_2} (\delta T + \delta W)\, dt = 0, \tag{10.18}$$

where T is the kinetic energy and W the work function. The kinetic energy of the entire bar is

$$T(t) = \tfrac{1}{2} \int_0^L m(x) \left[\frac{\partial y(x, t)}{\partial t}\right]^2 dx, \tag{10.19}$$

where $m(x)$ is the distributed mass and $y(x, t)$ the transverse displacement. The work function must include the effect of the bending moment, the axial force, and the transverse load. To evaluate the work done by the axial force we must calculate the change in the horizontal projection of an element of length ds,

$$ds - dx = \left\{ (dx)^2 + \left[\frac{\partial y(x, t)}{\partial x} \right]^2 (dx^2) \right\}^{1/2} - dx \cong \frac{1}{2} \left[\frac{\partial y(x, t)}{\partial x} \right]^2 dx,$$

where the assumption has been made that the displacements are sufficiently small that in the binomial expansion only the first two terms can be retained. We note that the force $P(x, t)$ acts against the change in the horizontal projection, so that the work is negative. Furthermore, it is assumed that the displacements are so small that they do not bring about a change either in the axial force $P(x, t)$ or the transverse distributed force $p(x, t)$. It follows that the work function is

$$W(t) = -\frac{1}{2} \int_0^L EI(x) \left[\frac{\partial^2 y(x, t)}{\partial x^2} \right]^2 dx - \frac{1}{2} \int_0^L P(x, t) \left[\frac{\partial y(x, t)}{\partial x} \right]^2 dx$$

$$+ \int_0^L p(x, t) y(x, t) \, dx, \tag{10.20}$$

and we note that the $\frac{1}{2}$ coefficient multiplying the first integral is due to the linear relation between the bending moment and the curvature, whereas the $\frac{1}{2}$ coefficient in the second integral comes from the binomial expansion; the transverse displacement does not affect the axial force. Similarly, the last integral has no such coefficient, because the distance traveled by the force $p(x, t)$ is simply $y(x, t)$ and, furthermore, this transverse motion does not affect $p(x, t)$.

The variation in kinetic energy is

$$\delta T = \int_0^L m \frac{\partial y}{\partial t} \delta \left(\frac{\partial y}{\partial t} \right) dx = \int_0^L m \frac{\partial y}{\partial t} \frac{\partial}{\partial t} (\delta y) \, dx,$$

so

$$\int_{t_1}^{t_2} \delta T \, dt = \int_{t_1}^{t_2} \left[\int_0^L m \frac{\partial y}{\partial t} \frac{\partial}{\partial t} (\delta y) \, dx \right] dt = \int_0^L \left[\int_{t_1}^{t_2} m \frac{\partial y}{\partial t} \frac{\partial}{\partial t} (\delta y) \, dt \right] dx$$

$$= \int_0^L \left[m \frac{\partial y}{\partial t} \delta y \Big|_{t_1}^{t_2} - \int_{t_1}^{t_2} \frac{\partial}{\partial t} \left(m \frac{\partial y}{\partial t} \right) \delta y \, dt \right] dx$$

$$= - \int_{t_1}^{t_2} \int_0^L m \frac{\partial^2 y}{\partial t^2} \delta y \, dx \, dt, \tag{10.21}$$

because, by definition, $\delta y(x, t)$ is zero at $t = t_1$ and $t = t_2$.

The virtual work can be written

$$\delta W = -\int_0^L EI \frac{\partial^2 y}{\partial x^2} \delta\left(\frac{\partial^2 y}{\partial x^2}\right) dx - \int_0^L P \frac{\partial y}{\partial x} \delta\left(\frac{\partial y}{\partial x}\right) dx + \int_0^L p \, \delta y \, dx$$

$$= -\int_0^L EI \frac{\partial^2 y}{\partial x^2} \frac{\partial^2}{\partial x^2} (\delta y) \, dx - \int_0^L P \frac{\partial y}{\partial x} \frac{\partial}{\partial x} (\delta y) \, dx + \int_0^L p \, \delta y \, dx$$

$$= -EI \frac{\partial^2 y}{\partial x^2} \frac{\partial}{\partial x} (\delta y) \Big|_0^L + \frac{\partial}{\partial x}\left(EI \frac{\partial^2 y}{\partial x^2}\right) \delta y \Big|_0^L - \int_0^L \frac{\partial^2}{\partial x^2}\left(EI \frac{\partial^2 y}{\partial x^2}\right) \delta y \, dx$$

$$-P \frac{\partial y}{\partial x} \delta y \Big|_0^L + \int_0^L \frac{\partial}{\partial x}\left(P \frac{\partial y}{\partial x}\right) \delta y \, dx + \int_0^L p \, \delta y \, dx$$

$$= -EI \frac{\partial^2 y}{\partial x^2} \delta\left(\frac{\partial y}{\partial x}\right) \Big|_0^L + \left[\frac{\partial}{\partial x}\left(EI \frac{\partial^2 y}{\partial x^2}\right) - P \frac{\partial y}{\partial x}\right] \delta y \Big|_0^L$$

$$-\int_0^L \left[\frac{\partial^2}{\partial x^2}\left(EI \frac{\partial^2 y}{\partial x^2}\right) - \frac{\partial}{\partial x}\left(P \frac{\partial y}{\partial x}\right) - p\right] \delta y \, dx. \tag{10.22}$$

Introducing (10.21) and (10.22) in (10.18) we obtain

$$-\int_{t_1}^{t_2} \left\{ \int_0^L \left[m \frac{\partial^2 y}{\partial t^2} + \frac{\partial^2}{\partial x^2}\left(EI \frac{\partial^2 y}{\partial x^2}\right) - \frac{\partial}{\partial x}\left(P \frac{\partial y}{\partial x}\right) - p\right] \delta y \, dx \right.$$

$$\left. - EI \frac{\partial^2 y}{\partial x^2} \delta\left(\frac{\partial y}{\partial x}\right) \Big|_0^L + \left[\frac{\partial}{\partial x}\left(EI \frac{\partial^2 y}{\partial x^2}\right) - P \frac{\partial y}{\partial x}\right] \delta y \Big|_0^L \right\} dt = 0. \tag{10.23}$$

The integral must vanish for any arbitrary values of δy and $\delta(\partial y/\partial x)$, so we can set these variations equal to zero at $x = 0$ and $x = L$ and different from zero throughout the domain. Therefore, we must have

$$\frac{\partial^2}{\partial x^2}\left(EI \frac{\partial^2 y}{\partial x^2}\right) - \frac{\partial}{\partial x}\left(P \frac{\partial y}{\partial x}\right) + m \frac{\partial^2 y}{\partial t^2} = p, \tag{10.24}$$

which is the differential equation of motion. Furthermore, if we consider (10.24), because of the arbitrary nature of the variations, we can write

$$\left[EI \frac{\partial^2 y}{\partial x^2}\right] \delta\left(\frac{\partial y}{\partial x}\right) \Big|_0^L = 0,$$

$$\left[\frac{\partial}{\partial x}\left(EI \frac{\partial^2 y}{\partial x^2}\right) - P \frac{\partial y}{\partial x}\right] \delta y \Big|_0^L = 0, \tag{10.25}$$

which takes into account the possibility that either

$$EI \frac{\partial^2 y}{\partial x^2} = 0 \quad \text{or} \quad \frac{\partial y}{\partial x} = 0 \quad \text{at } x = 0, L \tag{10.26}$$

and that

$$\frac{\partial}{\partial x}\left(EI\frac{\partial^2 y}{\partial x^2}\right) - P\frac{\partial y}{\partial x} = 0 \quad \text{or} \quad y = 0 \qquad \text{at } x = 0, L. \qquad (10.27)$$

Equations (10.26) and (10.27) represent the boundary conditions. The first of equations (10.26) implies the vanishing of the bending moment and the second of (10.26) indicates the vanishing of the slope of the deflection curve at either end. The first of equations (10.27) means that the vertical force is zero and the second that the deflection is zero at either end. Hence the boundary conditions are of both the natural and geometric types.

10–4 VIBRATION OF ROTATING BARS

Let us consider the free transverse vibration problem of a uniform bar rotating about a vertical axis with a constant angular velocity Ω. The bar is fixed at the end $x = 0$ and free at the end $x = L$ (Figure 10.2). We ignore the

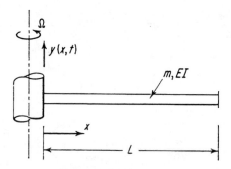

FIGURE 10.2

longitudinal elastic motion of the bar, so at any point x the axial force is due to the centrifugal force†

$$P(x, t) = P(x) = \int_x^L m\Omega^2 \xi \, d\xi = \tfrac{1}{2}m\Omega^2 L^2\left(1 - \frac{x^2}{L^2}\right), \qquad (10.28)$$

so, for no transverse load, (10.24) reduces to

$$EI\frac{\partial^4 y}{\partial x^4} - \tfrac{1}{2}m\Omega^2 L^2 \frac{\partial}{\partial x}\left[\left(1 - \frac{x^2}{L^2}\right)\frac{\partial y}{\partial x}\right] + m\frac{\partial^2 y}{\partial t^2} = 0. \qquad (10.29)$$

† The hub radius is assumed negligible compared with the length L.

The associated boundary conditions are

$$y = 0 \quad \text{and} \quad \frac{\partial y}{\partial x} = 0 \qquad \text{at } x = 0 \tag{10.30}$$

$$EI \frac{\partial^2 y}{\partial x^2} = 0 \quad \text{and} \quad \frac{\partial}{\partial x}\left(EI \frac{\partial^2 y}{\partial x^2}\right) = 0 \qquad \text{at } x = L, \tag{10.31}$$

because $P(x)$ is zero at $x = L$. We note that these are the same boundary conditions as for a cantilever beam that does not rotate. Equations (10.29) through (10.31) represent a free vibration problem. We shall now proceed to obtain the natural modes. The corresponding eigenvalue problem consists of the differential equation

$$EI \frac{d^4 Y}{dx^4} - \tfrac{1}{2} m\Omega^2 L^2 \frac{d}{dx}\left[\left(1 - \frac{x^2}{L^2}\right)\frac{dY}{dx}\right] = \omega^2 m Y \tag{10.32}$$

and the boundary conditions

$$Y = 0 \quad \text{and} \quad \frac{dY}{dx} = 0 \qquad \text{at } x = 0 \tag{10.33}$$

$$\frac{d^2 Y}{dx^2} = 0 \quad \text{and} \quad \frac{d^3 Y}{dx^3} = 0 \qquad \text{at } x = L \tag{10.34}$$

Equation (10.32) can be written

$$L[Y] = \omega^2 m Y, \tag{10.35}$$

where L is an operator given by

$$L = EI \frac{d^4}{dx^4} - \tfrac{1}{2} m\Omega^2 L^2 \frac{d}{dx}\left[\left(1 - \frac{x^2}{L^2}\right)\frac{d}{dx}\right]. \tag{10.36}$$

We shall now verify that the operator L is self-adjoint. To this end we consider two comparison functions u and v, satisfying all the boundary conditions of the problem, and write

$$\begin{aligned}
\int_0^L u L[v]\, dx &= \int_0^L u\left\{EI \frac{d^4 v}{dx^4} - \tfrac{1}{2} m\Omega^2 L^2 \frac{d}{dx}\left[\left(1 - \frac{x^2}{L^2}\right)\frac{dv}{dx}\right]\right\} dx \\
&= u\left(EI \frac{d^3 v}{dx^3}\right)\Big|_0^L - \frac{du}{dx}\left(EI \frac{d^2 v}{dx^2}\right)\Big|_0^L + \int_0^L \frac{d^2 u}{dx^2} EI \frac{d^2 v}{dx^2}\, dx \\
&\quad - \tfrac{1}{2} m\Omega^2 L^2 \left[u\left(1 - \frac{x^2}{L^2}\right)\frac{dv}{dx}\Big|_0^L - \int_0^L \frac{du}{dx}\left(1 - \frac{x^2}{L^2}\right)\frac{dv}{dx}\, dx\right] \\
&= EI \int_0^L \frac{d^2 u}{dx^2}\frac{d^2 v}{dx^2}\, dx + \tfrac{1}{2} m\Omega^2 L^2 \int_0^L \left(1 - \frac{x^2}{L^2}\right)\frac{du}{dx}\frac{dv}{dx}\, dx.
\end{aligned} \tag{10.37}$$

We must note that the result is symmetric in u and v, so the operator L is indeed self-adjoint.

We shall seek to set up the eigenvalue problem by means of Galerkin's method. To this end we assume a solution in the form of a series

$$Y(x) = \sum_{j=1}^{n} a_j u_j(x), \tag{10.38}$$

where a_j are coefficients to be determined and $u_j(x)$ are comparison functions. We choose as comparison functions for the rotating bar the eigenfunctions of the cantilever beam obtained by setting $\Omega = 0$. Following the procedure described in Section 6–6 we obtain the eigenvalue problem

$$[k]\{a\} = \omega^2[m]\{a\}, \tag{10.39}$$

where the k_{rj} are obtained from

$$k_{rj} = k_{jr} = \int_0^L u_r L[u_j]\, dx$$

$$= EI \int_0^L \frac{d^2 u_r}{dx^2} \frac{d^2 u_j}{dx^2}\, dx + \tfrac{1}{2} m \Omega^2 L^2 \int_0^L \left(1 - \frac{x^2}{L^2}\right) \frac{du_r}{dx} \frac{du_j}{dx}\, dx. \tag{10.40}$$

The coefficients are symmetric, because L is self-adjoint. The coefficients m_{rj} are simply

$$m_{rj} = m_{jr} = m \int_0^L u_r u_j\, dx, \tag{10.41}$$

so both matrices $[k]$ and $[m]$ are symmetric.

The force $P(x)$ has a tendency to increase the potential energy of the system, so a "stiffer" system is obtained, with the tendency of increasing the natural frequencies as compared to the natural frequencies of the nonrotating bar.

10–5   EFFECT OF A CONSTANT AXIAL FORCE ON THE TRANSVERSE VIBRATION OF A UNIFORM FREE–FREE BAR

Another example of the effect of an axial force upon the bending vibration of a bar is the free-free bar subjected to a compressive force P_0 at one end, say $x = 0$ (Figure 10.3). For simplicity let us assume that the bar is uniform,

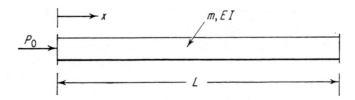

FIGURE 10.3

and if we consider only the steady-state longitudinal response to the constant force P_0, the longitudinal force at any point is†

$$P(x, t) = P(x) = -P_0\left(1 - \frac{x}{L}\right). \tag{10.42}$$

Hence the differential equation (10.24), for no transverse load, reduces to

$$EI \frac{\partial^4 y}{\partial x^4} + P_0 \frac{\partial}{\partial x}\left[\left(1 - \frac{x}{L}\right)\frac{\partial y}{\partial x}\right] + m \frac{\partial^2 y}{\partial t^2} = 0, \tag{10.43}$$

and the associated boundary conditions (10.26) and (10.27) lead to

$$EI \frac{\partial^2 y}{\partial x^2} = 0 \quad \text{and} \quad \frac{\partial}{\partial x}\left(EI \frac{\partial^2 y}{\partial x^2}\right) + P_0 \frac{\partial y}{\partial x} = 0 \qquad \text{at } x = 0 \tag{10.44}$$

$$EI \frac{\partial^2 y}{\partial x^2} = 0 \quad \text{and} \quad \frac{\partial}{\partial x}\left(EI \frac{\partial^2 y}{\partial x^2}\right) = 0 \qquad \text{at } x = L. \tag{10.45}$$

We note that the differential equation (as well as the boundary conditions) is homogeneous. One could set up the eigenvalue problem as in Section 10–4 and attempt a series solution. The operator L in this case is

$$L = EI \frac{d^4}{dx^4} + P_0 \frac{d}{dx}\left[\left(1 - \frac{x}{L}\right)\frac{d}{dx}\right], \tag{10.46}$$

and upon selecting two comparison functions u and v and integrating by parts we obtain

$$\int_0^L u L[v] \, dx = EI \int_0^L \frac{d^2 u}{dx^2} \frac{d^2 v}{dx^2} \, dx - P_0 \int_0^L \left(1 - \frac{x}{L}\right)\frac{du}{dx}\frac{dv}{dx} \, dx, \tag{10.47}$$

so in this case, too, the operator L is self-adjoint. Hence, we must conclude that the eigenfunctions resulting from the solution of the eigenvalue problem associated with (10.43) to (10.45) are orthogonal.

† Differentiate (7.73) with respect to x, multiply the result by EA, and ignore the time-dependent harmonics.

At this point we shall abandon this approach, in which we seek eigen-functions that satisfy the differential equation and the boundary condi-tions exactly, and seek a solution of the eigenvalue problem by one of the approximate methods discussed in Chapter 6. Except for the second of the boundary conditions, (10.44), which is a natural boundary condition, all the boundary conditions are the same as for the free-free bar. Hence we can use one of the methods assuming a solution in terms of a series of admissible functions. We shall choose the assumed-modes method, and to this end we use as admissible functions the normal modes $\phi_i(x)$ of the uniform free-free bar. Hence we assume a solution in the form

$$y(x, t) = \sum_{i=0}^{n} \phi_i(x)q_i(t), \tag{10.48}$$

where $q_i(t)$ $(i = 0, 1, \ldots, n)$ are generalized coordinates to be determined. Consequently the kinetic energy expression can be written

$$T = \tfrac{1}{2} \int_0^L m\left(\frac{\partial y}{\partial t}\right)^2 dx = \tfrac{1}{2} \int_0^L m\left[\sum_{i=0}^{n} \phi_i\dot{q}_i\right]\left[\sum_{j=0}^{n} \phi_j\dot{q}_j\right] dx$$

$$= \tfrac{1}{2} \sum_{i=0}^{n} \sum_{j=0}^{n} \dot{q}_i\dot{q}_j \int_0^L m\phi_i\phi_j \, dx = \tfrac{1}{2} \sum_{i=0}^{n} \dot{q}_i^2, \tag{10.49}$$

because the functions ϕ_i are such that

$$\int_0^L m\phi_i\phi_j \, dx = \delta_{ij}. \tag{10.50}$$

The potential energy expression is

$$V = \tfrac{1}{2} \int_0^L EI\left(\frac{\partial^2 y}{\partial x^2}\right)^2 dx + \tfrac{1}{2} \int_0^L P\left(\frac{\partial y}{\partial x}\right)^2 dx$$

$$= \tfrac{1}{2} \int_0^L EI\left[\sum_{i=0}^{n} \phi_i''q_i\right]\left[\sum_{j=0}^{n} \phi_j''q_j\right] dx - \tfrac{1}{2} \int_0^L P_0\left(1 - \frac{x}{L}\right)\left[\sum_{i=0}^{n} \phi_i'q_i\right]\left[\sum_{j=0}^{n} \phi_j'q_j\right] dx$$

$$= \tfrac{1}{2} \sum_{i=0}^{n} \sum_{j=0}^{n} q_iq_j\left[\int_0^L EI\phi_i''\phi_j'' \, dx - P_0 \int_0^L \left(1 - \frac{x}{L}\right)\phi_i'\phi_j' \, dx\right]$$

$$= \tfrac{1}{2} \sum_{i=0}^{n} \sum_{j=0}^{n} k_{ij}q_iq_j, \tag{10.51}$$

where primes indicate differentiation with respect to x and

$$k_{ij} = \int_0^L EI\phi_i''\phi_j'' \, dx - P_0 \int_0^L \left(1 - \frac{x}{L}\right)\phi_i'\phi_j' \, dx$$

$$= \omega_i^2\delta_{ij} - P_0 \int_0^L \left(1 - \frac{x}{L}\right)\phi_i'\phi_j' \, dx. \tag{10.52}$$

In the above equation ω_i denotes the ith natural frequency of the free-free uniform bar with no axial force applied.

Introducing (10.49) and (10.51) in Lagrange's equations

$$\frac{d}{dt}\left(\frac{\partial T}{\partial \dot{q}_i}\right) - \frac{\partial V}{\partial q_i} = 0, \qquad i = 0, 1, \ldots, n, \tag{10.53}$$

we obtain a set of $n + 1$ ordinary differential equations that can be arranged in the matrix form

$$[I]\{\ddot{q}\} + [k]\{q\} = \{0\}, \tag{10.54}$$

where $[I]$ is the identity matrix and the elements in the matrix $[k]$ are given by (10.52).

The admissible functions ϕ_i $(i = 0, 1, 2, \ldots, n)$ satisfy the differential equation

$$\phi_i'''' = \beta_i^4 \phi_i, \qquad i = 0, 1, 2, \ldots, n, \tag{10.55}$$

and the boundary conditions

$$\begin{aligned} \phi_i'' &= 0 & \text{at } x = 0,L, \\ \phi_i''' &= 0 & \text{at } x = 0,L. \end{aligned} \tag{10.56}$$

Hence, the admissible functions are (see Section 5–10)

$$\begin{aligned} \phi_0 &= A_0, \\ \phi_1 &= A_1\left(\frac{x}{L} - \frac{1}{2}\right), \\ \phi_i &= A_i[(\cos \beta_i L - \cosh \beta_i L)(\sin \beta_i x + \sinh \beta_i x) \\ &\quad - (\sin \beta_i L - \sinh \beta_i L)(\cos \beta_i x + \cosh \beta_i x)], \qquad i = 2, 3, \ldots, n, \end{aligned} \tag{10.57}$$

where the constants A_i $(i = 0, 1, 2, \ldots, n)$ are such that (10.50) is satisfied and the values β_i are obtained from the solution of the characteristic equation

$$\cos \beta L \cosh \beta L = 1. \tag{10.58}$$

It can be easily shown that

$$A_0 = \frac{1}{\sqrt{mL}}, \qquad A_1 = \sqrt{\frac{12}{mL}}.$$

The constants A_i $(i = 2, 3, \ldots, n)$ are obtained from the following equations:

$$\int_0^L m\phi_i^2 \, dx = m\int_0^L \phi_i^2 \, dx = 1, \qquad i = 2, 3, \ldots, n. \tag{10.59}$$

The integral $\int \phi_i^2 \, dx$ can be evaluated by a complex procedure of integrations by parts, making use of (10.55). Omitting the details, we give the result

$$\int \phi_i^2 \, dx = \frac{x}{4\beta_i^4} [\beta_i^4 \phi_i^2 - 2\phi_i' \phi_i''' + (\phi_i'')^2] + \frac{3}{4\beta_i^4} \phi_i \phi_i'''$$

$$- \frac{1}{4\beta_i^4} \phi_i' \phi_i'', \qquad i = 2, 3, \ldots, n. \tag{10.60}$$

Hence the last of equations (10.57), in conjunction with (10.59) and (10.60), yields

$$\int_0^L m\phi_i^2 \, dx = \frac{mL}{4} \phi_i^2(L)$$

$$= \frac{mL}{4} A_i^2 [(\cos \beta_i L - \cosh \beta_i L)(\sin \beta_i L + \sinh \beta_i L)$$

$$- (\sin \beta_i L - \sinh \beta_i L)(\cos \beta_i L + \cosh \beta_i L)]^2$$

$$= 1, \qquad i = 2, 3, \ldots, n.$$

Using (10.58) the above equation yields

$$A_i = \pm \frac{1}{\sqrt{mL}} \frac{1}{\sin \beta_i L - \sinh \beta_i L}, \qquad i = 2, 3, \ldots, n. \tag{10.61}$$

Choosing the negative sign, it follows that the normalized admissible functions can be written

$$\phi_0 = \frac{1}{\sqrt{mL}},$$

$$\phi_1 = \sqrt{\frac{12}{mL}} \left(\frac{x}{L} - \frac{1}{2} \right), \tag{10.62}$$

$$\phi_i = \frac{1}{\sqrt{mL}} \Bigg[(\cos \beta_i x + \cosh \beta_i x)$$

$$- \frac{\cos \beta_i L - \cosh \beta_i L}{\sin \beta_i L - \sinh \beta_i L} (\sin \beta_i x + \sinh \beta_i x) \Bigg], \qquad i = 2, 3, \ldots, n.$$

It will be recalled that the corresponding values ω_i to be used in (10.52) are

$$\omega_0 = \omega_1 = 0, \qquad \omega_i = \beta_i^2 \sqrt{\frac{EI}{m}}, \qquad i = 2, 3, \ldots, n.$$

We note, however, that

$$k_{0i} = k_{i0} = 0, \qquad i = 0, 1, 2, \ldots, n, \tag{10.63}$$

so the matrix $[k]$ is singular. This is to be expected, because there are no forces in the transverse direction, so we must have

$$\int_0^L my \, dx = 0, \tag{10.64}$$

which indicates that $q_0 = 0$. Hence we can reduce the problem so that we have only n ordinary differential equations instead of $n + 1$. This can be done by simply ignoring the translational rigid-body mode, ϕ_0, and summing the series (10.48) from 1 to n.

The resulting eigenvalue problem is

$$[k]\{q\} = \Lambda[I]\{q\}, \tag{10.65}$$

where the coefficients k_{ij} are given by (10.52). To evaluate these coefficients we first note that (10.55) leads to

$$\int \phi_i' \phi_j' \, dx = \frac{1}{\beta_i^4 - \beta_j^4} (\phi_i''''\phi_j' - \phi_i'''\phi_j'' + \phi_i''\phi_j''' - \phi_i'\phi_j'''')$$

$$= \frac{1}{\beta_i^4 - \beta_j^4} (\beta_i^4 \phi_i \phi_j' - \phi_i'''\phi_j'' + \phi_i''\phi_j''' - \beta_j^4 \phi_i'\phi_j).$$

Let us denote the integral on the right side of (10.52) by Γ_{ij}. It can be shown that for $i \neq j$ the value of this integral is

$$\Gamma_{ij} = \int_0^L \left(1 - \frac{x}{L}\right)\phi_i' \phi_j' \, dx$$

$$= \left[\left(1 - \frac{x}{L}\right) \int \phi_i' \phi_j' \, dx\right]_0^L + \frac{1}{L}\int_0^L \left[\int \phi_i' \phi_j' \, dx\right] dx$$

$$= \left[\left(1 - \frac{x}{L}\right) \frac{1}{\beta_i^4 - \beta_j^4}(\beta_i^4 \phi_i \phi_j' - \phi_i'''\phi_j'' + \phi_i''\phi_j''' - \beta_j^4 \phi_i'\phi_j)\right]_0^L$$

$$+ \frac{1}{L}\frac{1}{\beta_i^4 - \beta_j^4} \int_0^L (\beta_i^4 \phi_i \phi_j' - \phi_i'''\phi_j'' + \phi_i''\phi_j''' - \beta_j^4 \phi_i'\phi_j) \, dx$$

$$= \left[\left(1 - \frac{x}{L}\right) \frac{1}{\beta_i^4 - \beta_j^4}(\beta_i^4 \phi_i \phi_j' - \phi_i'''\phi_j'' + \phi_i''\phi_j''' - \beta_j^4 \phi_i'\phi_j)\right]_0^L$$

$$+ \frac{1}{L}\frac{1}{(\beta_i^4 - \beta_j^4)^2}\left[-2(\beta_i^4 + \beta_j^4)\phi_i''\phi_j'' + (\beta_i^4 + 3\beta_j^4)\phi_i'''\phi_j'\right.$$

$$\left.+ (\beta_j^4 + 3\beta_i^4)\phi_i'\phi_j''' - 4\beta_i^4\beta_j^4\phi_i\phi_j\right]_0^L$$

$$= -\frac{1}{\beta_i^4 - \beta_j^4}[\beta_i^4\phi_i(0)\phi_j'(0) - \beta_j^4\phi_i'(0)\phi_j(0)]$$

$$- \frac{1}{L}\frac{4\beta_i^4\beta_j^4}{(\beta_i^4 - \beta_j^4)^2}[\phi_i(L)\phi_j(L) - \phi_i(0)\phi_j(0)], \quad i,j = 2, 3, \ldots, n. \tag{10.66}$$

For $i = j$ we obtain, after a series of integrations,

$$\Gamma_{ii} = \int_0^L \left(1 - \frac{x}{L}\right)(\phi_i')^2 \, dx$$

$$= \frac{1}{8}\left\{x\left(2 - \frac{x}{L}\right)\left[(\phi_i')^2 - 2\phi_i\phi_i'' + \frac{1}{\beta_i^4}(\phi_i''')^2\right]\right.$$

$$+ 2\left(1 - \frac{x}{L}\right)\left(3\phi_i\phi_i' - \frac{1}{\beta_i^4}\phi_i''\phi_i'''\right) + \frac{3}{L}\phi_i^2 - \frac{1}{L\beta_i^4}(\phi_i'')^2\right\}_0^L$$

$$= \frac{L}{8}\left\{[\phi_i'(L)]^2 + \frac{3}{L^2}[\phi_i^2(L) - \phi_i^2(0)] - \frac{6}{L}\phi_i(0)\phi_i'(0)\right\},$$

$$i = 2, 3, \ldots, n. \quad (10.67)$$

This allows us to write the eigenvalue problem (10.65) in the form

$$[\omega_i^2\delta_{ij} - P_0\Gamma_{ij}]\{q\} = \Lambda[\delta_{ij}]\{q\}, \quad (10.68)$$

which yields n eigenvalues Λ_r and associated eigenvectors $\{q^{(r)}\}$, so that the eigenfunctions are

$$Y^{(r)}(x) = \sum_{i=1}^n \phi_i(x)q_i^{(r)}, \quad r = 1, 2, \ldots, n. \quad (10.69)$$

We note that the eigenvalues Λ_r depend on the compressive force P_0. They can be obtained by solving the characteristic equation

$$|\omega_i^2\delta_{ij} - P_0\Gamma_{ij} - \Lambda\delta_{ij}| = 0. \quad (10.70)$$

It is quite conceivable that for certain values of P_0 at least one of the eigenvalues becomes negative or complex. In this case the corresponding mode is unstable and the linear theory, based on small displacements, ceases to be valid. Of course, before any eigenvalue becomes negative it must become zero. We shall call the values of P_0 for which the eigenvalues Λ_r are zero *critical values* and denote them P_{cr}. Instead of obtaining these critical values from the characteristic equation we shall follow a different route. We note that for $\Lambda = 0$ the eigenvalue problem, for the uniform bar, can be written

$$EI\frac{d^4y}{dx^4} + P_{cr}\frac{d}{dx}\left[\left(1 - \frac{x}{L}\right)\frac{dy}{dx}\right] = 0, \quad (10.71)$$

where y is subject to the boundary conditions

$$\frac{d^2y}{dx^2} = 0 \quad \text{and} \quad \frac{d^3y}{dx^3} + \frac{P_{cr}}{EI}\frac{dy}{dx} = 0 \quad \text{at } x = 0 \quad (10.72)$$

$$\frac{d^2y}{dx^2} = 0 \quad \text{and} \quad \frac{d^3y}{dx^3} = 0 \quad \text{at } x = L. \quad (10.73)$$

Equation (10.71) can be integrated immediately to yield

$$\frac{d^3y}{dx^3} + \frac{P_{cr}}{EI}\left(1 - \frac{x}{L}\right)\frac{dy}{dx} = 0, \tag{10.74}$$

where the constant of integration was taken zero to satisfy the second of either (10.72) or (10.73). Equation (10.74) describes the static buckling problem of a uniform bar under the action of distributed axial loads.†

Making the substitution

$$\frac{dy}{dx} = u, \qquad z = \left(\frac{P_{cr}L^2}{EI}\right)^{1/3}\left(1 - \frac{x}{L}\right), \tag{10.75}$$

(10.74) reduces to

$$\frac{d^2u}{dz^2} + zu = 0. \tag{10.76}$$

Equation (10.76) is related to a Bessel equation. Its solution is

$$u = C_1\text{Ai}(-z) + C_2\text{Bi}(-z), \tag{10.77}$$

where C_1 and C_2 are constants of integration and $\text{Ai}(-z)$ and $\text{Bi}(-z)$ are called *Airy functions* and are related to the Bessel functions by the expressions‡

$$\left.\begin{array}{l}\text{Ai}(-z) = \dfrac{1}{3}z^{1/2}[J_{1/3}(\zeta) + J_{-1/3}(\zeta)] \\[3mm] \text{Bi}(-z) = \left(\dfrac{z}{3}\right)^{1/2}[J_{1/3}(\zeta) - J_{-1/3}(\zeta)]\end{array}\right\} \qquad \zeta = \frac{2}{3}z^{3/2}. \tag{10.78}$$

Equation (10.77) is subject to the boundary conditions

$$\frac{du}{dz} = 0 \qquad \text{at } z = 0, \tag{10.79}$$

$$\frac{du}{dz} = 0 \qquad \text{at } z = z_{cr} = \left(\frac{P_{cr}L^2}{EI}\right)^{1/3} = \alpha_{cr}{}^{1/3}. \tag{10.80}$$

Substituting (10.79) and (10.80) in (10.77), we obtain

$$C_1\frac{d\text{Ai}(0)}{dz} + C_2\frac{d\text{Bi}(0)}{dz} = 0,$$

$$C_1\frac{d\text{Ai}(-\alpha_{cr}{}^{1/3})}{dz} + C_2\frac{d\text{Bi}(-\alpha_{cr}{}^{1/3})}{dz} = 0, \tag{10.81}$$

---

† See S. Timoshenko, *Theory of Elastic Stability*, McGraw-Hill, New York, 1936, p. 115.

‡ See M. Abramowitz and I. A. Stegun, eds., *Handbook of Mathematical Functions*, National Bureau of Standards, Washington, D.C., 1965, p. 446.

and since

$$\left.\begin{array}{l} \dfrac{d\mathrm{Ai}(-z)}{dz} = -\dfrac{1}{3}\,z[J_{-2/3}(\zeta) - J_{2/3}(\zeta)] \\[3mm] \dfrac{d\mathrm{Bi}(-z)}{dz} = \dfrac{z}{\sqrt{3}}\,[J_{-2/3}(\zeta) + J_{2/3}(\zeta)] \end{array}\right\} \qquad \zeta = \dfrac{2}{3}\,z^{3/2}, \qquad (10.82)$$

it can be easily shown that the critical values P_{cr} can be obtained from the equation

$$J_{2/3}(\zeta_{cr}) = 0, \qquad \zeta_{cr} = \frac{2}{3}\,z_{cr}{}^{3/2}, \qquad (10.83)$$

where $J_{2/3}(\zeta)$ is related to the Airy functions by

$$J_{2/3}(\zeta) = \frac{\sqrt{3}}{2z}\left[\sqrt{3}\,\frac{d\mathrm{Ai}(-z)}{dz} + \frac{d\mathrm{Bi}(-z)}{dz}\right]. \qquad (10.84)$$

The first zero of $J_{2/3}(\zeta)$ is $\zeta = 0$. The second and third zeros were obtained by interpolation from Table 10.11 of the '*Handbook of Mathematical Functions*. Ignoring the case $\zeta = z = 0$, the critical values are

$$\begin{aligned} z_{cr1} &= 2.9483, & P_{cr1} &= 25.6454\,\frac{EI}{L^2}, \\[3mm] z_{cr2} &= 4.5778, & P_{cr2} &= 95.9335\,\frac{EI}{L^2}. \end{aligned} \qquad (10.85)$$

For values of P_0 other than the critical values, the natural frequencies are finite and they tend to be lower than the natural frequencies of the bar without the axial force. Another effect of the axial thrust is to distort the mode shapes as compared with the natural modes of a free-free bar with no axial force. The distortion consists of a reduction of the deformation at the end at which the external compressive force is applied, which results in a movement of the nodal points of each mode toward that end.

It should be noted that the values of the compressive force larger than P_{cr1} are mainly of academic interest in the absence of a means of preventing buckling. Furthermore, as the displacements become large we can no longer justify the use of the linear theory, which is based on the small displacements assumption (see Section 10–3).

Some of the problems associated with the dynamic stability of an elastic bar under constant and pulsating thrusts, in which a simplified control system is incorporated to obtain directional stability, are discussed in a paper by Beal.†

† T. R. Beal, *Am. Inst. Aeron. Astron. J.*, 3, 486–494 (1965).

10–6 NATURAL MODES OF BARS UNDER COMBINED FLEXURE AND TORSION†

The shear center or center of flexure of a cross section is defined as the point through which the transverse load must pass so that there will be no torsional motion produced. If the cross section is symmetric, the shear center coincides with the mass center of the cross section.‡ For a nonsymmetric cross section the shear center does not coincide, in general, with the center of mass. When the locus of the mass centers does not coincide with the locus of the shear centers, the vibration will consist of a combination of flexure and torsion. Let us consider such a case and denote the distance between the mass

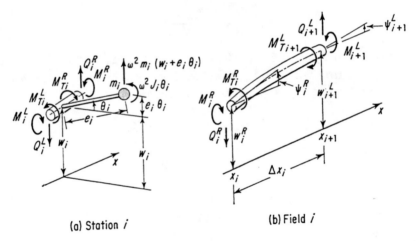

(a) Station i (b) Field i

FIGURE 10.4

center and shear center for any cross section as the eccentricity $e(x)$. We shall make the assumption that the locus of the shear centers of the bar in its undeformed position is a straight line. We shall further assume that the straight line passing through the shear center and mass center for any cross section remains a straight line making an angle $\theta(x)$ with respect to the initial position during deformation.

Because such problems occur for the most part in the case of nonuniform bars, we shall formulate the problem by means of the lumped method using

---

† This topic is referred to as *coupled vibration* in some texts, because of earlier terminology used in the field of aircraft vibration. We shall reserve the term *coupling* for the concept discussed earlier.

‡ We assume that the mass center is identical with the area center of the cross section.

transfer matrices similar to the methods discussed in Sections 6–11 and 6–12. Hence we shall divide the system into $n + 1$ lumped masses m_i at distances e_i from the shear centers (Figure 10.4). The shear centers in the undeformed position are located at points x_i called *stations*. The segments between stations, called *fields*, are assumed to be massless and to have uniform bending and torsional stiffness.

In the following w_i denotes transverse displacements, ψ_i bending angles, M_i bending moments, Q_i transverse shearing forces, θ_i torsional angles, and M_{Ti} torsional moments. The mass polar moment of inertia† of the mass m_i about an axis parallel to the shear center locus is denoted J_i. We shall use influence coefficients as defined in Section 6–12, and, in addition, we introduce the influence coefficient $a_i^{\theta M}$, defined as the angular torsional displacement at station $i + 1$ due to a unit torque, $M_{Ti+1}^L = 1$, when station i is regarded as fixed.

The relations between the quantities on the right of the station i and the ones on its left are

$$w_i^R = w_i^L = w_i, \qquad \psi_i^R = \psi_i^L = \psi_i, \tag{10.86}$$

$$M_i^R = M_i^L, \qquad \theta_i^R = \theta_i^L = \theta_i,$$

and, in addition, we have

$$Q_i^R = Q_i^L - \omega^2 m_i(w_i + e_i\theta_i) = -\omega^2 m_i w_i + Q_i^L - \omega^2 e_i m_i \theta_i, \tag{10.87}$$

$$M_{Ti}^R = M_{Ti}^L + \omega^2 m_i(w_i + e_i\theta_i)e_i + \omega^2 J_i\theta_i$$

$$= \omega^2 e_i m_i w_i + \omega^2(e_i^2 m_i + J_i)\theta_i + M_{Ti}^L. \tag{10.88}$$

Equations (10.86) through (10.88) can be arranged in the matrix form

$$
\begin{Bmatrix} w \\ \psi \\ M \\ Q \\ \theta \\ M_T \end{Bmatrix}_i^R = [T_S]_i \begin{Bmatrix} w \\ \psi \\ M \\ Q \\ \theta \\ M_T \end{Bmatrix}_i^L, \tag{10.89}
$$

† One may also consider including the rotatory inertia and shear deformation effects, as discussed in Section 10–7.

where

$$
\begin{Bmatrix} w \\ \psi \\ M \\ Q \\ \theta \\ M_T \end{Bmatrix}^R_i = \begin{Bmatrix} w^R_i \\ \psi^R_i \\ M^R_i \\ Q^R_i \\ \theta^R_i \\ M^R_{Ti} \end{Bmatrix}, \qquad \begin{Bmatrix} w \\ \psi \\ M \\ Q \\ \theta \\ M_T \end{Bmatrix}^L_i = \begin{Bmatrix} w^L_i \\ \psi^L_i \\ M^L_i \\ Q^L_i \\ \theta^L_i \\ M^L_{Ti} \end{Bmatrix}
\tag{10.90}
$$

are called state vectors and $[T_S]_i$ is the station transfer matrix for the station i. The station transfer matrix has the form

$$
[T_S]_i = \begin{bmatrix} 1 & 0 & 0 & 0 & 0 & 0 \\ 0 & 1 & 0 & 0 & 0 & 0 \\ 0 & 0 & 1 & 0 & 0 & 0 \\ -\omega^2 m_i & 0 & 0 & 1 & -\omega^2 e_i m_i & 0 \\ 0 & 0 & 0 & 0 & 1 & 0 \\ \omega^2 e_i m_i & 0 & 0 & 0 & \omega^2(e_i^2 m_i + J_i) & 1 \end{bmatrix}.
\tag{10.91}
$$

For the field i we have the relations (see Section 6–12)

$$
\begin{aligned}
w^L_{i+1} &= w^R_i + \Delta x_i \psi^R_i + a_i^{wM} M^L_{i+1} + a_i^{wQ} Q^L_{i+1} \\
&= w^R_i + \Delta x_i \psi^R_i + a_i^{wM} M^R_i + (a_i^{wQ} - \Delta x_i a_i^{wM}) Q^R_i, \\
\psi^L_{i+1} &= \psi^R_i + a_i^{\psi M} M^L_{i+1} + a_i^{\psi Q} Q^L_{i+1} \\
&= \psi^R_i + a_i^{\psi M} M^R_i + (a_i^{\psi Q} - \Delta x_i a_i^{\psi M}) Q^R_i, \\
M^L_{i+1} &= M^R_i - \Delta x_i Q^R_i, \\
Q^L_{i+1} &= Q^R_i,
\end{aligned}
\tag{10.92}
$$

and, in addition,

$$
\begin{aligned}
\theta^L_{i+1} &= \theta^R_i - a_i^{\theta M} M^L_{Ti+1} = \theta^R_i - a_i^{\theta M} M^R_{Ti}, \\
M^L_{Ti+1} &= M^R_{Ti}.
\end{aligned}
\tag{10.93}
$$

Hence, in a similar way, we have

$$
\begin{Bmatrix} w \\ \psi \\ M \\ Q \\ \theta \\ M_T \end{Bmatrix}^L_{i+1} = [T_F]_i \begin{Bmatrix} w \\ \psi \\ M \\ Q \\ \theta \\ M_T \end{Bmatrix}^R_i,
\tag{10.94}
$$

where the field transfer matrix $[T_F]_i$ is

$$[T_F]_i = \begin{bmatrix} 1 & \Delta x_i & a_i^{wM} & (a_i^{wQ} - \Delta x_i a_i^{wM}) & 0 & 0 \\ 0 & 1 & a_i^{\psi M} & (a_i^{\psi Q} - \Delta x_i a_i^{\psi M}) & 0 & 0 \\ 0 & 0 & 1 & -\Delta x_i & 0 & 0 \\ 0 & 0 & 0 & 1 & 0 & 0 \\ \hline 0 & 0 & 0 & 0 & 1 & -a_i^{\theta M} \\ 0 & 0 & 0 & 0 & 0 & 1 \end{bmatrix},$$

$$= \begin{bmatrix} 1 & \Delta x_i & (\Delta x_i)^2/2EI_i & -(\Delta x_i)^3/6EI_i & 0 & 0 \\ 0 & 1 & \Delta x_i/EI_i & -(\Delta x_i)^2/2EI_i & 0 & 0 \\ 0 & 0 & 1 & -\Delta x_i & 0 & 0 \\ 0 & 0 & 0 & 1 & 0 & 0 \\ \hline 0 & 0 & 0 & 0 & 1 & -\Delta x_i/GJ_{0i} \\ 0 & 0 & 0 & 0 & 0 & 1 \end{bmatrix}, \quad (10.95)$$

where EI_i is the flexural stiffness and GJ_{0i} the torsional stiffness.

We note from (10.91) and (10.95) that the terms rendering the flexural and torsional vibration dependent on each other are all in the station matrix $[T_S]_i$ and result from the equations of motion rather than the force deformation or moment-deformation relations.

From this point the procedure for obtaining the natural modes of vibration is the same as the one presented in Section 6–12.

10-7 EFFECT OF AXIAL FORCES ON THE NATURAL MODES OF VIBRATION OF A NONUNIFORM FREE–FREE BAR

The effect of axial forces on the transverse vibration of bars was discussed in Sections 10–3 through 10–5. When the bar is nonuniform it may be more practicable to adopt a lumped-parameter approach. We shall also use this opportunity to introduce the rotatory inertia and shear deformation effects upon the natural modes of vibration. The axial forces are such that they result in a constant longitudinal acceleration K. The approach used is similar to the one used in Section 10–6.

It is important to recognize the nature of the shear deformation effect. As in Section 5–2, let us define the transverse displacement as w, the angular

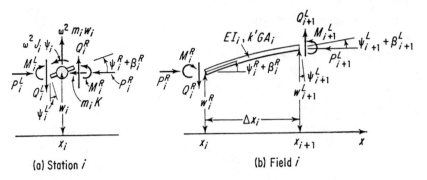

(a) Station i (b) Field i

FIGURE 10.5

displacement due to bending by ψ, and the angular distortion due to shear by β. The slope of the deflection curve is

$$\frac{\partial w}{\partial x} = \psi + \beta. \qquad (10.96)$$

However, an infinitesimal element rotates only by an amount ψ, because the shear in that case causes only distortion and no rotation. Hence the angular acceleration giving the rotatory inertia term is $\ddot{\psi}$. (Note that for finite elements of a bar, in which the elements length is not small relative to the bar height, the rotatory inertia due to shear deformation may not be negligible. This would require the addition of terms that are not included in the following analysis.) It also should be noted that the contribution of the axial forces P to the shearing forces at any point will be $P\psi$.

From Figure 10.5(a), representing station i, we obtain the following relations:

$$w_i^R = w_i^L = w_i, \qquad \psi_i^R = \psi_i^L = \psi_i,$$

$$M_i^R = M_i^L - \omega^2 J_i \psi_i = M_i^L - \omega^2 J_i \psi_i^L, \qquad (10.97)$$

$$Q_i^R = Q_i^L - \omega^2 m_i w_i = Q_i^L - \omega^2 m_i w_i^L,$$

where J_i is the mass moment of inertia of the mass m_i about an axis normal to the plane of the bar. In addition, we have the force equation in the horizontal direction,

$$P_i^R = P_i^L - m_i K, \qquad (10.98)$$

where K is the constant longitudinal acceleration.

The relations for the field i are obtained from Figure 10.5(b) in conjunction with the definition of the influence coefficients a_i^{wM}, a_i^{wQ}, $a_i^{\psi M}$, and $a_i^{\psi Q}$ given in Section 6–12. These relations are

$$w_{i+1}^L = w_i^R + \Delta x_i(\psi_i^R + \beta_i^R) + a_i^{wM} M_{i+1}^L + a_i^{wQ}(Q_{i+1}^L + P_{i+1}^L \psi_{i+1}^L),$$

$$\psi_{i+1}^L = \psi_i^R + a_i^{\psi M} M_{i+1}^L + a_i^{\psi Q}(Q_{i+1}^L + P_{i+1}^L \psi_{i+1}^L),$$

$$M_{i+1}^L = M_i^R - \Delta x_i Q_i^R - (w_{i+1}^L - w_i^R)P_i^R, \qquad (10.99)$$

$$Q_{i+1}^L + P_{i+1}^L \psi_{i+1}^L = Q_i^R + P_i^R \psi_i^R,$$

$$P_{i+1}^L = P_i^R.$$

The relation between the shearing force and the shear deformation is

$$\beta_i^R = a_i^{\beta Q}(Q_{i+1}^L + P_{i+1}^L \psi_{i+1}^L) = a_i^{\beta Q}(Q_i^R + P_i^R \psi_i^R), \qquad (10.100)$$

where the influence coefficient $a_i^{\beta Q}$ is defined as the shear deformation slope in field i due to a unit shearing force applied at station $i + 1$ when station i is regarded as fixed.

The axial force can be treated as a known quantity independent of the transverse motion, so that in effect we have only four variables, w_i, ψ_i, M_i, and Q_i. Equations (10.97) can be written

$$\left\{ \begin{array}{c} w \\ \psi \\ M \\ Q \end{array} \right\}_i^R = [T_s]_i \left\{ \begin{array}{c} w \\ \psi \\ M \\ Q \end{array} \right\}_i^L, \qquad (10.101)$$

where

$$\left\{ \begin{array}{c} w \\ \psi \\ M \\ Q \end{array} \right\}_i^R = \left\{ \begin{array}{c} w_i^R \\ \psi_i^R \\ M_i^R \\ Q_i^R \end{array} \right\}, \qquad \left\{ \begin{array}{c} w \\ \psi \\ M \\ Q \end{array} \right\}_i^L = \left\{ \begin{array}{c} w_i^L \\ \psi_i^L \\ M_i^L \\ Q_i^L \end{array} \right\} \qquad (10.102)$$

are the state vectors corresponding to the right and left sides of station i and

$$[T_s]_i = \begin{bmatrix} 1 & 0 & 0 & 0 \\ 0 & 1 & 0 & 0 \\ 0 & -\omega_i^2 J_i & 1 & 0 \\ -\omega^2 m_i & 0 & 0 & 1 \end{bmatrix} \qquad (10.103)$$

is the station transfer matrix associated with station i.

To write the relations for field i by means of a field transfer matrix we must rearrange (10.99) and make use of (10.100). The result is

$$w_{i+1}^L = w_i^R + \frac{\Delta x_i + \Delta x_i a_i^\beta{}^Q P_i^R + a_i^w{}^Q P_i^R}{1 + a_i^w{}^M P_i^R} \psi_i^R + \frac{a_i^w{}^M}{1 + a_i^w{}^M P_i^R} M_i^R$$

$$+ \frac{\Delta x_i a_i^\beta{}^Q - \Delta x_i a_i^w{}^M + a_i^w{}^Q}{1 + a_i^w{}^M P_i^R} Q_i^R,$$

$$\psi_{i+1}^L = \left[1 + a_i^\psi{}^Q P_i^R - \frac{a_i^\psi{}^M P_i^R(\Delta x_i + \Delta x_i a_i^\beta{}^Q P_i^R + a_i^w{}^Q P_i^R)}{1 + a_i^w{}^M P_i^R}\right]\psi_i^R$$

$$+ \frac{a_i^\psi{}^M}{1 + a_i^w{}^M P_i^R} M_i^R + \left[a_i^\psi{}^Q - \frac{a_i^\psi{}^M(\Delta x_i + \Delta x_i a_i^\beta{}^Q P_i^R + a_i^w{}^Q P_i^R)}{1 + a_i^w{}^M P_i^R}\right] Q_i^R,$$

$$\text{(10.104)}$$

$$M_{i+1}^L = -\frac{P_i^R(\Delta x_i + \Delta x_i a_i^\beta{}^Q P_i^R + a_i^w{}^Q P_i^R)}{1 + a_i^w{}^M P_i^R} \psi_i^R + \frac{1}{1 + a_i^w{}^M P_i^R} M_i^R$$

$$- \frac{\Delta x_i + \Delta x_i a_i^\beta{}^Q P_i^R + a_i^w{}^Q P_i^R}{1 + a_i^w{}^M P_i^R} Q_i^R,$$

$$Q_{i+1}^L = -(P_i^R)^2\left[a_i^\psi{}^Q - \frac{a_i^\psi{}^M(\Delta x_i + \Delta x_i a_i^\beta{}^Q P_i^R + a_i^w{}^Q P_i^R)}{1 + a_i^w{}^M P_i^R}\right]\psi_i^R - \frac{a_i^\psi{}^M P_i^R}{1 + a_i^w{}^M P_i^R} M_i^R$$

$$+ \left[1 - a_i^\psi{}^Q P_i^R + \frac{a_i^\psi{}^M P_i^R(\Delta x_i + \Delta x_i a_i^\beta{}^Q P_i^R + a_i^w{}^Q P_i^R)}{1 + a_i^w{}^M P_i^R}\right] Q_i^R,$$

where the influence coefficients have the values

$$a_i^w{}^Q = \frac{(\Delta x_i)^3}{3EI_i}, \qquad a_i^w{}^M = \frac{(\Delta x_i)^2}{2EI_i}, \qquad a_i^\psi{}^Q = \frac{(\Delta x_i)^2}{2EI_i},$$

$$a_i^\psi{}^M = \frac{\Delta x_i}{EI_i}, \qquad a_i^\beta{}^Q = \frac{1}{k'GA_i}. \qquad \text{(10.105)}$$

In the above expressions EI_i is the bending stiffness and $k'GA_i$ is the shearing stiffness [see (5.3)].

Equations (10.104) can be arranged in the matrix form

$$\left\{\begin{array}{c} w \\ \psi \\ M \\ Q \end{array}\right\}_{i+1}^L = [T_F]_i \left\{\begin{array}{c} w \\ \psi \\ M \\ Q \end{array}\right\}_i^R, \qquad \text{(10.106)}$$

where the elements of the field transfer matrix $[T_F]_i$ are the coefficients of w_i^R, ψ_i^R, M_i^R, and Q_i^R respectively.

The procedure for obtaining the natural modes is described in Section 6–12.

For a free-free bar subjected to a constant force P_0 at the end $x = 0$ we have

$$P_i^L = P_0, \qquad (10.107)$$

and considering a bar consisting of $n + 1$ lumps and n fields the forces become

$$P_i^R = P_{i+1}^L = P_0 - K \sum_{j=1}^{i} m_j = P_0 \left(1 - \frac{\sum_{j=1}^{i} m_j}{\sum_{j=1}^{n+1} m_j} \right). \qquad (10.108)$$

The boundary conditions are

$$\begin{aligned} M_1^L = 0, \qquad Q_1^L = P_0 \psi_1^L = P_0 \psi_1^R, \\ M_{n+1}^R = 0, \qquad Q_{n+1}^R = 0. \end{aligned} \qquad (10.109)$$

Problems

10.1 A uniform bar of length L, hinged at both ends, is elastically supported so that the additional restoring distributed force acting in the transverse direction is

$$F_{\text{rest}} = -k \frac{x}{L} \left(1 - \frac{x}{L} \right) w.$$

Check whether the operator L is self-adjoint. Let $k = 25 \times 10^2 EI/L^4$ and solve the corresponding eigenvalue problem for the transverse motion by means of an approximate method and obtain the first three modes and natural frequencies. Plot the normal modes.

10.2 A uniform circular plate, clamped at the boundary $r = a$, is elastically supported so that the restoring force is proportional to the displacement; the constant of proportionality is k. Derive a general expression for the normal modes and natural frequencies of the system.

10.3 A bar in transverse vibration is subjected at any point x to an axial force $P(x, t)$ and a distributed transverse force $p(x, t)$. Derive the differential equation of motion by summing forces on a differential element. Indicate the boundary conditions corresponding to all possible types of end supports.

10.4 Derive the eigenvalue problem for the transverse vibration of a rotating bar hinged at the end $x = 0$ and free at the end $x = L$ (see Figure 10.2 with the end $x = 0$ hinged instead of clamped). Check whether the operator L is self-adjoint. Solve the eigenvalue problem for a uniform bar by an approximate method and obtain the first three modes. Let

$$\Omega^2 = 9 \frac{EI}{mL^4}.$$

10.5 Formulate the boundary-value problem for the transverse motion of the system shown in Figure 10.6. Solve the eigenvalue problem corresponding to the uniform bar if

$$\alpha = 0, \qquad P(t) = P_0 = \frac{12EI}{L^2}.$$

Obtain the first three modes by an approximate method and compare your results with the free-free beam solution, $P_0 = 0$.

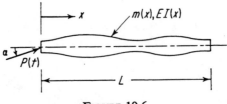

FIGURE 10.6

10.6 A nonsymmetrical bar is fixed at the end $x = 0$ and free at the end $x = L$ as shown in Figure 10.7, which represents a top view. The cross-sectional data are

$$A(x) = A\left(1 - \frac{1}{2}\frac{x}{L}\right), \qquad I(x) = I\left(1 - \frac{1}{2}\frac{x}{L}\right),$$

$$J(x) = J\left[1 - \frac{3}{4}\left(\frac{x}{L}\right)^2\right], \qquad e(x) = \sqrt{\frac{J}{8A}}\left(2\frac{x}{L} - 1\right),$$

where $A(x)$ and $I(x)$ are the cross-sectional area and moment of inertia, respectively, $J(x)$ is the mass polar moment of inertia about the center of mass, and $e(x)$ is the distance between the shear center and the mass center at any point x. Assume that the torsional stiffness is proportional to the polar moment of inertia, $GJ_0(x) = GJ(x)$ and, furthermore, let $G = \frac{3}{8}E$. The distributed mass is proportional to the cross-sectional area. Divide the bar into lumps so that there are six degrees of freedom and obtain the first two natural frequencies and natural modes. Plot the natural modes in the form of contour lines of equal deflection as seen from the top.

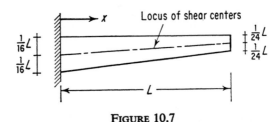

FIGURE 10.7

10.7 Solve Problem 10.5 by the lumped method of Section 10–7, including the rotatory inertia and shear deformation effects. Let $k'G = E/4$ and assume that the cross section is a square of side $L/10$. Use six lumps and obtain the first two natural frequencies and natural modes for

(a) $P_0 = 0$.

(b) $P_0 = 12EI/L^2$.

Plot the corresponding modes for both cases.

Selected Readings

Bisplinghoff, R. L., H. Ashley, and R. L. Halfman, *Aeroelasticity*, Addison-Wesley, Reading, Mass., 1957.

Pestel, E. L., and F. A. Leckie, *Matrix Methods in Elastomechanics*, McGraw-Hill, New York, 1963.

Seide, P., Effect of Constant Longitudinal Acceleration on the Transverse Vibration of Uniform Beams, Aerospace Corporation *TDR-169 (3560–30) TN-6*, October 1963.

Thomson, W. T., *Vibration Theory and Applications*, Prentice-Hall, Englewood Cliffs, N.J., 1965.

Timoshenko, S., *Vibration Problems in Engineering*, Van Nostrand, New York, 1937.

CHAPTER 11

RANDOM
VIBRATION

11-1 INTRODUCTION

In the physical world there are many systems for which the excitation can be related in a simple fashion to the factors producing the excitation. Furthermore, if the function describing the system excitation is obtained as the outcome of an experiment, and the experiment is repeated a large number of times under essentially the same conditions, the same record is produced every time. For that reason these functions are called *deterministic*. They are characterized by the fact that the value of the function can be given in advance for any time t. These functions can be divided into sinusoidal, periodic, and nonperiodic; an example of each is shown in Figure 11.1(a), (b), and (c), respectively. Hence deterministic functions can be regarded as consisting either of prescribed periodic functions or prescribed transients which lead to steady-state and transient responses. We had many opportunities to work with deterministic functions in earlier chapters.

On the other hand, there are cases in which, for a variety of reasons, it is not possible to predict exactly the outcome of an experiment. Furthermore, if such an experiment is repeated a number of times, different records are obtained, no matter how carefully the experimenter controls the conditions that might influence the outcome of the experiment. Perhaps the reason for not being able to predict exactly the outcome of the experiment is that we are not fully aware of all the contributing factors. It is also possible that, although we know all the factors involved, the number of factors affecting the outcome of the experiment is so large that it is not feasible to take them all into consideration. The functions describing such phenomena are called *random* and a typical random function is shown in Figure 11.1(*d*). Examples of random functions are the noise intensity caused by jet and rocket engines, or displacement-type excitation to which airplanes taxiing on rough runways are subjected, or noise in communication systems. An attempt to obtain

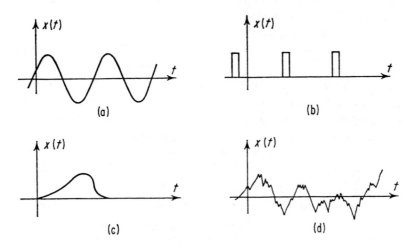

(a)

(b)

(c)

(d)

FIGURE 11.1

instantaneous responses to a random excitation leads to a degree of complexity which alerts one to the necessity of adopting new methods of analysis.

If the outcomes obtained by a repetition of the random experiment under practically the same conditions present no discernable pattern, then no meaningful analysis is possible. But in the physical world there are many random experiments which, when re-created a large number of times, do form a definite pattern, in the sense that their outcomes tend to center around certain average values. This characteristic of random experiments is called *statistical regularity*. The concept of statistical regularity cannot be proved mathematically and is a result of inductive reasoning.

In view of the concept of statistical regularity, one may wonder whether or not it is possible to construct a mathematical model to represent a repetitive experiment. Such a model should enable one to make predictions as to the chance of a certain outcome of the experiment occurring, when the experiment is repeated a large number of times. This leads, in a natural way, to the probability model as the mathematical model. Such a model allows one to make a *probabilistic* statement, as opposed to a deterministic statement, about the outcome of an experiment without the need to perform the experiment.

There are several ways of defining probability. The pure mathematicians adopt an *axiomatic* point of view in defining probability. On the other hand, engineers and physicists define probability as the *relative frequency of occurrence* of a particular outcome out of all possible outcomes, when the experiment is repeated a large number of times. The outcomes of random experiments are called *events*. Associated with each event there is a probability of occurrence.

The outcomes of an experiment can be represented by points called *sample points*. The set of points representing all possible outcomes of the experiment is called the *sample space*, or *event space*, of the experiment in question. A *real random variable* is a real-valued variable defined on a sample space. Random variables can be discrete or continuous. One can associate with the amplitude of a random variable a *probability density function* and statistical averages such as the *mean value* and *mean square value* (encountered in Chapter 1 in connection with deterministic variables). The collection of records of random variables describing an experiment is called a *random process*. For a given time, $t = t_1$, one can evaluate statistical averages over the process. A random process is said to be *stationary* if these averages do not depend on the time t_1.† When the averages over the process can be replaced by averages over a single record the process is said to be *ergodic*. In many cases it is possible to describe a random process adequately, although not completely, by means of the mean value and mean square value. This is particularly true if the random process is normal, or *Gaussian*. The *central limit theorem* states the condition a random process must satisfy to be Gaussian. Random processes are conveniently treated by means of power spectral density functions.

If the excitation is a random variable, the response of a system is a random variable. Furthermore, if the excitation process is Gaussian, the response process is Gaussian, *provided the system is linear*. Hence if the excitation of a linear system is a Gaussian process, the response mean value and response mean square value adequately describe the response.

In this chapter we first present an introduction to probability theory. The concept of a random variable and its probability distribution, as well as the expected value of a function of a random variable, are discussed. The idea of a random process is introduced and averages over the process and over a representative sample function are defined. A general relation between the response power spectral density and the excitation power spectral density is

---

† The exact definition is given in Section 11–4.

derived for a linear system. Finally, the response mean value and mean square value are obtained for a single-degree-of-freedom system, a multi-degree-of-freedom system, and a continuous system.

11-2 PROBABILITY

An important concept in probability theory is the *event*. Depending on the outcome of a certain experiment one can decide whether an event A has occurred or not. The concept of event can be conveniently discussed by means of terminology used in *set theory*.

A *set* is a collection of objects. The objects of a set A are called *elements* and denoted a_i $(i = 1, 2, \ldots, k)$. The numerals representing the number of dots on the six faces of a die can be looked upon as the elements of a set consisting of six elements, $k = 6$. All odd integers larger than 0 and smaller than 10 can form a set of five elements. A part of a larger set is called a *subset*. As an example, the set consisting of the numerals representing the number of dots on the faces of a die which is smaller than three is a subset consisting of two elements, as opposed to the set of six elements previously mentioned. If all the subsets considered are not arbitrary but subsets of a larger set S, then S is a *space*. Sets can be finite or infinite. Finite sets can be defined by the number of their elements or by listing the conditions to be satisfied by all their elements. If a set contains no elements it is said to be an *empty set* or *null set*. The *union* $A + B$ of two sets, A and B, is a set containing all the elements of A or B or both A and B. The *intersection* AB of two sets, A and B, is the set containing all the elements common to both A and B.

The outcomes of a well-defined experiment can be looked upon as the elements of the space S. Consistent with this definition events are subsets of S. An event A consisting of a single element is said to be *elementary*. The single performance of a well-defined experiment is called a *trial*. If the outcome of a trial is a_k, where a_k is one of the elements of the set A, then event A is said to have occurred. An elementary event occurs only when the outcome of the trial is its single element. One can repeat the experiment a large number of times and observe the ratio of the number of times event A occurred to the total number of trials. Consequently one can require the satisfaction of the following axioms:

Axiom I. To each event A, which is the subset of the space S, there is assigned a nonnegative, real number $\Pr(A)$ which is called the *probability* of event A,

$$\Pr(A) \geq 0. \tag{11.1}$$

Axiom II. The probability of an event certain to occur is unity,

$$\Pr(S) = 1. \tag{11.2}$$

Axiom III. The probability of *two mutually exclusive events*, A and B, is equal to the sum of the probabilities of A and B,

$$\Pr(A + B) = \Pr(A) + \Pr(B), \tag{11.3}$$

where $A + B$ is the union of A and B.

To evaluate the probability of a given event one can use the *relative frequency of occurrence* definition,

$$\Pr(A) = \frac{n}{N}, \qquad N \to \infty, \tag{11.4}$$

where n is the total number of times the event occurred during a total number of trials N.

An experiment and its outcomes can be conveniently represented as a space and its points. An elementary event defines a point in that space called a *sample point*. The set of points representing all the possible outcomes of an

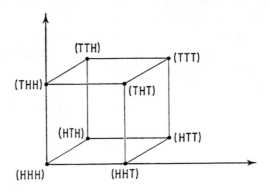

FIGURE 11.2

experiment is called a *sample space* of the experiment in question. As an example consider the experiment of tossing three coins simultaneously. A single outcome of the experiment is any combination of heads and tails. There are eight such possible outcomes; they are plotted in the sample space shown in Figure 11.2. It is not necessary, however, to plot the sample space in a three-dimensional plot. One could just as easily plot the eight sample points along a straight line.

If the coins are unbiased there is every reason to believe that the probability of any of the elementary events is 1/8. In other experiments there may be reasons for which the probability of the elementary events are not equal. For example, in rolling a loaded die it would be naïve to think (as some people might have found out) that the probability of any of the six faces showing is 1/6. In such cases one can perform a large number of experiments and assign probabilities to each elementary event according to the relative

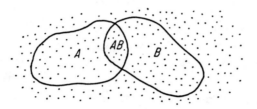

FIGURE 11.3

frequency definition. Hence to every elementary event a_i we can assign a probability p_i. If an event A consists of the elements a_i $(i = 1, 2, \ldots, k)$, then

$$\Pr(A) = \sum_{i=1}^{k} \Pr(a_i) = \sum_{i=1}^{k} p_i, \tag{11.5}$$

or, the probability of occurrence of an event A is equal to the sum of the probabilities of the sample points associated with the occurrence of A.

Next let us consider two events, A and B, of the same experiment. Figure 11.3 shows the sample space associated with the experiment. One can distinguish three regions in the sample space. The first region contains points belonging to event A only, the second one to event B only, and the third one to both events A and B. Of course, the latter is the region over which the first two overlap. The event that either A or B or both A and B occur is the union $A + B$ and the corresponding probability is denoted $\Pr(A + B)$. Furthermore, the intersection AB is the event that both A and B occur and $\Pr(AB)$ is the probability associated with that event. From Figure 11.3 we conclude that

$$\Pr(A + B) = \Pr(A) + \Pr(B) - \Pr(AB), \tag{11.6}$$

where $\Pr(AB)$ had to be subtracted, because the overlapping region has been counted twice. Of course, when events A and B are *mutually exclusive*, the intersection AB is zero and (11.6) reduces to (11.3). Equation (11.6) is known as the *addition theorem*.

As an illustration, let us consider the experiment of rolling an unbiased die and denote the event of less than four dots showing A and the event of an even number of dots showing B. The question is asked as to the probability of the union $A + B$ occurring. The probability of occurrence of A is $\Pr(A) = 1/2$ and the probability of B is $\Pr(B) = 1/2$. The intersection AB is the event of two dots showing and the corresponding probability is $\Pr(AB) = 1/6$. It follows from (11.6) that

$$\Pr(A + B) = 1/2 + 1/2 - 1/6 = 5/6.$$

This result can be easily checked, because the only way for the union $A + B$ not to occur is for the face with five dots to show.

The probability that event B will occur if it is given that event A is certain to occur is called *conditional probability* and is denoted $\Pr(B/A)$. Because A is certain to occur we are concerned only with the sample points belonging to A. In addition, we know that the sum of the probabilities of the elementary events in A must be 1, so we must readjust the probabilities associated with each point in A. This readjustment, however, must not affect the relative value of any two elementary events. To this end we multiply all the probabilities p_i of the elementary events by a constant c and denote the new probabilities π_i, so that

$$\pi_i = cp_i,$$

where π_i must be such that

$$\sum_A \pi_i = c \sum_A p_i = c \Pr(A) = 1.$$

This allows us to write

$$\pi_i = cp_i = \frac{p_i}{\Pr(A)}.$$

The conditional probability $\Pr(B/A)$ is obtained by considering only the points in A that also belong to B, the points in the common region. Thus

$$\Pr(B/A) = \sum_{AB} \pi_i = \frac{\sum_{AB} p_i}{\Pr(A)}.$$

But the numerator $\sum_{AB} p_i$ is precisely the probability that the intersection AB occurs, that is, $\Pr(AB)$. It follows that the conditional probability can be written

$$\Pr(B/A) = \frac{\Pr(AB)}{\Pr(A)}. \tag{11.7}$$

Equation (11.7) can be rewritten

$$\Pr(AB) = \Pr(A)\Pr(B/A), \tag{11.8}$$

which is the statement of the *multiplication theorem*. It is obvious that the probability that both events A and B occur can also be written

$$\Pr(AB) = \Pr(B)\Pr(A/B). \tag{11.9}$$

If events A and B are *statistically independent*, we have

$$\Pr(B/A) = \Pr(B), \qquad \Pr(A/B) = \Pr(A), \tag{11.10}$$

from which it follows that

$$\Pr(AB) = \Pr(A)\Pr(B), \qquad A,\ B \text{ independent.} \tag{11.11}$$

As an example, let us consider the experiment of tossing two unbiased coins simultaneously, and calculate the probability of obtaining one tail if it is given that one head is certain to occur. The four possible outcomes of the experiment are:

head	head
head	tail
tail	head
tail	tail

The event of obtaining at least one head is denoted A and the event of one tail and one head occurring, AB. Examining the four possible outcomes we conclude that $\Pr(A) = 3/4$ and $\Pr(AB) = 1/2$. Using (11.7) we obtain

$$\Pr(B/A) = \frac{1/2}{3/4} = 2/3.$$

11-3 RANDOM VARIABLES AND PROBABILITY DISTRIBUTIONS

In Section 11–2 we have shown that all the elementary events associated with a random experiment define a sample space. Each sample point corresponds to an elementary event and is assigned a probability. These points can be identified by a numerical-valued variable x. The numerical-valued variable x, defined on a sample space, is called a *random variable*. Random variables can be discrete or continuous.

Let us consider a real *discrete random variable* x, associated with a random experiment. This variable can take the possible values x_k. The complete set of probabilities $p(x_k)$ is called the *probability distribution* and is shown in Figure 11.4(*a*).

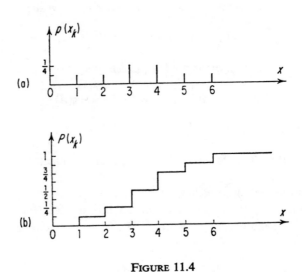

FIGURE 11.4

The set consisting of all the outcomes x_k for which $x_k \leq X$ is an event. The probability of this event is

$$\Pr(x_k \leq X) = \sum_{x_k \leq x} p(x_k). \qquad (11.12)$$

The function $P(x_k) = \Pr(x_k \leq X)$ is called the *probability distribution function* and is plotted in Figure 11.4(*b*). It is obvious that

$$P(\infty) = \Pr(x_k \leq +\infty) = 1. \qquad (11.13)$$

We shall be interested in random variables for which the probability density function is everywhere continuous. Such a random variable is called a *continuous random variable* and is shown in Figure 11.5(*a*). Let us consider a continuous random variable $x(t)$ and ask the question as to the chance that the amplitude of $x(t)$ at any particular time t will be smaller than a given value x. We shall call this the probability that $x(t)$ is smaller than x and denote it $\Pr(x(t) < x)$. From Figure 11.5(*a*) we see that the amplitude of the function $x(t)$ is smaller than x when t lies within the intervals $\Delta t_1, \Delta t_2, \Delta t_3, \ldots$. We shall assume that summing the values of the

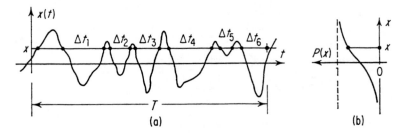

FIGURE 11.5

intervals Δt_i, dividing by the length of the total interval T, and letting $T \to \infty$, we obtain an estimate of the desired value

$$\Pr(x(t) < x) = \lim_{T \to \infty} \frac{1}{T} \sum_i \Delta t_i. \tag{11.14}$$

As we let x vary we obtain a function which is denoted $P(x)$,

$$P(x) = \Pr(x(t) < x). \tag{11.15}$$

The function $P(x)$, which is shown in Figure 11.5(b), is the *probability distribution function* associated with a continuous random variable. It is a monotonically increasing function which possesses the properties

$$P(-\infty) = 0, \qquad 0 \le P(x) \le 1, \qquad P(\infty) = 1. \tag{11.16}$$

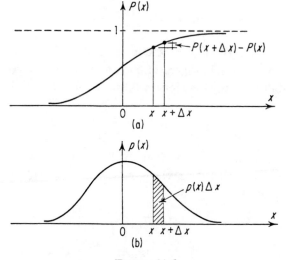

FIGURE 11.6

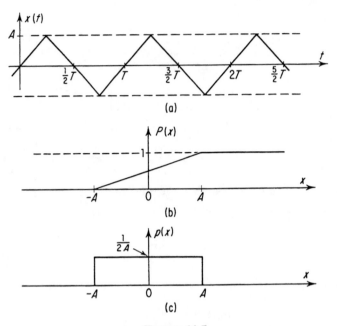

FIGURE 11.7

Next we consider another value $x + \Delta x$, to which corresponds the cumulative probability $P(x + \Delta x)$ and ask the question as to the probability that the amplitude of $x(t)$ lies between the values x and $x + \Delta x$. This probability is, of course, $P(x + \Delta x) - P(x)$. At this point we define the *probability density function* as

$$p(x) = \lim_{\Delta x \to 0} \frac{P(x + \Delta x) - P(x)}{\Delta x} = \frac{dP(x)}{dx}, \qquad (11.17)$$

and we conclude that $p(x)$ is just the slope of the function $P(x)$. The functions $P(x)$ and $p(x)$ are shown in Figure 11.6(*a*) and (*b*), respectively.

The probability density function, $p(x)$, can also be defined by

$$\Pr(x_1 < x < x_2) = \int_{x_1}^{x_2} p(x)\, dx, \qquad (11.18)$$

which means that the probability that $x(t)$ lies between the values x_1 and x_2 is equal to the area under the $p(x)$ curve between these values. The function $p(x)$ has the properties

$$p(x) \geq 0, \qquad p(-\infty) = 0, \qquad p(\infty) = 0,$$

$$P(x) = \int_{-\infty}^{x} p(\xi)\, d\xi, \qquad P(\infty) = \int_{-\infty}^{\infty} p(x)\, dx = 1. \qquad (11.19)$$

As an example let us consider the function shown in Figure 11.7(a). The probability density function is shown in Figure 11.7(c) and is known as the *rectangular* or *uniform* distribution.

It was pointed out in Section 11–1 that the vibration caused by a jet or rocket engine at a certain point of a flight vehicle can be regarded as a random variable. In such a case the amplitude, frequency, and phase angle vary randomly. When the random variable is the result of many factors, none of which contributes significantly to this result, the distribution approaches, under very general conditions, the *normal*, or *Gaussian*, *distribution* (see the *central limit theorem*, Section 11–6). The normal distribution is described by the expressions

$$P(x) = \frac{1}{\sqrt{2\pi}} \int_{-\infty}^{x} e^{-(1/2)\xi^2} \, d\xi,$$

$$p(x) = \frac{1}{\sqrt{2\pi}} e^{-(1/2)x^2}$$

(11.20)

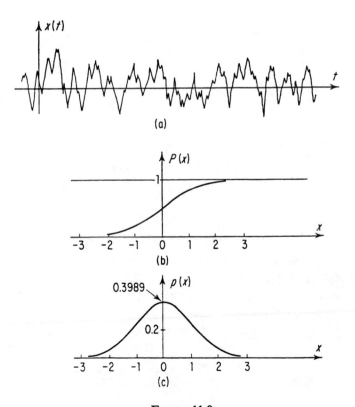

(a)

(b)

(c)

FIGURE 11.8

and can be seen in Figure 11.8. The above distribution, $P(x)$, is also known
as the *error function* and is tabulated in many mathematical handbooks.†

When the random variable is restricted to positive values of the argument,
it may tend to follow the *Rayleigh distribution*, defined by

$$P(x) = \begin{cases} 1 - e^{-(1/2)x^2}, & x > 0, \\ 0, & x < 0, \end{cases}$$

$$p(x) = \begin{cases} xe^{-(1/2)x^2}, & x > 0, \\ 0, & x < 0. \end{cases}$$

(11.21)

These are shown in Figure 11.9.

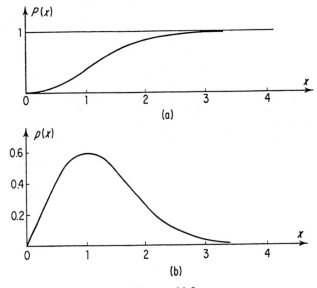

FIGURE 11.9

Let us consider the function $x(t)$ and assume that we know the probability
density function $p(t)$ associated with the variable t and are interested in
obtaining the probability density function $p(x)$. But the probability that x lies
within the interval between x_0 and $x_0 + \Delta x_0$ must be equal to the probability
that t lies within the intervals between t_1 and $t_1 + \Delta t_1$, t_2 and $t_2 + \Delta t_2$, t_3
and $t_3 + \Delta t_3$, etc., as can be easily seen from Figure 11.10. Hence we have

$$\Pr(x_0 < x < x_0 + \Delta x_0) = \sum_i \Pr(t_i < t < t_i + \Delta t_i). \tag{11.22}$$

† Most error functions in tables have a slightly different definition.

Equation (11.22) implies that

$$p(x_0) \, \Delta x_0 = \sum_i p(t_i) |\Delta t_i|, \tag{11.23}$$

where the absolute value of Δt_i must be used to account for the fact that to an increment Δx_0 may correspond a negative increment Δt_i, as is the case with Δt_2 and Δt_4 in Figure 11.10.

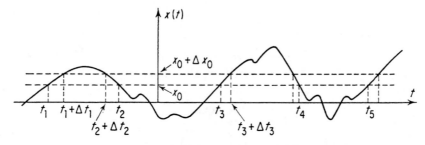

FIGURE 11.10

Letting x_0 assume the arbitrary value x and going to the limit, we can write the probability density function $p(x)$ in the form

$$p(x) = \sum_i \frac{p(t_i)}{|dx/dt_i|} = \sum_i \left[\frac{p(t)}{|dx/dt|} \right]_{t=t_i}, \tag{11.24}$$

where t_i are the values of t corresponding to $x(t) = x$. Note that (11.24) is valid provided the slope dx/dt is different from zero at all these points.

As an illustration consider the function

$$x(t) = A \sin \frac{2\pi t}{T}. \tag{11.25}$$

Assuming t has an equal chance of assuming any value from 0 to T, we have

$$p(t) = \begin{cases} \dfrac{1}{T}, & 0 < t < T, \\ 0, & t < 0, t > T, \end{cases} \tag{11.26}$$

which means that the variable t possesses a uniform distribution defined previously. Furthermore, for each value of x there are two values of t in the interval $0 < t < T$, and, because the absolute values of the slopes at these points are equal, we have

$$p(x) = 2 \frac{1}{T} \frac{1}{|dx/dt|} = \frac{2}{T} \frac{1}{(2\pi/T)A \cos (2\pi t/T)}$$

$$= \frac{1}{\pi} \frac{1}{A[1 - \sin^2 (2\pi t/T)]^{1/2}}. \tag{11.27}$$

Hence we can write

$$p(x) = \begin{cases} \dfrac{1}{\pi} \dfrac{1}{(A^2 - x^2)^{1/2}}, & |x| < A, \\ 0, & |x| > A. \end{cases} \tag{11.28}$$

The *mean value*,[†] or *expected value*, of the random variable x is given by[‡]

$$E[x] = \int_{-\infty}^{\infty} x p(x)\, dx = \bar{x}. \tag{11.29}$$

Next let us consider a real single-valued continuous function $g(x)$ of the random variable x. Then, by definition, the expected value of $g(x)$ is

$$E[g(x)] = \int_{-\infty}^{\infty} g(x) p(x)\, dx = \overline{g(x)}. \tag{11.30}$$

In the special case in which $g(x) = x^2$, we have

$$E[x^2] = \overline{x^2} = \int_{-\infty}^{\infty} x^2 p(x)\, dx, \tag{11.31}$$

where $E[x^2] = \overline{x^2}$ is called the *mean square value* of the random variable x. Its square root is called *root mean square* or *rms value*.

The *variance* of x, denoted by $\sigma^2(x)$, is defined as the mean square value of x about the mean,

$$\sigma^2(x) = E[(x - \bar{x})^2] = \int_{-\infty}^{\infty} (x - \bar{x})^2 p(x)\, dx = \overline{x^2} - (\bar{x})^2. \tag{11.32}$$

The positive square root of the variance, denoted $\sigma(x)$, is called *standard deviation* of x, and if the mean value is zero it is identical to the rms value of x.

Now let us consider two random variables $x(t)$ and $y(t)$ and define the joint, or *second-order*, probability distribution function $P(x, y)$ associated with the probability that $x(t) \le x$ and $y(t) \le y$,

$$P(x, y) = \Pr[x(t) < x, y(t) \le y]. \tag{11.33}$$

The probability density function $p(x, y)$ can be shown in a two-dimensional plot (see Figure 11.11). In terms of the function $p(x, y)$ we can write

$$P(x, y) = \int_{-\infty}^{x} \int_{-\infty}^{y} p(\xi, \eta)\, d\xi\, d\eta, \tag{11.34}$$

[†] Also called *average value*.
[‡] Note that this definition does not invalidate the definition of mean value given in Chapter 1.

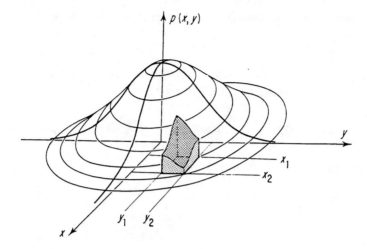

FIGURE 11.11

where ξ and η are dummy variables. We note that the probability that $x_1 < x < x_2$ and $y_1 < y < y_2$ is

$$\Pr(x_1 < x < x_2, y_1 < y < y_2) = \int_{x_1}^{x_2} \int_{y_1}^{y_2} p(x, y) \, dx \, dy. \quad (11.35)$$

It is represented by the shaded volume in Figure 11.11.

The probability density function $p(x, y)$ possesses the properties

$$p(x, y) \geq 0, \qquad \int_{-\infty}^{\infty} \int_{-\infty}^{\infty} p(x, y) \, dx \, dy = 1. \quad (11.36)$$

The probability that x lies within the interval $x_1 < x < x_2$ regardless of the value of y is

$$\Pr(x_1 < x < x_2, -\infty < y < \infty) = \int_{x_1}^{x_2} \left[\int_{-\infty}^{\infty} p(x, y) \, dy \right] dx = \int_{x_1}^{x_2} p(x) \, dx, \quad (11.37)$$

where

$$p(x) = \int_{-\infty}^{\infty} p(x, y) \, dy \quad (11.38)$$

is the first-order probability density of x alone. The probability density for a single variable from a collection of variables is called *marginal probability density*. Similarly, $p(y)$ is the first-order probability density of y alone,

$$p(y) = \int_{-\infty}^{\infty} p(x, y) \, dx. \quad (11.39)$$

When the two random variables x and y are *statistically independent*, then

$$p(x, y) = p(x)p(y). \tag{11.40}$$

Next let us consider the expected value of a real single-valued continuous function $g(x, y)$ of the random variables $x(t)$ and $y(t)$,

$$E[g(x, y)] = \int_{-\infty}^{\infty} \int_{-\infty}^{\infty} g(x, y)p(x, y) \, dx \, dy = \overline{g(x, y)}. \tag{11.41}$$

Consider the special case in which

$$g(x, y) = (x - \bar{x})(y - \bar{y}),$$
$$\bar{x} = \int_{-\infty}^{\infty} xp(x) \, dx, \qquad \bar{y} = \int_{-\infty}^{\infty} yp(y) \, dy, \tag{11.42}$$

and define the *covariance* $\rho(x, y)$ between x and y as

$$\rho(x, y) = E[(x - \bar{x})(y - \bar{y})] = \int_{-\infty}^{\infty} \int_{-\infty}^{\infty} (x - \bar{x})(y - \bar{y})p(x, y) \, dx \, dy$$
$$= E[xy] - E[x]E[y]. \tag{11.43}$$

When x and y are statistically independent, using (11.40) we obtain

$$\rho(x, y) = E[xy] - E[x]E[y] = \int_{-\infty}^{\infty} \int_{-\infty}^{\infty} xyp(x, y) \, dx \, dy - E[x]E[y]$$
$$= \int_{-\infty}^{\infty} xp(x) \, dx \int_{-\infty}^{\infty} yp(y) \, dy - E[x]E[y] = 0. \tag{11.44}$$

Hence, for statistically independent random variables, the covariance is zero.

The change of variable procedure, (11.24), can be extended to second- and higher-order probability density functions. This change of variables is most conveniently accomplished by using the Jacobian of the transformation in question.†

11–4   ENSEMBLE AVERAGES. STATIONARY RANDOM PROCESSES

Let us assume that as a result of the same experiment performed a large number of times under the same conditions, one obtains individual time histories $^k x(t)$. Each of the functions $^k x(t)$ is said to be a *sample function*.

† See A. Papoulis, *Probability, Random Variables, and Stochastic Processes*, McGraw-Hill, New York, 1965, p. 201.

The entire collection of random records, or *ensemble* of sample functions, obtained by repeating the same experiment under identical conditions, is called a *random process* and denoted $\{{}^kx(t)\}$. Figure 11.12 shows such a family of time histories.

Because of reasons beyond the control of the experimenter, the sample functions ${}^kx(t)$ are all different. As an example, the functions ${}^kx(t)$ may represent the amplitude of vibration of a certain component of an aircraft

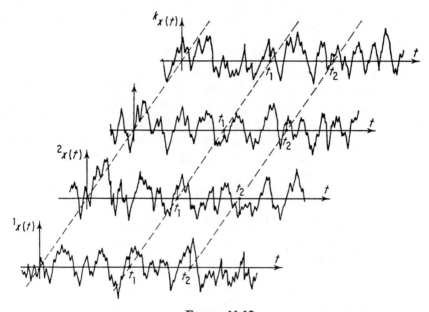

FIGURE 11.12

taxiing along a given stretch of a runway several times a day, every day. Factors such as the runway roughness, wind conditions, and tire pressure will very likely influence the level of vibration.

Let us denote by $\{{}^ky(t)\}$ another random process—the process associated with the random variable $y(t)$. In general, the random processes $\{{}^kx(t)\}$ and $\{{}^ky(t)\}$ involve a large number of sample functions. For a given value of time, $t = t_1$, we can calculate averages over the ensemble. For example, the *mean values* of $x(t)$ and $y(t)$ associated with time $t = t_1$ can be written

$$\mu_x(t_1) = \langle {}^kx(t_1) \rangle = \lim_{n \to \infty} \frac{1}{n} \sum_{k=1}^{n} {}^kx(t_1),$$

$$\mu_y(t_1) = \langle {}^ky(t_1) \rangle = \lim_{n \to \infty} \frac{1}{n} \sum_{k=1}^{n} {}^ky(t_1),$$

(11.45)

where quantities averaged over the ensemble are denoted by angular brackets. In general, the mean values calculated for time $t = t_2$ are different from the ones calculated for $t = t_1$:

$$\mu_x(t_1) \neq \mu_x(t_2), \qquad \mu_y(t_1) \neq \mu_y(t_2). \tag{11.46}$$

Next we define the *covariance functions* as

$$
\begin{aligned}
\rho_x(t_1, t_2) &= \langle [^kx(t_1) - \mu_x(t_1)][^kx(t_2) - \mu_x(t_2)] \rangle, \\
\rho_y(t_1, t_2) &= \langle [^ky(t_1) - \mu_y(t_1)][^ky(t_2) - \mu_y(t_2)] \rangle, \\
\rho_{xy}(t_1, t_2) &= \langle [^kx(t_1) - \mu_x(t_1)][^ky(t_2) - \mu_y(t_2)] \rangle,
\end{aligned}
\tag{11.47}
$$

where the averages are obtained in a way similar to that used to obtain (11.45). In general these functions will depend on the values t_1 and t_2.

To describe a random process completely we must use an infinite set of statistical quantities involving other values of time $t_3, t_4, t_5, \ldots$. But if $\{^kx(t)\}$ and $\{^ky(t)\}$ form a two-dimensional Gaussian distribution for a given value t, then $\{^kx(t)\}$ and $\{^ky(t)\}$ are separately Gaussian† and can be described completely by their mean values and covariance functions.

Next let us investigate how the mean values and covariance functions vary as the times t_1 and t_2 vary. To this end we let $t_1 = t$ and $t_2 = t + \tau$ and write

$$
\begin{aligned}
\rho_x(t, \tau) &= \langle [^kx(t) - \mu_x(t)][^kx(t + \tau) - \mu_x(t + \tau)] \rangle, \\
\rho_y(t, \tau) &= \langle [^ky(t) - \mu_y(t)][^ky(t + \tau) - \mu_y(t + \tau)] \rangle, \\
\rho_{xy}(t, \tau) &= \langle [^kx(t) - \mu_x(t)][^ky(t + \tau) - \mu_y(t + \tau)] \rangle,
\end{aligned}
\tag{11.48}
$$

and we note that now the covariance functions depend on t and τ. In the special case in which $\tau = 0$, we have

$$
\begin{aligned}
\rho_x(t, 0) &= \langle [^kx(t) - \mu_x(t)]^2 \rangle = \sigma_x^2(t), \\
\rho_y(t, 0) &= \langle [^ky(t) - \mu_y(t)]^2 \rangle = \sigma_y^2(t), \\
\rho_{xy}(t, 0) &= \langle [^kx(t) - \mu_x(t)][^ky(t) - \mu_y(y)] \rangle = \rho_{xy}(t),
\end{aligned}
\tag{11.49}
$$

where $\sigma_x^2(t)$ and $\sigma_y^2(t)$ are the *variances* of the random processes $\{^kx(t)\}$ and $\{^ky(t)\}$, respectively, and $\rho_{xy}(t)$ is the covariance between $\{^kx(t)\}$ and $\{^ky(t)\}$ for a given value of t. Hence, in general, $\sigma_x^2(t)$, $\sigma_y^2(t)$, and $\rho_{xy}(t)$ are functions of t.

The random processes $\{^kx(t)\}$ and $\{^ky(t)\}$ are said to be *weakly stationary*, both individually and jointly, if the quantities $\mu_x(t)$, $\mu_y(t)$, $\rho_x(t, \tau)$, $\rho_y(t, \tau)$, and $\rho_{xy}(t, \tau)$ do not depend on time t, which means that their values are not affected by a translation of the time origin. The random processes $\{^kx(t)\}$ and $\{^ky(t)\}$ are said to be *strongly stationary*, both individually and jointly,

† J. S. Bendat et al., *ASD TR 61–123*, Wright-Patterson AFB, Ohio, 1961, pp. 4–32.

if all possible statistical quantities of $\{^k x(t)\}$ and $\{^k y(t)\}$ are not affected by a translation of time t. In view of the statement made above with regard to the Gaussian distributions we see that for such distributions the two concepts coincide. A random process that is not stationary is said to be *nonstationary*. We shall confine ourselves only to stationary processes.† In fact, we shall be concerned only with weakly stationary processes and we shall refer to these processes simply as *stationary*. One may regard a stationary random process as somewhat analogous to the steady-state vibration in the case of deterministic functions.

It is not difficult to see that for stationary random processes the mean values are constant,

$$\mu_x = \langle {}^k x(t) \rangle = \text{const}, \qquad \mu_y = \langle {}^k y(t) \rangle = \text{const}, \qquad (11.50)$$

because, by definition, their values are independent of time. Furthermore, for the same reason, the values of the covariance functions depend only on τ, so equations (11.48) reduce to

$$\rho_x(\tau) = R_x(\tau) - \mu_x^2, \qquad \rho_y(\tau) = R_y(\tau) - \mu_y^2, \qquad \rho_{xy}(\tau) = R_{xy}(\tau) - \mu_x \mu_y, \qquad (11.51)$$

where the functions

$$R_x(\tau) = \langle [^k x(t)][^k x(t + \tau)] \rangle, \qquad R_y(\tau) = \langle [^k y(t)][^k y(t + \tau)] \rangle \qquad (11.52)$$

are called *autocorrelation functions* of the random processes $\{^k x(t)\}$ and $\{^k y(t)\}$ and the function

$$R_{xy}(\tau) = \langle [^k x(t)][^k y(t + \tau)] \rangle \qquad (11.53)$$

is called the *cross-correlation function* between $\{^k x(t)\}$ and $\{^k y(t)\}$. Note that when the mean values are zero we obtain

$$\rho_x(\tau) = R_x(\tau), \qquad \rho_y(\tau) = R_y(\tau), \qquad \rho_{xy}(\tau) = R_{xy}(\tau), \qquad (11.54)$$

which means that the correlation functions are identical with the covariance functions. This case is not without importance, because, for linear systems, one can look upon mean values as constant values superposed on the randomly varying ones.

† For a discussion of nonstationary processes the reader is referred to a report by J. S. Bendat et al., *Advanced Concepts of Stochastic Processes and Statistics for Flight Vehicle Vibration Estimation and Measurement, ASD TR 62–973*, Wright-Patterson AFB, Ohio, 1962.

If in addition to the means being zero, $\tau = 0$, equations (11.54) give

$$R_x(0) = \sigma_x^2, \qquad R_y(0) = \sigma_y^2, \qquad R_{xy}(0) = \rho_{xy}. \qquad (11.55)$$

For stationary random processes we have the relations

$$R_x(\tau) = \langle [^kx(t)][^kx(t + \tau)] \rangle = \langle [^kx(t - \tau)][^kx(t)] \rangle = R_x(-\tau),$$
$$R_y(\tau) = \langle [^ky(t)][^ky(t + \tau)] \rangle = \langle [^ky(t - \tau)][^ky(t)] \rangle = R_y(-\tau), \qquad (11.56)$$

so that the *autocorrelation functions are even functions of* τ. On the other hand, we have

$$R_{xy}(\tau) = \langle [^kx(t)][^ky(t + \tau)] \rangle = \langle [^kx(t - \tau)][^ky(t)] \rangle = R_{yx}(-\tau), \qquad (11.57)$$

and note that, in general, $R_{xy}(\tau)$ *is not* an even function of τ. Next consider

$$\langle ([^kx(t)] \pm [^kx(t + \tau)])^2 \rangle = \langle [^kx(t)]^2 \rangle \pm 2\langle [^kx(t)][^kx(t + \tau)] \rangle + \langle [^kx(t + \tau)]^2 \rangle$$
$$= 2\langle [^kx(t)]^2 \rangle \pm 2\langle [^kx(t)][^kx(t + \tau)] \rangle$$
$$= 2R_x(0) \pm 2R_x(\tau).$$

The above expression must be larger than zero, so we have

$$R_x(0) \geq |R_x(\tau)|. \qquad (11.58)$$

A similar expression can be obtained for $R_y(0)$, so we must conclude that

$$R_x(0) \geq |R_x(\tau)|, \qquad R_y(0) \geq |R_y(\tau)|, \qquad (11.59)$$

or the autocorrelation functions have an upper bound for $\tau = 0$. It can also be shown that

$$R_x(0)R_y(0) \geq |R_{xy}(\tau)|^2. \qquad (11.60)$$

One can check the stationarity assumption by calculating mean values and the correlation function associated with the random processes $\{^kx(t)\}$ and $\{^ky(t)\}$ using a large number of sample functions. The processes are stationary if the corresponding quantities are equal for various values of time.

11–5 TIME AVERAGES. ERGODIC RANDOM PROCESSES

All the averages discussed in Section 11–4 were ensemble averages. Their evaluation requires either a large number of sample functions or the probability distribution of the sample functions. It is possible, however, to select arbitrarily two sample functions $^kx(t)$ and $^ky(t)$ out of the random processes $\{^kx(t)\}$ and $\{^ky(t)\}$ and average over the time t. Such averages are called *time averages* or *temporal averages* and were encountered in Chapter 1. The *temporal means* of the functions $^kx(t)$ and $^ky(t)$ are defined by

$$^k\mu_x = \lim_{T \to \infty} \frac{1}{T} \int_{-T/2}^{T/2} {}^kx(t)\, dt,$$

$$^k\mu_y = \lim_{T \to \infty} \frac{1}{T} \int_{-T/2}^{T/2} {}^ky(t)\, dt,$$

(11.61)

and because the integrals are definite, the results are not functions of time. The *temporal covariance functions* are defined by

$$^k\rho_x(\tau) = \lim_{T \to \infty} \frac{1}{T} \int_{-T/2}^{T/2} [{}^kx(t) - {}^k\mu_x][{}^kx(t + \tau) - {}^k\mu_x]\, dt,$$

$$^k\rho_y(\tau) = \lim_{T \to \infty} \frac{1}{T} \int_{-T/2}^{T/2} [{}^ky(t) - {}^k\mu_y][{}^ky(t + \tau) - {}^k\mu_y]\, dt,$$

(11.62)

$$^k\rho_{xy}(\tau) = \lim_{T \to \infty} \frac{1}{T} \int_{-T/2}^{T/2} [{}^kx(t) - {}^k\mu_x][{}^ky(t + \tau) - {}^k\mu_y]\, dt.$$

The stationary random processes $\{^kx(t)\}$ and $\{^ky(t)\}$ are said to be *ergodic* if the statistical averages over the two sample spaces (ensemble averages) are equal to the time averages over the sample functions $^kx(t)$ and $^ky(t)$ evaluated over sample functions of long duration. Hence, if *regardless of which pair of functions* are selected from the corresponding stationary random processes, we have

$$^k\mu_x = \mu_x, \qquad ^k\mu_y = \mu_y,$$

$$^k\rho_x(\tau) = \rho_x(\tau), \qquad ^k\rho_y(\tau) = \rho_y(\tau), \qquad ^k\rho_{xy}(\tau) = \rho_{xy}(\tau),$$

(11.63)

then the random processes $\{^kx(t)\}$ and $\{^ky(t)\}$ are said to be *weakly ergodic*. If all the ensemble averages are deducible from corresponding temporal averages, the random processes are said to be *strongly ergodic*. Again for Gaussian processes the two concepts are equivalent.† It follows that *any ergodic process is by necessity a stationary process but a stationary process is not necessarily ergodic.*

The ergodicity assumption allows one to use only one sample function from each random process in calculating averages instead of using the entire ensemble. This, in effect, says that the chosen sample functions are representatives of the corresponding random processes. This is a relatively strong

† See J. S. Bendat et al., *ASD TR 61–123*, Wright-Patterson AFB, Ohio, 1961, pp. 4–46.

assumption and must be checked carefully before assuming that a process is ergodic.

Because for an ergodic process any functions $^k x(t)$ and $^k y(t)$ can be taken as representatives of the random processes $\{^k x(t)\}$ and $\{^k y(t)\}$, we shall drop the superscript k. Next we define various averages in terms of a notation commonly used in the field of random vibration. The *temporal means* of the representative functions $x(t)$ and $y(t)$ are defined by

$$E[x] = \bar{x} = \lim_{T \to \infty} \frac{1}{T} \int_{-T/2}^{T/2} x(t)\, dt,$$

$$(11.64)$$

$$E[y] = \bar{y} = \lim_{T \to \infty} \frac{1}{T} \int_{-T/2}^{T/2} y(t)\, dt,$$

where $\bar{x}$ and $\bar{y}$ do not depend on t. The *temporal mean square values* are given by

$$E[x^2] = \overline{x^2} = \lim_{T \to \infty} \frac{1}{T} \int_{-T/2}^{T/2} x^2(t)\, dt,$$

$$(11.65)$$

$$E[y^2] = \overline{y^2} = \lim_{T \to \infty} \frac{1}{T} \int_{-T/2}^{T/2} y^2(t)\, dt,$$

and *temporal correlation functions* have the form†

$$\psi_x(\tau) = \overline{x(t)x(t + \tau)} = \lim_{T \to \infty} \frac{1}{T} \int_{-T/2}^{T/2} x(t)x(t + \tau)\, dt,$$

$$\psi_y(\tau) = \overline{y(t)y(t + \tau)} = \lim_{T \to \infty} \frac{1}{T} \int_{-T/2}^{T/2} y(t)y(t + \tau)\, dt, \qquad (11.66)$$

$$\psi_{xy}(\tau) = \overline{x(t)y(t + \tau)} = \lim_{T \to \infty} \frac{1}{T} \int_{-T/2}^{T/2} x(t)y(t + \tau)\, dt,$$

where the first two are called *autocorrelation functions* and the third one is a *cross-correlation function*.

We note that for ergodic processes with zero mean values we have

$$\overline{x^2} = \sigma_x^2 = R_x(0), \qquad \overline{y^2} = \sigma_y^2 = R_y(0), \qquad (11.67)$$

---

† For finite pulses the autocorrelation functions can be obtained by averaging over the length of the pulse; otherwise the result is zero.

and regardless of whether the means are zero or not,

$$\psi_x(\tau) = R_x(\tau), \qquad \psi_y(\tau) = R_y(\tau), \qquad \psi_{xy}(\tau) = R_{xy}(\tau). \qquad (11.68)$$

It is not difficult to prove that the functions $\psi_x(\tau)$, $\psi_y(\tau)$, and $\psi_{xy}(\tau)$ possess the same properties as the functions $R_x(\tau)$, $R_y(\tau)$, and $R_{xy}(\tau)$ provided the processes $x(t)$ and $y(t)$ are stationary. These properties are expressed by (11.56), (11.57), (11.59), and (11.60).

Example 11.1

Consider the rectangular pulse

$$f(t) = \begin{cases} A, & |t| < t_1, \\ 0, & |t| > t_1, \end{cases} \qquad (a)$$

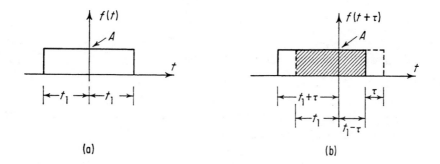

(a)   (b)

FIGURE 11.13

and calculate the autocorrelation function $\psi_f(\tau)$, Figure 11.13. The auto-correlation function $\psi_f(\tau)$ is defined by

$$\psi_f(\tau) = \lim_{T \to \infty} \frac{1}{T} \int_{-T/2}^{T/2} f(t)f(t + \tau)\, dt. \qquad (b)$$

We are dealing here with a pulse of finite duration, so it would be meaningless to let $T \to \infty$. We shall choose instead $T = 2t_1$.† From Figure 11.13 we see that for $0 < \tau < 2t_1$ the range of integration is defined by the shaded area,

$$\psi_f(\tau) = \frac{1}{2t_1} \int_{-t_1}^{t_1 - \tau} A^2 dt = \frac{A^2}{2t_1} [t_1 - \tau - (-t_1)] = A^2 \left(1 - \frac{\tau}{2t_1}\right). \qquad (c)$$

† See the footnote concerning finite pulses in connection with (11.66), p. 486.

For $\tau > 2t_1$ we obtain

$$\psi_f(\tau) = 0. \tag{d}$$

We also recall that $\psi_f(\tau)$ is an even function of τ,

$$\psi_f(-\tau) = \psi_f(\tau). \tag{e}$$

The autocorrelation function is plotted in Figure 11.14.

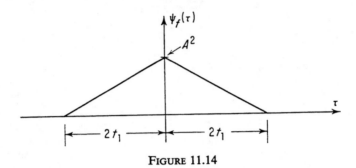

FIGURE 11.14

We note that the maximum occurs at $\tau = 0$,

$$\psi_f(0) = A^2. \tag{f}$$

This is always the case, unless $f(t)$ is a periodic function, in which case the maximum value is obtained for any integer multiple of the period.

11–6 NORMAL RANDOM PROCESS. CENTRAL LIMIT THEOREM

Let us consider the sum

$$x = \sum_{i=1}^{n} x_i, \tag{11.69}$$

where x_i are statistically independent random variables with respective probability densities $p_i(x)$. The variables x_i have finite mean values μ_i and finite variances σ_i^2, respectively. The *central limit theorem*† states that under very minor restrictions which can always be assumed to be satisfied in practice, the distribution of the sum random variable x approaches the normal

† See C. A. Bennett, and N. L. Franklin, *Statistical Analysis in Chemistry and the Chemical Industry*, Wiley, New York, 1954, p. 89. Also J. S. Bendat, *Principles and Applications of Random Noise Theory*, Wiley, New York, 1958, p. 96.

distribution function as $n \to \infty$ regardless of the shape of the densities $p_i(x)$. The mean value and the variance of the sum random variable are

$$\mu = E[x] = \sum_{i=1}^{n} \mu_i,$$

$$\sigma^2 = E[(x - \mu)^2] = \sum_{i=1}^{n} \sigma_i^2. \tag{11.70}$$

Under these conditions the normally distributed variable x, for large n, has the probability density function

$$p(x) = \frac{1}{\sqrt{2\pi}\sigma} \exp\left[-\frac{1}{2}\frac{(x - \mu)^2}{\sigma^2}\right]. \tag{11.71}$$

A necessary condition for the theorem to be valid is that no single random variable x_i contributes significantly to the sum. This assumption is generally ensured if the variables x_i are identically distributed.

If every sample function of a random process satisfies these conditions, the process is called a *normal, or Gaussian, random process*. Stationary Gaussian random processes whose power spectral density functions (see the definition in Chapter 1 or Section 11–7) are continuous are also ergodic.†

In practical applications these conditions are approximately satisfied by many random processes, so such processes can be considered, at least approximately, normal.

11-7 SPECTRAL DENSITY OF A STATIONARY RANDOM PROCESS

In Chapter 1 we indicated that the Fourier transform of a nonperiodic function may be regarded as its frequency decomposition. Hence, let us consider a stationary random process $\{f(t)\}$ and let $S_f(\omega)$ be the Fourier transform of the autocorrelation function, $R_f(\tau)$,‡

$$S_f(\omega) = \int_{-\infty}^{\infty} R_f(\tau)e^{-i\omega\tau}\,d\tau, \tag{11.72}$$

where $R_f(\tau) = \langle f(t)f(t + \tau)\rangle$, so that the autocorrelation function can be written as the inversion integral

$$R_f(\tau) = \frac{1}{2\pi}\int_{-\infty}^{\infty} S_f(\omega)e^{i\omega\tau}\,d\omega. \tag{11.73}$$

---

† See J. S. Bendat et al., *ASD TR 61–123*, Wright-Patterson AFB, Ohio, 1961, pp. 4–46.
‡ In defining $S(\omega)$, various texts have different factors multiplying the integral. The definition given here makes $S(\omega)$ the Fourier transform of $R(\tau)$.

A sufficient condition for the existence of $S_f(\omega)$ is that

$$\int_{-\infty}^{\infty} |R_f(\tau)| \, d\tau < \infty, \tag{11.74}$$

under which conditions $S_f(\omega)$ is bounded.† Recalling the definition of the autocorrelation function and letting $\tau = 0$, we obtain

$$R_f(0) = \langle [f(t)]^2 \rangle = \frac{1}{2\pi} \int_{-\infty}^{\infty} S_f(\omega) \, d\omega. \tag{11.75}$$

We conclude that by summing all the elements $S_f(\omega) \, d\omega$ and dividing the result by 2π, we obtain the mean square value of the process. Hence $S_f(\omega)$ can be interpreted as the *mean square spectral density*. This points out a way of obtaining the mean square value of a stationary random process when its mean square spectral density, $S_f(\omega)$, is known. The function $S_f(\omega)$ is also called the *power spectral density function*, owing to the analogy with electrical systems, for which the average power dissipated in a resistor by the frequency components of a voltage lying in a band between ω and $\omega + d\omega$ is proportional to $S_f(\omega) \, d\omega$.

If we consider its physical interpretation we must conclude that $S_f(\omega)$ *is always positive (or zero)*, $S_f(\omega) \geq 0$.

We have shown in Section 11–4 that $R_f(\tau)$ is an even function of τ, $R_f(\tau) = R_f(-\tau)$. It follows from (11.72) that

$$S_f(\omega) = \int_{-\infty}^{\infty} R_f(\tau) e^{-i\omega\tau} \, d\tau = \int_{-\infty}^{\infty} R_f(-\tau) e^{-i\omega\tau} \, d\tau$$

$$= -\int_{\infty}^{-\infty} R_f(\eta) e^{i\omega\eta} \, d\eta = S_f(-\omega), \tag{11.76}$$

where η is a dummy variable of integration. Hence we conclude that $S_f(\omega)$ *is an even function of* ω. Because $R_f(\tau)$ is an even function of τ, (11.72) gives

$$S_f(\omega) = \int_{-\infty}^{\infty} R_f(\tau) e^{-i\omega\tau} \, d\tau = \int_{-\infty}^{\infty} R_f(\tau) \cos \omega\tau \, d\tau$$

$$= 2 \int_{0}^{\infty} R_f(\tau) \cos \omega\tau \, d\tau, \tag{11.77}$$

so $S_f(\omega)$ *is a real function of* ω, because it is the integral of a real function. It follows that $S_f(\omega)$ is an even, real function of ω, so (11.73) reduces to

$$R_f(\tau) = \frac{1}{\pi} \int_{0}^{\infty} S_f(\omega) \cos \omega\tau \, d\omega. \tag{11.78}$$

† When $R_f(\tau)$ is a sinusoidal function, there may be some difficulty in obtaining the integral, which can be avoided by using delta functions.

Equations (11.77) and (11.78) are known as *Wiener-Khinchin equations*, and except for a factor of 2 they represent a Fourier cosine transform pair.

Again we can let $\tau = 0$ and write

$$R_f(0) = \frac{1}{\pi} \int_0^\infty S_f(\omega) \, d\omega, \tag{11.79}$$

so if we have at our disposal the function $S_f(\omega)$, obtained through data reduction, we can calculate the mean square value of the random process. The advantage of (11.79) over (11.75) is that it contains no negative frequencies.

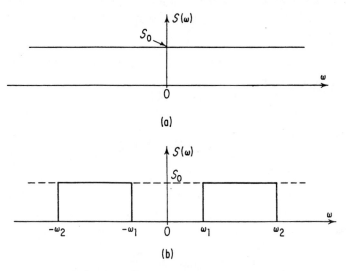

FIGURE 11.15

A random excitation whose power spectral density is flat over a band of frequencies considerably wider than the band of the system excited is called *white noise*. The name comes from the analogy with white light, which has a flat spectrum over the visible range. If the band of frequencies is infinitely long, one speaks of an *ideal white noise* [see Figure 11.15(a)]. This, however, is a physical impossibility, because it implies an infinite mean square. It also implies that $R_x(\tau) = 0$, $\tau \neq 0$, so that $x(t)$ and $x(t + \tau)$ are independent for $\tau \neq 0$. A more realistic power spectrum density is the *band-limited white noise* [see Figure 11.15(b)], which is flat over a band of frequencies extending from ω_1 to ω_2 which are called *lower cutoff* and *upper cutoff* frequencies, respectively. For the band-limited white noise the mean square value, $R_f(0)$, is simply $(1/\pi)(\omega_2 - \omega_1)S_0$.

Example 11.2

Calculate the autocorrelation function associated with the band-limited white noise spectral density shown in Figure 11.15(b).

Using (11.78) we obtain the autocorrelation function

$$R_f(\tau) = \frac{1}{\pi} \int_0^\infty S_f(\omega) \cos \omega\tau \, d\omega = \frac{S_0}{\pi} \int_{\omega_1}^{\omega_2} \cos \omega\tau \, d\omega$$

$$= \frac{S_0}{\pi} \frac{\sin \omega_2\tau - \sin \omega_1\tau}{\tau}, \tag{a}$$

where ω_2 is the upper cutoff frequency and ω_1 is the lower cutoff frequency. In the case in which $\omega_1 = 0$ we have the autocorrelation function

$$R_f(\tau) = \frac{S_0\omega_2}{\pi} \frac{\sin \omega_2\tau}{\omega_2\tau}, \tag{b}$$

which is plotted in Figure 11.16.

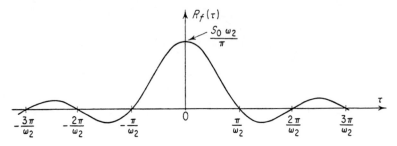

FIGURE 11.16

As $\omega_2 \to \infty$ the spectral density approaches the ideal white noise $S_f(\omega) = S_0$, and the autocorrelation function becomes

$$R_f(\tau) = S_0\delta(\tau), \tag{c}$$

where $\delta(\tau)$ is the unit impulse.

11-8 CORRELATION THEOREM. PARSEVAL'S THEOREM

Let us consider two nonperiodic functions $f_1(t)$ and $f_2(t)$ such that

$$\int_{-\infty}^\infty |f_1(t)| \, dt < \infty, \qquad \int_{-\infty}^\infty |f_2(t)| \, dt < \infty \tag{11.80}$$

and write the corresponding Fourier transform pairs

$$F_1(\omega) = \int_{-\infty}^{\infty} f_1(t)e^{-i\omega t}\, dt,$$

$$f_1(t) = \frac{1}{2\pi} \int_{-\infty}^{\infty} F_1(\omega)e^{i\omega t}\, d\omega,$$

$$\tag{11.81}$$

$$F_2(\omega) = \int_{-\infty}^{\infty} f_2(t)e^{-i\omega t}\, dt,$$

$$f_2(t) = \frac{1}{2\pi} \int_{-\infty}^{\infty} F_2(\omega)e^{i\omega t}\, d\omega.$$

$$\tag{11.82}$$

Next let us form the integral

$$\int_{-\infty}^{\infty} f_1(t)f_2(t + \tau)\, dt = \int_{-\infty}^{\infty} f_1(t)\left[\frac{1}{2\pi} \int_{-\infty}^{\infty} F_2(\omega)e^{i\omega(t + \tau)}\, d\omega\right] dt$$

$$= \frac{1}{2\pi} \int_{-\infty}^{\infty} F_2(\omega)e^{i\omega\tau}\left[\int_{-\infty}^{\infty} f_1(t)e^{i\omega t}\, dt\right] d\omega$$

$$= \frac{1}{2\pi} \int_{-\infty}^{\infty} F_2(\omega)F_1(-\omega)e^{i\omega\tau}\, d\omega$$

$$= \frac{1}{2\pi} \int_{-\infty}^{\infty} F_1^*(\omega)F_2(\omega)e^{i\omega\tau}\, d\omega, \tag{11.83}$$

where $F_1^*(\omega) = F_1(-\omega)$ is the complex conjugate of $F_1(\omega)$. Equation (11.83) resembles in structure an inverse Fourier transform formula, so we are led to the conclusion that

$$F_1^*(\omega)F_2(\omega) = \int_{-\infty}^{\infty}\left[\int_{-\infty}^{\infty} f_1(t)f_2(t + \tau)\, dt\right]e^{-i\omega\tau}\, d\tau \tag{11.84}$$

is the Fourier transform of the integral

$$\int_{-\infty}^{\infty} f_1(t)f_2(t + \tau)\, dt. \tag{11.85}$$

The reciprocal relations, (11.83) and (11.84), which constitute a Fourier transform pair, are called the *correlation theorem*. In the special case in which $\tau = 0$, (11.83) reduces to

$$\int_{-\infty}^{\infty} f_1(t)f_2(t)\, dt = \frac{1}{2\pi} \int_{-\infty}^{\infty} F_1^*(\omega)F_2(\omega)\, d\omega. \tag{11.86}$$

Equation (11.86) represents a relation concerning the integral of the product of two nonperiodic functions and is called *Parseval's theorem.*

If the function $f_2(t)$ is identical to the function $f_1(t)$, (11.86) becomes

$$\int_{-\infty}^{\infty} f_1^2(t)\, dt = \frac{1}{2\pi} \int_{-\infty}^{\infty} |F_1(\omega)|^2\, d\omega, \tag{11.87}$$

which is known as *Parseval's relation.*†

11–9 SPECTRAL DENSITIES OF SAMPLE FUNCTIONS. ERGODIC PROCESSES

The temporal autocorrelation function $\psi_f(\tau)$ of a sample function $f(t)$ is defined as‡

$$\psi_f(\tau) = \lim_{T \to \infty} \frac{1}{T} \int_{-T/2}^{T/2} f(t)f(t+\tau)\, dt. \tag{11.88}$$

By the correlation theorem (11.83) discussed in Section 11–8 we can write immediately

$$\psi_f(\tau) = \lim_{T \to \infty} \frac{1}{2\pi T} \int_{-\infty}^{\infty} F^*(\omega)F(\omega)e^{i\omega\tau}\, d\omega$$

$$= \lim_{T \to \infty} \frac{1}{2\pi T} \int_{-\infty}^{\infty} |F(\omega)|^2 e^{i\omega\tau}\, d\omega. \tag{11.89}$$

But a Fourier transform, $G_f(\omega)$, can be written for the autocorrelation function in the form

$$G_f(\omega) = \int_{-\infty}^{\infty} \psi_f(\tau)e^{-i\omega\tau}\, d\tau \tag{11.90}$$

with the corresponding inversion formula

$$\psi_f(\tau) = \frac{1}{2\pi} \int_{-\infty}^{\infty} G_f(\omega)e^{i\omega\tau}\, d\omega. \tag{11.91}$$

Comparing (11.89) and (11.91) we must have

$$G_f(\omega) = \lim_{T \to \infty} \frac{1}{T} |F(\omega)|^2. \tag{11.92}$$

---

† In many texts (11.87) is called *Parseval's theorem.*
‡ See the footnote concerning finite pulses in connection with (11.66), p. 486.

Letting $\tau = 0$ in (11.89) we conclude that $G_f(\omega)$ is the power spectral density of the temporal mean square value multiplied by 2π.†

If $f(t)$ is one sample function of an ergodic process, we must have

$$G_f(\omega) = S_f(\omega), \tag{11.93}$$

or the power spectral density of the random process is equal to the power spectral density of any of its sample functions. Hence, all the properties of $S_f(\omega)$ shown in Section 11–7 are also properties of $G_f(\omega)$ if the process is ergodic.

Example 11.3

Find the mean square spectral density associated with the rectangular pulse of Example 11.1.

In Example 11.1 we obtained the autocorrelation function of the rectangular pulse. Its expression is

$$\psi_f(\tau) = \begin{cases} A^2\left(1 - \dfrac{\tau}{2t_1}\right), & |\tau| < 2t_1, \\ 0, & |\tau| > 2t_1, \end{cases} \tag{a}$$

where A is the amplitude and $2t_1$ the duration of the pulse.

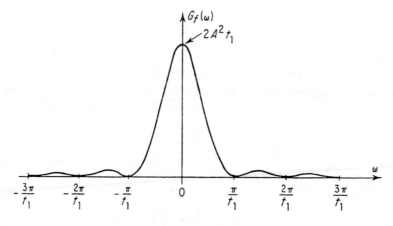

FIGURE 11.17

† For finite pulses, $G_f(\omega)$ is often called the *energy density spectrum* of $f(t)$.

Using (11.90) we obtain the mean square spectral density

$$G_f(\omega) = \int_{-\infty}^{\infty} \psi_f(\tau) e^{-i\omega\tau} \, d\tau = 2 \int_0^{\infty} \psi_f(\tau) \cos \omega\tau \, d\tau,$$

$$= 2A^2 \int_0^{2t_1} \left(1 - \frac{\tau}{2t_1}\right) \cos \omega\tau \, d\tau = 2A^2 t_1 \left(\frac{\sin \omega t_1}{\omega t_1}\right)^2, \tag{b}$$

which is plotted in Figure 11.17.

11–10 RESPONSE OF LINEAR SYSTEMS TO STATIONARY RANDOM EXCITATION. GENERAL RELATIONS

In Chapter 1 we have shown that the relation between the response $x(t)$ and the excitation $f(t)$, for a linear system,[†] can be written in the form of the convolution integral,

$$x(t) = \int_0^t f(\tau) h(t - \tau) \, d\tau, \tag{11.94}$$

in which $h(t)$ is the impulsive response. Equation (11.94) holds true for any linear system and for any type of excitation as long as it is defined only for $t > 0$ and is zero for $t < 0$. If we wish to consider functions $f(t)$ defined for $t < 0$ we must modify (11.94). It is not difficult to show that to account for functions $f(t)$ of negative argument the convolution integral should be written

$$x(t) = \int_{-\infty}^t f(\tau) h(t - \tau) \, d\tau. \tag{11.95}$$

But, by definition, $h(t - \tau)$ is zero for $t < \tau$, and because τ is the variable of integration, the upper limit of the integral in (11.95) can be changed to any value larger than t without affecting the result. Hence let us write

$$x(t) = \int_{-\infty}^{\infty} f(\tau) h(t - \tau) \, d\tau, \tag{11.96}$$

where the upper limit is chosen as infinity, to preserve the symmetry of the integral. This allows us to write

$$x(t) = \int_{-\infty}^{\infty} h(\tau) f(t - \tau) \, d\tau. \tag{11.97}$$

[†] For a discussion of nonlinear systems, see A. Papoulis, *Probability, Random Variables, and Stochastic Processes*, McGraw-Hill, New York, 1965, p. 305.

Let us denote by $X(\omega)$ the Fourier transform of $x(t)$, so that

$$X(\omega) = \int_{-\infty}^{\infty} x(t)e^{-i\omega t}\, dt = \int_{-\infty}^{\infty} f(\tau)\left[\int_{-\infty}^{\infty} h(t-\tau)e^{-i\omega t}\, dt\right] d\tau$$

$$= \int_{-\infty}^{\infty} f(\tau)e^{-i\omega\tau}\left[\int_{-\infty}^{\infty} h(\lambda)e^{-i\omega\lambda}\, d\lambda\right] d\tau = H(\omega)F(\omega), \qquad (11.98)$$

where $H(\omega)$ is the Fourier transform of the impulsive response. This relation was derived in Chapter 1 [see (1.109)], where $H(\omega)$ was identified as the complex frequency response.

Let us say that $f(t)$ is a sample function representing a stationary excitation random process $\{f(t)\}$ and $x(t)$ denotes the corresponding response. The mean value of the response process, averaged over the ensemble, is

$$\langle x(t)\rangle = \left\langle \int_{-\infty}^{\infty} h(\tau)f(t-\tau)\, d\tau \right\rangle, \qquad (11.99)$$

and because we can change the order of averaging and integration we can write†

$$\langle x(t)\rangle = \int_{-\infty}^{\infty} h(\tau)\langle f(t-\tau)\rangle\, d\tau. \qquad (11.100)$$

But for stationary random processes $\langle f(t-\tau)\rangle = \langle f(t)\rangle = $ const, independent of τ, so

$$\langle x(t)\rangle = \langle f(t)\rangle \int_{-\infty}^{\infty} h(\tau)\, d\tau = \langle f(t)\rangle H(0), \qquad (11.101)$$

where

$$H(0) = \int_{-\infty}^{\infty} h(\tau)\, d\tau \qquad (11.102)$$

is merely the complex frequency response $H(\omega)$ with zero substituted for ω. It follows that *if the excitation mean value is zero, the response mean value is also zero.*

---

† Note that the averaging is done with respect to t and the variable of integration is τ. This is not necessarily sufficient, however. In a more rigorous analysis, we must insist that the integral $\int_{-\infty}^{\infty} |h(\tau)\langle f(t-\tau)\rangle|\, d\tau$ exists for all τ. The conditions of validity are given on pages 65 and 66 of W. B. Davenport, Jr., and W. L. Root, *Introduction to Random Signals and Noise*, McGraw-Hill, New York, 1958.

Next let us explore the response autocorrelation function of a stationary random process. First let us write

$$x(t) = \int_{-\infty}^{\infty} h(\tau_1)f(t - \tau_1)\, d\tau_1,$$

$$x(t + \tau) = \int_{-\infty}^{\infty} h(\tau_2)f(t + \tau - \tau_2)\, d\tau_2,$$

(11.103)

so the autocorrelation function of the response is obtained by averaging over the ensemble,†

$$
\begin{aligned}
R_x(\tau) &= \langle x(t)x(t + \tau)\rangle \\
&= \left\langle \int_{-\infty}^{\infty} h(\tau_1)f(t - \tau_1)\, d\tau_1 \int_{-\infty}^{\infty} h(\tau_2)f(t + \tau - \tau_2)\, d\tau_2 \right\rangle \\
&= \left\langle \int_{-\infty}^{\infty}\int_{-\infty}^{\infty} h(\tau_1)h(\tau_2)f(t - \tau_1)f(t + \tau - \tau_2)\, d\tau_1\, d\tau_2 \right\rangle \\
&= \int_{-\infty}^{\infty}\int_{-\infty}^{\infty} h(\tau_1)h(\tau_2)\langle f(t - \tau_1)f(t + \tau - \tau_2)\rangle\, d\tau_1\, d\tau_2.
\end{aligned}
$$

(11.104)

But for stationary processes

$$
\begin{aligned}
\langle f(t - \tau_1)f(t + \tau - \tau_2)\rangle &= \langle f(t)f(t + \tau + \tau_1 - \tau_2)\rangle \\
&= R_f(\tau + \tau_1 - \tau_2),
\end{aligned}
$$

(11.105)

where the latter is recognized as the autocorrelation function of the excitation. Hence

$$R_x(\tau) = \int_{-\infty}^{\infty}\int_{-\infty}^{\infty} h(\tau_1)h(\tau_2)R_f(\tau + \tau_1 - \tau_2)\, d\tau_1\, d\tau_2.$$

(11.106)

Equation (11.106) may not be convenient for computational purposes, and so we will attempt to express $R_x(\tau)$ in a different form. To this end we consider the mean square spectral density of the response

$$
\begin{aligned}
S_x(\omega) &= \int_{-\infty}^{\infty} R_x(\tau)e^{-i\omega\tau}\, d\tau \\
&= \int_{-\infty}^{\infty} e^{-i\omega\tau}\left\{\int_{-\infty}^{\infty} h(\tau_1)\left[\int_{-\infty}^{\infty} h(\tau_2)R_f(\tau + \tau_1 - \tau_2)\, d\tau_2\right] d\tau_1\right\} d\tau.
\end{aligned}
$$

(11.107)

But

$$R_f(\tau + \tau_1 - \tau_2) = \frac{1}{2\pi}\int_{-\infty}^{\infty} S_f(\omega)e^{i\omega(\tau + \tau_1 - \tau_2)}\, d\omega,$$

(11.108)

† See the footnote in connection with (11.100), p. 497.

so introducing (11.108) in (11.107), interchanging the order of integration, and rearranging, we can write

$$S_x(\omega) = \int_{-\infty}^{\infty} e^{-i\omega\tau} \left\{ \int_{-\infty}^{\infty} h(\tau_1) \right.$$

$$\left. \times \left[\int_{-\infty}^{\infty} h(\tau_2) \left(\frac{1}{2\pi} \int_{-\infty}^{\infty} S_f(\omega) e^{i\omega(\tau + \tau_1 - \tau_2)} \, d\omega \right) d\tau_2 \right] d\tau_1 \right\} d\tau$$

$$= \int_{-\infty}^{\infty} e^{-i\omega\tau} \left\{ \frac{1}{2\pi} \int_{-\infty}^{\infty} S_f(\omega) e^{i\omega\tau} \right.$$

$$\left. \times \left[\int_{-\infty}^{\infty} h(\tau_1) e^{i\omega\tau_1} \left(\int_{-\infty}^{\infty} h(\tau_2) e^{-i\omega\tau_2} \, d\tau_2 \right) d\tau_1 \right] d\omega \right\} d\tau$$

$$= \int_{-\infty}^{\infty} e^{-i\omega\tau} \left\{ \frac{1}{2\pi} \int_{-\infty}^{\infty} S_f(\omega) H(\omega) H(-\omega) e^{i\omega\tau} \, d\omega \right\} d\tau$$

$$= \int_{-\infty}^{\infty} e^{-i\omega\tau} \left\{ \frac{1}{2\pi} \int_{-\infty}^{\infty} S_f(\omega) |H(\omega)|^2 e^{i\omega\tau} \, d\omega \right\} d\tau, \tag{11.109}$$

because $H(-\omega) = H^*(\omega)$ is the complex conjugate of the complex frequency response $H(\omega)$. From (11.107) and (11.109) we conclude that

$$S_x(\omega) = |H(\omega)|^2 S_f(\omega), \tag{11.110}$$

and

$$R_x(\tau) = \frac{1}{2\pi} \int_{-\infty}^{\infty} S_x(\omega) e^{i\omega\tau} \, d\omega$$

$$= \frac{1}{2\pi} \int_{-\infty}^{\infty} |H(\omega)|^2 S_f(\omega) e^{i\omega\tau} \, d\omega \tag{11.111}$$

are a Fourier transform pair. Equation (11.110) is an algebraic expression relating the mean square spectral density of the excitation and response, whereas (11.111) gives the response autocorrelation function in the form of an inverse Fourier transform.

In particular, for $\tau = 0$, we obtain

$$R_x(0) = \langle [x(t)]^2 \rangle = \frac{1}{2\pi} \int_{-\infty}^{\infty} |H(\omega)|^2 S_f(\omega) \, d\omega, \tag{11.112}$$

which allows us to calculate the mean square value of the response process.

Hence, for a stationary process, it is of vital importance to know the power spectral density, $S_f(\omega)$, of the excitation process.

11–11   RESPONSE OF LINEAR SYSTEMS TO ERGODIC EXCITATION. GENERAL RELATIONS

Let us consider again the relation between the response $x(t)$ of a random process and the corresponding excitation $f(t)$ for a linear system. The relation is

$$x(t) = \int_{-\infty}^{\infty} h(\tau)f(t - \tau)\, d\tau. \tag{11.113}$$

The *response temporal mean value* can be written

$$E[x(t)] = \bar{x} = E\left[\int_{-\infty}^{\infty} h(\tau)f(t - \tau)\, d\tau\right] = \int_{-\infty}^{\infty} h(\tau)E[f(t - \tau)]\, d\tau. \tag{11.114}$$

But for an ergodic process, which is by definition a stationary process, the expected value of $f(t - \tau)$ does not depend on τ, so (11.114) can be written

$$E[x(t)] = E[f(t)]\int_{-\infty}^{\infty} h(\tau)\, d\tau = E[f(t)]H(0), \tag{11.115}$$

where $H(0)$ is defined by (11.102). Hence, if the excitation mean value is zero, the response mean value is also zero.

The *response temporal autocorrelation function* is given by

$$E[x(t)x(t + \tau)] = \psi_x(\tau) = \lim_{T \to \infty} \frac{1}{T}\int_{-T/2}^{T/2} x(t)x(t + \tau)\, dt. \tag{11.116}$$

In a way similar to the one used in Section 11–10 it can be shown that

$$\psi_x(\tau) = \int_{-\infty}^{\infty}\int_{-\infty}^{\infty} h(\tau_1)h(\tau_2)\psi_f(\tau + \tau_1 - \tau_2)\, d\tau_1\, d\tau_2, \tag{11.117}$$

where

$$\psi_f(\tau) = \lim_{T \to \infty} \frac{1}{T}\int_{-T/2}^{T/2} f(t)f(t + \tau)\, dt \tag{11.118}$$

is the excitation temporal autocorrelation function associated with the excitation sample function $f(t)$.

The *response temporal mean square spectral density* is obtained by writing the Fourier transform of $\psi_x(\tau)$,

$$\begin{aligned}
G_x(\omega) &= \int_{-\infty}^{\infty} \psi_x(\tau)e^{-i\omega\tau}\, d\tau \\
&= \int_{-\infty}^{\infty} e^{-i\omega\tau}\left[\int_{-\infty}^{\infty}\int_{-\infty}^{\infty} h(\tau_1)h(\tau_2)\psi_f(\tau + \tau_1 - \tau_2)\, d\tau_1\, d\tau_2\right] d\tau.
\end{aligned} \tag{11.119}$$

Using the same procedure as in Section 11–10, the above leads to

$$G_x(\omega) = |H(\omega)|^2 G_f(\omega),\tag{11.120}$$

so the response temporal autocorrelation function becomes

$$\psi_x(\tau) = \frac{1}{2\pi}\int_{-\infty}^{\infty} G_x(\omega)e^{i\omega\tau}\, d\omega = \frac{1}{2\pi}\int_{-\infty}^{\infty} |H(\omega)|^2 G_f(\omega)e^{i\omega\tau}\, d\omega.\tag{11.121}$$

When $\tau = 0$, (11.121) reduces to the *response temporal mean square value*

$$E[x^2(t)] = \overline{x^2} = \psi_x(0) = \frac{1}{2\pi}\int_{-\infty}^{\infty} |H(\omega)|^2 G_f(\omega)\, d\omega.\tag{11.122}$$

If the excitation random process is ergodic, it follows that the response random process is also ergodic. For an ergodic random process the ensemble averages must equal the corresponding time averages

$$\begin{aligned}
E[x(t)] &= \langle x(t)\rangle, \\
\psi_x(\tau) &= R_x(\tau), \\
G_x(\omega) &= S_x(\omega),
\end{aligned}\tag{11.123}$$

where the time averages are obtained by using any of the sample functions of the random process.

11–12 RESPONSE OF A SINGLE-DEGREE-OF-FREEDOM SYSTEM TO RANDOM EXCITATION

The system shown in Figure 11.18 travels with a uniform velocity v on a rough surface. We shall assume that the velocity v is such that the lower ends of the spring and damper are imparted a vertical motion $y(t)$.

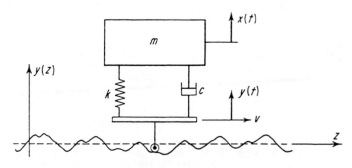

FIGURE 11.18

The corresponding equation of motion is

$$\ddot{x}(t) + 2\zeta\omega_n\dot{x}(t) + \omega_n^2 x(t) = \omega_n^2 f(t), \tag{11.124}$$

where

$$f(t) = \frac{2\zeta}{\omega_n}\dot{y}(t) + y(t) \tag{11.125}$$

is an equivalent displacement excitation. The excitation $f(t)$ can be assumed to be an ergodic random process, so the response $x(t)$ will also be an ergodic process. Furthermore, if the excitation process is Gaussian, the response process will also be Gaussian.† Let us assume that the process is Gaussian, in which case it is fully described by the mean value and autocorrelation function. Assuming that the excitation process is given, we shall be interested in calculating the response mean value and the response autocorrelation function. The process is ergodic, so these quantities can be calculated by using time averages.

For a stationary process the mean value is a constant. We can measure the displacement $y(t)$ with respect to a reference line in a way that the mean value excitation is zero,

$$E[f(t)] = 0, \tag{11.126}$$

from which it follows immediately that the response mean value is zero,

$$E[x(t)] = 0. \tag{11.127}$$

The excitation autocorrelation function can be obtained by integrating numerically

$$\psi_f(\tau) = \frac{1}{T}\int_{-T/2}^{T/2} f(t)f(t + \tau)\,dt \tag{11.128}$$

over a sample function $f(t)$ of very large duration. This allows us to calculate the excitation mean square spectral density

$$G_f(\omega) = \int_{-\infty}^{\infty} \psi_f(\tau)e^{-i\omega\tau}\,d\tau. \tag{11.129}$$

The response mean square spectral density is obtained by using (11.120),

$$G_x(\omega) = |H(\omega)|^2 G_f(\omega), \tag{11.130}$$

where

$$H(\omega) = \frac{1}{1 - (\omega/\omega_n)^2 + i(2\zeta\omega/\omega_n)} \tag{11.131}$$

---

† This is true if the system is linear. If the system is not linear, the response to a Gaussian excitation is not Gaussian. For further details see A. Papoulis, *Probability, Random Variables, and Stochastic Processes*, McGraw-Hill, New York, 1965, p. 476.

is the complex frequency response of the system. The response autocorrelation function is obtained by writing the inverse Fourier transform of $G_x(\omega)$,

$$\psi_x(\tau) = \frac{1}{2\pi} \int_{-\infty}^{\infty} |H(\omega)|^2 G_f(\omega) e^{i\omega\tau} \, d\omega. \tag{11.132}$$

Letting $\tau = 0$ in (11.132), we obtain the response mean square value,

$$\overline{x^2} = E[x^2(t)] = \psi_x(0) = \frac{1}{2\pi} \int_{-\infty}^{\infty} |H(\omega)|^2 G_f(\omega) \, d\omega. \tag{11.133}$$

As an illustration let us assume that $f(t)$ is such that its mean square spectral density is the ideal white noise,

$$G_f(\omega) = G_0 = \text{const.} \tag{11.134}$$

Introducing (11.134) in (11.133) we obtain the response mean square value

$$\overline{x^2} = \frac{G_0}{2\pi} \int_{-\infty}^{\infty} \frac{d\omega}{[1 - (\omega/\omega_n)^2]^2 + (2\zeta\omega/\omega_n)^2}. \tag{11.135}$$

The above integral can be evaluated by means of the residue theorem[†] in conjunction with Jordan's lemma.[‡] To this end we introduce the complex variable

$$\Omega = \omega + i\lambda, \tag{11.136}$$

and instead of considering a real line integral we shall consider a complex integral over a closed contour, which can be evaluated by means of the residue theorem. The integral to be considered is

$$\oint \frac{d\Omega}{[1 - (\Omega/\omega_n)^2]^2 + (2\zeta\Omega/\omega_n)^2}, \tag{11.137}$$

where the contour is shown in Figure 11.19.

Next let us write

$$\int_{-\infty}^{\infty} = \lim_{R \to \infty} \left[\oint - \int \right], \tag{11.138}$$

† F. B. Hildebrand, *Advanced Calculus for Engineers*, Prentice-Hall, Englewood Cliffs, N.J., 1960, p. 523.
‡ F. B. Hildebrand, op. cit., p. 529.

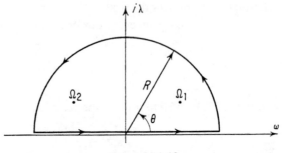

FIGURE 11.19

where the symbols are self-explanatory. According to Jordan's lemma, however, the integral over the semicircle vanishes as $R \to \infty$,

$$\lim_{R \to \infty} \oint \frac{d\Omega}{[1 - (\Omega/\omega_n)^2]^2 + (2\zeta\Omega/\omega_n)^2} = \lim_{R \to \infty} \int_0^\pi \frac{iRe^{i\theta}\, d\theta}{[1 - (Re^{i\theta}/\omega_n)^2]^2 + (2\zeta Re^{i\theta}/\omega_n)^2}$$

$$= 0. \tag{11.139}$$

It follows from (11.138) that

$$\int_{-\infty}^\infty \frac{d\omega}{[1 - (\omega/\omega_n)^2]^2 + (2\zeta\omega/\omega_n)^2} = \lim_{R \to \infty} \oint \frac{d\Omega}{[1 - (\Omega/\omega_n)^2]^2 + (2\zeta\Omega/\omega_n)^2}$$

$$= 2\pi i \sum \text{Res}, \tag{11.140}$$

where $\sum \text{Res}$ denotes the sum of the residues of the function

$$|H(\Omega)|^2 = \frac{\omega_n^4}{(\omega_n^2 - \Omega^2)^2 + (2\zeta\Omega\omega_n)^2} \tag{11.141}$$

at the poles enclosed by the contour consisting of the straight line and the semicircle. Of course, as the radius of the semicircle tends to infinity, it becomes obvious that we must sum the poles of $|H(\Omega)|^2$ in the upper half of the complex plane Ω.

Let us consider an underdamped system ($\zeta < 1$) so that there are two poles in the upper half-plane. These poles are

$$\Omega_1 = (i\zeta + \sqrt{1 - \zeta^2})\omega_n, \qquad \Omega_2 = (i\zeta - \sqrt{1 - \zeta^2})\omega_n. \tag{11.142}$$

But

$$\text{Res} \, (\Omega = \Omega_1) = (\Omega - \Omega_1)|H(\Omega)|^2 \Big|_{\Omega = \Omega_1} = \frac{\omega_n}{8i\zeta\sqrt{1 - \zeta^2}(i\zeta + \sqrt{1 - \zeta^2})}, \tag{11.143}$$

$$\text{Res} \, (\Omega = \Omega_2) = (\Omega - \Omega_2)|H(\Omega)|^2 \Big|_{\Omega = \Omega_2} = -\frac{\omega_n}{8i\zeta\sqrt{1 - \zeta^2}(i\zeta - \sqrt{1 - \zeta^2})},$$

so the response mean square becomes

$$\overline{x^2} = \frac{G_0}{2\pi} \int_{-\infty}^{\infty} \frac{d\omega}{[1 - (\omega/\omega_n)^2]^2 + (2\zeta\omega/\omega_n)^2}$$

$$= \frac{G_0}{2\pi} 2\pi i [\text{Res}\,(\Omega = \Omega_1) + \text{Res}\,(\Omega = \Omega_2)]$$

$$= \frac{G_0\omega_n}{8\zeta\sqrt{1 - \zeta^2}} \left(\frac{1}{i\zeta + \sqrt{1 - \zeta^2}} - \frac{1}{i\zeta - \sqrt{1 - \zeta^2}}\right) = \frac{G_0\omega_n}{4\zeta}. \quad (11.144)$$

Although the white noise is only an idealization, the result is not as unreasonable as it may seem. From Figure 11.20 we see that even if the

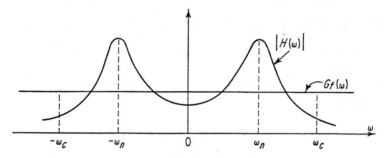

FIGURE 11.20

excitation spectral density is in the form of a band-limited white noise, with lower cutoff frequency zero and upper cutoff frequency much larger than the natural frequency of the undamped system, $\omega_c \gg \omega_n$, one obtains a good approximation of the response mean square.

Because the random process is Gaussian, the response mean value and mean square value are sufficient to determine the response probability density function, so it is possible to evaluate the probability that the response will exceed a certain displacement.

11–13 RESPONSE OF A DISCRETE SYSTEM TO RANDOM EXCITATION

Let us consider an n-degree-of-freedom system subjected to ergodic random excitations $F_i(t)$. The corresponding equations of motion, in matrix form, are

$$[m]\{\ddot{x}(t)\} + [c]\{\dot{x}(t)\} + [k]\{x(t)\} = \{F(t)\}, \quad (11.145)$$

where $[m]$, $[c]$, and $[k]$ are symmetric inertia, damping, and stiffness matrices, respectively. The column matrices $\{F(t)\}$ and $\{x(t)\}$ represent the random

excitations and responses, respectively. For simplicity we also assume that the classical modal matrix $[u]$, associated with the undamped system, can be used to represent a linear transformation uncoupling the system of equations.†
Hence let the solution of (11.145) be

$$\{x(t)\} = [u]\{q(t)\}, \tag{11.146}$$

where $q_r(t)$ $(r = 1, 2, \ldots, n)$ are generalized coordinates corresponding to the random responses $x_i(t)$ $(i = 1, 2, \ldots, n)$. The standard modal analysis leads to the uncoupled set of equations

$$\ddot{q}_r(t) + 2\zeta_r\omega_r\dot{q}_r(t) + \omega_r^2 q_r(t) = f_r(t), \qquad r = 1, 2, \ldots, n, \tag{11.147}$$

where $f_r(t)$ $(r = 1, 2, \ldots, n)$ are generalized random forces given by

$$f_r(t) = \{u^{(r)}\}^T\{F(t)\}, \tag{11.148}$$

where $\{u^{(r)}\}$ is the rth normal mode of the undamped system.
  Letting

$$Q_r(\omega) = \int_{-\infty}^{\infty} q_r(t)e^{-i\omega t}\, dt, \tag{11.149}$$

$$F_r(\omega) = \int_{-\infty}^{\infty} f_r(t)e^{-i\omega t}\, dt \tag{11.150}$$

be the Fourier transforms of $q_r(t)$ and $f_r(t)$, respectively, we can transform both sides of (11.147) and write

$$Q_r(\omega)(-\omega^2 + 2i\omega\zeta_r\omega_r + \omega_r^2) = F_r(\omega), \qquad r = 1, 2, \ldots, n, \tag{11.151}$$

which can be solved for $Q_r(\omega)$ to obtain

$$Q_r(\omega) = \frac{H_r(\omega)F_r(\omega)}{\omega_r^2}, \qquad r = 1, 2, \ldots, n, \tag{11.152}$$

where $$H_r(\omega) = \frac{1}{1 - (\omega/\omega_r)^2 + i(2\zeta\omega/\omega_r)}, \qquad r = 1, 2, \ldots, n \tag{11.153}$$

is the complex frequency response for the rth normal coordinate.
  From (11.146) we see that the motion of any point mass i can be written

$$x_i(t) = \sum_{r=1}^{n} u_i^{(r)}q_r(t). \tag{11.154}$$

† This is possible if the damping matrix $[c]$ is of a special form. For details see Chapter 9

Similarly, let the motion of point j be

$$x_j(t) = \sum_{s=1}^{n} u_j^{(s)} q_s(t), \tag{11.155}$$

where s is a dummy index replacing r for reasons that will become obvious shortly. The Fourier transforms of $x_i(t)$ and $x_j(t)$ are, respectively,

$$X_i(\omega) = \sum_{r=1}^{n} u_i^{(r)} Q_r(\omega) = \sum_{r=1}^{n} u_i^{(r)} \frac{H_r(\omega) F_r(\omega)}{\omega_r^2},$$

$$X_j(\omega) = \sum_{s=1}^{n} u_j^{(s)} Q_s(\omega) = \sum_{s=1}^{n} u_j^{(s)} \frac{H_s(\omega) F_s(\omega)}{\omega_s^2}. \tag{11.156}$$

Using (11.83), of the correlation theorem, we can calculate the temporal cross-correlation function

$$\overline{x_i(t) x_j(t + \tau)} = \lim_{T \to \infty} \frac{1}{T} \int_{-T/2}^{T/2} x_i(t) x_j(t + \tau) \, dt$$

$$= \lim_{T \to \infty} \frac{1}{2\pi T} \int_{-\infty}^{\infty} X_i^*(\omega) X_j(\omega) e^{i\omega\tau} \, d\omega$$

$$= \lim_{T \to \infty} \frac{1}{2\pi T} \sum_{r=1}^{n} \sum_{s=1}^{n} \frac{u_i^{(r)} u_j^{(s)}}{\omega_r^2 \omega_s^2} \int_{-\infty}^{\infty} H_r^*(\omega) H_s(\omega) F_r^*(\omega) F_s(\omega) e^{i\omega\tau} \, d\omega. \tag{11.157}$$

Letting

$$\psi_{frs}(\tau) = \lim_{T \to \infty} \frac{1}{T} \int_{-T/2}^{T/2} f_r(t) f_s(t + \tau) \, dt \tag{11.158}$$

be a generalized excitation cross-correlation function we can define its Fourier transform as

$$G_{frs}(\omega) = \int_{-\infty}^{\infty} \psi_{frs}(\tau) e^{-i\omega\tau} \, d\tau$$

$$= \lim_{T \to \infty} \frac{1}{T} \int_{-\infty}^{\infty} \left[\int_{-\infty}^{\infty} f_r(t) f_s(t + \tau) \, dt \right] e^{-i\omega\tau} \, d\tau. \tag{11.159}$$

From (11.84), of the correlation theorem, however, we conclude that

$$G_{frs}(\omega) = \lim_{T \to \infty} \frac{1}{T} F_r^*(\omega) F_s(\omega), \tag{11.160}$$

so (11.157) can be written

$$\overline{x_i(t) x_j(t + \tau)} = \frac{1}{2\pi} \sum_{r=1}^{n} \sum_{s=1}^{n} \frac{u_i^{(r)} u_j^{(s)}}{\omega_r^2 \omega_s^2} \int_{-\infty}^{\infty} H_r^*(\omega) H_s(\omega) G_{frs}(\omega) e^{i\omega\tau} \, d\omega. \tag{11.161}$$

Letting $\tau = 0$, (11.161) becomes

$$\overline{x_i(t)x_j(t)} = \frac{1}{2\pi} \sum_{r=1}^{n} \sum_{s=1}^{n} \frac{u_i^{(r)}u_j^{(s)}}{\omega_r^2\omega_j^2} \int_{-\infty}^{\infty} H_r^*(\omega)H_s(\omega)G_{frs}(\omega)\,d\omega, \quad (11.162)$$

which for $i = j$ reduces to the response mean square value

$$\overline{x_i^2(t)} = \frac{1}{2\pi} \sum_{r=1}^{n} \sum_{s=1}^{n} \frac{u_i^{(r)}u_i^{(s)}}{\omega_r^2\omega_s^2} \int_{-\infty}^{\infty} H_r^*(\omega)H_s(\omega)G_{frs}(\omega)\,d\omega, \quad (11.163)$$

associated with the random response $x_i(t)$. Note that $\overline{x_i^2(t)}$ is not a function of t.

For a Gaussian process it is sufficient to know the values $\overline{x_i^2(t)}$ $(i = 1, 2, \ldots, n)$, to determine the probability density functions associated with every coordinate $x_i(t)$, that is, if the excitation mean values are zero.

11-14 RESPONSE OF CONTINUOUS SYSTEMS TO RANDOM EXCITATION

The approach used for discrete systems can be used to obtain the response of continuous systems to random excitation. Using the notation of Chapter 9 we can write the differential equation of motion of a damped continuous system in the form

$$L[w(P, t)] + \frac{\partial}{\partial t} C[w(P, t)] + M(P)\frac{\partial^2 w(P, t)}{\partial t^2} = f(P, t), \quad (11.164)$$

where $f(P, t)$ denotes an ergodic distributed random excitation. The displacement $w(P, t)$ must satisfy (11.164) throughout domain D, and, in addition, it must satisfy the associated boundary conditions. We shall make the assumption that the operator C, representing the system damping, is a linear combination of the operator L and the mass distribution M, so modal analysis can be used to obtain a solution of (11.164). To this end we assume the solution of (11.164) as an infinite series of the normal modes $w_r(P)$, of the undamped system, multiplied by corresponding generalized coordinates $q_r(t)$,

$$w(P, t) = \sum_{r=1}^{\infty} w_r(P)q_r(t), \quad (11.165)$$

where the orthonormal set of functions $w_r(P)$ $(r = 1, 2, \ldots)$ satisfy the relations

$$\int_D m(P)w_r(P)w_s(P)\, dD(P) = \delta_{rs},$$

$$\int_D w_r(P)L[w_s(P)]\, dD(P) = \omega_r^2 \delta_{rs}, \qquad (11.166)$$

$$\int_D w_r(P)C[w_s(P)]\, dD(P) = 2\zeta_r \omega_r \delta_{rs},$$

where δ_{rs} is the Kronecker delta and ω_r denotes the rth natural frequency of the undamped system. This leads to the set of equations

$$\ddot{q}_r(t) + 2\zeta_r \omega_r \dot{q}_r(t) + \omega_r^2 q_r(t) = f_r(t), \qquad r = 1, 2, \ldots, \qquad (11.167)$$

where the generalized random forces $f_r(t)$ are given by

$$f_r(t) = \int_D w_r(P)f(P, t)\, dD(P), \qquad r = 1, 2, \ldots. \qquad (11.168)$$

Let $Q_r(\omega)$ be the Fourier transforms of $q_r(t)$ and let

$$
\begin{aligned}
F_r(\omega) &= \int_{-\infty}^{\infty} f_r(t)e^{-i\omega t}\, dt = \int_{-\infty}^{\infty} \left[\int_D w_r(P)f(P, t)\, dD(P) \right] e^{-i\omega t}\, dt \\
&= \int_D w_r(P)\left[\int_{-\infty}^{\infty} f(P, t)e^{-i\omega t}\, dt \right] dD(P) \\
&= \int_D w_r(P)F(P, \omega)\, dD(P) \qquad (11.169)
\end{aligned}
$$

be the Fourier transform of the generalized random force $f_r(t)$, where $F(P, \omega)$ is the Fourier transform of the distributed random force $f(P, t)$. Recalling the definition (11.153) of the complex frequency response $H_r(\omega)$, we can transform (11.167) and obtain the relation

$$Q_r(\omega) = \frac{H_r(\omega)F_r(\omega)}{\omega_r^2}, \qquad r = 1, 2, \ldots. \qquad (11.170)$$

The Fourier transforms of the responses at the points P and P' are

$$
\begin{aligned}
W(P, \omega) &= \sum_{r=1}^{\infty} w_r(P)Q_r(\omega) = \sum_{r=1}^{\infty} w_r(P)\frac{H_r(\omega)F_r(\omega)}{\omega_r^2}, \\
W(P', \omega) &= \sum_{s=1}^{\infty} w_s(P')Q_s(\omega) = \sum_{s=1}^{\infty} w_s(P')\frac{H_s(\omega)F_s(\omega)}{\omega_s^2},
\end{aligned}
\qquad (11.171)
$$

so that by (11.83), of the correlation theorem, we can calculate the cross-correlation function

$$\overline{w(P, t)w(P', t + \tau)} = \lim_{T \to \infty} \frac{1}{2\pi T} \int_{-\infty}^{\infty} W^*(P, \omega)W(P', \omega)e^{i\omega\tau}\, d\omega$$

$$= \lim_{T \to \infty} \frac{1}{2\pi T} \sum_{r=1}^{\infty} \sum_{s=1}^{\infty} \frac{w_r(P)w_s(P')}{\omega_r^2\omega_s^2} \int_{-\infty}^{\infty} H_r^*(\omega)$$

$$\times\, H_s(\omega)F_r^*(\omega)F_s(\omega)e^{i\omega\tau}\, d\omega, \qquad (11.172)$$

which resembles (11.157) for discrete systems. The expressions for $F_r(\omega)$ and $F_s(\omega)$ are different, however.

Next we define a generalized excitation cross-correlation function as

$$\psi_{frs}(\tau) = \lim_{T \to \infty} \frac{1}{T} \int_{-T/2}^{T/2} f_r(t)f_s(t + \tau)\, dt = \frac{1}{2\pi} \int_{-\infty}^{\infty} G_{frs}(\omega)e^{i\omega\tau}\, d\omega, \quad (11.173)$$

where $G_{frs}(\omega)$ is the Fourier transform of $\psi_{frs}(\tau)$. Using (11.84), it can be shown that

$$G_{frs}(\omega) = \lim_{T \to \infty} \frac{1}{T} F_r^*(\omega)F_s(\omega)$$

$$= \lim_{T \to \infty} \frac{1}{T} \int_D \int_D w_r(P)w_s(P')F_r^*(P, \omega)F_s(P', \omega)\, dD(P)\, dD(P'). \quad (11.174)$$

Hence, (11.172) becomes

$$\overline{w(P, t)w(P', t + \tau)} = \frac{1}{2\pi} \sum_{r=1}^{\infty} \sum_{s=1}^{\infty} \frac{w_r(P)w_s(P')}{\omega_r^2\omega_s^2} \int_{-\infty}^{\infty} H_r^*(\omega)H_s(\omega)G_{frs}(\omega)e^{i\omega\tau}\, d\omega, \tag{11.175}$$

which for $\tau = 0$ reduces to

$$\overline{w(P, t)w(P', t)} = \frac{1}{2\pi} \sum_{r=1}^{\infty} \sum_{s=1}^{\infty} \frac{w_r(P)w_s(P')}{\omega_r^2\omega_s^2} \int_{-\infty}^{\infty} H_r^*(\omega)H_s(\omega)G_{frs}(\omega)\, d\omega. \tag{11.176}$$

If the two points coincide, $P = P'$, we obtain the response mean square value

$$\overline{w^2(P, t)} = \frac{1}{2\pi} \sum_{r=1}^{\infty} \sum_{s=1}^{\infty} \frac{w_r(P)w_s(P)}{\omega_r^2\omega_s^2} \int_{-\infty}^{\infty} H_r^*(\omega)H_s(\omega)G_{frs}(\omega)\, d\omega, \tag{11.177}$$

which, for a Gaussian process with zero mean value, determines the probability density function for any point P.

Problems

11.1 Use (11.24) and prove that the plot of the probability density function $p(x)$ shown in Figure 11.7(c) is correct. The function $x(t)$ is shown in Figure 11.7(a).

11.2 Calculate the mean value and the mean square value of the function

$$x(t) = A \sin \frac{2\pi}{T} t$$

by using (11.29) and (11.31). The probability density of $x(t)$ is given by (11.28).

11.3 Calculate the temporal mean value and mean square value of the function

$$x(t) = A \sin \frac{2\pi}{T} t.$$

Note that for periodic functions temporal averages can be obtained by averaging over the period T.

11.4 Calculate and plot the temporal autocorrelation function of the function

$$x(t) = A \sin \frac{2\pi}{T} t.$$

Obtain the temporal mean square value of $x(t)$ from the autocorrelation function.

11.5 Calculate the autocorrelation function corresponding to the ideal white noise spectral distribution.

11.6 Calculate the autocorrelation function and the mean square spectral density of the unit step function.

11.7 Calculate and plot the autocorrelation function of the pulse-width modulated wave shown in Figure 11.21.

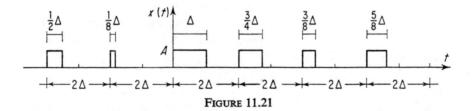

FIGURE 11.21

11.8 Calculate and plot the autocorrelation function of the periodic function shown in Figure 11.22. Calculate and plot the mean square spectral density of $x(t)$.

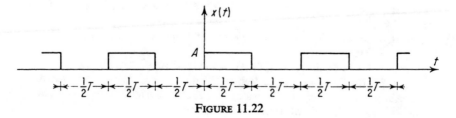

FIGURE 11.22

11.9   Calculate and plot the cross-correlation function between the function of Problem 11.8 and the function $y(t) = A \sin (2\pi/T)t$.

11.10   Prove (11.117) and (11.120).

11.11   A carriage travels with a uniform velocity on a rough surface as shown in Figure 11.23. The mass is uniform and its center is placed so that it is at an equal

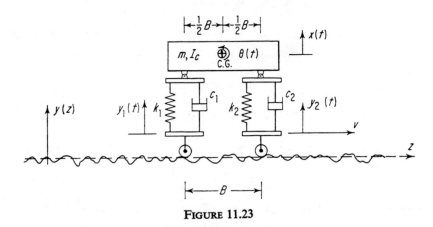

FIGURE 11.23

distance from both hinges. Assume that in addition to the vertical translation $x(t)$ the mass undergoes a small angular rotation $\theta(t)$ and derive the corresponding equations of motion. Derive general equations for the response autocorrelation functions and the cross-correlation function between $x(t)$ and $\theta(t)$ by assuming that damping is sufficiently small that one can ignore the coupling of the natural modes produced by damping. Obtain the response mean square values. The excitation random process can be assumed to be ergodic.

11.12   Consider a general continuous system subjected to a concentrated force of randomly time-dependent amplitude $f(t)$. The excitation processes can be assumed to be ergodic and written

$$f(P, t) = f(t)\delta(P - P_0).$$

Use the analysis of Section 11–14 and derive general equations for the response cross-correlation function and mean square value.

Selected Readings

Aseltine, J. A., *Transform Method in Linear System Analysis*, McGraw-Hill, New York, 1958.

Bendat, J. S., *Principles and Applications of Random Noise Theory*, Wiley, New York, 1958.

Bendat, J. S., L. D. Enochson, G. H. Klein, and A. G. Piersol, The Application of Statistics to the Flight Vehicle Vibration Problems, *ASD TR 61–123*, Aeronautical Systems Division, Air Force Systems Command, Wright-Patterson AFB, Ohio, 1961.

Crandall, S. H., ed., *Random Vibration*, M.I.T. Press, Cambridge, Mass., 1958.

Crandall, S. H., and W. D. Mark, *Random Vibration in Mechanical Systems*, Academic Press, New York, 1963.

Davenport, Jr., W. B., and W. L. Root, *Introduction to Random Signals and Noise*, McGraw-Hill, New York, 1958.

Hoel, P. G., *Introduction to Mathematical Statistics*, 3rd ed., Wiley, New York, 1962.

Hurty, W. C., and M. F. Rubinstein, *Dynamics of Structures*, Prentice-Hall, Englewood Cliffs, N.J., 1964.

Lee, Y. W., *Statistical Theory of Communication*, Wiley, New York, 1960.

Papoulis, A., *Probability, Random Variables, and Stochastic Processes*, McGraw-Hill, New York, 1965.

Thomson, W. T., *Vibration Theory and Applications*, Prentice-Hall, Englewood Cliffs, N.J., 1965.

Thomson, W. T., "Continuous Structures Excited by Correlated Random Forces," *Intern. J. Mech. Sci.* **4**, 109–114 (1962).

ELEMENTS OF
MATRIX
ALGEBRA

A-1 GENERAL CONSIDERATIONS

Matrices are sets of elements consisting of real or complex numbers and arranged in a rectangular formation of m rows and n columns. Rectangular arrays are frequently encountered in engineering. The motion of a system described by n independent coordinates can be represented by a rectangular matrix consisting of one column and n rows. The moments of inertia of a rigid body or the state of stress at a point may be represented by sets of nine numbers arranged in matrices of three rows and three columns. Sets of algebraic equations and coordinate transformations are readily represented by matrices.

Not any set of numbers arranged in matrix form may be called a matrix. To be able to regard a set of numbers as a matrix one must recall that (1) the idea of a matrix implies the treatment of its elements taken as a whole and in their proper arrangement, and (2) matrices must obey certain rules of addition and multiplication.

A-2   PRINCIPAL TYPES OF MATRICES AND THEIR NOTATION

(a) Rectangular Matrices

A rectangular matrix is denoted by square brackets and is of the form

$$\begin{bmatrix} 5 & -3 & 4 & 1 \\ 2 & 0 & 0 & 5 \\ 1 & 13 & 2 & -5 \end{bmatrix}. \tag{A.1}$$

In general one writes

$$\begin{bmatrix} a_{11} & a_{12} & a_{13} & \cdots & a_{1j} & \cdots & a_{1n} \\ a_{21} & a_{22} & \cdots\cdots\cdots\cdots & a_{2j} & \cdots & a_{2n} \\ \vdots & \vdots & & \vdots & & \vdots \\ a_{i1} & a_{i2} & \cdots\cdots\cdots\cdots & a_{ij} & \cdots & a_{in} \\ \vdots & \vdots & & \vdots & & \vdots \\ a_{m1} & a_{m2} & \cdots\cdots\cdots\cdots & a_{mj} & \cdots & a_{mn} \end{bmatrix}. \tag{A.2}$$

The above is denoted symbolically in the form $[a_{ij}]$ or, even simpler, $[a]$. If a matrix has m rows and n columns it is called *a matrix of order* (m, n) or simply *an* $m \times n$ *(m by n) matrix*.

(b) Column and Row Matrices

A matrix consisting of a single row is denoted

$$\lfloor x_1 \quad x_2 \quad \cdots \quad x_j \quad \cdots \quad x_n \rfloor = \lfloor x \rfloor. \tag{A.3}$$

A matrix consisting of a single column has the following notation:

$$\begin{Bmatrix} y_1 \\ y_2 \\ \vdots \\ y_i \\ \vdots \\ y_n \end{Bmatrix} = \{y\}. \tag{A.4}$$

(c) Square, Diagonal, and Unit Matrices

In the special but frequent case when $m = n$, the matrix is a square matrix,

$$[a] = \begin{bmatrix} a_{11} & a_{12} & \cdots & a_{1n} \\ a_{21} & a_{22} & \cdots & a_{2n} \\ \vdots & \vdots & & \vdots \\ a_{n1} & a_{n2} & \cdots & a_{nn} \end{bmatrix}. \tag{A.5}$$

If all off-diagonal elements in a square matrix are equal to zero ($a_{ij} = 0$ for $i \neq j$), the matrix reduces to a diagonal matrix

$$[a] = \begin{bmatrix} a_{11} & 0 & \cdots & 0 \\ 0 & a_{22} & \cdots & 0 \\ \vdots & \vdots & & \vdots \\ 0 & 0 & \cdots & a_{nn} \end{bmatrix}. \tag{A.6}$$

A diagonal matrix with all the diagonal elements equal to unity ($a_{ii} = 1$) is called a *unit matrix* or an *identity matrix*,

$$[I] = \begin{bmatrix} 1 & 0 & \cdots & 0 \\ 0 & 1 & \cdots & 0 \\ \vdots & \vdots & & \vdots \\ 0 & 0 & \cdots & 1 \end{bmatrix}. \tag{A.7}$$

(d) Transposed Matrices

The transpose of matrix $[a]$ is denoted $[a]^T$. It is obtained by taking every row in the transposed matrix identical to the corresponding column in the original matrix. Evidently if the original matrix is an $m \times n$ matrix its transpose is an $n \times m$ matrix,

$$[a]^T = \begin{bmatrix} a_{11} & a_{12} & \cdots & a_{1n} \\ a_{21} & a_{22} & \cdots & a_{2n} \\ \vdots & \vdots & & \vdots \\ a_{m1} & a_{m2} & \cdots & a_{mn} \end{bmatrix}^T = \begin{bmatrix} a_{11} & a_{21} & \cdots & a_{m1} \\ a_{12} & a_{22} & \cdots & a_{m2} \\ \vdots & \vdots & & \vdots \\ a_{1n} & a_{2n} & \cdots & a_{mn} \end{bmatrix}. \tag{A.8}$$

The transpose of a row matrix is a column matrix and vice versa,

$$\lfloor x \rfloor^T = \{x\}, \tag{A.9}$$

$$\{z\}^T = \lfloor z \rfloor. \tag{A.10}$$

(e) Symmetric and Skew-Symmetric Matrices

A symmetric matrix is equal to its transpose $a_{ij} = a_{ji}$, or

$$[a] = [a]^T. \tag{A.11}$$

A skew-symmetric matrix is defined by

$$a_{ij} = -a_{ji}, \qquad a_{ii} = 0. \tag{A.12}$$

Note that both the symmetric and the skew-symmetric matrices must be square.

(f) Null Matrices

A matrix whose elements are all zero is called a *null matrix*. It is denoted by [0], {0}, or $\lfloor 0 \rfloor$, according to whether it is a rectangular, column, or row matrix.

A-3 BASIC MATRIX OPERATIONS

(a) Equality of Matrices

Two matrices are equal if all elements of one matrix are equal to the corresponding elements of the other matrix. Hence

$$[a] = [b] \tag{A.13}$$

if and only if

$$a_{ij} = b_{ij} \tag{A.14}$$

for each pair of subscripts i and j.
Evidently the two matrices must have the same order.

(b) Addition and Subtraction of Matrices

Addition and subtraction of two matrices can be performed only if the two matrices have the same order. The corresponding elements of the second matrix are added or subtracted to obtain the resulting matrix.
Hence the matrix equation

$$[c] = [a] \pm [b] \tag{A.15}$$

implies that

$$c_{ij} = a_{ij} \pm b_{ij} \tag{A.16}$$

for any pair of subscripts i and j.
The matrix addition (or subtraction) *is a commutative process,*

$$[a] + [b] = [b] + [a], \tag{A.17}$$

and, furthermore, *it is an associative process,*

$$([a] + [b]) + [c] = [a] + ([b] + [c]). \tag{A.18}$$

(c) Scalar Multiplier

When a matrix is multiplied by a scalar the resulting matrix consists of the corresponding elements of the original matrix multiplied by the scalar. If the scalar is s,

$$[c] = s[a] \tag{A.19}$$

implies

$$c_{ij} = sa_{ij} \tag{A.20}$$

for any pair of subscripts i and j.

(d) Multiplication of Matrices

In general the matrix multiplication is *not a commutative process*.

In addition, two matrices may be multiplied only if they are conformable, which means the number of columns of the first matrix is equal to the number of rows of the second one. An example of a matrix multiplication is

$$\begin{array}{ccc} [a] & [b] & = & [c]. \\ m \times n & n \times p & & m \times p \end{array} \tag{A.21}$$

The above means that if $[a]$ is an $m \times n$ matrix and $[b]$ is an $n \times p$ matrix, the resulting matrix $[c]$ is an $m \times p$ matrix.

In general the multiplication is not commutative, so one must specify the relative position of the matrices to be multiplied. The multiplication (A.21) may be described in two ways: (1) the matrix $[a]$ postmultiplied by $[b]$ is equal to $[c]$, or (2) the matrix $[b]$ premultiplied by $[a]$ is equal to $[c]$.

The rule of matrix multiplication is

$$c_{ij} = \sum_{k=1}^{n} a_{ik}b_{kj}, \tag{A.22}$$

where k is a dummy index. An example is

$$\begin{bmatrix} a_{11} & a_{12} & a_{13} \\ a_{21} & a_{22} & a_{23} \end{bmatrix} \begin{bmatrix} b_{11} & b_{12} & b_{13} & b_{14} \\ b_{21} & b_{22} & b_{23} & b_{24} \\ b_{31} & b_{32} & b_{33} & b_{34} \end{bmatrix}$$

$$= \begin{bmatrix} a_{11}b_{11} + a_{12}b_{21} + a_{13}b_{31} & a_{11}b_{12} + a_{12}b_{22} + a_{13}b_{32} \\ a_{21}b_{11} + a_{22}b_{21} + a_{23}b_{31} & a_{21}b_{12} + a_{22}b_{22} + a_{23}b_{32} \end{bmatrix}$$

$$\begin{bmatrix} a_{11}b_{13} + a_{12}b_{23} + a_{13}b_{33} & a_{11}b_{14} + a_{12}b_{24} + a_{13}b_{34} \\ a_{21}b_{13} + a_{22}b_{23} + a_{23}b_{33} & a_{21}b_{14} + a_{22}b_{24} + a_{23}b_{34} \end{bmatrix}.$$

In the special case in which one of the matrices is a square matrix and the other is a unit matrix the multiplication is commutative,

$$[a][I] = [I][a] = [a]. \tag{A.23}$$

The multiplication of matrices *is an associative process.* If

$$[a][b] = [d], \tag{A.24}$$

$$[b][c] = [e], \tag{A.25}$$

$$\underset{m \times n}{[a]} \; \underset{n \times p}{[b]} \; \underset{p \times q}{[c]} \; = \; \underset{m \times p}{[d]} \; \underset{p \times q}{[c]} \; = \; \underset{m \times n}{[a]} \; \underset{n \times q}{[e]} \; = \; \underset{m \times q}{[f]}. \tag{A.26}$$

Note that the above operations imply

$$d_{ij} = \sum_{k=1}^{n} a_{ik} b_{kj}, \tag{A.27}$$

$$e_{ks} = \sum_{j=1}^{p} b_{kj} c_{js}, \tag{A.28}$$

$$f_{is} = \sum_{j=1}^{p} d_{ij} c_{js} = \sum_{k=1}^{n} \sum_{j=1}^{p} a_{ik} b_{kj} c_{js}, \tag{A.29}$$

$$f_{is} = \sum_{k=1}^{n} a_{ik} e_{ks} = \sum_{k=1}^{n} \sum_{j=1}^{p} a_{ik} b_{kj} c_{js}. \tag{A.30}$$

The *distributive law holds true* for continuous products of matrices. Hence if

$$[b] = [g] + [h], \tag{A.31}$$

then $$[f] = [a][b][c] = [a][g][c] + [a][h][c]. \tag{A.32}$$

In general, the matrix product

$$[a][b] = [0] \tag{A.33}$$

does not imply that either [a] or [b] or both [a] and [b] are null matrices. An example is

$$\begin{bmatrix} 1 & 1 \\ 1 & 1 \end{bmatrix} \begin{bmatrix} 1 & -1 \\ -1 & 1 \end{bmatrix} = \begin{bmatrix} 0 & 0 \\ 0 & 0 \end{bmatrix}.$$

Hence matrix algebra differs from ordinary algebra on two counts:
(1) matrix multiplication is not commutative, and (2) the fact that the product
of two matrices is equal to a null matrix cannot be construed to mean that
either multiplicand (or both) is a null matrix. Both these rules hold true in
ordinary algebra.

A–4 DETERMINANT OF A SQUARE MATRIX. SINGULAR MATRIX

The determinant of the matrix $[a]$ is denoted $|a|$. When the determinant
of a matrix vanishes ($|a| = 0$), the matrix $[a]$ is called *singular*. Evidently the
matrix $[a]$ must be square.

A–5 ADJOINT OF A MATRIX

Let the determinant of matrix $[a]$ be

$$|a| = |a_{ij}| = \begin{vmatrix} a_{11} & a_{12} & \cdots & a_{1n} \\ a_{21} & a_{22} & \cdots & a_{2n} \\ \vdots & \vdots & & \vdots \\ a_{n1} & a_{n2} & \cdots & a_{nn} \end{vmatrix}, \tag{A.34}$$

and denote by M_{rs} the minor determinant corresponding to the element a_{rs},
where M_{rs} is obtained by striking out of the determinant $|a|$ the rth row and
sth column. The signed minor corresponding to the element a_{rs} is called the
cofactor A_{rs},

$$A_{rs} = (-1)^{r+s} M_{rs}. \tag{A.35}$$

By definition $[A_{ji}]$ is the adjoint of $[a_{ij}]$. It is obtained by transposing the
matrix of the cofactors of $[a_{ij}]$,

$$[A_{ji}] = [(-1)^{i+j} M_{ij}]^T. \tag{A.36}$$

A–6 INVERSE, OR RECIPROCAL, OF A MATRIX

The inverse, or reciprocal, of a matrix $[a]$ is denoted $[a]^{-1}$ and defined by

$$[a][a]^{-1} = [a]^{-1}[a] = [I]. \tag{A.37}$$

Consider the product

$$
[a_{ij}][A_{ji}] =
\begin{bmatrix}
a_{11} & a_{12} & \cdots & a_{1n} \\
a_{21} & a_{22} & \cdots & a_{2n} \\
\vdots & \vdots & & \vdots \\
a_{n1} & a_{n2} & \cdots & a_{nn}
\end{bmatrix}
$$

$$
\times
\begin{bmatrix}
M_{11} & -M_{21} & \cdots & (-1)^{1+n}M_{n1} \\
-M_{12} & M_{22} & \cdots & (-1)^{2+n}M_{n2} \\
\vdots & \vdots & & \vdots \\
(-1)^{1+n}M_{1n} & (-1)^{2+n}M_{2n} & \cdots & M_{nn}
\end{bmatrix}
$$

$$
= \left[\sum_{j=1}^{n} (-1)^{i+j} a_{kj} M_{ij} \right]. \tag{A.38}
$$

But

$$
\sum_{j=1}^{n} (-1)^{i+j} a_{kj} M_{ij} =
\begin{vmatrix}
a_{11} & a_{12} & \cdots & a_{1n} \\
a_{21} & a_{22} & \cdots & a_{2n} \\
\vdots & \vdots & & \vdots \\
a_{n1} & a_{n2} & \cdots & a_{nn}
\end{vmatrix}
= |a| \qquad \text{if } i = k. \tag{A.39}
$$

If $i \neq k$, however,

$$
\sum_{j=1}^{n} (-1)^{i+j} a_{kj} M_{ij} =
\begin{vmatrix}
a_{11} & a_{12} & \cdots & a_{1n} \\
a_{21} & \cdots\cdots\cdots\cdots & a_{2n} \\
\vdots & & \vdots \\
a_{k1} & a_{k2} & \cdots & a_{kn} \\
\vdots & \vdots & & \vdots \\
a_{k1} & a_{k2} & \cdots & a_{kn} \\
\vdots & \vdots & & \vdots \\
a_{n1} & a_{n2} & \cdots & a_{nn}
\end{vmatrix}
= 0, \tag{A.40}
$$

because (A.40) is the determinant obtained from matrix $[a]$ by replacing the elements in the ith row by the corresponding ones in the kth row and keeping the kth row intact. The result is a determinant with the ith and kth rows identical.

It follows that (A.38) may be rewritten in a more compact way,

$$
[a_{ij}][A_{ji}] = [a][A_{ji}] = |a|[I], \tag{A.41}
$$

where $[I]$ is of the same order as $[a]$ and $[A_{ji}]$.

Premultiply throughout (A.41) by $[a]^{-1}$ and obtain

$$
[a]^{-1}[a_{ij}][A_{ji}] = [a]^{-1}[a][A_{ji}] = [I][A_{ji}] = [A_{ji}]
$$
$$
= |a|[a]^{-1}[I] = |a|[a]^{-1}. \tag{A.42}
$$

Dividing through by $|a|$, which is a scalar, the inverse of $[a]$ is obtained in the form

$$[a]^{-1} = \frac{[A_{ji}]}{|a|}, \tag{A.43}$$

or the inverse of matrix $[a]$ is obtained by dividing its adjoint matrix $[A_{ji}]$ by its determinant $|a|$.

As the order of the matrix $[a]$ increases, the above formula becomes impracticable, and it is advisable to seek other ways of computing the inverse of a matrix. To this end let us consider a set of n algebraic equations in n unknowns x_j,

$$\sum_{j=1}^{n} a_{ij}x_j = b_i, \qquad i = 1, 2, \ldots, n. \tag{A.44}$$

Equation (A.44) can be written in the matrix form (see Section A–8)

$$[a]\{x\} = [I]\{b\}, \tag{A.45}$$

where $[I]$ is the identity or unit matrix. Premultiplying both sides of (A.45) by $[a]^{-1}$, we obtain

$$\{x\} = [a]^{-1}\{b\}, \tag{A.46}$$

where the column matrix $\{x\}$ represents the solution of the algebraic equations (A.44). The right side of (A.46) contains the inverse matrix $[a]^{-1}$, so it follows that all the *elementary operations* used in the solution of a set of algebraic equations can be used to compute the inverse of a matrix. From these elementary operations we shall use only the multiplication of an equation by a number other than zero and the addition of two equations. This method of computing the inverse of a matrix is intimately related to the *Gauss-Jordan reduction method* for the solution of a set of algebraic equations. The method consists of performing identical operations on the corresponding rows of the matrices $[a]$ and $[I]$ in such a way that $[a]$ is reduced to a unit matrix. The original unit matrix becomes the reciprocal $[a]^{-1}$. This is shown schematically by

$$\begin{array}{cc} [a], & [I], \\ \vdots & \vdots \\ [I], & [a]^{-1}. \end{array} \tag{A.47}$$

As an illustration, let us obtain the reciprocal of a 3×3 matrix, which is sufficiently simple to handle and yet retains all the characteristic features of the problem. First we arrange the matrices $[a]$ and $[I]$ side by side:

$$[a] = \begin{bmatrix} 5 & 3 & 1 \\ 3 & 6 & 2 \\ 1 & 2 & 3 \end{bmatrix}, \qquad [I] = \begin{bmatrix} 1 & 0 & 0 \\ 0 & 1 & 0 \\ 0 & 0 & 1 \end{bmatrix}. \qquad (A.48)$$

Next divide the first row by 5 and the second row by 3, leave the third row intact, and subtract the first resulting row from the second and third resulting rows. These operations yield

$$\begin{bmatrix} 1 & \frac{3}{5} & \frac{1}{5} \\ 0 & \frac{7}{5} & \frac{7}{15} \\ 0 & \frac{7}{5} & \frac{14}{5} \end{bmatrix}, \qquad \begin{bmatrix} \frac{1}{5} & 0 & 0 \\ -\frac{1}{5} & \frac{1}{3} & 0 \\ -\frac{1}{5} & 0 & 1 \end{bmatrix}.$$

Now multiply the second and third rows by $\frac{5}{7}$ and subtract the second resulting row from the third resulting one, to obtain

$$\begin{bmatrix} 1 & \frac{3}{5} & \frac{1}{5} \\ 0 & 1 & \frac{1}{3} \\ 0 & 0 & \frac{5}{3} \end{bmatrix}, \qquad \begin{bmatrix} \frac{1}{5} & 0 & 0 \\ -\frac{1}{7} & \frac{5}{21} & 0 \\ 0 & -\frac{5}{21} & \frac{5}{7} \end{bmatrix}.$$

Multiply the third row by $\frac{3}{5}$; then subtract $\frac{1}{3}$ times the third resulting row from the second and $\frac{1}{5}$ times the third resulting row from the first, so that

$$\begin{bmatrix} 1 & \frac{3}{5} & 0 \\ 0 & 1 & 0 \\ 0 & 0 & 1 \end{bmatrix}, \qquad \begin{bmatrix} \frac{1}{5} & \frac{1}{35} & -\frac{3}{35} \\ -\frac{1}{7} & \frac{2}{7} & -\frac{1}{7} \\ 0 & -\frac{1}{7} & \frac{3}{7} \end{bmatrix}.$$

Finally $\frac{3}{5}$ times the second row is subtracted from the first row, to give

$$[I] = \begin{bmatrix} 1 & 0 & 0 \\ 0 & 1 & 0 \\ 0 & 0 & 1 \end{bmatrix}, \qquad [a]^{-1} = \frac{1}{7}\begin{bmatrix} 2 & -1 & 0 \\ -1 & 2 & -1 \\ 0 & -1 & 3 \end{bmatrix}. \qquad (A.49)$$

A-7 TRANSPOSITION AND RECIPROCATION OF PRODUCTS OF MATRICES

It was shown in Section A–3 that

$$\underset{m \times n}{[a]} \quad \underset{n \times p}{[b]} \quad = \quad \underset{m \times p}{[c]} \qquad (A.50)$$

implies the operation

$$c_{ij} = \sum_{k=1}^{n} a_{ik} b_{kj}. \tag{A.51}$$

Now consider the product $[b]^T[a]^T$. Any term a_{ik} in the matrix $[a]$ is equal to the term a_{ki} in the matrix $[a]^T$. Similarly, any term b_{kj} in the matrix $[b]$ is equal to the term b_{jk} in the matrix $[b]^T$. But

$$\sum_{k=1}^{n} b_{jk} a_{ki} = c_{ji}, \tag{A.52}$$

from which we conclude that

$$[b]^T[a]^T = [c]^T. \tag{A.53}$$

In general, it can be shown that if

$$[a][b][c] \cdots [s] = [t], \tag{A.54}$$

then

$$[t]^T = [s]^T \cdots [c]^T[b]^T[a]^T, \tag{A.55}$$

or the transpose of a continuous product of matrices is equal to the continuous product of the transposed matrices *in reversed order*.

Consider again the product

$$[a][b] = [c], \tag{A.56}$$

and premultiply both sides by $[b]^{-1}[a]^{-1}$. The result is

$$[b]^{-1}[a]^{-1}[a][b] = [b]^{-1}[I][b] = [b]^{-1}[b] = [I]$$
$$= [b]^{-1}[a]^{-1}[c]. \tag{A.57}$$

Now postmultiply (A.57) by $[c]^{-1}$. The result is

$$[I][c]^{-1} = [c]^{-1} = [b]^{-1}[a]^{-1}[c][c]^{-1}$$
$$= [b]^{-1}[a]^{-1}[I] = [b]^{-1}[a]^{-1}. \tag{A.58}$$

Hence

$$[c]^{-1} = [b]^{-1}[a]^{-1}. \tag{A.59}$$

In general, if matrix $[t]$ is as given by (A.54), its reciprocal is

$$[t]^{-1} = [s]^{-1} \cdots [c]^{-1}[b]^{-1}[a]^{-1}, \tag{A.60}$$

or the reciprocal of a continuous product of matrices is equal to the continuous product of the reciprocals *in reversed order*.

A-8 LINEAR TRANSFORMATIONS

Consider a set of linear equations

$$y_1 = a_{11}x_1 + a_{12}x_2 + a_{13}x_3 + \cdots + a_{1n}x_n = \sum_{j=1}^{n} a_{1j}x_j,$$

$$y_2 = a_{21}x_1 + a_{22}x_2 + \cdots\cdots\cdots + a_{2n}x_n = \sum_{j=1}^{n} a_{2j}x_j, \qquad \text{(A.61)}$$

$$\vdots \qquad \vdots \qquad \vdots \qquad\qquad \vdots \qquad \vdots$$

$$y_n = a_{n1}x_1 + a_{n2}x_2 + \cdots\cdots\cdots + a_{nn}x_n = \sum_{j=1}^{n} a_{nj}x_j.$$

These equations may be written in the condensed form

$$y_i = \sum_{j=1}^{n} a_{ij}x_j = a_{ij}x_j, \qquad i = 1, 2, \ldots, n, \qquad \text{(A.62)}$$

in which a repeated index implies summation.

Equations (A.62) may be cast in a matrix form,

$$\{y\} = [a]\{x\}, \qquad \text{(A.63)}$$

where the column matrices $\{y\}$ and $\{x\}$ may be interpreted as n-dimensional vectors. Equation (A.63) implies that a matrix $[a]$ operating on a vector $\{x\}$ transforms the vector $\{x\}$ into the vector $\{y\}$.

Coordinate transformations of this type are of vital importance in the study of vibrations.

Selected Readings

Frazer, R. A., W. J. Duncan, and A. R. Collar, *Elementary Matrices*, Cambridge Univ. Press, New York, 1957.

Hildebrand, F. B., *Methods of Applied Mathematics*, Prentice-Hall, Englewood Cliffs, N.J., 1960.

Hohn, F. E., *Elementary Matrix Algebra*, Macmillan, New York, 1958.

ELEMENTS OF
LAPLACE
TRANSFORMATION

B–1 INTEGRAL TRANSFORMATIONS. GENERAL DISCUSSION

If a function $f(t)$ is defined by a differential equation and some initial conditions (or boundary conditions), under certain circumstances it may be simpler to obtain a solution of the problem by translating the problem for $f(t)$ into a problem for $\bar{f}(s)$ given by the integral transformation

$$\bar{f}(s) = \int_a^b f(t)K(s, t) \, dt, \tag{B.1}$$

where $K(s, t)$ is a given function of s and t called the *kernel* of the transformation. When the limits a and b are finite, $\bar{f}(s)$ is a finite transform.

If the problem consists of an ordinary differential equation, the dependent variable $f(t)$ is a function of only one variable, t. In such a case an integral transformation converts the differential equation into an algebraic equation in terms of the transformed function $\bar{f}(s)$, where s is a parameter. In the process initial conditions (or boundary conditions) are accounted for. The algebraic equation can be solved for $\bar{f}(s)$ without much difficulty and the function $f(t)$ is obtained by performing an inverse transformation.

Integral transform methods can also be used to solve boundary-value problems defined by partial differential equations. In such a case one transformation reduces the number of independent variables by one. Hence, if two independent variables are involved, instead of solving a partial differential equation one must solve an ordinary differential equation, which, in general, is a considerably easier task. The main difficulty in using integral transform methods is to carry out the inverse transformation, which, for the most part, involves evaluation of an integral. The transform and its inverse are called a *transform pair*, and for many transforms in use there are tables of transform pairs available. In many cases it is possible to find the inverse transformation in the corresponding transform tables, thus eliminating the need to evaluate an inversion integral.

Two of the most widely used integral transformations are the Laplace and Fourier transforms. The first one is discussed in this appendix and the second one will be presented in Appendix C.

B-2   THE LAPLACE TRANSFORMATION. DEFINITION

The Laplace transform method provides a most convenient means of solving linear ordinary differential equations with constant coefficients. This type of equation arises frequently in the study of oscillations of discrete linear systems. The method can be used successfully also in the case of linear partial differential equations describing the response of continuous systems.

The same type of problem was treated by Heaviside by means of his operational calculus. Much of Heaviside's work was based on his intuition and the mathematical treatment was often obscure. The Laplace transform method, although similar to Heaviside's operational calculus, is mathematically rigorous. The method accounts automatically for the initial conditions and it provides a great deal of insight into the physical system.

Let a function $f(t)$ be given for all values of time larger than zero, $t > 0$. By definition, the Laplace transformation of $f(t)$ is given by

$$\mathscr{L}f(t) = \bar{f}(s) = \int_0^\infty e^{-st} f(t)\, dt. \qquad (B.2)$$

We note that the kernel of the transformation is $K(s, t) = e^{-st}$, where s is a subsidiary variable which in general is a complex quantity. The complex plane $s = x + iy$ will be referred to as the s plane.

The function $f(t)$ is subject to certain restrictions, because the integral, (B.1), must converge. If $f(t)$ is such that

$$|e^{-st} f(t)| < Ce^{-(s-a)t}, \qquad \text{Re } s > a, \qquad (B.3)$$

where C is a constant and Re s denotes the real part of s, then the Laplace transformation of $f(t)$ exists. Condition (B.3) implies that $f(t)$ does not increase more rapidly than Ce^{at} with increasing t. Such a function $f(t)$ is said to be of *exponential order* and is denoted $f(t) = O(e^{at})$.

Another condition of the existence of Laplace transformation is that $f(t)$ be piecewise continuous, which means that in a given interval it has a finite number of finite discontinuities and no infinite discontinuity.

Fortunately most functions describing physical phenomena satisfy these conditions.

B-3   FIRST SHIFTING THEOREM
(in the Complex Plane)

Consider

$$F(t) = f(t)e^{at}, \tag{B.4}$$

where a may be a real or complex number, and evaluate its Laplace transformation:

$$\bar{F}(s) = \int_0^\infty [f(t)e^{at}]e^{-st}\, dt$$

$$= \int_0^\infty f(t)e^{-(s-a)t}\, dt = \bar{f}(s-a). \tag{B.5}$$

It follows that

$$\mathscr{L}f(t)e^{at} = \bar{f}(s-a). \tag{B.6}$$

Hence the effect of multiplying $f(t)$ by e^{at} in the real domain is to shift the transform of $f(t)$ by an amount a in the s domain. Since the s domain is the complex plane, this theorem is also called the *complex shifting theorem*.

B-4   TRANSFORMATION OF DERIVATIVES

Suppose the Laplace transformation of $f(t)$ exists, denote $\lim_{t\to 0} f(t) = f(0)$, and calculate the Laplace transformation of the first derivative of $f(t)$, $df(t)/dt$. Integrating by parts we can write

$$\mathscr{L}\frac{df(t)}{dt} = \int_0^\infty e^{-st}\frac{df(t)}{dt}\, dt = e^{-st}f(t)\Big|_0^\infty$$

$$- \int_0^\infty (-se^{-st})f(t)\, dt = -f(0) + s\bar{f}(s). \tag{B.7}$$

In general, for the nth derivative of $f(t)$ we obtain

$$\mathcal{L}\frac{d^n f(t)}{dt^n} = \mathcal{L}f^{(n)}(t) = -f^{(n-1)}(0) - sf^{(n-2)}(0)$$
$$-s^2 f^{(n-3)}(0) - \cdots - s^{n-1}f(0) + s^n \bar{f}(s), \qquad (B.8)$$

where the following notation was adopted:

$$\frac{d^{(n-j)}f(t)}{dt^{n-j}}\bigg|_{t=0} = f^{(n-j)}(0). \qquad (B.9)$$

Equation (B.8) is valid only if $f(t)$ and all its derivatives through the $(n-1)$st are continuous.

B-5 TRANSFORMATION OF ORDINARY DIFFERENTIAL EQUATIONS

Consider the differential equation governing the motion $x(t)$ of a mass on a spring and viscous damper in parallel. The mass is subjected to an exciting force $F(t)$. The equation is

$$m\frac{d^2 x(t)}{dt^2} + c\frac{dx(t)}{dt} + kx(t) = F(t). \qquad (B.10)$$

One can transform both sides of (B.10). Using results obtained in Section B-4 we write

$$m[s^2\bar{x}(s) - sx(0) - \dot{x}(0)] + c[s\bar{x}(s) - x(0)] + k\bar{x}(s) = \bar{F}(s), \qquad (B.11)$$

where $x(0)$ and $\dot{x}(0)$ are the initial displacement and velocity. Equation (B.11), solved for $\bar{x}(s)$, yields

$$\bar{x}(s) = \frac{1}{ms^2 + cs + k}\bar{F}(s) + \frac{ms + c}{ms^2 + cs + k}x(0)$$
$$+ \frac{m}{ms^2 + cs + k}\dot{x}(0). \qquad (B.12)$$

Equation (B.12) is called the *subsidiary equation of the differential equation*. To obtain the response $x(t)$ one must evaluate the inverse transformation of $\bar{x}(s)$. Note that the Laplace transformation method provides automatically for initial conditions.

B–6   THE INVERSE TRANSFORMATION

Equation (B.12) gives an expression $\bar{x}(s)$, which is a function of s. One must now evaluate the inverse transformation, which is denoted symbolically

$$\mathscr{L}^{-1}\bar{x}(s) = x(t), \tag{B.13}$$

meaning that the inverse transformation of $\bar{x}(s)$ is $x(t)$.

In general the operation $\mathscr{L}^{-1}\bar{f}(s)$ involves the evaluation of the integral

$$f(t) = \mathscr{L}^{-1}\bar{f}(s) = \frac{1}{2\pi i}\int_{\gamma-i\infty}^{\gamma+i\infty} e^{st}\bar{f}(s)\,ds, \tag{B.14}$$

where the path of integration is a line parallel to the imaginary axis crossing the real axis at $\text{Re } s = \gamma$ and extending from $-\infty$ to $+\infty$.

In many cases, however, one can go a long way in using the Laplace transformation without having to recourse to line integrals. This is the case when Jordan's lemma can be used to replace the line integral by a closed contour integral, which, in turn, can be evaluated by means of the residue theorem.† In other cases it may be even simpler to obtain the inverse transformation from tables of Laplace transform pairs. Quite often, although the function is not in the tables, it can be reduced to a form to be found in tables. Hence, to increase the possibility of using tables, one must study ways to render seemingly complicated functions into a form listed in tables. To this end we shall consider the partial-fractions method.

B–7   METHOD OF PARTIAL FRACTIONS

Suppose that one is interested in finding the inverse transformation of the function

$$\bar{f}(s) = \frac{A(s)}{B(s)}, \tag{B.15}$$

where both $A(s)$ and $B(s)$ are polynomials in s. Note that (B.12) is of this type. In general $B(s)$ is a polynomial of higher order than $A(s)$.

Let $B(s)$ be a polynomial of nth order with n roots a_k such that it may be written

$$B(s) = (s - a_1)(s - a_2)\cdots(s - a_k)\cdots(s - a_n), \tag{B.16}$$

† See F. B. Hildebrand, *Advanced Calculus for Engineers*, Prentice-Hall, Englewood Cliffs, N.J., 1960, p. 569 (Prob. 34); see also Section 11–12 of this book.

and consider first the case in which all n roots are distinct. Equation (B.15) can be written

$$\bar{f}(s) = \frac{A(s)}{B(s)} = \frac{C_1}{s - a_1} + \frac{C_2}{s - a_2} + \cdots + \frac{C_k}{s - a_k} + \cdots + \frac{C_n}{s - a_n} \quad \text{(B.17)}$$

where C_k are coefficients given by

$$C_k = \lim_{s \to a_k} [(s - a_k)\bar{f}(s)] = \frac{A(s)}{B'(s)} \bigg|_{s = a_k} \quad \text{(B.18)}$$

in which B' is the derivative of B with respect to s. Because†

$$\mathscr{L}^{-1} \frac{1}{s - a_k} = e^{a_k t}, \quad \text{(B.19)}$$

(B.17) can be readily inverted to obtain

$$f(t) = \mathscr{L}^{-1}\bar{f}(s) = C_1 e^{a_1 t} + C_2 e^{a_2 t} + \cdots + C_n e^{a_n t}$$

$$= \sum_{k=1}^{n} C_k e^{a_k t} = \sum_{k=1}^{n} \lim_{s \to a_k} [(s - a_k)\bar{f}(s)e^{st}] = \sum_{k=1}^{n} \frac{A(s)}{B'(s)} e^{st} \bigg|_{s = a_k}. \quad \text{(B.20)}$$

The points $a_1, a_2, \ldots, a_n$ are called *simple poles* of $\bar{f}(s)$. Equation (B.20) may be used for functions other than ratios of two polynomials, provided the function has simple poles. Poles are points at which the function $\bar{f}(s)$ becomes infinite.

Consider now the case in which $B(s)$ has a multiple root of order k. In other words, $\bar{f}(s)$ has a pole of order k as opposed to poles of first order, or simple poles, examined previously. Hence consider

$$B(s) = (s - a_1)^k (s - a_2)(s - a_3) \cdots (s - a_n). \quad \text{(B.21)}$$

The partial-fractions expansion in this case is of the form

$$\bar{f}(s) = \frac{A(s)}{B(s)} = \frac{C_{11}}{(s - a_1)^k} + \frac{C_{12}}{(s - a_1)^{k-1}} + \cdots + \frac{C_{1k}}{s - a_1}$$

$$+ \frac{C_2}{s - a_2} + \frac{C_3}{s - a_3} + \cdots + \frac{C_n}{s - a_n}. \quad \text{(B.22)}$$

One can easily show that, in general, for the repeated root

$$C_{1r} = \frac{1}{(r - 1)!} \frac{d^{r-1}}{ds^{r-1}} [(s - a_1)^k \bar{f}(s)]_{s = a_1}, \quad r = 1, 2, \ldots, k. \quad \text{(B.23)}$$

† Note that $\mathscr{L}1 = 1/s$. Hence, by the *first shifting theorem*, $\mathscr{L}e^{a_k t} = 1/(s - a_k)$. Equation (B.19) follows directly from this result.

The simple poles of $\bar{f}(s)$ are treated, as previously, by means of (B.18). One can easily show that

$$\mathscr{L}^{-1} \frac{1}{(s - a_1)^r} = \frac{t^{r-1}}{(r - 1)!} e^{a_1 t}, \tag{B.24}$$

so that the inverse of (B.22) becomes

$$f(t) = \left[C_{11} \frac{t^{k-1}}{(k - 1)!} + C_{12} \frac{t^{k-2}}{(k - 2)!} + \cdots + C_{1k} \right] e^{a_1 t}$$
$$+ C_2 e^{a_2 t} + C_3 e^{a_3 t} + \cdots + C_n e^{a_n t}. \tag{B.25}$$

Equation (B.25) can be shown to be equal to

$$f(t) = \frac{1}{(k - 1)!} \frac{d^{k-1}}{ds^{k-1}} [(s - a_1)^k \bar{f}(s) e^{st}]_{s = a_1}$$
$$+ \sum_{i=2}^{n} [(s - a_i)\bar{f}(s)e^{st}]_{s = a_i}. \tag{B.26}$$

B–8   SECOND SHIFTING THEOREM (in the Real Domain)

Consider the Laplace transformation

$$\bar{f}(s) = \int_0^\infty e^{-s\lambda} f(\lambda) \, d\lambda \tag{B.27}$$

and let

$$\lambda = t - a, \tag{B.28}$$

so that

$$\bar{f}(s) = \int_0^\infty e^{-s(t-a)} f(t - a) \, dt = e^{as} \int_0^\infty e^{-st} [f(t - a)u(t - a)] \, dt, \tag{B.29}$$

or

$$e^{-as} \bar{f}(s) = \int_0^\infty e^{-st} [f(t - a)u(t - a)] \, dt. \tag{B.30}$$

Theorem. *If $\bar{f}(s)$ is the Laplace transform of $f(t)$,*

$$\mathscr{L}^{-1} e^{-as} \bar{f}(s) = f(t - a)u(t - a). \tag{B.31}$$

B-9   THE CONVOLUTION INTEGRAL. BOREL'S THEOREM

Consider two functions $f_1(t)$ and $f_2(t)$ both defined only for $t > 0$. Assume that $f_1(t)$ and $f_2(t)$ possess the Laplace transformations $\bar{f}_1(s)$ and $\bar{f}_2(s)$, respectively, and consider the integral

$$x(t) = \int_0^t f_1(\tau)f_2(t - \tau)\, d\tau = \int_0^\infty f_1(\tau)f_2(t - \tau)\, d\tau. \qquad (B.32)$$

The function $x(t)$ given by these integrals, sometimes denoted $x(t) = f_1(t)*f_2(t)$, is called the *convolution*, or *Faltung*, of the functions f_1 and f_2 over the interval $(0, \infty)$. The above is true because $f_2(t - \tau) = 0$ for $\tau > t$ or $t - \tau < 0$. Transforming both sides of (B.32) we obtain

$$\bar{x}(s) = \int_0^\infty e^{-st}\left[\int_0^\infty f_1(\tau)f_2(t - \tau)\, d\tau\right] dt$$

$$= \int_0^\infty f_1(\tau)\, d\tau \int_0^\infty e^{-st}f_2(t - \tau)\, dt$$

$$= \int_0^\infty f_1(\tau)\, d\tau \int_\tau^\infty e^{-st}f_2(t - \tau)\, dt. \qquad (B.33)$$

Note that the limit in the second integral was changed, because $f_2(t - \tau) = 0$ for $t < \tau$.

Next denote $t - \tau = \lambda$ in the second integral and, noting that for $t = \tau$ we have $\lambda = 0$, write

$$\bar{x}(s) = \int_0^\infty f_1(\tau)\, d\tau \int_\tau^\infty e^{-st}f_2(t - \tau)\, dt$$

$$= \int_0^\infty f_1(\tau)\, d\tau \int_0^\infty e^{-s(\tau + \lambda)}f_2(\lambda)\, d\lambda$$

$$= \int_0^\infty e^{-s\tau}f_1(\tau)\, d\tau \int_0^\infty e^{-s\lambda}f_2(\lambda)\, d\lambda = \bar{f}_1(s)\bar{f}_2(s). \qquad (B.34)$$

From (B.32) and (B.34) it follows that

$$x(t) = \mathscr{L}^{-1}\bar{x}(s) = \mathscr{L}^{-1}\bar{f}_1(s)\bar{f}_2(s)$$

$$= \int_0^t f_1(\tau)f_2(t - \tau)\, d\tau = \int_0^t f_1(t - \tau)f_2(\tau)\, d\tau. \qquad (B.35)$$

This is true because it does not matter which of the functions $f_1(t)$ and $f_2(t)$ is shifted. The integrals in (B.35) are called *convolution integrals*.†

---

† A special case of the convolution integral is discussed in Section 1–7, without any reference to Laplace transformation.

Theorem. *The inverse transformation of the product of two transforms is equal to the convolution of their inverse transformations.*

Selected Readings

Churchill, R. V., *Operational Mathematics*, 2nd ed., McGraw-Hill, New York, 1958.

Hildebrand, F. B., *Advanced Calculus for Engineers*, Prentice-Hall, Englewood Cliffs, N.J., 1960.

Thomson, W. T., *Laplace Transformation*, Prentice-Hall, Englewood Cliffs, N.J., 1960.

Tranter, C. J., *Integral Transforms in Mathematical Physics*, Methuen, London, 1959.

ELEMENTS OF
FOURIER
TRANSFORMATIONS

C–1 GENERAL DEFINITIONS

Appendix B contains a general discussion of integral transforms as well as an introduction to the Laplace transformation. Other types of integral transformations widely used are the Fourier transformations. The general discussion of Section B–1 applies equally well to the Fourier transforms.

We shall be concerned with several varieties of Fourier transformations: *Fourier sine transform*, defined by

$$\bar{F}_s(p) = \int_0^\infty f(x) \sin px \, dx. \qquad (C.1)$$

Fourier cosine transform, defined by

$$\bar{F}_c(p) = \int_0^\infty f(x) \cos px \, dx. \qquad (C.2)$$

Complex Fourier transform, defined by

$$\bar{F}(p) = \int_{-\infty}^\infty f(x) e^{ipx} \, dx. \qquad (C.3)$$

The conditions for the existence of the above transformations will be

536

discussed later. When the limits of the integral are finite the transforms are
called finite. We shall discuss the finite transformations:

Finite sine transform, defined by

$$\bar{f}_s(p) = \int_0^\pi f(x) \sin px \, dx. \tag{C.4}$$

Finite cosine transform, defined by

$$\bar{f}_c(p) = \int_0^\pi f(x) \cos px \, dx. \tag{C.5}$$

In the case of finite transforms, p is a positive integer and the inverse
transformation consists of summations rather than integrations.

C-2 FOURIER INTEGRAL FORMULA

Let a function be represented by a Fourier series of period $2\pi\lambda$,

$$f(x) = \tfrac{1}{2}a_0 + \sum_{n=1}^\infty \left(a_n \cos \frac{nx}{\lambda} + b_n \sin \frac{nx}{\lambda}\right). \tag{C.6}$$

The coefficients a_0, a_n, and b_n are called Fourier coefficients and are given by

$$a_0 = \frac{1}{\pi\lambda} \int_{-\pi\lambda}^{\pi\lambda} f(\xi) \, d\xi,$$

$$a_n = \frac{1}{\pi\lambda} \int_{-\pi\lambda}^{\pi\lambda} f(\xi) \cos \frac{n\xi}{\lambda} \, d\xi, \tag{C.7}$$

$$b_n = \frac{1}{\pi\lambda} \int_{-\pi\lambda}^{\pi\lambda} f(\xi) \sin \frac{n\xi}{\lambda} \, d\xi.$$

Introducing (C.7) in (C.6) we obtain

$$f(x) = \frac{1}{2\pi\lambda} \int_{-\pi\lambda}^{\pi\lambda} f(\xi) \, d\xi$$

$$+ \frac{1}{\pi\lambda} \sum_{n=1}^\infty \int_{-\pi\lambda}^{\pi\lambda} f(\xi) \left[\cos \frac{n\xi}{\lambda} \cos \frac{nx}{\lambda} + \sin \frac{n\xi}{\lambda} \sin \frac{nx}{\lambda}\right] d\xi$$

$$= \frac{1}{2\pi\lambda} \int_{-\pi\lambda}^{\pi\lambda} f(\xi) \, d\xi + \frac{1}{\pi\lambda} \sum_{n=1}^\infty \int_{-\pi\lambda}^{\pi\lambda} f(\xi) \cos \frac{n(x - \xi)}{\lambda} \, d\xi. \tag{C.8}$$

Next let $n/\lambda = \alpha$ and $1/\lambda = \Delta\alpha$ and take the limit by letting $\lambda \to \infty$, to obtain

$$
f(x) = \lim_{\substack{\Delta\alpha \to 0 \\ \lambda \to \infty}} \left[\frac{1}{2\pi} \Delta\alpha \int_{-\pi\lambda}^{\pi\lambda} f(\xi)\, d\xi + \frac{1}{\pi} \sum_{n=1}^{\infty} \Delta\alpha \int_{-\pi\lambda}^{\pi\lambda} f(\xi) \cos n\, \Delta\alpha(x - \xi)\, d\xi \right]
$$

$$
= \frac{1}{\pi} \int_0^\infty \left[\int_{-\infty}^\infty f(\xi) \cos \alpha(x - \xi)\, d\xi \right] d\alpha, \tag{C.9}
$$

which is the *Fourier integral formula*.

Note that the above is true only if the integral $\int_{-\infty}^\infty |f(\xi)|\, d\xi$ exists. In addition in any given interval (a, b), the function $f(x)$ must satisfy *Dirichlet's conditions*: (1) the function $f(x)$ must have only a finite number of maxima and minima in the interval (a, b); and (2) the function $f(x)$ must have only a finite number of finite discontinuities and no infinite discontinuities in the interval (a, b). For example, the function $f(x) = \sin(1/x)$ has an infinite number of maxima and minima in the neighborhood $x = 0$, so it does not satisfy Dirichlet's conditions.

C-3 INVERSION FORMULAS

In (C.9) let $\alpha = p$ and expand to obtain

$$
f(x) = \frac{1}{\pi} \int_0^\infty \left[\int_{-\infty}^\infty f(\xi) \cos p\xi\, d\xi \right] \cos xp\, dp
$$

$$
+ \frac{1}{\pi} \int_0^\infty \left[\int_{-\infty}^\infty f(\xi) \sin p\xi\, d\xi \right] \sin xp\, dp. \tag{C.10}
$$

If $f(x)$ is an odd function,

$$
\int_{-\infty}^\infty f(\xi) \cos p\xi\, d\xi = 0, \tag{C.11}
$$

$$
\int_{-\infty}^\infty f(\xi) \sin p\xi\, d\xi = 2 \int_0^\infty f(\xi) \sin p\xi\, d\xi = 2\bar{F}_s(p). \tag{C.12}
$$

Therefore, the inversion formula for the Fourier sine transform is

$$
f(x) = \frac{2}{\pi} \int_0^\infty \bar{F}_s(p) \sin xp\, dp. \tag{C.13}
$$

Similarly, if $f(x)$ is an even function we obtain the inversion formula for the Fourier cosine transform,

$$
f(x) = \frac{2}{\pi} \int_0^\infty \bar{F}_c(p) \cos xp\, dp. \tag{C.14}
$$

For the inversion formula of the complex Fourier transform we return to (C.9). The integral

$$I(\alpha, x) = \int_{-\infty}^{\infty} f(\xi) \sin \alpha(x - \xi) \, d\xi \qquad (C.15)$$

is an odd function of α. Hence

$$\int_{-\infty}^{\infty} I(\alpha, x) \, d\alpha = \int_{-\infty}^{\infty} \left[\int_{-\infty}^{\infty} f(\xi) \sin \alpha(x - \xi) \, d\xi \right] d\alpha = 0. \qquad (C.16)$$

Multiply (C.16) by i and subtract the result from (C.9) to obtain

$$\begin{aligned}
f(x) &= \frac{1}{\pi} \int_{0}^{\infty} \left[\int_{-\infty}^{\infty} f(\xi) \cos \alpha(x - \xi) \, d\xi \right] d\alpha \\
&= \frac{1}{2\pi} \int_{-\infty}^{\infty} \left\{ \int_{-\infty}^{\infty} f(\xi)[\cos \alpha(x - \xi) - i \sin \alpha(x - \xi)] \, d\xi \right\} d\alpha \\
&= \frac{1}{2\pi} \int_{-\infty}^{\infty} e^{-i\alpha x} \left[\int_{-\infty}^{\infty} f(\xi)e^{i\alpha\xi} \, d\xi \right] d\alpha.
\end{aligned} \qquad (C.17)$$

Let $\alpha = p$ in (C.17), recall (C.3), and write the inversion formula for the complex Fourier transform,

$$f(x) = \frac{1}{2\pi} \int_{-\infty}^{\infty} \bar{F}(p)e^{-ipx} \, dp. \qquad (C.18)$$

C-4 FOURIER TRANSFORMS OF DERIVATIVES OF FUNCTIONS

To apply the Fourier transformation method to differential equations it is necessary to develop expressions for transforms of derivatives. Denote by $\bar{F}^{(r)}(p)$ the Fourier transform of the rth derivative of $f(x)$, perform an integration by parts, and obtain

$$\bar{F}^{(r)}(p) = \int_{-\infty}^{\infty} \frac{d^r f(x)}{dx^r} e^{ipx} \, dx = e^{ipx} \frac{d^{r-1}f(x)}{dx^{r-1}} \Big|_{-\infty}^{\infty} - ip\bar{F}^{(r-1)}(p). \qquad (C.19)$$

If the function $f(x)$ is such that its $(r-1)$st derivative vanishes as $x \to \pm\infty$, it follows that

$$\bar{F}^{(r)}(p) = -ip\bar{F}^{(r-1)}(p). \qquad (C.20)$$

Assuming all the derivatives of $f(x)$ through the $(r-1)$st vanish as $x \to \pm\infty$, we obtain

$$\bar{F}^{(r)}(p) = (-ip)^r \bar{F}(p), \tag{C.21}$$

where $\bar{F}(p)$ is the complex Fourier transform of $f(x)$.

For the Fourier cosine transform of the rth derivative of $f(x)$ we obtain

$$\bar{F}_c^{(r)}(p) = \int_0^\infty \frac{d^r f(x)}{dx^r} \cos px \, dx = -a_{r-1} + p\bar{F}_s^{(r-1)}(p), \tag{C.22}$$

where
$$\lim_{x \to \infty} \frac{d^{r-1} f(x)}{dx^{r-1}} = 0, \qquad \lim_{x \to 0} \frac{d^{r-1} f(x)}{dx^{r-1}} = a_{r-1} \tag{C.23}$$

and $\bar{F}_s^{(r-1)}(p)$ is the Fourier sine transform of the $(r-1)$st derivative of $f(x)$. From (C.22) we obtain

$$\bar{F}_c^{(r)}(p) = -a_{r-1} - p^2 \bar{F}_c^{(r-2)}(p). \tag{C.24}$$

It must be noted from (C.22) and (C.24) that the formula obtained when r is an even number will be different than the formula obtained when r is an odd number. Hence the general formulas can be written

$$\bar{F}_c^{(2r)}(p) = -\sum_{n=0}^{r-1} (-1)^n a_{2r-2n-1} p^{2n} + (-1)^r p^{2r} \bar{F}_c(p), \tag{C.25}$$

$$\bar{F}_s^{(2r+1)}(p) = -\sum_{n=0}^{r} (-1)^n a_{2r-2n} p^{2n} + (-1)^r p^{2r+1} \bar{F}_s(p). \tag{C.26}$$

Similarly, for the Fourier sine transforms we obtain

$$\bar{F}_s^{(2r)}(p) = -\sum_{n=1}^{r} (-1)^n a_{2r-2n} p^{2n-1} + (-1)^r p^{2r} \bar{F}_s(p), \tag{C.27}$$

$$\bar{F}_s^{(2r+1)}(p) = -\sum_{n=1}^{r} (-1)^n a_{2r-2n+1} p^{2n-1} + (-1)^r p^{2r+1} \bar{F}_c(p). \tag{C.28}$$

Evidently the choice of the transform to be used depends on the boundary conditions of the problem.

C-5   FINITE FOURIER TRANSFORMS

The previous three types of transforms may be used when one or both boundaries are at infinity. Assuming that a boundary is at infinity is a mathematical expediency that eliminates the need of prescribing the corresponding boundary conditions. Many times the domains are of finite size, in which case if a transform method is to be used the transform must be *finite*. To this end we wish to define the *finite sine transform* as

$$\bar{f}_s(p) = \int_0^\pi f(x) \sin px \, dx, \tag{C.29}$$

where p is a positive integer. The choice of π as the upper limit will become self-evident shortly.

Next, assume that $f(x)$ can be expanded in a Fourier sine series,

$$f(x) = \sum_{p=1}^\infty a_p \sin px, \tag{C.30}$$

where the coefficients a_p are determined by introducing (C.30) in (C.29) and writing

$$\bar{f}_s(p) = \int_0^\pi \left(\sum_{r=1}^\infty a_r \sin rx \right) \sin px \, dx$$

$$= \sum_{r=1}^\infty a_r \int_0^\pi \sin rx \sin px \, dx = \sum_{r=1}^\infty a_r \frac{\pi}{2} \delta_{rp} = \frac{\pi}{2} a_p. \tag{C.31}$$

Hence the inversion formula for the finite sine transform is

$$f(x) = \frac{2}{\pi} \sum_{p=1}^\infty \bar{f}_s(p) \sin px. \tag{C.32}$$

Similarly, one can define the *finite cosine transform* as

$$\bar{f}_c(p) = \int_0^\pi f(x) \cos px \, dx, \tag{C.33}$$

which has the inversion formula

$$f(x) = \frac{1}{\pi} \bar{f}_c(0) + \frac{2}{\pi} \sum_{p=1}^\infty \bar{f}_c(p) \cos px, \tag{C.34}$$

where

$$\bar{f}_c(0) = \int_0^\pi f(x) \, dx. \tag{C.35}$$

Again the decision as to the use of the sine or the cosine transformation is dictated by the boundary conditions. The variable range is $0 \le x \le \pi$, and within this range the function $f(x)$ must satisfy Dirichlet's conditions. At any point of discontinuity the left side of (C.32) and (C.34) should be replaced by the average $\frac{1}{2}[f(x + 0) + f(x - 0)]$.

Next, let the range be $0 \leq x \leq L$ and change the parameter p accordingly. The finite sine transform is redefined in the form

$$\bar{f}_s(n) = \int_0^L f(x) \sin \frac{n\pi x}{L} \, dx, \tag{C.36}$$

where n is a positive integer. It is not difficult to show that the corresponding inversion formula is

$$f(x) = \frac{2}{L} \sum_{n=1}^{\infty} \bar{f}_s(n) \sin \frac{n\pi x}{L}. \tag{C.37}$$

In a similar way, we redefine the finite cosine transform as

$$\bar{f}_c(n) = \int_0^L f(x) \cos \frac{n\pi x}{L} \, dx \tag{C.38}$$

with the inversion formula

$$f(x) = \tfrac{1}{2}\bar{f}_c(0) + \frac{2}{L} \sum_{n=1}^{\infty} \bar{f}_c(n) \cos \frac{n\pi x}{L}, \tag{C.39}$$

where

$$\bar{f}_c(0) = \int_0^L f(x) \, dx. \tag{C.40}$$

The new independent variable x, which now has units of length rather than radians, must satisfy Dirichlet's conditions. The finite transformation as given by (C.36) and (C.38) are most conveniently used when the transformation of the spatial variable is desired.

C-6 FINITE FOURIER TRANSFORMS OF DERIVATIVES OF FUNCTIONS

Using integration by parts, the following finite sine transforms, of even- and odd-order derivatives, respectively, can be obtained:

$$\bar{f}_s^{(2r)}(n) = -\sum_{j=1}^{r} (-1)^j \left(\frac{n\pi}{L}\right)^{2j-1} [(-1)^{n+1}f^{(2r-2j)}(L) + f^{(2r-2j)}(0)]$$
$$+ (-1)^r \left(\frac{n\pi}{L}\right)^{2r} \bar{f}_s(n), \tag{C.41}$$

$$\bar{f}_s^{(2r+1)}(n) = -\sum_{j=1}^{r} (-1)^j \left(\frac{n\pi}{L}\right)^{2j-1} [(-1)^{n+1}f^{(2r-2j+1)}(L) + f^{(2r-2j+1)}(0)]$$
$$+ (-1)^{r+1} \left(\frac{n\pi}{L}\right)^{2r+1} \bar{f}_c(n). \tag{C.42}$$

Similarly, the finite cosine transforms of even- and odd-order derivatives are, respectively,

$$\bar{f}_c^{(2r)}(n) = -\sum_{j=0}^{r-1} (-1)^j \left(\frac{n\pi}{L}\right)^{2j} [(-1)^{n+1} f^{(2r-2j-1)}(L) + f^{(2r-2j-1)}(0)]$$

$$+ (-1)^r \left(\frac{n\pi}{L}\right)^{2r} \bar{f}_c(n), \tag{C.43}$$

$$\bar{f}_c^{(2r+1)}(n) = -\sum_{j=0}^{r} (-1)^j \left(\frac{n\pi}{L}\right)^{2j} [(-1)^{n+1} f^{(2r-2j)}(L) + f^{(2r-2j)}(0)]$$

$$+ (-1)^r \left(\frac{n\pi}{L}\right)^{2r+1} \bar{f}_s(n). \tag{C.44}$$

C-7 CONVOLUTION THEOREMS

Let $\bar{F}(p)$ and $\bar{G}(p)$ be the complex Fourier transforms of $f(x)$ and $g(x)$, respectively, and define the convolution of the functions $f(x)$ and $g(x)$ as

$$h(x) = f(x)*g(x) = \int_{-\infty}^{\infty} f(x - \eta)g(\eta)\, d\eta. \tag{C.45}$$

The function $f(x - \eta)$ can be written, however, as the inversion integral

$$f(x - \eta) = \frac{1}{2\pi} \int_{-\infty}^{\infty} \bar{F}(p)e^{-ip(x-\eta)}\, dp, \tag{C.46}$$

so that

$$h(x) = \int_{-\infty}^{\infty} g(\eta)\left[\frac{1}{2\pi} \int_{-\infty}^{\infty} \bar{F}(p)e^{-ip(x-\eta)}\, dp\right] d\eta$$

$$= \frac{1}{2\pi} \int_{-\infty}^{\infty} \bar{F}(p)\left[\int_{-\infty}^{\infty} g(\eta)e^{ip\eta}\, d\eta\right]e^{-ipx}\, dx$$

$$= \frac{1}{2\pi} \int_{-\infty}^{\infty} \bar{F}(p)\bar{G}(p)e^{-ipx}\, dp. \tag{C.47}$$

Theorem. *If $\bar{F}(p)$ and $\bar{G}(p)$ are the Fourier transforms of $f(x)$ and $g(x)$, respectively, the inverse transformation of the product of $\bar{F}(p)$ and $\bar{G}(p)$ is the convolution integral (C.45).*

Similar theorems exist for the other kinds of Fourier transforms.

Selected Readings

Sneddon, I. N., *Fourier Transforms*, McGraw-Hill, New York, 1951.

Titchmarsh, E. C., *Introduction to the Theory of Fourier Integrals*, Oxford Univ. Press, New York, 1948.

Tranter, C. J., *Integral Transforms in Mathematical Physics*, Methuen, London, 1959.

INDEX